The Biochemistry of the Nucleic Acids

The Biochemistry of the Nucleic Acids

NINTH EDITION

R. L. P. Adams   R. H. Burdon
A. M. Campbell
D. P. Leader   R. M. S. Smellie

Department of Biochemistry, University of Glasgow

LONDON NEW YORK
CHAPMAN AND HALL

First published, 1950 by Methuen and Co. Ltd
Second edition, 1953 Third edition, 1957
Fourth edition, 1960 Fifth edition, 1965
Sixth edition, 1969 Seventh edition, 1972

Published by Chapman and Hall Ltd
11 New Fetter Lane, London EC4P 4EE
First published as a Science Paperback, 1972
Eighth edition, 1976 Ninth edition, 1981

Published in the USA by Chapman and Hall
in association with Methuen, Inc.
733 Third Avenue, New York NY 10017

Filmset by Enset Ltd.
Midsomer Norton, Bath

Printed in Great Britain by
Fletcher and Son Ltd., Norwich

ISBN 0 412 22680 4 (cased)
ISBN 0 412 22690 1 (paperback)

British Library Cataloguing in Publication Data

The biochemistry of the nucleic acids.—9th ed.
1. Nucleic acids
I. Title II. Adams, R.L.P.
574.87'328 QP620 80-42129

ISBN 0-412-22680-4
ISBN 0-412-22690-1 Pbk

Contents

Preface

When the first edition of this book was published in 1950, it set out to present an elementary outline of the state of knowledge of nucleic acid biochemistry at that time and it was the first monograph on the subject to appear since Levene's book on Nucleic Acids in 1931. The fact that a ninth edition is required after thirty years and that virtually nothing of the original book has been retained is some measure of the speed with which knowledge has advanced in this field.

'The Child's Guide to the Nucleic Acids' as it is known within the Department in Glasgow is still intended primarily as an introduction to the subject for advanced undergraduates in biochemistry and molecular biology, for graduates embarking upon studies in the field of nucleic acids, for chemists seeking to find some understanding of the more biological aspects of the subject and for biologists who require some knowledge of the chemical and molecular aspects.

The first seven editions emerged from the pen of the late J.N. Davidson who died in September 1972 shortly after completing the seventh edition. The eighth edition was revised extensively by four of his colleagues who recognized the need for a book giving a reasonably comprehensive coverage of the field at an up to date but elementary level. In doing so an attempt was made to retain something of the character and structure of the earlier editions while at the same time introducing new ideas and concepts and eliminating some of the more out-dated material.

Progress between 1976 and 1980 has been even more rapid than in the previous four years and the ninth edition has undergone very extensive revision not only in the content of individual chapters but also in general organization and layout. With a large amount of additional material to present the book has grown in size but every effort has been made to keep the increase within bounds by excluding non-essential detail. In a field in which new developments are occurring so rapidly it is inevitable that new knowledge will accumulate more quickly than it can be embodied in a new edition but we have endeavoured to incorporate into this edition material published up until the date of completion of the manuscript in September 1980.

It is a pleasure to express our thanks to those who have allowed us to reproduce figures, diagrams and plates. In particular we should like to thank

Dr Lesley Coggins for Plates I and V
Drs D.E. Oling and A.L. Oling for Plate II
Dr P.R. Cook and Dr S.J. McCready for Plate III
Dr U. Scheer and Dr John Sommerville for Plate VI
Dr S.L. McKnight and Dr O.L. Miller for Plate VII
Dr J.A. Lake for the model used in Plate VIII

We are particularly grateful to Mr Ian Ramsden of the Department of Medical Illustration and Photography in the West Medical Building of Glasgow University and to Miss J.M. Gillies, Mrs J. Greenwood and Mrs C.M. Dow for valuable secretarial assistance.

R.L.P.A.
R.H.B.
A.M.C.
D.P.L.
R.M.S.S.

September 1980

Abbreviations and nomenclature

The abbreviations employed in this book are those approved by the Commission on Biochemical Nomenclature (CBN) of the International Union of Pure and Applied Chemistry (IUPAC) and the International Union of Biochemistry (IUB).

Nucleosides

A	adenosine
G	guanosine
C	cytidine
U	uridine
ψ	5-ribosyluracil (pseudouridine)
I	inosine
X	xanthine
T	ribosylthymine (ribothymidine)
N	unspecified nucleoside
R	unspecified purine nucleoside
Y	unspecified pyrimidine nucleoside
dA	2′-deoxyribosyladenine
dG	2′-deoxyribosylguanine
dC	2′-deoxyribosylcytosine
dT	2′-deoxyribosylthymine (thymidine)

Minor nucleosides (when in sequence)

m^1A	1-methyladenosine
m^6_2A	N^6-dimethyladenosine
iA	N^6-isopentenyladenosine
m^5C	5-methylcytidine
ac^4C	N^4-acetylcytidine
m^1G	1-methylguanosine
m^2G	N^2-methylguanosine
m^2_2G	N^2-dimethylguanosine
m^1I	1-methylinosine
Cm	2′-*O*-methylcytidine
Gm	2′-*O*-methylguanosine

Um	2′-*O*-methyluridine
hU	5,6-dihydrouridine

Nucleotides

AMP	adenosine 5′-monophosphate
GMP	guanosine 5′-monophosphate
CMP	cytidine 5′-monophosphate
UMP	uridine 5′-monophosphate
dAMP	2′-deoxyribosyladenine 5′-monophosphate
dGMP	2′-deoxyribosylguanine 5′-monosphosphate
dCMP	2′-deoxyribosylcytosine 5′-monophosphate
dTMP	2′-deoxyribosylthymine 5′-monosphosphate
2′-AMP, 3′-AMP, 5′-AMP etc.	2′-, 3′- and 5′-phosphates of adenosine etc.
ADP etc.	5′-(pyro) diphosphates of adenosine etc.
ATP etc.	5′-(pyro) triphosphates of adenosine etc.
ddTTP	2′, 3′-dideoxyribosylthymine 5′- triphosphate
araCTP	1-β-D-arabinofuranosylcytosine 5′-triphosphate

Polynucleotides

DNA	deoxyribonucleic acid
cDNA	complementary DNA
mtDNA	mitochondrial DNA
RNA	ribonucleic acid
mRNA	messenger RNA
rRNA	ribosomal RNA
tRNA	transfer RNA
nRNA	nuclear RNA
hnRNA	heterogeneous nuclear RNA
snRNA	small nuclear RNA
Alanine tRNA or $tRNA^{Ala}$, etc.	transfer RNA that normally accepts alanine
Alanyl-$tRNA^{Ala}$ or Ala-$tRNA^{Ala}$ or Ala-tRNA	transfer RNA that normally accepts alanine with alanine residue covalently linked
poly(N), or $(N)_n$ or $(rN)_n$	polymer of ribonucleotide N
poly (dN) or $(dN)_n$	polymer of deoxyribonucleotide N

poly(N-N′), or r(N-N′)$_n$ or (rN-rN′)$_n$	copolymer of N—N′-N—N′ in regular, alternating, *known* sequence
poly(A)·poly(B) or (A)$_n$(B)$_n$	two chains, generally or completely associated
poly(A), poly(B) or (A)$_n$, (B)$_n$	two chains, association unspecified or unknown
poly(A) + poly(B) or (A)$_n$ + (B)$_n$	two chains, generally or completely unassociated

Miscellaneous

RNase, DNase	ribonuclease, deoxyribonuclease
P_i, PP_i	inorganic orthophosphate and pyrophosphate

Amino acids

Ala	Alanine
Arg	Arginine
Asn	Asparagine
Asp	Aspartic acid
Cys	Cysteine
Gln	Glutamine
Glu	Glutamic acid
Gly	Glycine
His	Histidine
Ile	Isoleucine
Leu	Leucine
Lys	Lysine
Met	Methionine
fMet	Formylmethionine
Phe	Phenylalanine
Pro	Proline
Ser	Serine
Thr	Threonine
Trp	Tryptophan
Tyr	Tyrosine
Val	Valine

Further details of the recommendations of the Commission on Biochemical Nomenclature are printed in the *J. Biol. Chem.* **246,** 4894 (1971), *Biochim. Biophys. Acta,* **247**, 1 (1971), *Biochemistry,* **5,** 1445 (1966), *Arch. Biochem. Biophys.*, **115,** 1 (1966), *J. mol. Biol.*, **55,** 299 (1971), and *Progress in Nucleic Acid Research and Molecular Biology,* **22,** (1979).

In naming enzymes, the recommendations of the Commission on Enzymes of the International Union of Biochemistry (1972) are followed as far as possible. The numbers recommended by the Commission are inserted in the text after the name of each enzyme.

1 Introduction

The fundamental investigations which led to the discovery of the nucleic acids were made by Friedrich Miescher [1] (1844–95), who may be regarded as the founder of our knowledge of the chemistry of the cell nucleus. In early work carried out in 1868, in the laboratory of Hoppe-Seyler in Tübingen, he isolated the nuclei from pus cells obtained from discarded surgical bandages and showed the presence in them of an unusual phosphorus-containing compound that he called 'nuclein' and which we now know to have been nucleoprotein. Miescher's investigations were continued in Basel, where he spent most of his working life and where he became interested in salmon sperm as a source of nuclear material. In 1872 he showed that isolated sperm heads contained an acidic compound, now recognized as nucleic acid, and a base to which the name 'protamine' was given. It was subsequently shown that nucleic acids were normal constituents of all cells and tissues and Miescher's investigations of the nucleic acids were continued by Altman, who in 1889 described a method for the preparation of protein-free nucleic acids from animal tissues and from yeast. The work was continued later by Kossel in Heidelberg, Jones in Baltimore, Levene in New York, Hammarsten in Stockholm, Gulland in Nottingham and many others [2–7].

One of the best animal sources of nucleic acid was found to be the thymus gland, and it is not surprising therefore that much of the early work was concentrated on nucleic acid from this source. On hydrolysis it was found to yield the purine bases adenine and guanine, the pyrimidine bases cytosine and thymine, a deoxypentose and phosphoric acid. The nucleic acid from yeast on the other hand yielded on hydrolysis adenine, guanine, cytosine, uracil, a pentose sugar and phosphoric acid. Yeast nucleic acid therefore differed from thymus nucleic acid in containing uracil in place of thymine and a pentose in place of a deoxypentose. Since most nucleic acids from animal sources appeared to resemble thymus nucleic acid, and since the only other nucleic acid which had at that time (1920) been prepared in reasonable quantities from a plant source appeared to be very similar to yeast nucleic acid, the impression grew that deoxypentose nucleic acid was characteristic of animal tissues, and pentosenucleic acid was characteristic of plant tisues [5]. Thus Jones, in 1920, stated categorically: 'we come to understand quite clearly that there are only two nucleic acids in nature, one obtainable from the nuclei of animal cells and the other from the nuclei of plant cells' [6].

It was not long before the validity of this concept was questioned. It had been known for many years that pentose derivatives were present in animal tissues. For example, the so-called β-nucleoprotein, which was originally prepared from mammalian pancreas by O. Hammarsten [7] in 1894, was known to contain a pentose sugar, and Jorpes [8] eventually prepared from this material a nucleic acid of the pentose type which he showed to resemble yeast nucleic acid and to be abundant in pancreatic tissue. The presence of pentosenucleic acids in mammary tissue was also suggested by the work of Odenius [9] and of Mandel and Levene [10]. Pentosenucleotide derivatives were also demonstrated in chick embryo pulp by Calvery [11], in spleen and the liver by Jones and Perkins [12] and by Thomas and Berariu [13], and in sea urchin eggs by Blanchard [14]. It thus appeared probable that pentosenucleic acids were normal constituents of animal tissues as well as of plant cells, and Jones and Perkins [12] expressed the view: 'the distinction between plant and animal nucleic acids will in future not be so definitely drawn'.

Final proof that ribonucleic acid is a general constituent of animal, plant and bacterial cells was not forthcoming until the early 1940's as a consequence of the ultraviolet spectrophotometric studies of Caspersson [15], the histochemical observations of Brachet [16] and the chemical analyses of Davidson [17, 18].

It took a surprisingly long time also to establish the nature of the sugars in deoxypentose and pentose nucleic acids. Eventually, however, the deoxypentose from a range of sources was shown to be D(−)-2-deoxyribose [19–23] and the pentose was shown to be D(−)-ribose [24–30] so that the two types of nucleic acid are now known as deoxyribonucleic acid (DNA) and ribonucleic acid (RNA) respectively. Looking back it seems remarkable that confirmation of the nature of these sugars was not obtained until the mid 1950's around the same time that the double-helical structure for DNA was being put forward by Watson and Crick [31].

These advances established the biology of the nucleic acids on a new foundation. The use of new techniques in cytochemistry and cell fractionation showed that DNA and RNA are normal constituents of all cells, whether plant or animal, DNA being confined mainly to the nucleus while RNA is found also in the cytoplasm. [17, 18, 32–35.]

The development of techniques of subcellular fractionation and for the isolation of nuclei [36–41] made possible chemical measurements of the distribution of DNA and RNA amongst the subcellular fractions of various cell types, and led ultimately to the recognition of RNA in the nuclear, ribosomal and soluble fractions of cells (see Chapter 5) and to the demonstration of the constancy in the average amount of DNA per nucleus in the somatic cells of any given species [42].

The advent of isotope techniques led to the demonstration that both DNA and RNA could be synthesized *de novo* in most tissues from low-molecular-weight precursors [44] and from an early stage a correlation began to emerge between the rate of cell division in a tissue and the rate of uptake of isotopes into the DNA of that tissue [44]. So far as the synthesis of RNA was concerned, cell-fractionation studies combined with measurements of the incorporation of labelled precursors into the RNAs of different subcellular fractions [43, 44], demonstrated that there were considerable differences in the metabolic activities of the various classes of RNAs, nuclear RNA showing specially high levels of isotope incorporation.

Many of the earliest contributions to our understanding of the structure of nucleic acids arose out of the work of Levene and Jacobs [3, 45, 46] who established the presence of D-ribose, hypoxanthine and phosphorus in inosinic acid from muscle and later the presence of ribonucleotides of adenine, guanine, cytosine and uracil in yeast nucleic acid. They also recognized the occurrence of thymine in place of uracil in thymus nucleic acid.

The presence of the bases in approximately equimolar proportions led to the development of the tetranucleotide hypothesis for both DNA and RNA, in which both nucleic acids were considered to be polymeric structures containing equivalent amounts of mononucleotides derived from each of the four purine and pyrimidine bases linked together in repeating units. This concept of nucleic acid structure survived until the late 1940s despite the fact that evidence for it was not strong [47] and it was only when methods for the quantitative analysis of nucleic acids had been developed [48, 49] that the tetranucleotide hypothesis was finally abandoned as a consequence of the demonstration that the various nucleotides did not necessarily occur in equimolar proportions [50].

In the early 1950s Chargaff [51] drew attention to certain regularities in the composition of DNA, namely that the sum of the purines was equal to the sum of the pyrimidines, that the sum of the amino bases (adenine and cytosine) was equal to the sum of the keto bases (guanine and thymine) and that adenine and thymine, and guanine and cytosine, were present in equivalent amounts (Chapters 2 and 3). These observations were to be of crucial importance in the subsequent interpretation of X-ray crystallographic analyses.

The elucidation of the detailed structure of nucleosides and nucleotides can largely be attributed to Todd and his collaborators (for review see [52]), who established the nature of the glycosidic linkage between the sugar residues and the purine or pyrimidine bases and the nature of the phosphate ester bonds. Their work taken together with the studies of Cohn and his colleagues [53] provided final confirmation of the nature of the 3′, 5′-internucleotide linkage in both DNA and RNA and made it possible for clear concepts as to the primary structure of the two types of nucleic acids to be put forward (Chapters 2, 3 and 5).

At the time that these new developments were taking place in the understanding of the chemical structures of nucleotides and polynucleotides, progress was also being made in X-ray crystallographic studies of DNA. Arising from the work of Astbury [54], Pauling and Corey [55], Wilkins and his colleagues [56] and Franklin and Gosling [57], Watson and Crick [31] proposed their now famous double-helical structure made up of specifically hydrogen-bonded base pairs (Chapter 2) which suggested 'a possible copying mechanism for the genetic material'.

The elucidation of the primary structures of DNA and RNA taken together with the early studies on the incorporation of labelled precursors into the nucleic acids provided the impetus for researches into the detailed mechanisms of biosynthesis of both types of nucleic acid. The almost accidental discovery of polynucleotide phosphorylase by Grunberg-Manago, Ortiz and Ochoa in 1955 [59] was a major step forward making available for the first time an enzyme with the capacity to form polyribonucleotides, either as homopolymers or heteropolymers, using ribonucleoside 5′-diphosphates as the substrates. This system provided vital clues as to possible mechanisms of polymerization of nucleotide monomers and also made available for the first time a range of biosynthetic polyribonucleotides whose structures and interactions could be investigated and compared with those of their naturally-occurring counterparts. In subsequent years the range of synthetic polynucleotides synthesized by polynucleotide phosphorylase was to prove invaluable in the elucidation of the genetic code (Chapter 12).

Within a few years of the discovery of polynucleotide phosphorylase, an enzyme from *E.coli*, catalysing the synthesis of DNA-like polymers from deoxyribonucleoside 5′-triphosphates, was identified by Kornberg and his collaborators [60]. This enzyme was named DNA polymerase and is now recognized as one of a family of enzymes concerned in the replication and repair of DNA molecules [58] (Chapters 8 and 9).

RNA polymerases, catalysing the synthesis of polyribonucleotides from ribonucleoside 5′-triphosphates were identified almost simultaneously by several groups around 1960 (for review see [61–64]) so that within the space of a very few years understanding of the whole field of nucleic acid biosynthesis and function underwent a complete transformation.

The subsequent chapters of this book are concerned with more detailed consideration of the various aspects of structure, function and biosynthesis of nucleic acids that have been touched upon above and with the further developments that have occurred in the field. Other treatments of these topics, some by now almost historical, are to be found in the following books [58, 65–82].

REFERENCES

1 Miescher, F. (1879), *Die Histochemischen und Physiologischen Arbeiten*, Leipzig.
2 Fruton, J.S. (1972), *Molecules and Life*, Wiley-Interscience, New York.
3 Levene, P.A. and Bass, L.W. (1931), *Nucleic Acids*, Chemical Catalog Co., New York.
4 Altmann, R. (1889), *Nucleinsauren, Arch.Anat.Physiol.*, 524.
5 Levene, P.A. (1921), *J. Biol. Chem.*, **48**, 119.
6 Jones, W. (1920), *The Nucleic Acids*, Longmans, London.
7 Hammarsten, O. (1894), *Hoppe-Seyler's Ztschr.*, **19**, 19.
8 Jorpes, E. (1924), *Biochem. Ztschr.*, **151**, 227.
(1928), *Acta Med. Scand.*, **68**, 253, 503
(1934), *Biochem. J.*, **28**, 2102.
9 Odenius, R. (1900), *Jahresber. Fortschr. Thierchem.*, **30**, 39.
10 Mandel, J.A. and Levene, P.A. (1905), *Hoppe-Seyler's Ztschr.*, **46**, 155.
11 Calvery, H.O. (1928), *J. Biol. Chem.*, **77**, 489, 497.
12 Jones, W. and Perkins, M.E. (1924–5), *J. Biol. Chem.*, **62**, 290.
13 Thomas, P. and Berariu, C. (1924), *Compt. Rend. Soc. Biol.*, **91**, 1470.
14 Blanchard, K.C. (1935), *J. Biol. Chem.*, **108**, 251.
15 Caspersson, T. (1950), *Cell Growth and Cell Function*, Norton, New York.
16 Brachet, J. (1950), *Chemical Embryology*, Interscience, New York.
17 Davidson, J.N. and Waymouth, C. (1944), *Biochem. J.*, **38**, 39.
18 Davidson, J.N. and Waymouth, C. (1944–5), *Nutrition Abs. Rev.*, **14**, 1.
19 Levene, P.A. and Mori, T. (1929), *J. Biol. Chem.*, **83**, 803.
20 Vischer, E. Zamenhof, S. and Chargaff, E. (1949), *J. Biol. Chem.*, **177**, 429.
21 Chargaff, E., Vischer, E., Doniger, R., Green, C. and Misani, F. (1949), *J. Biol. Chem.*, **177**, 405.
22 Jones, A.S. and Laland, S.G. (1954), *Acta Chem. Scand.*, **8**, 603.
23 Laland, S.G. and Overend, W.G. (1954), *Acta Chem. Scand.*, **8**, 192.
24 Levene, P.A. and Jacobs, W.A. (1909), *Ber. Deutsch. Chem. Ges.*, **42**, 2102, 2469, 2474, 2703.
25 Barker, G.R. and Guilland, J.M. (1943), *J. Chem. Soc.*, 625.
Barker, G.R., Farrar, K.R. and Gulland, J.M. (1947), *J. Chem. Soc.*, 21.

26 Davidson, J.N. and Waymouth, C. (1944), *Biochem. J.*, **38,** 375.
27 Vischer, E. and Chargaff, E. (1948), *J. Biol. Chem.*, **176,** 715.
28 Schwerdt, C.E. and Loring, H.S. (1947), *J. Biol. Chem.*, **167,** 593.
29 Ada, G.L. and Gottschalk, A., (1956), *Biochem. J.*, **62,** 686.
30 MacDonald, D.L. and Knight, C.A. (1953), *J. Biol. Chem.*, **202,** 45.
31 Watson, J.D. and Crick, F.H.C. (1953), *Nature,* **171,** 737
32 Feulgen, R. and Rossenbeck, H. (1924), *Hoppe-Seyler's Ztschr.*, **135,** 203.
33 Kiesel, A. and Belozerski, A.N. (1934), *Hoppe-Seyler's Ztschr.*, **229,** 160.
34 Belozerski, A.N. (1936), *Biochimia,* **1,** 253.
(1939), *Compt. Rend. Acad. Sci. URSS,* **25,** 751,
35 Behrens, M. (1938), *Hoppe-Seyler's Ztschr.* **253,** 185.
36 Claude, A. (1946), *J. Exp. Med.*, **84,** 51.
37 Hogeboom, C., Schneider, W.C. and Palade, G.E. (1948), *J. Biol. Chem.*, **172,** 619.
38 Dounce, A.L. (1943), *J. Biol. Chem.*, **147,** 685.
(1945), *ibid.*, **151,** 221, 235.
39 Mirsky, A.E. and Pollister, A.W. (1943), *J. Gen. Physiol.*, **30,** 117.
40 Behrens, M. (1938), *Aberhalden's Hanbuch der Biologische Arbeitsmethoden,* Sect. 5, Part 10, p. 1363.
41 Dounce, A.L. (1955), in *The Nucleic Acids,* (ed. E. Chargaff and J.N. Davidson), Academic Press, New York, Vol. 2, p. 93.
42 Vendrely, R. (1955) *The Nucleic Acids,* (ed. E. Chargaff and J.N. Davidson, Academic Press, New York, Vol. 2, p. 155.
43 Marshak, A. and Calvet, F. (1949), *J. Cell Comp. Physiol.*, **34,** 451.
44 Smellie, R.M.S. (1955), in *The Nucleic Acids,* (ed. E. Chargaff and J.N. Davidson), Academic Press, New York, Vol. 2, p. 393.
45 Levene, P.A. and Jacobs, W.A. (1908), *Ber. Deutsch. Chem. Ges.*, **41,** 2703.
(1909), *ibid.*, **42,** 335.
46 Levene, P.A. and Jacobs, W.A. (1912), *J. Biol. Chem.*, **12,** 411.
47 Gulland, J.H. (1947), *Symp.Soc.Exp.Biol.*, **1,** 1.
48 Vischer, E. and Chargaff, E. (1947), *J. Biol. Chem.*, **168,** 781.
49 Wyatt, G.R. (1955) in *The Nucleic Acids,* (ed. E. Chargaff and J.N. Davidson), Academic Press, New York, Vol. 1, p. 243.
50 Chargaff, E. (1950), *Experientia,* **6,** 201.
51 Chargaff, E. (1955), in *The Nucleic Acids* (ed. E. Chargaff and J.N. Davidson), Academic Press, New York, Vol. 1, p. 307.
52 Brown, D.M. and Todd, A.R. (1955), in *The Nucleic Acids,* (ed. E. Chargaff and J.N. Davidson), Academic Press, New York, Vol. 1, p. 409.
53 Cohn, W.E. (1956) in *Currents in Biochemical Research,* (ed. D.E. Green), Interscience, New York, p. 460.
54 Astbury, W.T. (1947), *Symp. Soc. Exp. Biol.*, **1,** 66.
55 Pauling, L. and Corey, R.B. (1953), *Proc. Nat. Acad. Sci.*, **39,** 84.
56 Wilkins, M.F.H., Stokes, A.R. and Wilson, H.R. (1953), *Nature,* **171,** 738.
57 Franklin, R.E. and Gosling, R.G. (1953), *Nature,* **171,** 740; **172,** 156.
58 Kornberg, A. (1980), *DNA Replication.* Freeman, San Francisco.
59 Grunberg-Manago, M., Ortiz, P.J. and Ochoa, S. (1956), *Biochim. Biophys. Acta, 20,* 269.

60 Lehmann, I.R., Bessman, M.J., Simms, E.S. and Kornberg, A. (1958), *J. Biol. Chem.*, **233,** 163.
61 Smellie, R.M.S. (1963), in *Progress in Nucleic Acid Research* (ed. J.N. Davidson and W.E. Cohn), Academic Press, New York, Vol. 1, p. 27.
62 Hurwitz, J. and August, J.T. (1963) in *Progress in Nucleic Acid Research* (ed. J.N. Davidson and W.E. Cohn), Academic Press, New York, Vol. 1, p. 59.
63 Jordan, D.O. (1960), *Chemistry of the Nucleic Acids,* Butterworths, London.
64 Potter, V.R. (1960), *Nucleic Acid Outlines,* Burgess Publishing Company, Minneapolis.
65 Steiner, R.F. and Beers, R.F. (1961), *Polynucleotides,* Elsevier, Amsterdam.
66 Allen, F.W. (1962), *Ribonucleoproteins and Ribonucleic Acids,* Elsevier, Amsterdam.
67 Perutz, M.F. (1962), *Proteins and Nucleic Acids,* Elsevier, Amsterdam.
68 Chargaff, E. (1963), *Essays on Nucleic Acids,* Elsevier, Amsterdam.
69 Davidson, J.N. and Cohn, W.E. (eds.) (1963–73), *Progress in Nucleic Acid Research and Molecular Biology,* Academic Press, New York, Vols. 1–13.
70 Cohn, W.E. (ed.) (1974–79), *Prog. Nucleic Acid Res. Mol. Biol.* Academic Press, New York, Vols. 14–22.
71 Michelson, A.M. (1963), *The Chemistry of Nucleosides and Nucleotides,* Academic Press, New York.
72 Synthesis and Structure of Macromolecules (1963), *Cold Spring Harb. Symp. Quant. Biol.* Vol. 28.
73 Taylor, J.H. (ed.) (1963), *Molecular Genetics,* Part I; (1967), Part II. Academic Press, New York.
74 Vogel, H.J., Bryson, V. and Lampen, J.O. (eds.) (1963), *Informational Macromolecules,* Academic Press, New York.
75 Steiner, R.F. (1965), *The Chemical Foundations of Molecular Biology,* Van Nostrand, Princeton.
76 Jukes, T.H. (1966), *Molecules and Evolution,* Columbia University Press, New York.
77 Kendrew, J. (1966), *The Thread of Life,* Bell, London.
78 Watson, J.D. (1968), *The Double Helix,* Atheneum, New York.
79 Cantoni, G.L. and Davies, D.R. (eds.) (1966), *Procedures in Nucleic Acid Research,* Harper and Row, New York.
80 Grossman, L. and Moldave, K. (eds.) (1967), Part A; (1968), Part B; (1971) Parts C and D, *Nucleic Acids,* being Vols. 12, 20 and 21 of *Methods in Enzymology* (ed. S.P. Colowick and N.O. Kaplan), Academic Press, New York.
81 Parish, J.H. (1972), *Principles and Practice of Experiments with Nucleic Acids,* Longmans, London.
82 Lewin, B. (1974), *Gene Expression,* Vols. I and II. Wiley, London.

The structure of nucleic acids

2

2.1 GENERAL

Before any account is given of the structure of the nucleic acids proper, it is essential to understand the structures of their component parts. Complete hydrolysis of the nucleic acids yields pyrimidine and purine bases, a sugar component, and phosphoric acid. Partial hydrolysis yields compounds known as nucleosides and nucleotides. Each of these component parts will be discussed in turn.

2.2 PYRIMIDINE BASES

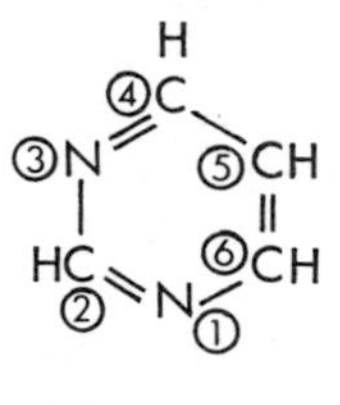

pyrimidine

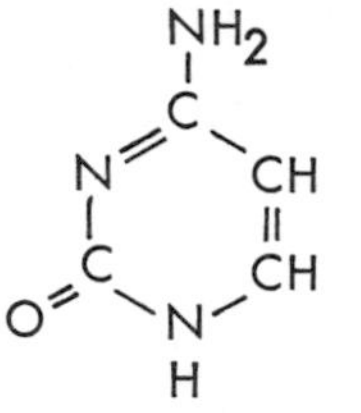

cytosine

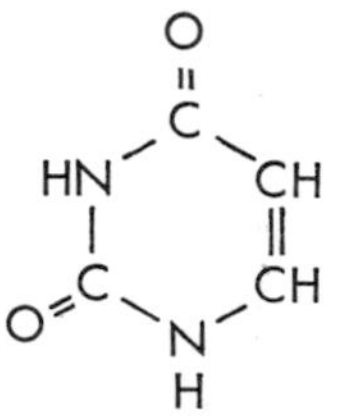

uracil

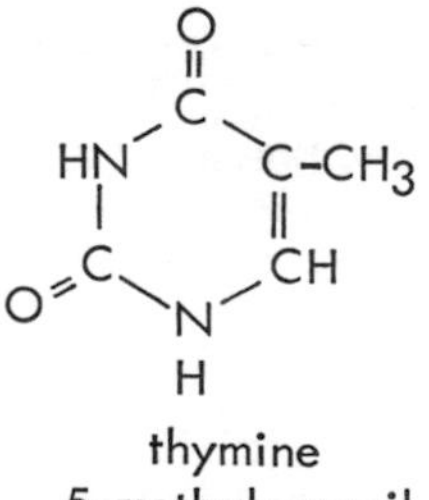

thymine
5-methyl-uracil

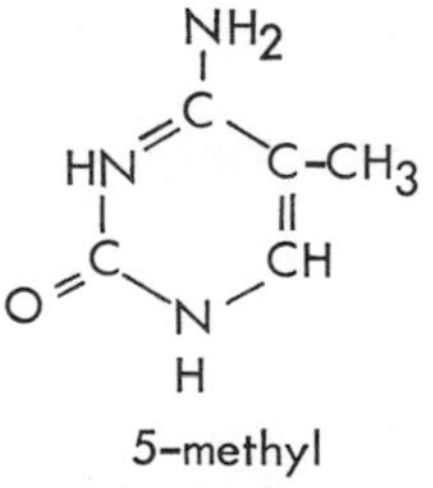

5-methyl
cytosine

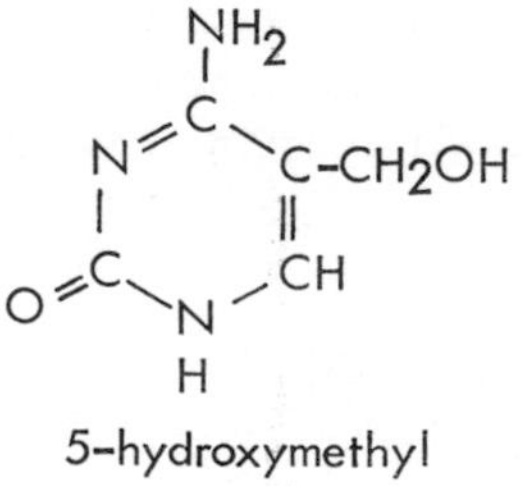

5-hydroxymethyl
cytosine

The pyrimidine bases are derivatives of the parent compound pyrimidine, and the bases found in the nucleic acids are cytosine found in both RNA and DNA, uracil found in RNA and thymine and 5-methylcytosine found in DNA. In certain of the coliphages cytosine is replaced by 5-hydroxymethylcytosine or glucosylated derivatives of 5-hydroxymethylcytosine.

The pyrimidine bases can undergo keto-enol tautomerism as shown below for uracil:

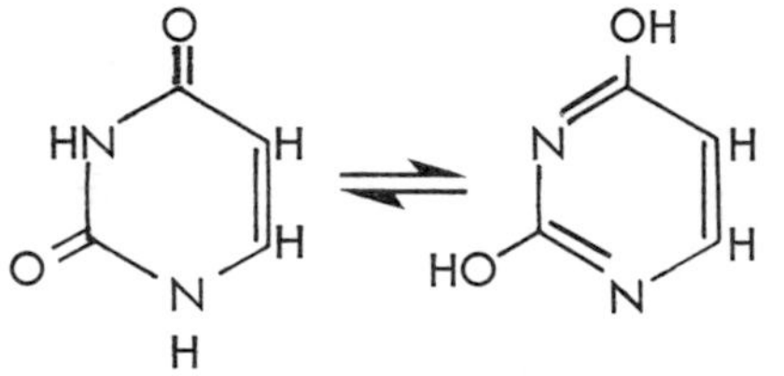

Lactam Lactim

2.3 PURINE BASES

Both types of nucleic acids contain the purine bases, adenine and guanine. They are derivatives of the parent compound purine which is formed by the fusion of a pyrimidine ring and an imidazole ring. It should be noted that the style of numbering of the pyrimidine ring in the purines differs from that used for pyrimidines themselves.

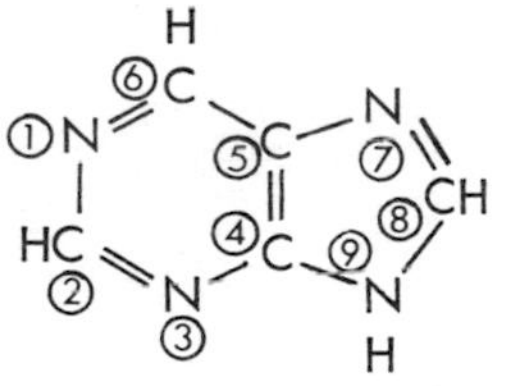

Purine

Adenine and guanine have the following structures:

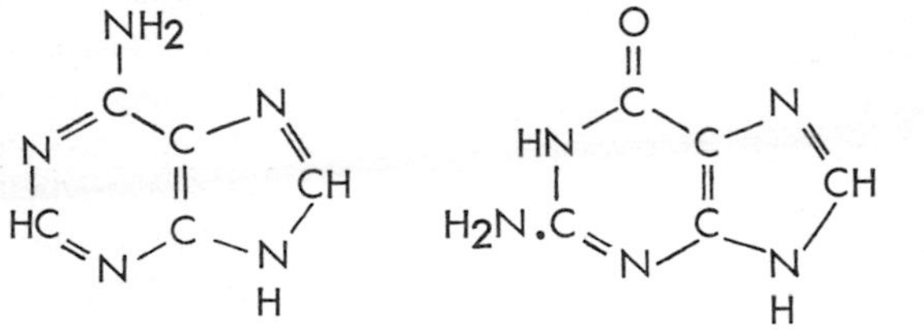

6 - aminopurine
adenine

2-amino 6 - hydroxy purine
guanine

As in the pyrimidines so with the purines the bases can exist in two tautomeric forms as shown below for guanine:

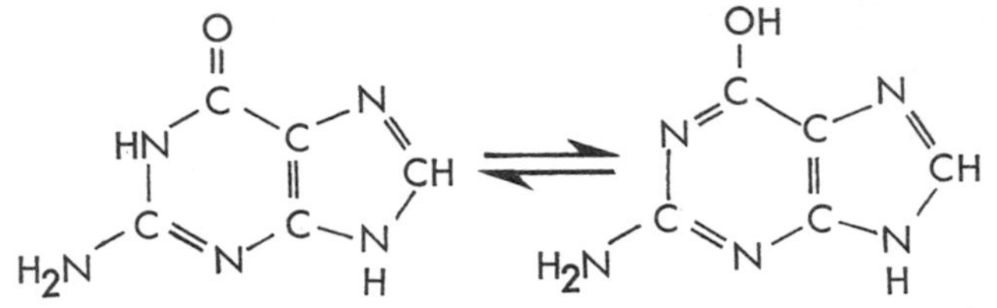

Tautomeric forms of guanine

Other naturally occurring purine derivatives include hypoxanthine, xanthine, and uric acid.

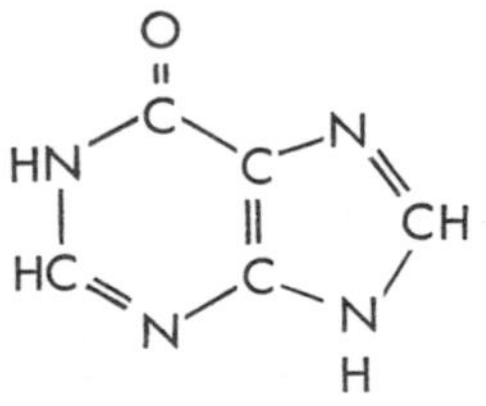

hypoxanthine
6-hydroxypurine

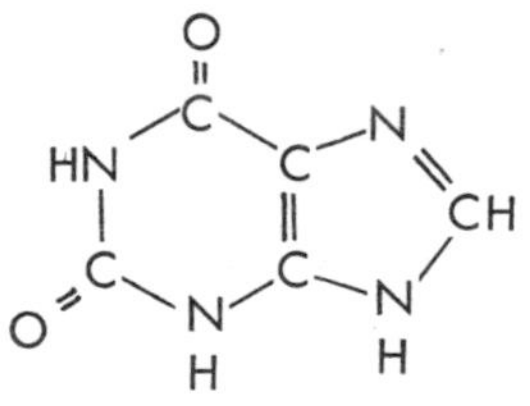

xanthine
2,6 dihydroxypurine

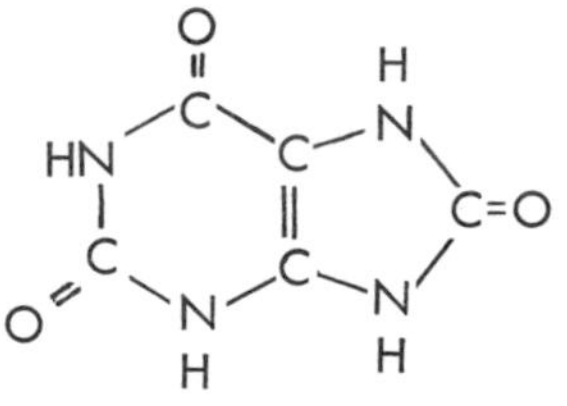

uric acid
2,6,8-trihydroxypurine

Table 2.1 Some of the more important minor bases in RNA.

1-methyladenine	dihydrouracil
2-methyladenine	5-hydroxyuracil
6-methyladenine	5-carboxymethyluracil
6,6-dimethyladenine	5-methyluracil (thymine)
6-isopentenyladenine	5-hydroxymethyluracil
2-methylthio-6-isopentenyladenine	2-thiouracil
6-hydroxymethylbutenyladenine	3-methyluracil
6-hydroxymethylbutenyl-2-methylthioadenine	5-methylamino-2-thiouracil
1-methylguanine	5-methyl-2-thiouracil
2-methylguanine	5-uracil-5-hydroxyacetic acid
2,2-dimethylguanine	3-methylcytosine
7-methylguanine	4-methylcytosine
2,2,7-trimethylguanine	5-methylcytosine
hypoxanthine	5-hydroxymethylcytosine
1-methylhypoxanthine	2-thiocytosine
xanthine	4-acetylcytosine
6-aminoacyladenine	
7-(4,5-cis-dihydroxyl-l-clyclopenten-3-ylaminomethyl)-7-deazaguanosine(Q)	

Certain 'minor bases' are also found in small amounts in some nucleic acids [1–4]. For example, 'transfer' RNA (tRNA) which is discussed on pages 114 to 116 contains a wide variety of methylated bases, including thymine [3]. These unusual bases comprise less than 5 per cent of the total base content of the tRNA and vary in relative amounts from species to species. Some of the minor bases in RNA are listed in Table 2.1.
The phenylalanine transfer RNA from yeast contains a most unusual base known as Wye base (Yt), the structure of which is shown below [5,6].

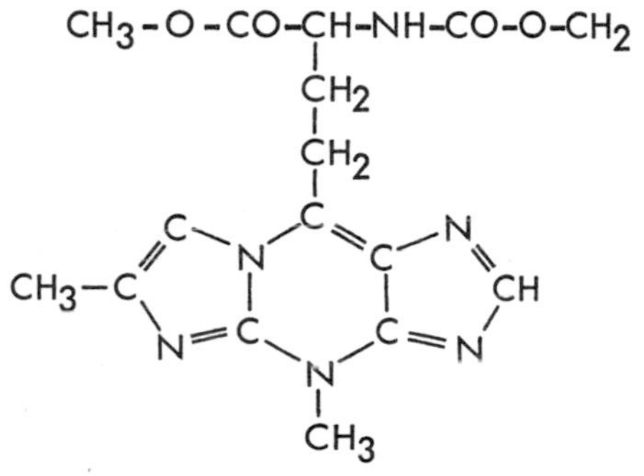

The chemistry of the pyrimidines and purines has been reviewed by Bendich [7] and by Ulbricht [53].

2.4 PENTOSE AND DEOXYPENTOSE SUGARS

The main sugar component of RNA is D-ribose which in polynucleotides occurs in the β-D-ribofuranose form. In DNA this sugar is replaced by 2-deoxyribose also in the β-D-furanose form. This apparently small difference between the two types of nucleic acid has wide-ranging effects on both their chemistry and structure since the presence of the bulky hydroxyl group on the 2 position of the sugar not only limits the range of possible secondary structures available to the RNA molecule but also makes it more susceptible to chemical and enzymic degradation.

Some RNAs, notably ribosomal RNAs, contain very small amounts of 2-*O*-methyl ribose.

Glucose occurs glycosidically linked to hydroxymethyclytosine in the DNA from certain strains of bacteriophage (p. 70).

When the pentose sugars occur in nucleic acids or nucleotides the carbon atoms are numbered as 1′, 2′, 3′ etc., to avoid confusion with the numbering of the ring atoms of the bases.

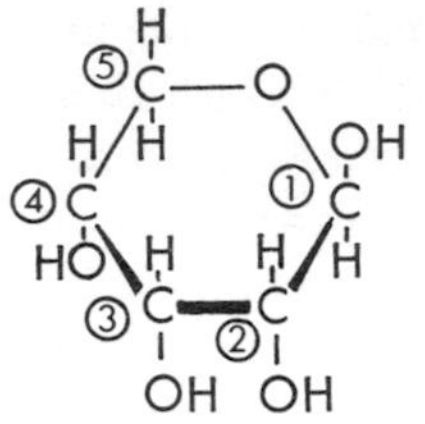

β-D-ribopyranose 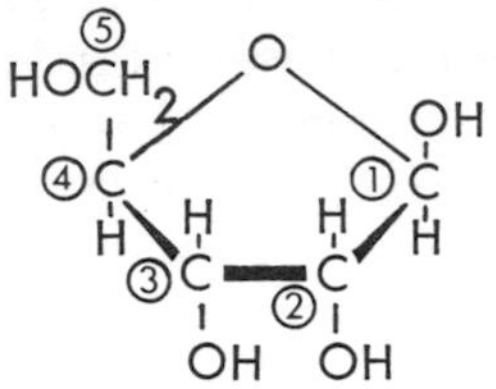 β-D-ribofuranose

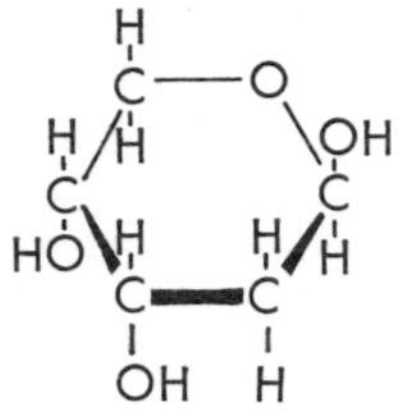

β-D-2-deoxyribopyranose 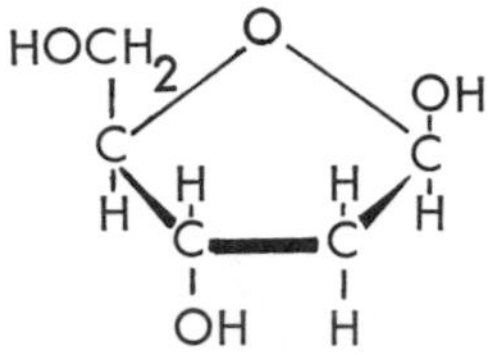 β-D-2-deoxyribofuranose

2.5 NUCLEOSIDES

When a purine or a pyrimidine base is linked to ribose or deoxyribose the resulting compound is known as a nucleoside. Thus adenine condenses with ribose to form the nucleoside adenosine, guanine forms guanosine, cytosine forms cytidine, and uracil forms uridine. These ribonucleosides can be formed on partial hydrolysis of RNA. The ribonucleoside from hypoxanthine is named inosine. The nucleosides derived from 2-deoxyribose are known as deoxyribonucleosides-deoxyadenosine, deoxyguanosine, deoxycytidine, deoxythymidine and so on.

In addition to the nucleosides listed above several others are found in very small amounts in certain classes of nucleic acids. These are listed in Table 2.2.

Table 2.2 Minor nucleosides in RNA [9] (see also Table 2.1).

1-ribosylthymine	2′-*O*-methyluridine
5-ribosyluracil (pseudouridine)	2′-*O*-methylcytidine
2-ribosylguanine	2′-*O*-methylpseudouridine
2′-*O*-methyladenosine	2′-*O*-methyl-4-methylcytidine
2′-*O*-methylguanosine	

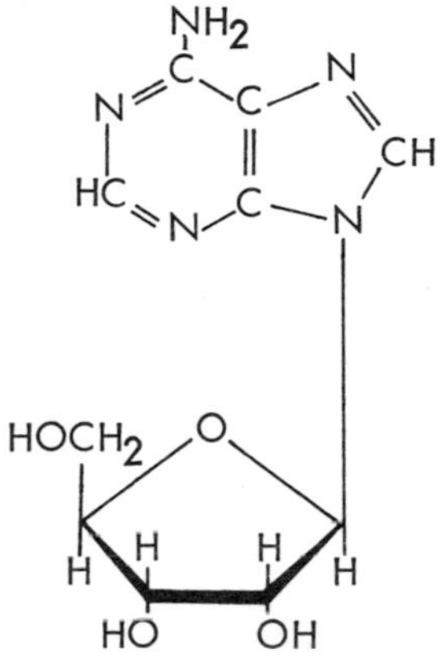

adenosine
9-β-D-ribofuranosyl
adenine

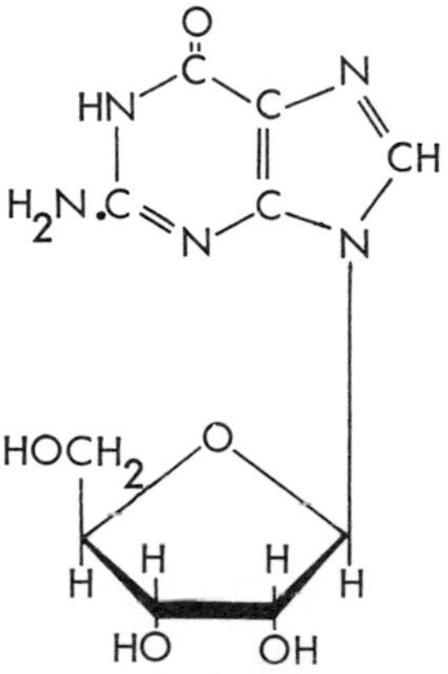

guanosine
9-β-D-ribofuranosyl
guanine

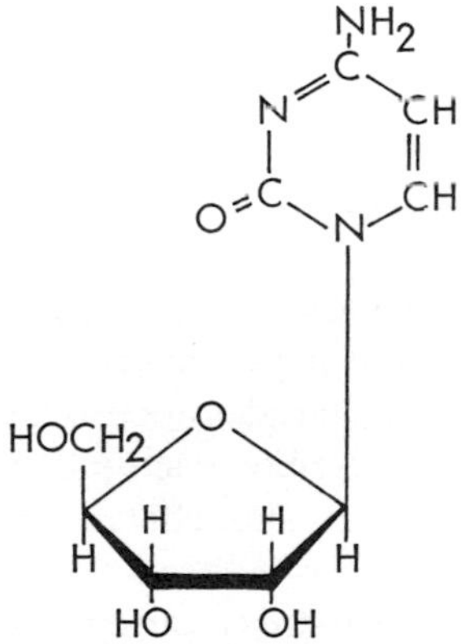

cytidine
1-β-D-ribofuranosyl
cytosine

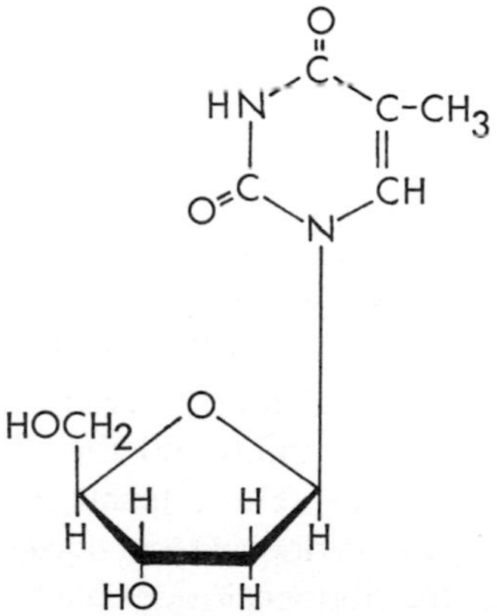

thymidine
1-β-D-2-deoxyribo-
furanosylthymine

The nucleoside 5-ribosyluracil has been obtained in small amounts from the digestion products of RNA particularly tRNA and has been named pseudouridine.

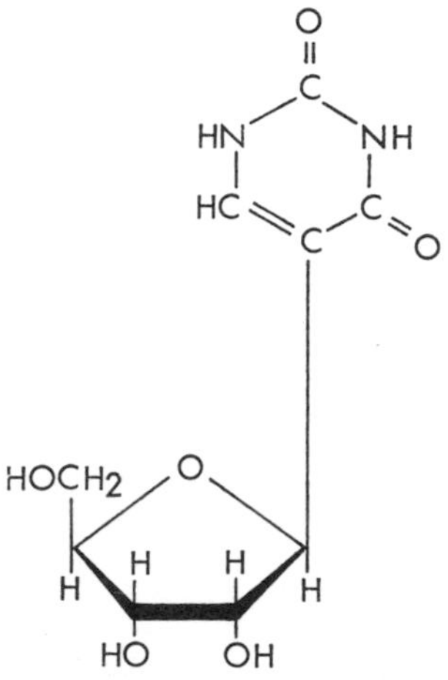

pseudouridine (ø)
5-β-D-ribofuranosyluracil

2.6 NUCLEOTIDES

The structures of nucleotides are dealt with in more specialised terms by Hutchinson [10], by Michelson [11]. They are all phosphoric acid esters of the nucleosides. Those derived from ribonucleosides are usually referred to as ribonucleotides and those from deoxyribonucleosides as deoxyribonucleotides. These terms are sometimes abbreviated to riboside, ribotide, deoxyriboside and deoxyribotide.

Since the ribonucelosides have three free hydroxyl groups on the sugar ring, three possible ribonucleoside monophosphates can be formed. Adenosine, for example, can give rise to three monophosphates (adenylic acids), adenosine 5′-phosphate, adenosine 3′-phosphate and adenosine 2′-phosphate.

In the same way guanosine, cytidine and uridine can give rise to three guanosine monophosphates (guanylic acids), three cytidine monophosphates (cytidylic acids), and three uridine monophosphates (uridylic acids) respectively. They are frequently referred to by the abbreviations shown in the table of abbreviations at the beginning of the book.

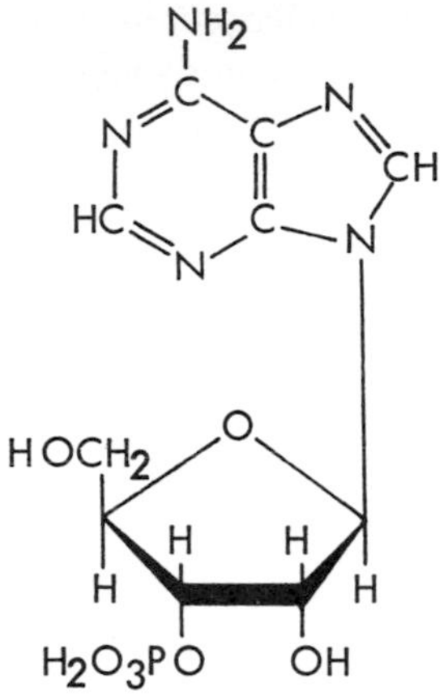

adenosine
3' - phosphate

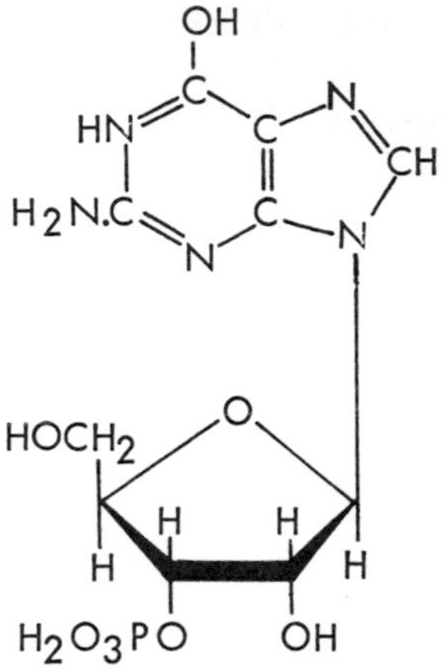

guanosine
3' - phosphate

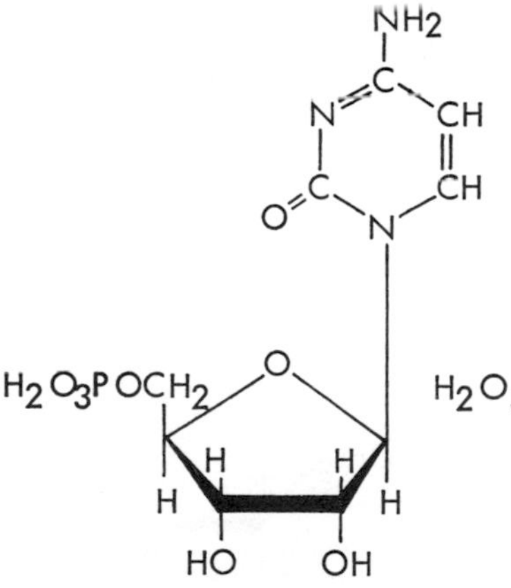

cytidine
5' - phosphate

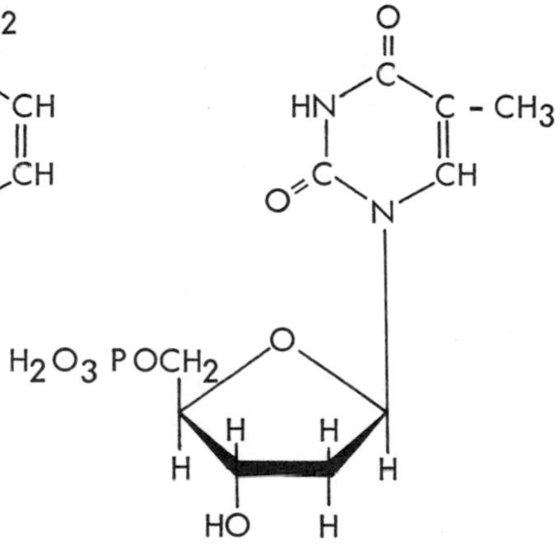

thymidine
5' - phosphate

The ribonucleoside 5′-phosphates may be further phosphorylated at position 5′ to yield 5′-di- and-tri-phosphates. Thus adenosine 5′-phosphate (AMP) yields adenosine 5′-diphosphate (ADP) and adenosine 5′-triphosphate (ATP). Adenosine 5′- and guanosine 5′-tetraphosphate have also been described.

Similarly the other ribonucleoside 5′-phosphates yield such di- and tri-phosphates as GDP, CDP, UDP, GTP, CTP and UTP. The 5′-monophosphates of adenosine, guanosine, cytidine, and uridine together with the corresponding di- and tri-phosphates all occur in the free state in the cell as do the deoxyribonucleoside triphosphates which are referred to as dAMP, dADP, dATP, dTMP, dTDP, dTTP, etc.

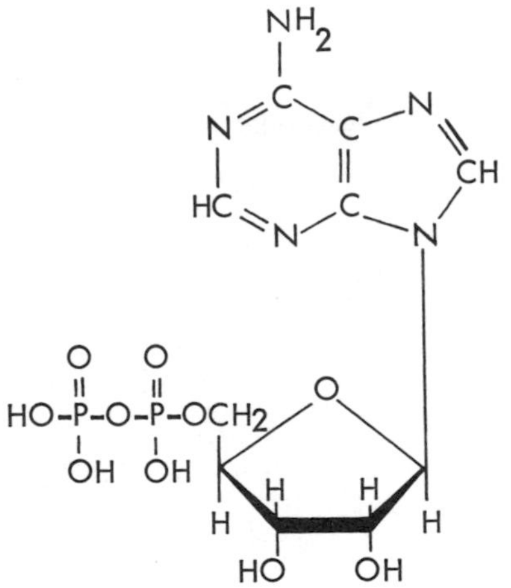

adenosine diphosphate (ADP)

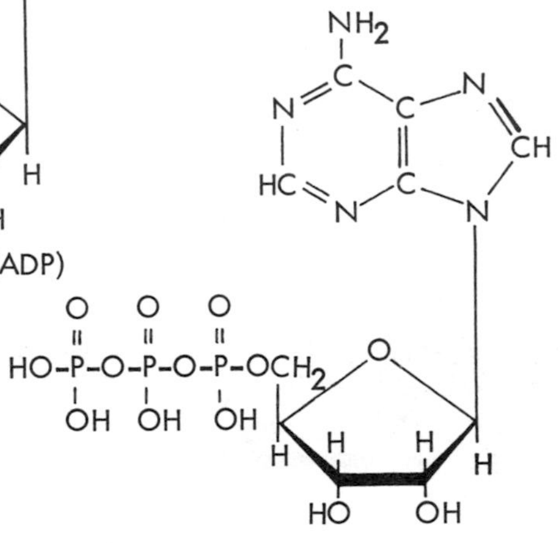

adenosine triphosphate (ATP)

Ribonucleoside 3′, 5′-diphosphates and 2′, 3′ cyclic monophosphates can be formed on hydrolysis of RNA molecules, and ribonucleoside 3′, 5′ cyclic monophosphates occur in many tissues where they play multiple roles in the regulation of metabolic pathways, as do guanosine tetraphosphate (ppGpp) and adenosine hexaphosphate (pppAppp) (see [50] for review).

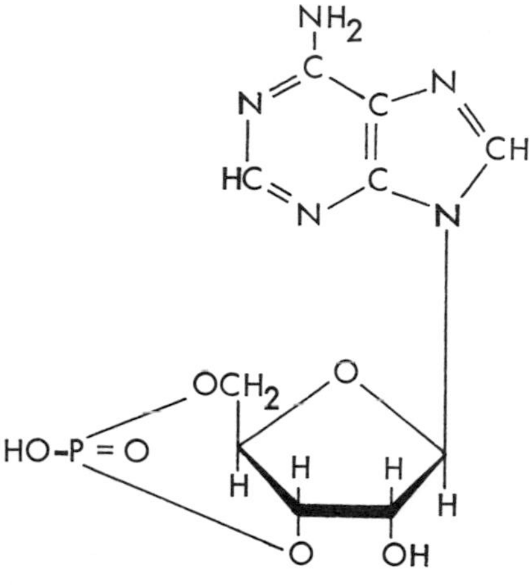

adenosine
3':5'-cyclic phosphate (cAMP)

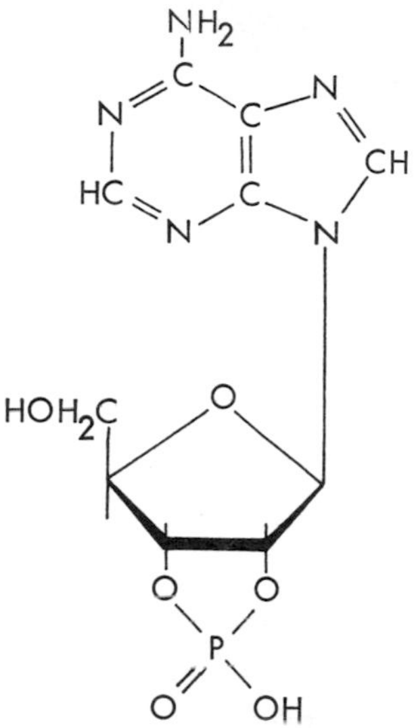

adenosine
2':3'-cyclic phosphate

2.7 THE PRIMARY STRUCTURE OF THE NUCLEIC ACIDS

The internucleotide bond in both DNA and RNA is the phosphodiester linkage. In the case of DNA where C-4′ in the sugar is occupied in ring formation and C-2′ carries no hydroxyl group, only the hydroxyl groups at positions 3′ and 5′ are available for internucleotide linkages. The primary structure of the polynucleotide chain in DNA is shown in Fig. 2.1.

In the case of RNA where there is a hydroxyl group at the 2′ position, it is possible to postulate a 2′,5′ linkage rather than the 3′,5′ linkage which occurs in DNA. However, hydrolysis with phosphodiesterase from snake venom yields nucleoside 5′-monophosphates and hydrolysis with phosphodiesterase from spleen yields the nucleoside 3′-monophosphates, suggesting that the internucleotide linkage in most RNA is identical to that in DNA. An unusual nucleotide linkage can be shown to occur in 2′,5′ linked oligo (A) induced by the antiviral agent interferon [12].

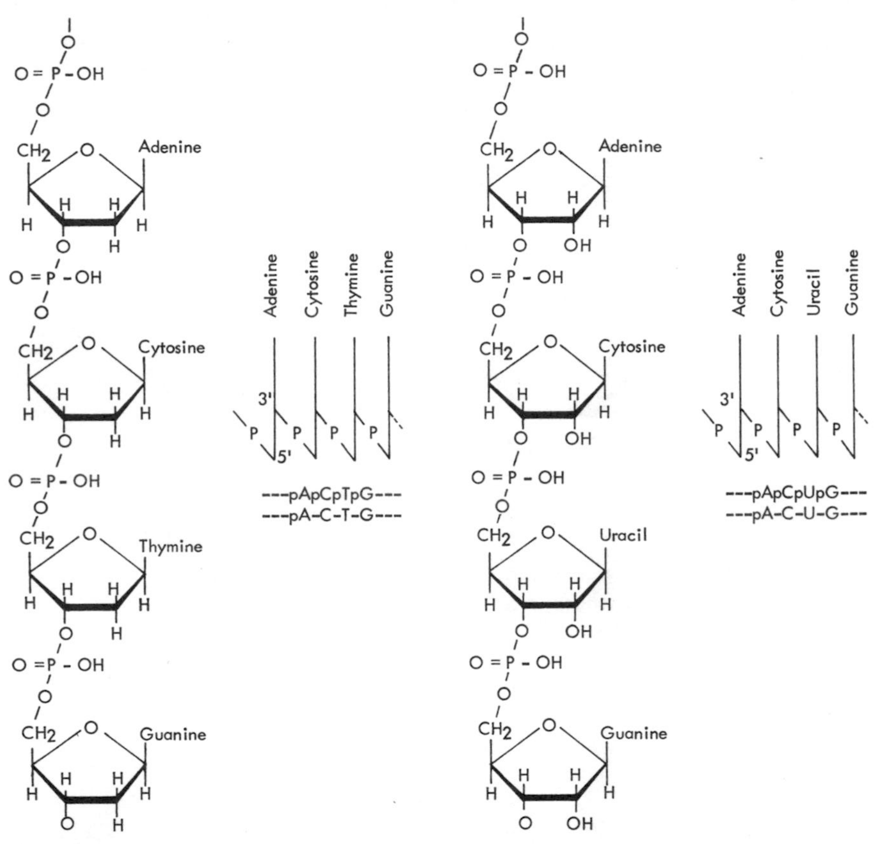

Fig. 2.1 A section of the polynucleotide chain in DNA (on the left) and RNA (on the right). The shorthand notations are shown alongside.

2.8 SHORTHAND NOTATION

The representation of polynucleotide chains by complete formulae is clumsy and it has become customary to use the schematic system illustrated in Fig. 2.1 where the chain, shown in full on the left, is abbreviated as on the right. The vertical line denotes the carbon chain of the sugar with the base attached at C-1′. The diagonal line from the middle of the vertical line indicates the phosphate link at C-3′ while that at the end of the vertical line remote from the base denotes the phosphate link at C-5′. This system may be used for either RNA or DNA.

To simplify further the representation of specific polynucleotides in shorthand notation, the following system originally suggested by Heppel,

Oritz and Ochoa [13] and now embodied in the rules of the CBN is commonly employed. A phosphate group is denoted by p; when placed to the right of the nucleoside symbol, the phosphate is esterified at C-3′ of the ribose moiety; when placed to the left of the nucleoside symbol, the phosphate is esterified at the C-5′ of the ribose moiety. Thus, UpUp or U-Up is a dinucleotide with one phosphate monoesterified at C-3′ of a uridine residue and a phosphodiester bond between C-5′ of that same uridine residue and C-3′ of the other uridine group. UpU or U-U would be the dinucleoside monophosphate, uridylyl (3′ 5′) uridine. The letter p *between* nucleoside residues may be replaced by a hyphen.

The following examples illustrate the method:

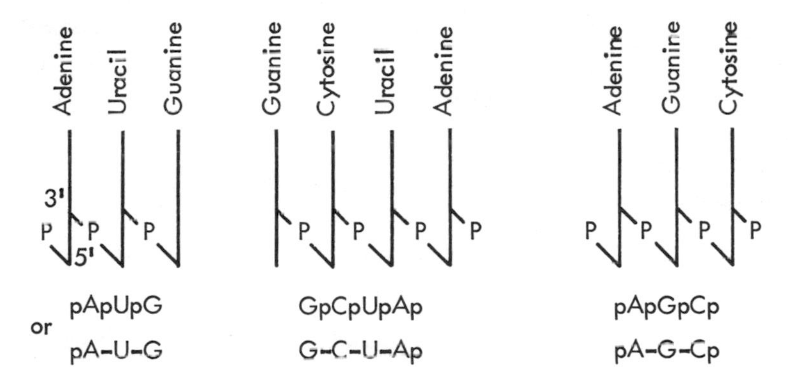

pApUpG
or
pA-U-G

GpCpUpAp
G-C-U-Ap

pApGpCp
pA-G-Cp

The letters A, G, C, U and T represent adenosine, guanosine, cytidine, uridine, and ribothymidine respectively. The prefix d (e.g. dA) may be used to indicate a deoxyribonucleoside.

There is now agreement on nomenclature for cyclic-terminal nucleotides. They may be represented by using the symbol cyclic-p or >p to indicate a 2′,3′-phosphoryl group or by means of the symbol p. Thus, U-cyclic-p or U>p is uridine 2′,3′-phosphate and UpU-cyclic-p or UpU>p is the cyclic-terminal dinucleotide.

2.9 ALKALINE HYDROLYSIS OF RNA

Alkaline digestion of RNA yields both the nucleoside 2′ and 3′-monophosphates. These isomers are readily converted into a mixture of both under acid conditions, but are stable without interconversion in alkaline solution. Interconversion involves the formation of a cyclic intermediate, the nucleoside 2′,3′-phosphate (II in Fig. 2.2), which yields on hydrolysis a mixture of the 2′- and 3′-phosphates. For example,

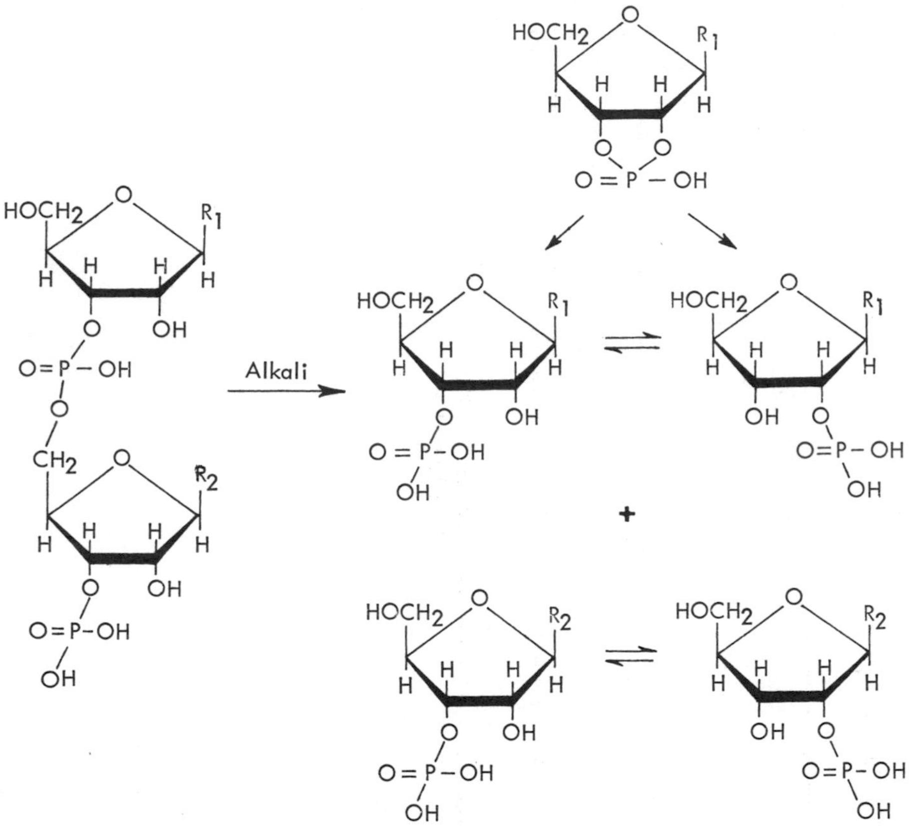

Fig. 2.2 Hydrolysis of a trinucleotide by alkali. R represents a purine or pyrimidine base.

the trinucleotide I shown in Fig. 2.2 yields the cyclic nucleoside 2′:3′-phosphates (II) which then give rise to a mixture of nucleoside 2′- and 3′-phosphates (III and IV).

Hydrolysis by pancreatic ribonuclease also proceeds via the formation of cyclic phosphates (Chapter 6). On the other hand, treatment with venom phosphodiesterase would yield nucleoside 5′phosphates (see Chapter 6).

2′-*O*-Methylribose is found in rRNA and tRNA [24]. The inter-nucleotide bond adjacent to a ribose residue methylated at the 2-position is of course resistant to hydrolysis by alkali and by pancreatic ribonuclease since cyclization between the 2′- and 3′-positions is not possible. Moreover, pancreatic RNase cannot degrade such internucleotide bonds which would otherwise be susceptible to its nucleolytic action.

An unusual primary structure is found in the case of poly ADP-ribose, a compound which is synthesized in large amounts in eukaryotic cell nuclei and frequently found attached to nuclear proteins. It has the structure

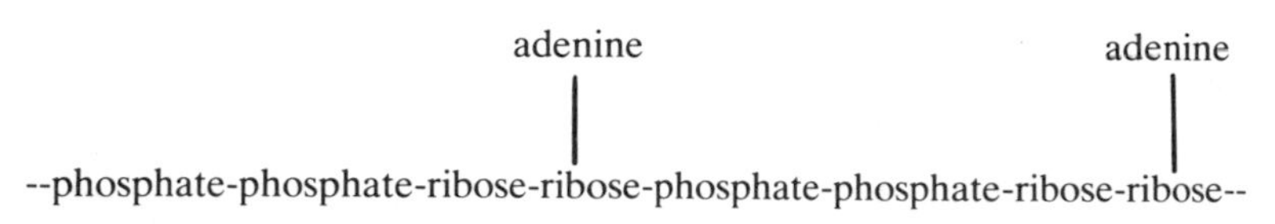

the two ribose moieties being linked by a 2′-1′ bond [48]. A further unusual primary structure is found on the 5′ end of many mRNA molecules and known as a 'cap' (Chapter 5). It has the general structure 7-Me-GpppNpNp- [51].

2.11 THE SECONDARY STRUCTURE OF DNA

X-ray diffraction has been extensively employed in the study of the molecular architecture of DNA by Astbury [14] and later by Franklin and Gosling [15] and on a very extensive scale by Wilkins and his colleagues [16–18]. Using early information obtained by this technique and chemical observations, Watson and Crick [19–21] in 1953 put forward the view that the DNA molecule is double stranded and in the form of a right-handed helix with the two polynucleotide chains wound round the same axis and held together by hydrogen bonds between the bases.

By making scale models they were able to show that the bases could fit in if they were arranged in pairs of one pyrimidine and one purine. The equivalence of adenine and thymine, and guanine and cytosine in most naturally occurring DNA molecules, first observed by Chargaff [26] suggested that the most likely hydrogen-bonding configuration of base pairs be of these two types (Fig. 2.3). Other base-pairing arrangements have since been suggested [25] but have only been shown to occur in RNA in unusual situations where they are involved in tertiary structure stabilization (Chapter 3).

The most important consequence of the base-pairing configuration found in DNA is that the order in which the bases occur in one chain automatically determines the order in which they occur in the other complementary chain. Apart from this essential condition, there are no restrictions on the sequence of the bases along the chains. The pairs of bases are flat and may be stacked one above the other like a pile of plates

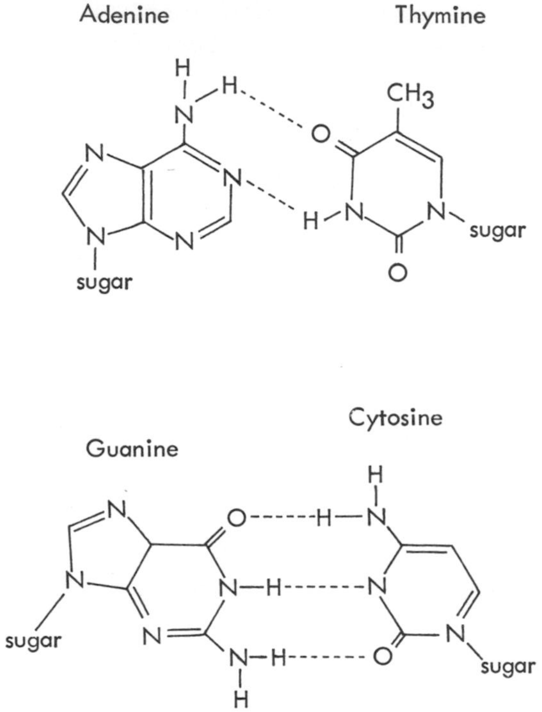

Fig. 2.3 The normal base-pairing arrangement found in DNA.

so that the molecule is readily represented as a spiral staircase with the base pairs forming the treads (Fig. 2.4). The two polynucleotide chains are of opposite polarity in the sense that the internucleotide linkage in one strand is 3′→5′ while in the other it is 5′→3′. The two helices are right handed and cannot be separated without unwinding. The pitch of the duplex is 3.4 nm and since there are 10 base pairs in each turn of the helix, there is a distance of 0.34 nm between each base pair. The diameter of the helix is 2 nm.

While the basic model put forward by Watson and Crick remains close to the accepted structure of the DNA molecule in solution, the more refined X-ray diffraction studies of Wilkins and his colleagues have shown that DNA fibres can have three possible structures (Table 2.3) with the B structure corresponding closely to the original Watson–Crick model. The A and C structures are also right-handed double helices, but differ in the pitch and in the number of bases per turn. In addition, the bases are not flat but are tilted in both A and C conformations. Their biological

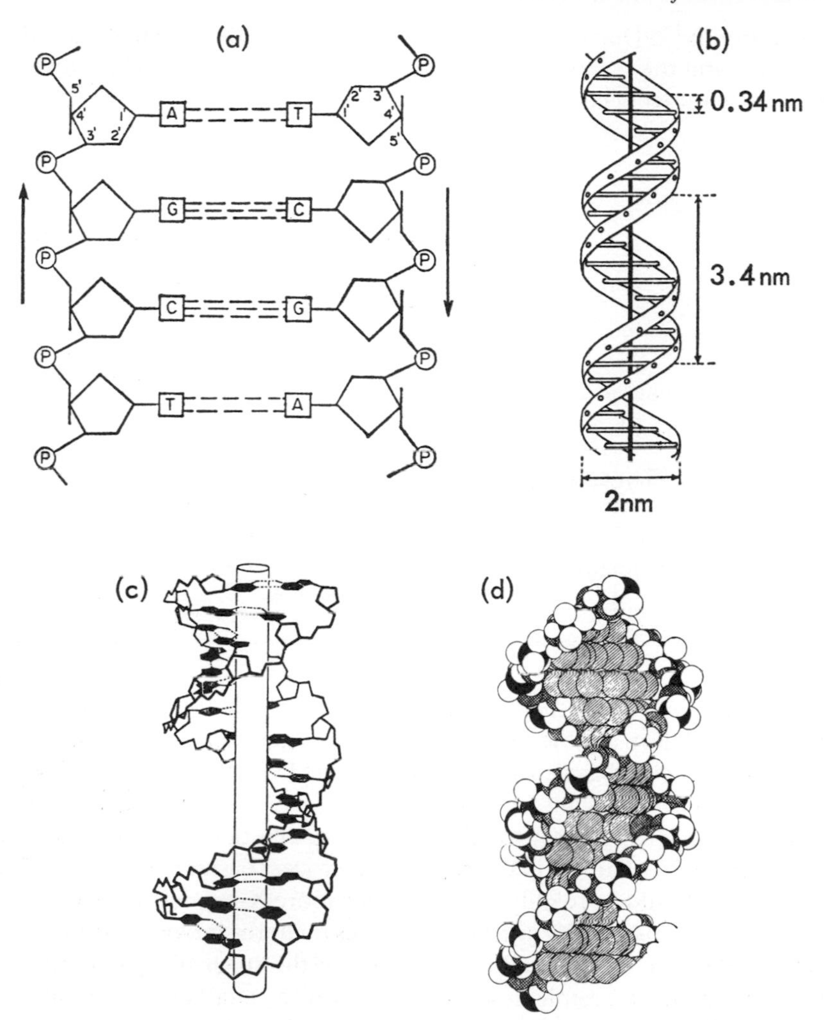

Fig. 2.4 Various ways of diagrammatic representation of DNA: (a) showing polarity and base pairing but not helical twist, (b) showing helical twist and helix parameters but not base pairs, (c) showing helix and base pairs, (d) space filling accurate representation.

significance is not clear but the A conformation is believed to be very close to the structure adopted by double-stranded RNA and by DNA–RNA hybrids [27]. Because of the presence of the extra 2′-hydroxyl group, RNA appears to be unable to adopt the B conformation. Thus when it is engaged as a template for making RNA (Chapter 10), the DNA molecule must adopt the A conformation. Various organic solvents [28] and proteins [29] have been suggested to force the DNA from the B to the A form, and the transition from the B conformation to the C conformation appears to occur in concentrated salt solutions [22] and ethylene glycol [21].

Table 2.3 The different forms of DNA.

Form	Pitch (nm)	Residues per turn	Inclination of base pair from horizontal
A			
Na salt 75 per cent relative humidity	2.8	11	20°
B			
Na salt 92 per cent relative humidity	3.4	10	0°
C			
Li salt 66 per cent relative humidity	3.1	9.3	6°
DNA–RNA hybrid	2.8	11	20°

Since the late 1970s there have been several challenges to the B structure of the DNA double helix. It has been suggested that the fibre diffraction data could as readily fit a side by side (SBS) model which would have the helix changing from a right- to a left-handed sense at regular intervals [32]. The main attraction of this model is that it would make the unwinding of the duplex in the replication process easier to envisage. The discussion should be readily resolved when more X-ray data from crystals become available. X-ray diffraction studies on fibres provide information which relates to the general organization of the sugar phosphate chains but no detail at atomic resolution. However, analysis of crystals of defined DNA fragments can allow the unequivocal positioning of each atom in the helical framework and is a wholly definitive technique. Large oligonucleotides are not readily crystallized so the data currently available relates to small molecules which may well have an atypical structure because the end effects which do not normally make a substantial contribution to polynucleotide conformation may play a dominant role. Crystalline dpApTpApT has been shown to adopt a right-handed helical configuration with two base pairs in each segment [33]. On the other hand dCpGpCpGpCpG crystallizes as a left-handed double helix with twelve base pairs per turn known as Z DNA [34]. It is therefore clear that there is scope for sequence-dependent structural changes along the length of chromosomal DNA.

There have been many experimental attempts to relate the structure of DNA in solution to the structure in crystals or fibres, using techniques such as circular dichroism [22] and low-angle X-ray scattering [23, 24]. One of the best experimental systems for the analysis of secondary structure in solution is circular superhelical DNA (Section 2.12) and experimental and theoretical analysis on these molecules suggests that the solution structure is close to, but not entirely compatible with the B structure of DNA. The molecules are slightly unwound to give 10.4 rather than an integral 10 base pairs per turn [35]. An additional and separate phenomenon is the dynamic secondary structure of DNA. The double helix is not a totally fixed or rigid molecule but undergoes considerable internal deformation in a continuous manner. This aspect of DNA secondary structure is best shown by tritium-exchange experiments [36] which indicate that small segments of the double helix can swing apart and protrude into the external medium in a manner closely dependent on the environment of the DNA molecule. Whereas X-ray diffraction experiments and circular dichroism show the average conformations that the DNA molecule as a whole can assume, tritium-exchange experiments show the amount of deformation and twisting of these structures that can occur in localized areas. Fluorescence experiments have also confirmed that internal distortions of the DNA molecule occur to a hitherto unsuspected degree [39].

Multistranded structures for DNA have frequently been invoked, especially where the component strands are synthetic polynucleotides. One such situation where tetraplexes could occur in chromosomal DNA has been shown to be theoretically feasible if the sequence contains an inverted repeat. Tetraplex DNA has long been suggested as a necessary preliminary structure for genetic recombination [37].

The accepted shorthand method of describing two DNA strands has been to have the strand of 5′–3′ polarity on the top line of the sequence with the complementary strand of opposite polarity lying below.

Thus for example:

5′-AGGTC-3′
3′-TCCAG-5′

2.11 THE SECONDARY STRUCTURE OF RNA

While RNA molecules do not possess the regular interstrand hydrogen-bonded structure characteristic of DNA, they have the capacity to form double-helical regions. These helices can be formed between two separate RNA chains, but are more frequently found between two segments of the same chain folded back on itself. The secondary structure is similar to the A form of DNA with tilted bases, since the 2′-OH hinders B structure formation. The helical regions formed in this manner are seldom regular as the segments on the chain brought into opposition do not have entirely complementary sequences so non-bonded residues 'loop out' of the structure (Fig. 2.5). In some RNA molecules, such as tRNA, in the region of 70% of the bases are involved in secondary structure interactions (Chapter 5).

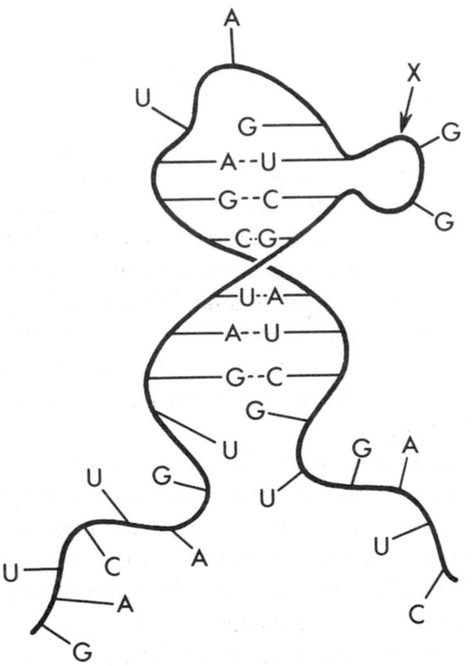

Fig. 2.5 A possible secondary structure of RNA illustrating a helical region with complementary base pairing [38]. A looped-out portion of the helix is shown at X.

Structures of this type frequently show unusual base pairing in addition to the expected A.U and G.C pairs and G.U pairing can be observed. Helix stability appears to require at least three conventional base pairs. The end of a hairpin loop appears always to have a minimum of three nucleotides.

2.12 THE TERTIARY STRUCTURE OF DNA

It is now clear that most of the DNA from animal and bacterial sources is restrained in the cell in a superhelical configuration (Plate 1). Small viral

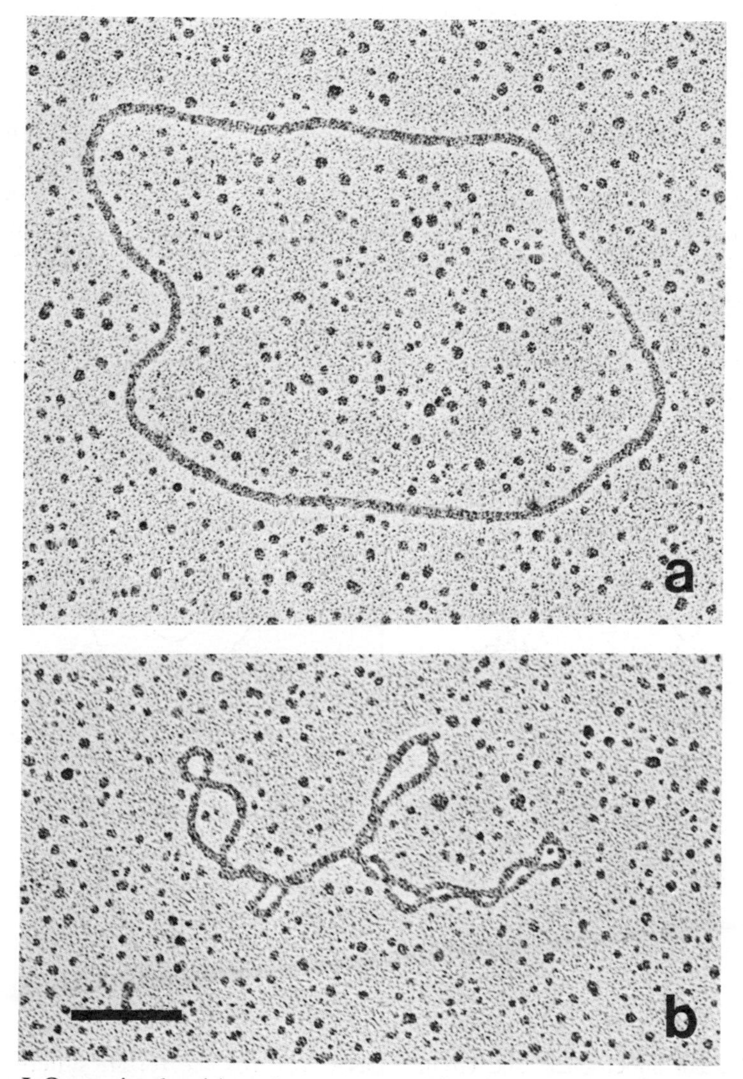

Plate I Open-circular (a) and supercoiled (b) forms of PM2 virus DNA. Bar represents 0.2 μm. (By courtesy of Dr Lesley Coggins).

DNA molecules can readily be isolated as supercoiled circles and the larger linear and circular molecules are divided into superhelical quasi-circular domains [40]. Supercoils are essentially covalently closed circles of DNA which have very different physical, chemical and biological properties from circular DNA molecules which have a break in one or other strand.

Covalently closed cyclic DNA molecules have no breaks at all in either strand, with the consequence that any change in the secondary structure of the DNA which affects the number of residues per turn of the helix must have concomitant effects on the tertiary structure of the molecule. Such DNA can best be envisaged as a simple cyclic structure which has been hypothetically opened, had the two strands of the double helix unwound by a few turns, and resealed. The resulting molecule could try to rewind the two strands back to their normal structure but would be unable to do so because of the covalent closure, with the result that the circular duplex itself will take on 'superhelical' or 'supercoiled' turns to compensate (Fig. 2.6). The number and nature of the supercoiled turns will, of course, depend on the difference between the secondary structure of the DNA when it was sealed and the secondary structure under the conditions of observation.

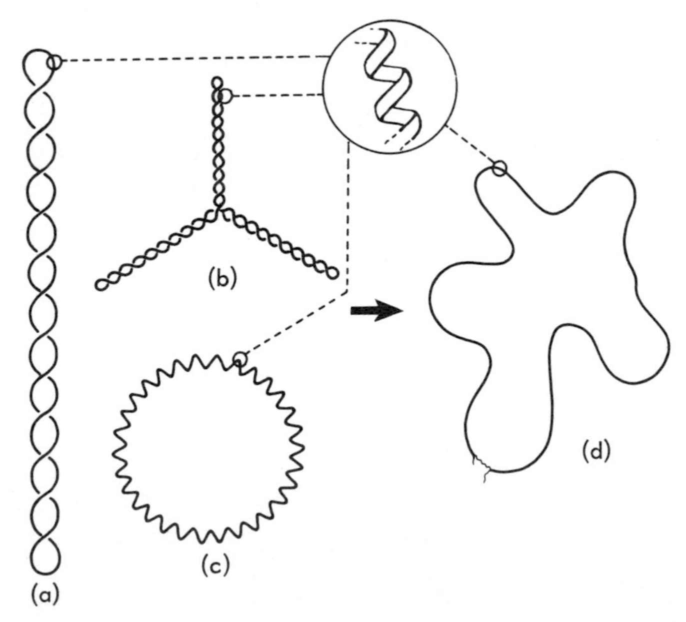

Fig. 2.6 Superhelical and circular DNA. (a), (b) and (c) are different conformations of superhelical DNA. A nick in one of the two component DNA strands leads to an open circular molecule (d).

Covalently closed double-stranded DNA differs very considerably from linear or open cyclic molecules of the same size and base composition. The two strands of the DNA are unable to separate and thus a high temperature is required to disrupt the structure of the supercoiled DNA. When these DNA molecules eventually do melt, the two strands cannot separate, but the entire molecule collapses into a compact, fast-sedimenting complex of the two interwound random coils. It has been suggested [41] that the secondary structure of such DNA can differ detectably from that of the open circular form because of the tertiary restraints on the molecule. SV40 DNA has been shown to have at least one region of unwound DNA [42] and polyoma to have at least three [43]. The intrinsic viscosity is low, and the sedimentation rate and electrophoretic mobility about 20 per cent faster than in the open circular molecules of the same size. The tertiary structure of the superhelix is most commonly represented by a straight interwound superhelix (Fig. 2.6). Other forms such as toroidal and branched structures have been shown to exist in solution [40].

The superhelix density of covalently closed cyclic DNA is defined as the number of superhelical turns per 10 base pairs. Superhelix density is determined by 'partial denaturation' of the DNA either with alkali [45] or with an intercalating dye [44]. The binding of the dye molecules to the DNA results in an increase in the number of residues per turn in the duplex and thus a decrease in the number of superhelical turns. The sedimentation rate of the DNA therefore drops to a minimum and then rises as the binding of further dye molecules causes the DNA to take on superhelical turns of the opposite sense. Frequently, superhelix density is estimated by gel electrophoresis in the presence of increasing amounts of an enzyme which abolishes superhelical turns [47]. The superhelix density of most covalently closed DNA molecules is -0.06 in neutral caesium chloride [44, 45]. Superhelix density is affected by both temperature and ionic strength [46]. Electron micrographs of various supercoiled molecules are shown in Plate I.

2.13 THE TERTIARY STRUCTURE OF RNA

X-ray crystallographic data of many of the small RNA molecules has shown that it is possible for extensive folding of the partially duplex arms across each other to occur with subsequent hydrogen bonding of the bases not involved in secondary structure stabilizing the folds (Fig. 2.7). The type of hydrogen bonding involved is frequently not that found in the conventional Watson–Crick base pair and unusual orientations of a pair, or a trio of bases are common (Fig. 2.8) [49]. Stacking interactions are also important in the maintenance of tertiary structure.

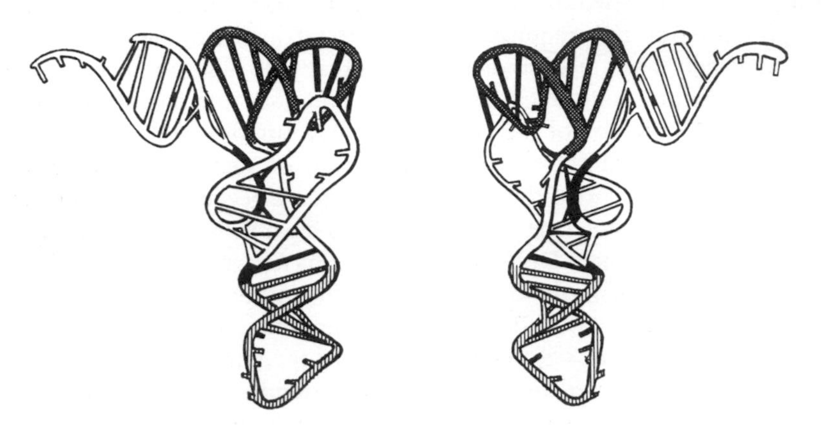

Fig. 2.7 Two different views of the same molecule of yeast $tRNA^{Phe}$ showing base-paired areas and tertiary structure. After Rich [52].

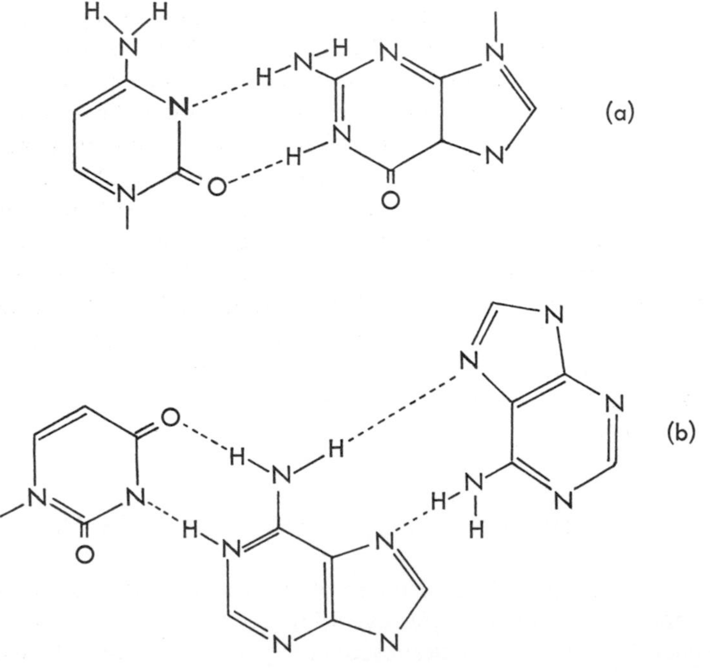

Fig. 2.8 Some examples of unusual hydrogen bonds stabilizing RNA tertiary structure: (a) a *trans* rather than *cis* base pair, (b) a complex of two adenines and one uracil.

REFERENCES

1 Adler, M., Weissman, B. and Gutman, A.B. (1958), *J. Biol. Chem.*, **230,** 717.
2 Littlefield, J. W. and Dunn, D.B. (1958), *Biochem. J.*, **70,** 642.
3 Smith, J.D. and Dunn, D.B. (1959), *Biochem. J.*, **72,** 294.
4 Davis, F.F., Carlucci, A.F. and Roubein, I.F. (1959), *J. Biol. Chem.*, **234,** 1525.
5 Nakanishi, K., Blobstein, S., Funamizu, M., Furutachi, N., Van Lear, G., Grunberger, D., Lanks, K.W. and Weinstein, I.B. (1971), *Nature New Biol.*, **234,** 107.
6 Thiebe, R., Zachau, H.G., Baczymskyj, L., Biemann, K. and Sonnenbichler, J. (1971), *Biochim. Biophys. Acta*, **240,** 163.
7 Bendich, A. (1955), in *The Nucleic Acids* (ed. E. Chargaff and J.N. Davidson), Academic Press, New York, Vol. 1, p. 81.
8 Bloomfield, V.A., Crothers, D.M., Tinoco, I. (1974), *Physical Chemistry of Nucleic Acids*, Harper & Row, New York.
9 Nichimura, S. (1972), in *Progress in Nucleic Acids Research and Molecular Biology* (eds. J.N. Davidson and W.E. Cohn), Academic Press, New York, Vol. 12, p. 49.
10 Hutchinson, D.W. (1964), *Nucleosides and Coenzymes*, Methuen, London.
11 Michelsen, A.M. (1963), *The Chemistry of Nucleosides and Nucleotides*, Academic Press, New York.
12 Kerr, I.M. and Brown, R.E. (1978), *Proc. Natl. Acad. Sci. USA*, **75,** 256–260.
13 Heppel, L.A., Ortiz, P.J. and Ochoa, S. (1967), *J. Biol. Chem.*, **229,** 679.
14 Astbury, W.T. (1974), *Symp. Soc. Exp. Biol.*, **1,** 66.
15 Franklin, R. and Gosling, R.G. (1953), *Nature*, **171,** 7400, **172,** 156.
16 Langridge, R., Wilson, H.R., Hooper, C.W., Wilkins, M.H.F. and Hamilton, L.D. (1960), *J. Mol. Biol.*, **3,** 547.
17 Fuller, W., Wilkins, M.F.H., Wilson, H.R. and Hamilton, L.D. (1965), *J. Mol. Biol.*, **12,** 60.
18 Davies, D.R. (1967), *Annu. Rev. Biochem.*, **36,** 321.
19 Watson, J.D. and Crick, F.H.C. (1953), *Nature*, **171,** 737, 964.
20 Watson, J.D. (1968), *The Double Helix*, Atheneum, New York.
21 Olby, R. (1964), *The Path to the Double Helix*, Macmillan.
22 Tunis-Schneider, M.J.B. and Maestre, M.F. (1970), *J. Mol. Biol.*, **52,** 521.
23 Bram, S. (1971), *J. Mol. Biol.*, **58,** 277.
24 Bram, S. (1973), *Cold Spring Harbour Symp. Quant. Biol.*, **38,** 83.
25 Arnott, S. (1970), *Science*, **167,** 1694.
26 Chargaff, E. (1963), *Essays on Nucleic Acids*, Elsevier/North Holland.
27 Tunis, M.J.B. and Hearst, J.E. (1958), *Biopolymers*, **6,** 128.
28 Brahms, J. and Mommaerts, W.H.F.M. (1964), *J. Mol. Biol.*, **10,** 73.
29 Shih, T.Y. and Fasman, G.D. (1971), *Biochem.*, **10,** 1675.
30 Feughelman, M., Langridge, R., Seeds, W.E., Stokes, A.R., Wilson, H.R., Hooper, C.W., Wilkins, M.F.H., Barclay, R.K. and Hamilton, L.D. (1955), *Nature*, **175,** 834.
31 Green, G. and Mahler, H.F. (1971), *Biochem.*, **10,** 2200.

32 Rodley, G.A., Scobie, R.S., Bates, R.H.T. and Lewitt, R.M. (1976), *Proc. Natl. Acad. Sci. USA,* **73,** 2959–2963.
33 Viswamitra, M.A., Kennard, O., Jones, P., Sheldrick, G.M., Salisbury, S., Flavello, L. and Shakked, Z. (1978), *Nature,* **273,** 687–690.
34 Wang, A.H., Quigley, G.J., Kolpak, F.J., Crawford, J.L., von Boom, J.H., van der Mare, G. and Rich, A. (1979), *Nature,* **282,** 680–686.
35 Wang, J.C. (1979), *Proc. Natl. Acad. Sci. USA,* **76,** 200–203.
36 McConnell, B. and Von Hippel, P.H. (1970), *J. Mol. Biol.,* **50,** 297–308.
37 Morgan, A.R. (1979), *Trends in Biochemical Sciences,* **4,** N244–N248.
38 Fresco, J.R. (1963), in *Informatinal Macromolecules* (ed. H.J. Vogel, V. Bryson and J.O. Lampen), Academic Press, New York, p. 121.
39 Wahl, P., Paoletti, J. and Pecq, J.B. (1970), *Proc. Natl. Acad. Sci. USA,* **65,** 417–421.
40 Campbell, A.M. (1978), *Trends in Biochemical Sciences,* **3,** 104–108.
41 Campbell, A.M. and Lochhead, D.S. (1971), *Biochem. J.,* **123,** 661.
42 Beard, P., Morrow, J.F. and Berg, P. (1973), *J. Virol.,* **12,** 1303.
43 Monjardino, J. and James, A.W. (1975), *Nature,* **225,** 249.
44 Wang, J.C. (1974), *J. Mol. Biol.,* **89,** 783.
45 Pulleyblank, D.E. and Morgan, A.R. (1975), *J. Mol. Biol.,* **91,** 1.
46 Wang, J.C. (1968), *J. Mol. Biol.,* **43,** 25.
47 Keller, W. (1975), *Proc. Natl. Acad. Sci. USA,* **72,** 4876–4880.
48 Hayaishi, O. (1976), *Trends in Biochemical Sciences,* **1,** 9–10.
49 Robertus, J.D., Lodno, J.G., Finch, J.T., Rhodes, D., Brown, R.S., Clark, B.F.C. and Oklug, A. (1974), *Nature,* **250,** 546–550.
50 Travers, A. (1980), *Nature,* **283,** 16.
51 Shatkin, A.J. (1976), *Cell,* **9,** 645–653.
52 Rich, A. and Schimmel, P.R. (1977), *Act. Chem. Res.,* **10,** 385.
53 Ulbricht, T.L.U. (1964), *Purines, Pyrimidines and Nucleotides,* Pergamon, London.

Chemical analysis of DNA

3

3.1 BASE COMPOSITION ANALYSIS

The common monomeric units of DNA are the four deoxyribonucleotides containing the bases adenine, cytosine, guaninc, and thymine. Many DNAs, however, contain small amounts of other bases, e.g. 5-methylcytosine, which is particularly abundant in wheat germ DNA (Table 3.1). In a few bacteriophages, one of the common pyrimidine bases is completely replaced by a different pyrimidine base, e.g. in T2, T4, and T6 5-hydroxymethylcytosine completely replaces cytosine, and in PBS 1 (a bacteriophage which attacks *Bacillus subtilis*) uracil replaces thymine.

Table 3.1 Molar proportions of bases (as moles of nitrogenous constituents per 100 g-atoms P) in DNAs from various sources (data from various authors).

Source of DNA	Adenine	Guanine	Cytosine	Thymine	5-Methyl-cytosine
Bovine thymus	28.2	21.5	21.2	27.8	1.3
Bovine spleen	27.9	22.7	20.8	27.3	1.3
Bovine sperm	28.7	22.2	20.7	27.2	1.3
Rat bone marrow	28.6	21.4	20.4	28.4	1.1
Herring testes	27.9	19.5	21.5	28.2	2.8
Paracentrotus lividus	32.8	17.7	17.3	32.1	1.1
Wheat germ	27.3	22.7	16.8	27.1	6.0
Yeast	31.3	18.7	17.1	32.9	—
Esch. coli	26.0	24.9	25.2	23.9	—
Mb. tuberculosis	15.1	34.9	35.4	14.6	—
ϕX174	24.3	24.5	18.2	32.3	—

Methods for determining the molar proportions of bases by hydrolysis and chromatography are discussed in detail by Bendich [1].

The results of the analysis of a number of DNAs as shown in Table 3.1 reveal wide variations in the molar proportions of bases in DNAs from different species although the DNAs from the different organs and tissues of any one species are essentially the same. Extensive tables showing the molar proportions of bases have been published [2].

It was Chargaff [3] who first drew attention to certain regularities in the composition of DNA. The sum of the purines is equal to the sum of the pyrimidines; the sum of the amino bases (adenine and cytosine) is equal to the sum of the keto (oxo) bases (guanine and thymidine); adenine and thymine are present in equimolar amounts, and guanine and cytosine are also found in equimolar amounts. This equivalence of A and T and of G and C is of the utmost importance in relation to the formation of the DNA helix (Chapter 2) and may be referred to as Chargaff's rule. There are two major deviations from the rule in Table 3.1. (a) In wheat germ DNA guanosine and cytosine are not present in equimolar amounts but this is explained by the scarcity of cytosine being compensated by the presence of 5-methylcytosine. (b) in ØX174 DNA and in the DNA of several similar small coliphages the adenine is not equimolar with thymine nor is guanine with cytosine. This is because ØX174 DNA is single-stranded.

DNAs fall into two main classes, the 'A-T rich type' in which adenine and thymine are in excess, and the much rarer 'G-C rich type' in which guanine and cytosine predominate.

The physical properties of DNA are strongly influenced by the percentage of G + C in the molecule, and the buoyant density of DNA in concentrated CsCl solutions is no exception. G + C rich DNA has a higher buoyant density than A + T rich DNA [2] and there is a linear relationship between the buoyant densities (ρ) of different DNAs and their G + C contents (see Fig. 3.1). This can be expressed by the relationship:

$$\rho = 1.660 + 0.098\ (\mathrm{GC})$$

where GC is the mole fraction of (G + C). The relative (G + C) content can also be determined from the thermal denaturation temperature which is discussed later (p. 38) [5], and from the ultraviolet spectrum of the DNA [6].

Several dyes and antibiotics have been shown to have strong binding specificity for either AT or GC base pairs. Thus distamycin and netropsin are AT specific and actinomycin is GC specific [57,58]. This property can be used to improve the separation of DNA molecules which differ in their GC content on CsCl gradients, since dye binding alters the buoyant

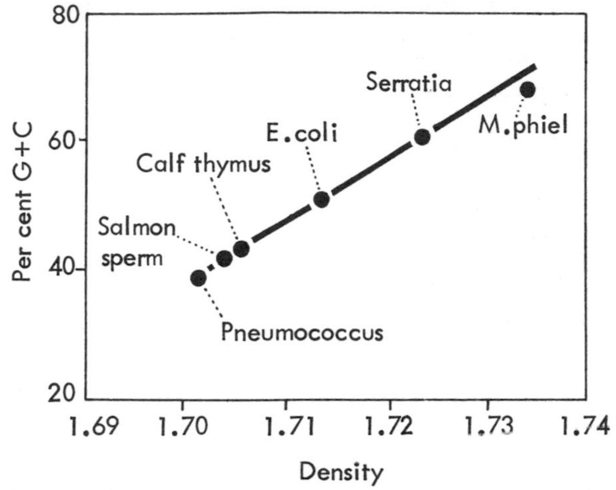

Fig.3.1 Relationship of density to content of guanine plus cytosine in DNAs from various sources [4].

density of the molecules. In addition, dyes such as the AT specific malachite green, or the GC specific phenyl neutral red, can be immobilized on polyacrylamide columns and used to fractionate DNA molecules of differing base composition [59].

On the basis of measurements of this sort the relative (G + C) contents of the DNAs from a wide variety of sources have been determined and are shown in Table 3.2. While mammalian DNAs show a (G + C) content between 40 and 45 per cent, the range of bacterial DNAs is much wider (30–75 per cent). The significance of these variations in base content has been discussed in relation to the taxonomy of bacteria [7, 8] and protozoa [9] and to the evolution of various organisms [10].

As will be evident later (p. 227), biosynthetic polydeoxyribonucleotides containing only A and T or only G and C can be prepared enzymically. Poly(dA-dT) occurs naturally as a satellite component in the DNA of the crab (*Cancer*) where it comprises 10–30 per cent of the total DNA. The land crab (*Gercarcinus*) has two satellite DNAs, poly(dA-dT) comprising 18 per cent of the total DNA and poly(dG)·poly(dC) comprising 3 per cent. Similar materials are found in other Crustacea [11].

3.2 MOLECULAR WEIGHT OF DNA

The molecular weights of DNA molecules are very difficult to determine accurately by the methods of classical chemistry since they range from 10^6 to more than 10^{10} (Table 3.3). Conventional analytical equilibrium ultra-

Table 3.2 The relative (G + C) content of DNAs from various sources [11,12].

Source of DNA	per cent (G + C)
Dictyostelium (slime mould)	22
M. pyogenes	34
Vaccinia virus	36
Bacillus cereus	37
B. megaterium	38
Haemophilus influenzae	39
Saccharomyces cerevisiae	39
Calf thymus	40
Rat liver	40
Bull sperm	41
Diplococcus pneumoniae	42
Wheat germ	43
Chicken liver	43
Mouse spleen	44
Salmon sperm	44
B. subtilis	44
T1 bacteriophage	46
Esch. coli	51
T7 bacteriophage	51
T3 bacteriophage	53
Neurospora crassa	54
Pseudomonas aeruginosa	68
Sarcina lutea	72
Micrococcus lysodeikticus	72
Herpes simplex virus	72
Mycobacterium phlei	73

centrifugation is unsatisfactory since the available instruments are not stable at speeds low enough to balance centrifugal forces against diffusion forces. The best absolute methods of DNA molecular weight determination are light scattering on low-angle instruments [12–14], equilibrium analytical ultracentrifugation in caesium chloride gradients [15], and viscoelastic relaxation [16]. Most other methods have an empirical basis and rely on the few light scattering experiments which have been performed for their calibration. These more empirical methods include the measurement of intrinsic viscosity [17], sedimentation rate [17], electron microscopy [18], and autoradiography [19]. Generally a combination of two or more of these methods can be used [20]. The advent of the laser has led to the development of a new technique for determining diffusion coefficients of large molecules from the Doppler shift in the wavelength of the scattered laser light. This technique together with conventional

Table 3.3 DNA molecular weights.

Source	Mol.Wt.	Length	Number of kilobase pairs (kb)
Bacteriophage ∅X 174	1.6×10^6	0.6 μm	—
Polyoma virus	3×10^6	1.5 μm	4.5
Mouse mitochondria	9.5×10^6	4.9 μm	14
Bacteriophage λ	33×10^6	17 μm	50
Bacteriophage T2 or T4	1.3×10^8	67 μm	200
Mycoplasma PPLO strain H-39	4×10^8	200 μm	600
H. influenzae chromosome	8×10^8	400 μm	1200
E. coli chromosome	3×10^9	1.5 mm	4500
Drosophila melanogaster chromosome	43×10^9	20 mm	70000

analytical velocity centrifugation to determine sedimentation coefficients should lead to reliable absolute molecular weight determinations [21]. The determination of the molecular weight of DNA has also been complicated by the difficulties experienced in the preparation of whole DNA molecules since high molecular weight DNA is very susceptible to hydrodynamic shearing forces and to the action of contaminating nucleases. Thus all molecular weight determinations on an unknown DNA must be carried out in nuclease-free (heat sterilized) conditions, without pipetting or any other manipulation which puts shear stress on the DNA; any methods which might themselves shear the DNA, such as viscosity determinations, must be used with caution. The size of large DNA molecules from individual chromosomes of eukaryotic cells has recently been determined by viscoelastic relaxation, and it appears that in *Drosophila* at least the DNA is contained in one complete molecule per chromosome [16]. The molecular weights of DNAs from a variety of sources are shown in Table 3.3.

In more recent years it has become customary to describe the size of DNA molecules in terms of kilobase pairs, rather than daltons of molecular weight (Chapter 2). A molecular weight of one million corresponds approximately to 1.5 kilobase pairs.

3.3 DNA DENATURATION. THE HELIX–COIL TRANSITION

When double-stranded DNA molecules are subjected to extremes of temperature or pH, the hydrogen bonds in the double helix are ruptured and the DNA collapses into two single-stranded molecules. If heat is used as the denaturant, the temperature at which this collapse occurs is known as the T_m or transition temperature. The absorption at 260 nm of any polynucleotide is due to that of its component bases. However, this absorption tends to be suppressed in the double-stranded DNA molecule where the bases are stacked above one another and inhibited from swinging out freely in solution by hydrogen bonds, and consequently it is very much lower than that of equimolar amounts of the component bases or nucleotides free in solution. This inhibition of absorption is to some extent relieved when the DNA molecule goes through the helix–coil transition and the bases are no longer so rigidly stacked although they still interact to some extent. As a consequence of this the absorption of DNA solutions rises by about 20–30 per cent as the DNA undergoes the melting process, and the transition process is usually measured by ultraviolet spectroscopy (Fig. 3.2). The increase in absorbance on melting is known as the *hyperchromic effect.*

Such a transition can be characterized by several factors:

(1) *The nature of the DNA.* Homogeneous DNA, such a viral DNA, melts over a short temperature range but heterogeneous DNA melts over a longer range.

(2) *The (G + C) content of the DNA.* The melting temperature of any DNA can be related to its (G + C) content since this base pair confers

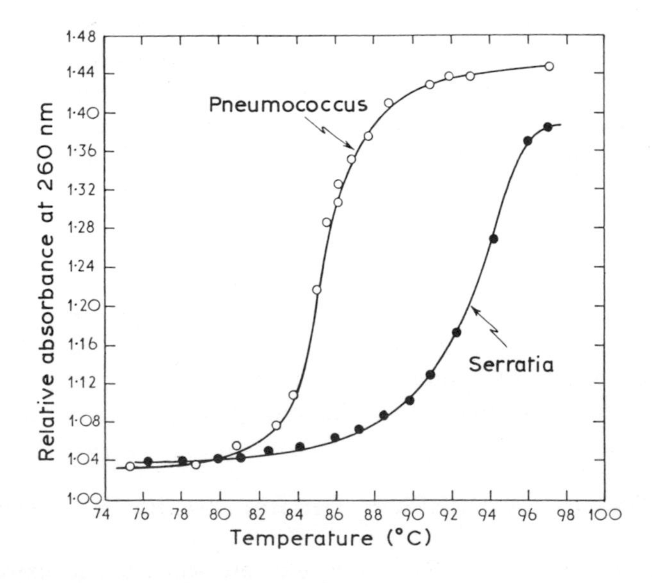

Fig. 3.2 Increase in absorbance at 260nm with rising temperature for DNA from two different sources [4].

extra stability on the molecule. The relationship can be expressed by the equation

$$\% \text{GC} = (T_m - 69.3)\, 2.44$$

in standard saline citrate (SSC, 0.15 M-NaCl. 0.015 M-citrate, pH 7) and by the equation

$$\% \text{GC} = (T_m - 53.9)\, 2.44$$

in one-tenth molar SSC [22]. In DNA molecules such as those from bacteriophage λ, in which some regions are richer in (G+C) than others, the transitions of both regions can be observed.

(3) *The nature of the solvent.* In low concentrations of counterion the transition temperature is low and its width broad. At higher concentrations of counterion the T_m is raised and with width of the transition becomes sharp. However, if 'denaturing' salts such as sodium perchlorate are present, addition of more salt will lower the T_m since the rupture of apolar bonds caused by the anion will overcome the ionic stabilization of the cation.

The helix–coil transition is also associated with a change in density of the DNA molecule, the single-stranded DNA being more dense than the equivalent double-stranded form with the same (G+C) content.

3.4 THE RENATURATION OF DNA. C_0t VALUE ANALYSIS

When two DNA strands are returned from the extreme conditions which caused them to melt to their original state they may reassociate to form a double helix again. However, whether or not they do so depends on a variety of factors. In principle the process should follow simple second-order kinetics but, because of the high probability of two sequences which were not preciously matched forming an imperfect union which is then difficult to dissociate, the degree of renaturation can vary from less than 1 per cent according to the nature of the sample and the conditions of renaturation.

(1) *The nature of the sample.* Simple sequence DNA molecules such as $d(G)_n \cdot d(C)_n$ have no difficulty finding the appropriate sequence with which to reanneal and so so without difficulty. However, if a eukaryotic DNA sample from a large genome is reannealed, clearly some of the sequences will have to encounter many other non-complementary sequences in solution before they find the correct partner. Thus if most of the DNA from eukaryotic cells is quickly cooled to a low temperature, so that the diffusion of the DNA in solution is inhibited. few of the single strands will reassociate.

(2) *The temperature of the reassociation process.* At very low temperatures (4°C) not only is diffusion limited but, for a DNA molecule which has become mismatched with a strand having only a few complementary bases, the opportunities to break away and continue its search for the correct complementary strands are reduced. Consequently DNA which is heated beyond the T_m and then quickly cooled to a low temperature will be denatured whereas solutions maintained at high temperatures below the T_m may renature.

(3) *The size of the DNA fragments.* Large string-like fragments of single-stranded DNA encounter diffusion problems and frequently cannot reanneal correctly since this effectively reduces their opportunities for finding complementary strands. For this reason DNA is often sheared for reannealing experiments.

(4) *The ionic strength of the solution.* Two highly charged DNA molecules are likely to repel one another, and the presence of salt is necessary to mask this repulsion in renaturation.

(5) *The concentration of the DNA.* Clearly, at higher concentrations the probability of two complementary strands encountering each other is raised.

(6) *The time allowed for reannealing experiments.* If renaturation is allowed under ideal conditions, two DNA samples of identical concentration should take different times to reanneal according to the genome size [see (1)]. For this reason the term $C_o t$ *value* has been defined for the study of reannealing of DNA and also for the study of the formations of DNA–RNA hybrids. C_o represents the DNA concentration in moles of

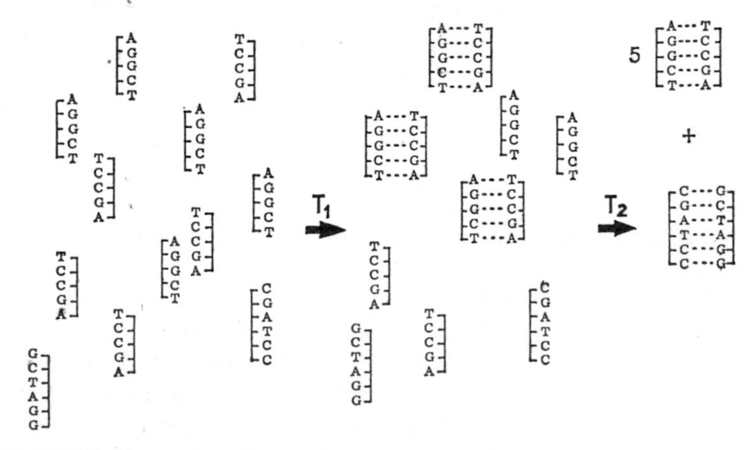

Fig. 3.3 Renaturation of complementary strands of DNA with time. After time T_1 most of the strands which are frequently repeated have found complementary partners but the unique strand has not. After a longer time, T_2, all strands have found complementary sequences.

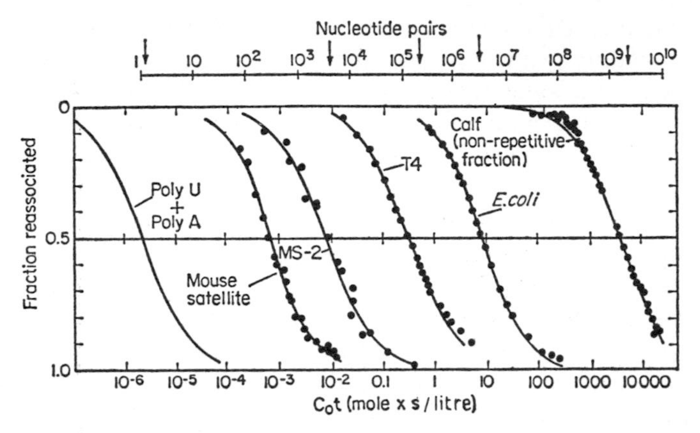

Fig. 3.4 The rate of reassociation of double-stranded polynucleotides from various sources showing how the rate decreases with the complexity of the organism and its genome (from [23]).

nucleotide/litre and t represents time in seconds. This is again monitored by ultraviolet spectroscopy or, more frequently, by hydroxyapatite chromatography since this can be used to separate double- and single-stranded DNA molecules and can be used at higher concentrations and hence for shorter times. Digestion of the remaining single-stranded DNA with a single-strand specific nuclease is also a technique employed. Fig. 3.3 shows the reannealing process and Fig. 3.4 the representation of $C_o t$ values for a variety of DNA molecules [23, 24].

3.5 ELECTRON MICROSCOPY OF DNA

Most electron micrographs of DNA are made by modifications of the monolayer technique originally developed by Kleinschmidt [25] where the molecules are bound to a film of denatured protein on the surface of an aqueous solution and consequently brought from a three dimensional to a two dimensional conformation by adsorption to the polypeptide net. The protein–DNA complex is then itself adsorbed to a solid film of collodion or parlodion, dried, and contrasted with heavy metals by shadowing or staining. The protein which has been most successfully employed for adsorption is cytochrome c which is rich in basic residues to bind to the phosphate groups of the DNA.

Single-stranded DNA tends to fold back on itself in solution forming intramolecular base pairs and consequently displays a condensed structure if the conventional methodology is used. This is overcome by the use of formamide [26, 27] which denatures the short base pairs regions of the folded single strands, or by the replacement of the cytochrome c with DNA unwinding proteins which have high affinity for single-stranded DNA and bind co-operatively to the sugar phosphate chain, thus stiffening it and spreading it out.

Denaturation mapping of DNA is used to determine the fluctuations in AT rich sequences along the molecule. As progressively greater amounts of formamide are used in the aqueous phase of the spreading medium, so bubbles of single-stranded DNA appear in the electron microscope, indicating areas of early denaturation which are assumed to be rich in AT sequences. Simple DNA molecules can be shown to have a characteristic denaturation pattern, and this can be employed as a frame of reference for the study of the biological characteristics of the molecule such as whether its replication is unidirectional or bidirectional (Chapter 8) [26].

Heteroduplex mapping of DNA is used to study the regions of homology beween two DNA molecules. The molecules are denatured and reannealed and the single-stranded regions in the heteroduplexes give an indication of the regions of non-homology between the two types. The technique can be used to detect deletions and insertions in viral mutants down to the level of 40 base pairs [27] but cannot readily detect mutations involving only a very small number of base pairs.

3.6 RESTRICTION MAPPING OF DNA

Large DNA molecules of differing sizes are readily separated by gel electrophoresis in 1–2% agarose. The DNA in the gel can then be visualized by staining with ethidium bromide and photographed under ultraviolet light. If only trace amounts of DNA are present then the radioactivity in the molecules can be detected by autoradiography with photographic film being placed on top of the gel and subsequently developed. The sensitivity of this technique is greatly enhanced for weak β emitters such as tritium if a scintillant is used to collect the radioactive energy and transfer it to photographic film in a technique known as fluorography. The size of the DNA molecules is determined by reference to fragments of known molecular weight which are run in neighbouring gel tracks. The general availability of restriction endonucleases (Chapter 6) has made it possible to use these highly sequence specific enzymes as tools to map long segments of DNA with respect to the cutting site of the nucleases. The shorter pieces produced by digestion with several nucleases may then be sequenced (Section 3.8) and placed in the correct order with respect to each other on the original large piece of DNA. This technique is known as restriction mapping. A simplified example is shown in Fig. 3.5.

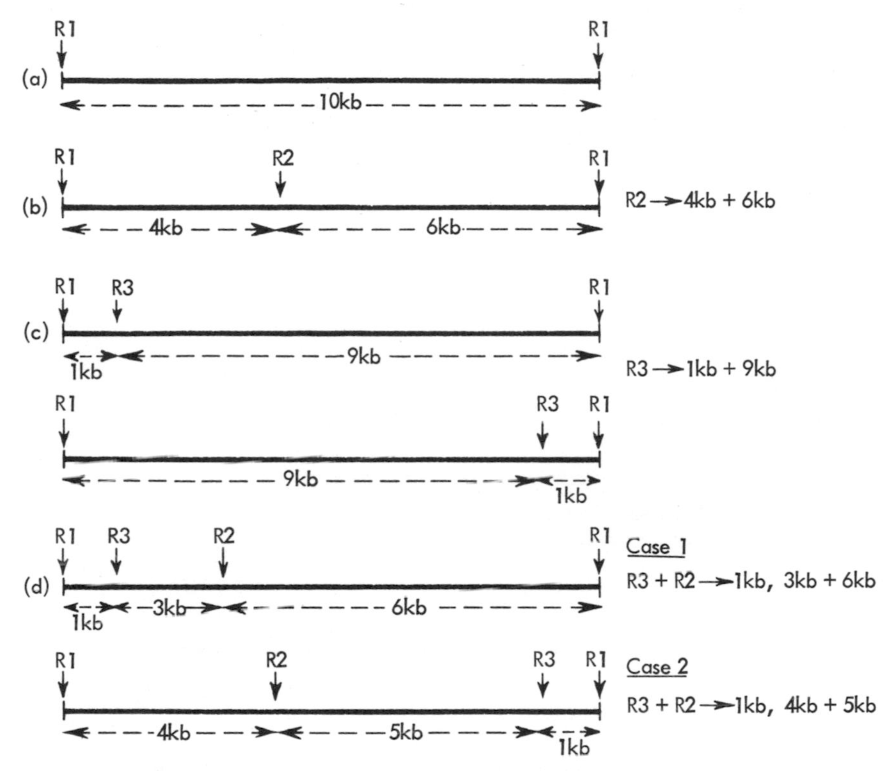

Fig. 3.5 Restriction mapping of a hypothetical DNA fragment 10 kilobases long produced by digestion with restriction enzyme R1. (a) Total fragment. (b) Fragment with restriction enzyme site R2 placed. All other restriction sites must now be located with reference to this site. (c) Two possible alternatives for restriction site R3 in relation to site R2. (d) The resolution of the two alternative R3 sites by the use of both enzymes R2 and R3 together.

One of the most classic examples of restriction mapping occurs in the case of the histone genes of the sea urchin *Psammechinus miliaris*. Birnstiel and his colleagues were able to show [28] that these genes occurred in a serially repeated unit of about 6 kilobases (kb). One site for the restriction enzyme EcoRI and one site for the restriction enzyme HindIII occurred in the 6kb segment. However, the enzyme Hpa I cut the segment in two places, yielding fragments of 4.7kb and 1.2kb. The question of whether the site of action of EcoRI was in the 4.7 or the 1.2kb fragments could be readily resolved by double digestion with both enzymes which yielded fragments of 2.8kb, 2.0kb and 1.2kb, thus placing the EcoRI site in the larger of the two Hpa I fragments.

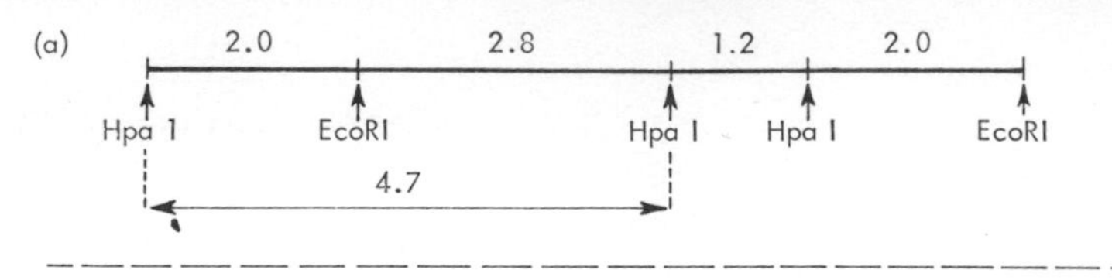

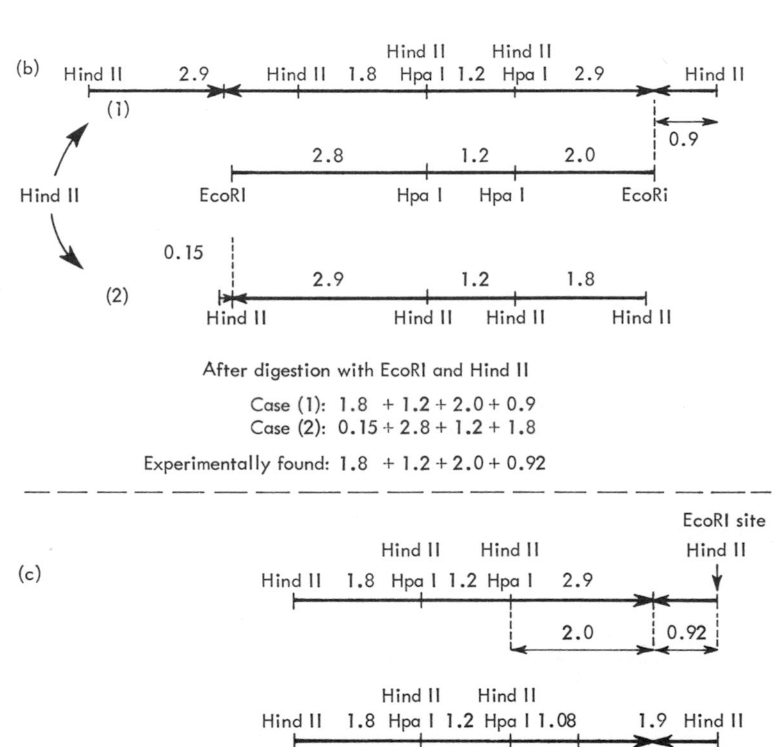

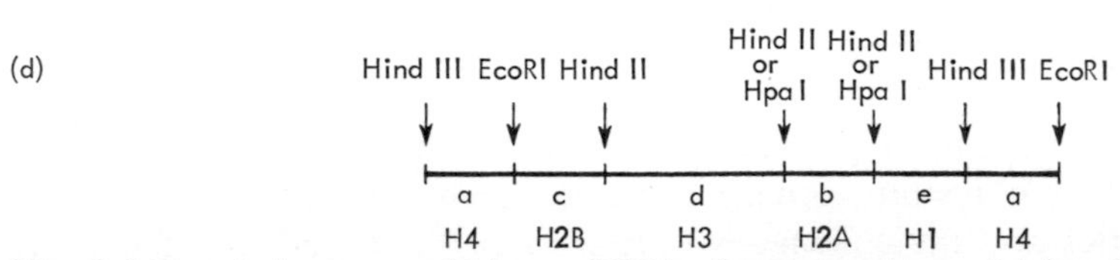

Fig. 3.6 Restriction map of histone DNA of *psammechinus*. (a) Double digestion with EcoRI and HpaI. (b) Orientation of the HindII cleavage pattern in respect to the known map (a). (c) Positioning of the HindIII cleavage site with respect to the EcoRI sit by double digestion with HindIII and HindII. The fact that four fragments of 1.8, 1.2, 1.08 and 1.9kb are obtained proves that HindIII cuts within the HindII 2.9kb fragment and therefore is to the 'left' of the EcoRI cleavage site. All values are given in kilobase (pairs) = kb (d) Location of the five histone mRNA coding sequences (from [28]).

To refine the physical map further, the enzyme HindII was used. This enzyme has the ability to cut the same sites as Hpa I which has specificity for the sequence GTTAAC, but in addition will cut at other sites since it will cut at GTTAAC or GTCGAC. HindII yielded three fragments of size 2.9kb, 1.8kb and 1.2kb. To resolve the question of whether the additional HindII site was close to the EcoRI site or 0.9kb distant from it, double digestion with both enzymes was performed. If the two sites were close together, the pattern would be expected to be very similar to that found by restricting with HindII alone, but if they were 0.9kb apart, four substantial fragments of 1.8kb, 1.2kb, 2.0kb and 0.9kb would be expected. The latter turned out to be the experimental result and the physical map shown in Fig. 3.6 could then be constructed. To map the position of the HindIII site with respect to the others, double digestion with EcoRI was used. This indicated that the two sites were 0.9kb apart. If the HindIII site mapped to the 'right' of the EcoRI site then it might be expected to be virtually coincidental with the extra HindII site and double digestion with HindIII and HindII should lead to a very similar pattern to that of HindII alone. The fact that double digestion led to four fragments of 1.9kb, 1.8kb, 1.2kb and 1.1kb therefore indicated that the HindIII site mapped 0.9kb to the 'left' of the EcoRI site. In this way the large serial repeat was broken into a smaller number of fragments whose physical relationship to each other in the DNA sequence was known. It must be emphasized that 'right' and 'left' in gene mapping are used in a purely arbitrary sense once the position of certain sites has been chosen to be represented in this manner.

3.7 HYBRIDIZATION OF RNA TO RESTRICTION FRAGMENTS BY THE SOUTHERN BLOT TECHNIQUE

DNA restriction fragments separated on agarose gels by the method described in Section 3.6 can be readily denatured and transferred to strips of cellulose nitrate filter. Radioactive RNA can then be applied to the filter where it will bind to the complementary DNA strands (Chapter 14). The filter is then washed and the radioactivity is detected by autoradiography or fluorography. In this way the DNA fragment which carries complementary sequences to any particular species of RNA can be detected and selected from a wide variety of products of restriction nuclease digestion. The technique is known as the Southern blot method [29], after its originator. The equivalent technique where the primary electrophoresis is performed with the RNA which is subsequently hybridized to radioactive DNA is known as northern blotting to indicate that the reverse process is occurring.

Fig. 3.6 shows how this technique can be effectively applied together with restriction mapping to indicate the relative positions of the sea urchin genes coding for each of the five histone messenger RNAs. Each messenger RNA hybridizes to only one of the restriction fragments produced by digestion with the combinations of nucleases described in Section 3.6 and consequently the physical position of each gene can be determined.

3.8 NUCLEOTIDE SEQUENCE ANALYSIS OF DNA

All contemporary methods of DNA sequencing are based on the high resolving power of polyacrylamide-gel electrophoresis. By this technique it is possible to separate two oligonucleotides which differ in size by a single nucleotide residue, the larger fragment moving more slowly because of its interaction with the gel matrix. In all sequencing methods a series of DNA fragments are produced with one end in common and the

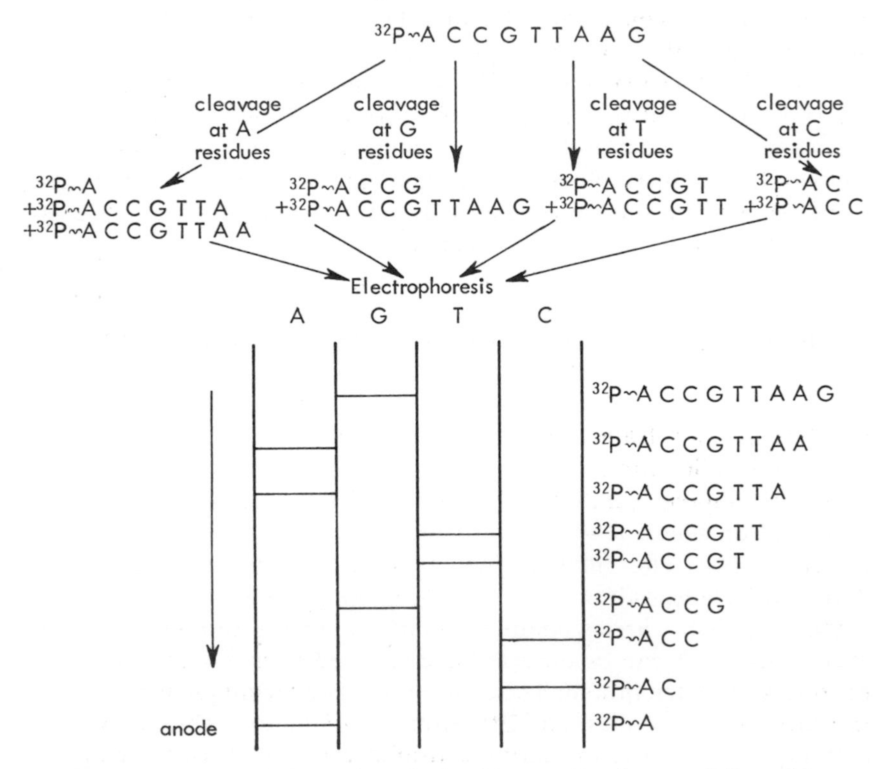

Fig. 3.7 Sequence information obtained by specific cleavage followed by electrophoresis.

other end varying in position. A minimum of four types of series of fragments is produced, each group being terminated by or cleaved at one of the four possible bases in DNA by specific enzymic or chemical means. The sequence can then be read directly off the gel autoradiograph as shown in the simplified diagram in Fig. 3.7.

The electrophoretic technique used for the separation of the strands of differing length is usually an 8–10% polyacrylamide gel though higher gel concentrations which were used in earlier methods are still occasionally employed [30–32]. The pH is generally 8.3 so that all oligonucleotides move to the anode, and a denaturant such as 7 M-urea is always present in the polymerizing solution to ensure that no intermolecular or intramolecular double-stranded regions can form. The resolving power of the gel depends firstly on the distance travelled by two neighbouring polynucleotide fragments and consequently it is customary to sequence long fragments of DNA on a series of gels which are run for differing time-periods or occasionally, at differing acrylamide concentrations. Thus, if the electrophoresis is carried out for a short period of time, the smaller fragments will be well resolved while the larger ones will be grouped close together towards the cathode. Electrophoresis for a longer time period improves the resolution of these larger fragments while the smaller ones move right off the end of the gel and are lost [33, 34]. The resolving power of the technique is also improved by the use of very thin gels of only 0.5 mm thickness since this leads to minimal scatter of the radioactive emissions during autoradiography and consequently sharper bands on the X-ray film. Thin gels also have the advantage of economy on materials and of reduced electrophoresis time since they can be run at a higher voltage without an unnacceptable generation of heat. In experiments designed to optimise these two main resolving parameters it is possible to determine sequences several hundred nucleotides long.

There are two basic ways in which DNA fragments ending in or cleaved at each of the four DNA bases are produced for gel sequence analysis. Those pioneered by the Sanger group [34–37] involve enzymatic copying of single-stranded DNA fragments and those pioneered by Maxam and Gilbert and their collaborators [30, 33] relate to chemical cleavage of single- or double-stranded DNA molecules at specific bases.

3.8.1 The Sanger dideoxy method of DNA sequence analysis [36]

This sequencing method requires a single-stranded DNA template and a primer (which is usually a restriction fragment) (Chapter 6) to direct the start of DNA synthesis to a specific point. Polymerization of deoxyribonucleotide triphosphates, at least one of which is labelled in the α position with ^{32}P is effected by DNA polymerase (Chapter 8). In each of four reaction mixtures an inhibitor of elongation which is itself incorporated into the growing chain is included. Most commonly these are the four 2′,3′ dideoxyribonucleotide triphosphates which lead to inevitable chain termination as the lack of a 3′-hydroxyl group makes phosphate ester formation with the subsequent triphosphate impossible. The ratio of inhibitor to substrate is carefully selected to optimize the range of fragment size produced in this manner with the competitive inhibitor only occasionally substituting for the correct deoxyribonucleotide triphosphate and thus terminating the chain in each of the four reaction mixtures. Thus the reaction mixture containing the inhibitor 2′,3′dideoxyATP will at the end of the reaction, contain a wide range of fragment sizes all having the same 5′ termini but different 3′ termini each ending in 2′,3′-deoxyadeosine while the mixture containing the inhibitor 2′3′dideoxy CTP will have a similar range of fragments each terminating at the 3′ end in 2′,3′dideoxy cytidine. Variations on the basic technique involve the enzyme used in the polymerization and the type of chain-terminating inhibitor used. The enzyme reverse transcriptase (Chapter 8) has a much lower K_m for dideoxyribonucleotide triphosphates and is used with much lower concentrations of inhibitor [40]. Arabinose [38] or ribose [39] nucleotide triphosphates can replace the dideoxyribonucleotide triphosphate inhibitors, the latter being treated with alkali subsequent to polymerization to give cleavage at the point of ribosubstitution [38, 39].

The majority of contemporary DNA sequence analyses involve double- rather than single-stranded DNA molecules and a disadvantage of this method is the need for prior separation of the two double strands to permit the polymerization to function correctly. It is possible to separate two DNA single strands of identical size on polyacrylamide or agarose gels after denaturation [30, 41–44] though the physical basis behind this phenomenon is not clearly understood. It is assumed that each complementary strand possesses a different capacity to fold back upon itself forming base pairs which are not exclusively Watson–Crick in character, rather in the manner in which RNA molecules can fold back on themselves, and that the consequent difference in secondary and tertiary structure leads to differing electrophoretic mobility. However, such separations frequently lead to poor yields and are time consuming. A biological method of producing a single strand for sequencing is to insert the double-stranded DNA into the replicative form DNA of the bacteriophage M13 by means of restriction endonucleases [45, 60]. The DNA is then isolated as the single-stranded form from the progeny bacteriophage particles (Chapter 4). Another method involves degradation of the double strand with exonuclease III followed by the use of the dideoxy method with a primer specific to only one strand.

3.8.2 The Maxam and Gilbert chemical method of DNA sequence analysis

In this method the DNA fragment which is to be sequenced, is first made radioactive at a unique 5′ or 3′ terminus. The commonest method for achieving this is to treat the DNA with alkaline phosphatase which removes the unlabelled phosphate at the 5′ end, and then to attach a radioactive ^{32}P phosphate label onto the 5′-hydroxyl group using bacteriophage T4-induced polynucleotide kinase and α-labelled ^{32}P-ATP [46–48]. However, it is possible instead to label the 3′ termini of the fragment utilising α-^{32}P-labelled ribonucleotide triphosphates and the enzyme terminal transferase (Chapter 8). If the newly synthesized ribonucleotide chain is then hydrolysed with alkali or piperidine [49, 50] then the ^{32}P label will remain at the 3′ end in two possible locations on either side of the initial ribonucleotide added. DNA polymerases (Chapter 8) may also be used to label the 3′ terminus with α-^{32}P-labelled deoxyribonucleotide triphosphates. Commonly, the piece of DNA to be sequenced is the product of digestion with a restriction endonuclease of known sequence specificity. If this nuclease has created a 'staggered' cut such that a 5′ template of known base sequence protrudes then DNA polymerase supplied with the appropriate deoxyribonucleotide triphosphates labelled in the α position will copy this sequence and complete the complementary strand at the 3′ end. Alternatively, if the restriction endonuclease has produced a fragment with no protruding ends, the DNA polymerase will, under the correct experimental conditions, exchange the existing 3′ terminal nucleotide for a radioactive one labelled in the α position [51–53].

Once the DNA fragment to be sequenced has been made radioactive by one of these techniques, it is necessary to separate the two labelled ends from each other. While this can be performed by denaturation followed by electrophoresis as discussed above, it is also possible to use a restriction nuclease of differing specificity to that used in the initial digestion to make an asymetric cut in the double-stranded fragment. The two labelled ends are then readily separated by electrophoresis by virtue of their presence on DNA molecules of differing size. While this second technique is experimentally simpler and leads to higher yields, the first technique has the advantage that if the entire length of the fragment can be produced in the two complementary single-stranded forms, it is possible to use the sequence of one strand to confirm that of the other. The DNA is then eluted from the gels and treated with the appropriate chemicals which react readily with both single- and double-stranded forms.

A variety of differing chemical reactions is used to cleave the chains where a specific base occurs. Ideally, each chain should only be cut once in a single place so the treatment must be designed to optimise this process. Cleavage at guanine residues is effected by methylation with dimethyl sulphate at the N7 position. This leads to instability of the glycosidic linkage which is then readily hydrolysed with piperidine [30, 33]. Since there are no readily available methods for cleavage at adenine residues alone, these are determined by hydrolysis of both purine nucleotide linkages with acid. The adenine residues can then be determined by reference to those bands which appear in the A+G gel track but not in the G gel track (Fig. 3.8). In earlier experimental protocols, adenine residues were also split by methylation before hydrolysis but this led to less well-defined resolution [30]. It is possible to obtain chain scission at both adenine and cytosine residues by the use of alkali [33]. In general, pyrimidine residues are hydrolysed by the use of hydrazine followed by piperidine. In 2M-NaCl, the reaction with thymine is inhibited and cleavage occurs only at cytosine so that the two again be differentiated by comparison of the gel tracks as shown in Fig. 3.8. Other base-specific cleavage methods have been developed [54] but are as yet in less general use.

One difference between the chemical method of DNA sequencing and the enzymatic method is that the chains which are produced and read on the sequencing gel do not in fact terminate in the base for which the cleavage system is specific, but in the base before that. While this is a largely academic distinction for sequencing purposes, it can be confusing to students. It must also be emphasized that the fragments to the distal side of the cleavage are also present on the sequencing gel, but are not visible on autoradiography as the fragment was labelled at only one terminus.

The major advantage of the chemical method of DNA sequencing is that it can be used with double-stranded DNA. More recently a method which utilizes the 5′ end labelling methodology originally developed for chemical sequencing together with the separation of double-stranded but singly labelled molecules by asymmetric restriction nuclease cutting as described above has been reported. This method involves the formation of 'nicks' in the double-stranded fragment with pancreatic deoxyribonuclease (Chapter 6) followed by extension of the 3′ hydroxyl groups thus exposed by DNA polymerase 'nick translation' (Chapter 8) utilizing the dideoxyribonucleotide triphosphate system and thus combines some of the better features of both techniques [55].

While the first direct readout sequencing method to be described in the literature the Sanger 'plus minus' method of DNA sequencing [35], undoubtedly played an important part in the development of DNA sequencing technology, its importancc is now largely historical since it has been superceded by the dideoxy technique and is not in general use. However, this method, together with those described above, has been used for the sequence analysis of the entire length of several small viral and bacteriophage DNA molecules (Chapter 4).

The sequencing of RNA molecules is described in Chapter 5. However, at the present stage of development of sequencing technology, one of the best methods of determining RNA sequences is to use the enzyme reverse transcriptase to make a complementary DNA copy of the RNA, and to use the DNA sequencing methods instead.

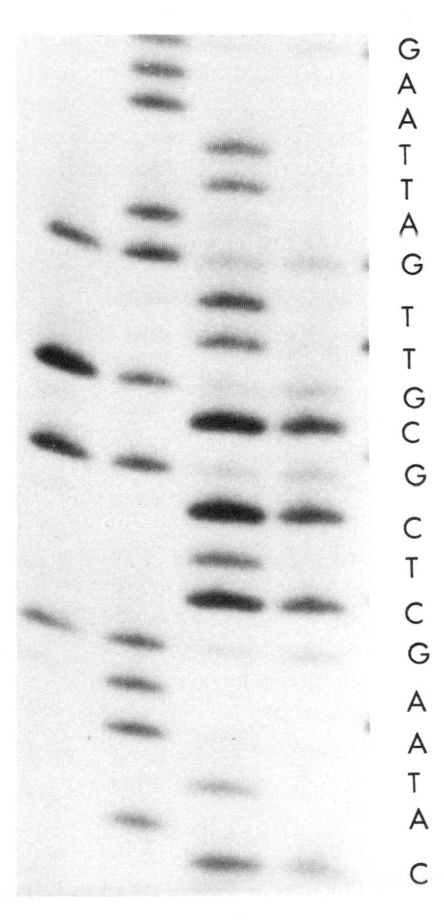

Fig. 3.8 The Gilbert Maxam method of DNA sequencing. Autoradiograph from DNA with the sequence: 32pCATAAGCTCGCGTTGATTAAG . . . determined by the Gilbert Maxam method. Bands which appear in the G and the G + A tracks are assigned to G. Bands which are in the A track alone are assigned to A. C and T can be similarly differentiated.

3.9 THE ALLOCATION OF SPECIFIC GENES TO SPECIFIC CHROMOSOMES

There are two main methods of investigation which determine which chromosome codes for a specific gene function. If the gene occurs at high frequency with several repeating units at each point of occurrence then it is possible to reanneal radioactive RNA or cDNA corresponding to this sequence to a preparation of metaphase chromosomes which have been denatured and to then determine which chromosomes have bound the radioactive material by a combination of light microscopy and autoradiography [61]. Such a procedure is not reliable in the case of structural genes which occur in a single or small number of copies, and these genes are usually allocated by techniques of somatic cell fusion [62]. If the enzyme which has the gene product has different characteristics in two separate species, then the two types of enzyme can be differentiated by isoenzyme electrophoretic analysis. The two cell types are fused under conditions in which only hybrid cells will grow because gene deficiencies in the two parent cells limit their growth potential in a selected medium (Chapter 7). The hybrid cells tend to lose inessential chromosomes speedily and a population of cloned hybrids each with a different complement of chromosomes can be isolated. In general, chromosomes are lost faster from the species which initially had the larger number of chromosomes so that in a fusion between human cells with a haploid number of 23 chromosomes and chinese-hamster cells with a haploid number of 11 or 12 chromosomes depending on the cell line employed, human chromosomes will be selectively lost. The population of clones can then be tested for the presence of both the human isoenzyme and the remaining human chromosomes and a correlation can be made between the presence of a human chromosome and the human isoenzyme activity. In this way it has been possible to allocate several hundred enzyme functions to specific human chromosomes [62].

REFERENCES

1 Bendich, A. (1957), *Methods in Enzymology,* Vol. III, Sect, V, p. 715 (S.P. Colowick and N.O. Kaplan, Eds.).
2 Sober, H.A. (Ed.) (1968), *Handbook of Biochemistry,* pp. H.11, H-30.
3 Chargaff, E. (1963), *Essays on Nucleic Acids.* Elsevier, Amsterdam.
4 Doty, P. (1961), *Harvey Lecturers,* **55,** 103.
5 Mandel, M. and Marmur, J. (1968), *Methods in Enzymology,* Vol. 12, Part B, p. 195 (L. Grossman and E. Moldave, Eds.). Academic Press, New York.

6 Felsenfeld, G. (1968), *Methods in Enzymology,* Vol. 12, Part B, p. 247 (L. Grossman and K. Moldave, Eds.). Academic Press, New York.
7 Marmur, J. (1963), *Ann. Rev. Microbiol.,* **17,** 329.
8 Sueoka, N. (1964), *The Bacteria,* p. 419 (I.C. Gunsalus and R.Y. Stanier, Eds.).
9 Schildkraut, C.L., Mandel, M., Levisohn, S., Smith-Sonneborn, J.E. and Marmur, J. (1962), *Nature,* **196,** 795.
10 Freese, E. (1962), *J. Theor. Biol.,* **3,** 82.
11 Laskowski, M. (1972), *Progr. Nucleic Acid Res. Mol. Biol.,* Vol. 12, p. 161. (J.N. Davidson and W.E. Cohn), Academic Press, New York.
12 Harpst, J.A., Krasna, A.I. and Zimm, B.H. (1968), *Biopolymers,* **6,** 595.
13 Krasna, A.I., Dawson, J.R. and Harpst, J.A. (1970), *Biopolymers,* **9,** 1017.
14 Krasna, A.I. (1970), *Biopolymers,* **9,** 1029.
15 Schmidt, V.W. and Hearst, J.E. (1969), *J. Mol. Biol.,* **44,** 143.
16 Kavenoff, R. and Zimm, B.H. (1963), *Chromosoma,* **41,** 1.
17 Crothers, D.M. and Zimm, B.H. (1965), *J. Mol, Biol.,* **12,** 525.
18 Lang, D. (1970), *J. Mol. Biol.,* **54,** 557.
19 Leighton, S.B. and Rubenstein, I. (1969), *J. Mol. Biol.,* **46,** 313.
20 Freifelder, D. (1970), *J. Mol. Biol.,* **54,** 567.
21 Dublin, S.B., Benedek, G.B., Bancroft, F.C. and Freifelder, D. (1970), *J. Mol. Biol.,* **54,** 547.
22 Mandel, M. and Marmur, J. (1976), in *Methods in Enzymology,* **12B,** (ed. Grossman, L. & Moldave, K.), pp. 195–206.
23 Britten, R.J. and Kohne, D.E. (1968), *Science,* **161,** 529.
24 Britten, R.J., Graham, D.E. and Neufeld, B.R. (1974), *Methods in Enzymology,* Vol. XXIX, Part E, p. 363 (ed. L. Grossman and K. Moldave), Academic Press, New York.
25 Kleinschmidt, A.K. (1968), in *Methods in Enzymology,* Vol. **12B** (ed. L. Grossman and K. Moldave), Academic Press, New York, pp. 361–377.
26 Inman, R.B. (1974) in *Methods in Enzymology* **12E** (ed. L. Grossman and K. Moldave), Academic Press, New York, pp. 451–458.
27 Davis, R.W., Simon, M. and Davidson, N. (1971), in *Methods in Enzymology* **12D,** (ed. L. Grossman and K. Moldave), Academic Press, New York, pp. 413–428.
28 Shaffner, W., Gross, K., Telford, J. and Birnstiel, M. (1976), *Cell,* **8,** 471–478.
29 Southern, E.M. (1965), *J. Mol. Biol.,* **98,** 503–517.
30 Maxam, A.M. and Gilbert, W. (1977), *Proc. Natl. Acad. Sci., U.S.A.,* **74,** 560–564.
31 Peacock, A.C. and Dingman, C.W. (1967), *Biochemistry,* **6,** 1818–1829.
32 Maniatis, T., Jeffrey, A. and Van Desande, H. (1975), *Biochemistry,* **14,** 3787–3795.
33 Maxam, A.M. and Gilbert, W. (1980), in *Methods in Enzymology,* (ed Wil, R.), **68,** p. 499, Academic Press, London and New York.
34 Sanger, F. and Coulson, A.R. (1978), *FEBS Lett.,* **87,** 107–110.
35 Sanger, F. and Coulson, A.R. (1975), *J. MOL. Biol.,* **94,** 441–448.
36 Sanger, F., Nicklen, S. and Coulson, A.R. (1977), *Proc. Natl. Acad. Sci., USA,* **74,** 5463–5467.

37 Air, G.M., Sanger, F. and Coulson, A.R. (1976), *J. Mol. Biol.*, **108,** 519–533.
38 Atkinson, M.R., Deutscher, M.P., Kornberg, A., Russell, A.F. and Moffat, J.G. (1969), *Biochemistry,* **8,** 4897–4904.
39 Barnes, W.M. (1976), *J. Mol. Biol.*, **119,** 83–89.
40 McGeoch, D.J. and Turnbull, N.T. (1978), *Nucl. Acids Res.*, **5,** 4007–4024.
41 Hayward, G.S. (1972), *Virology,* **49,** 342–349.
42 Flint, S.J., Gallimore, P.H. and Sharp, P.S. (1975), *J. Mol. Biol.*, **96,** 47–59.
43 Perlman, D. and Huberman, J.A. (1977), *Anal. Biochem.*, **83,** 666–670.
44 Szalay, A.A., Grohman, K. and Sinsheimer, R.L. (1977), *Nucl. Acids Res.*, **4,** 1569–1579.
45 Kaguni, J. and Reciy, D.S. (1979), *J. Mol. Biol.*, **135,** 863–878.
46 Can de Sande, J.H., Kleepe, K. and Khorana, H.G. (1973), *Biochemistry,* **12,** 5050–5056.
47 Lillehaug, J.R., Kleepe, R.K. and Keppe, K. (1976), *Biochemistry,* **15,** 1858–1864.
48 Berkner, K.L. and Folk, W.R. (1977), *J. Biol. Biochem.*, **252,** 3176–3181.
49 Kossel, H. and Roychoudhury, R. (1971), *Eur. J. Biochem.*, **22,** 271–278.
50 Chang, J.C., Temple, G.F., Poon, R., Neuman, K.H. and Kan, Y.W. (1977), *Proc. Natl. Acad. Sci. USA,* **74,** 5145–5152.
51 Donelson, J. and Wu, R. (1972), *J. Biol. Chem.*, **247,** 4654–4661.
52 Soeda, E., Kimura, G. and Miura, K. (1978), *Proc. Natl. Acad. Sci. USA,* **75,** 162–170.
53 Schwarz, E., Scherer, G., Hobom, G. and Kossel, H. (1978), *Nature,* **272,** 410–414.
54 Friedman, T. and Brown, D.M. (1978), *Nucl. Acids Res.*, **5,** 615–627.
55 Maat, J. and Smith, A.J.H. (1978), *Nucl. Acids Res.*, **5,** 4537–4545.
56 Smith, A.J.H. (1979), *Nucl. Acids Res.*, **6,** 831–848.
57 Gultman, T., Votavova, H. and Pivec, C. (1976), *Nucl. Acids Res.*, **3,** 835–845.
58 Binsteil, M., Telford, J., Weinberg, G. and Stafford, D. (1974), *Proc. Natl. Acad. Sci. USA,* **71,** 2900–2909.
59 Bunermann, H. and Muller, W. (1978), Nucl. Acids Res., **5,** 1059–1079.
60 Averson, S., Gait, M.J. and Young, M. (1980), *Nucl. Acids Res.*, **8,** 1731–1743.
61 Gall, J.G. and Pardue, M.L. (1971), in *Methods in Enzymology,* Vol. XXI (ed. Grossman, L. and Moldave, K.), p. 470–480.
62 Ringertz, N.R. and Savage, R.E. (1976), *Cell Hybrids,* p. 224–244, Academic Press, New York.

DNA and chromosomes 4

4.1 INTRODUCTION

There is a basic difference in the occurrence and organization of DNA and RNA molecules in both animal and bacterial cells. Whereas RNA molecules are found in diverse numbers and types throughout the cell, each type being designed for a specific function or set of linked functions (Chapter 5), DNA molecules are highly restricted in both number and location. Most DNA molecules contain the information for a wide range of unrelated functions and as a consequence they are generally very large. Furthermore, since the function of DNA molecules is to carry genetic information from generation to generation but not necessarily to express more than a small percentage of their potential at any point in time, it is necessary for any particular cell type to carry only a single or a few copies of each genetic message. In bacterial cells most of the DNA is represented by a single large chromosome of very high molecular weight. In animal cells, the DNA molecules are even larger and restricted to a comparatively small number of chromosomes according to the particular species of animal involved. In both cases, smaller DNA molecules which code for highly specialized functions can occasionally be found located outside the main cellular DNA complement, but these comprise a tiny fraction of the total cellular DNA.

One consequence of this method of organization of the genetic material is that two genes which contain dissimilar information with variant temporal requirements for expression can be located physically in close proximity on the same DNA molecule and the control of their function is therefore dependent on proteins which separate them physically or organizationally from each other. The multifunctional nature of DNA molecules undoubtedly makes them harder to study than RNA molecules though the advent of restriction endonuclease technology (Chapter 6) has made it possible for physical separations which do not occur in nature to be achieved in the laboratory, so that the portions of the chromosome which code for specific gene functions can now be separated from neighbouring regions.

Historically much of the early understanding of gene structure and function has come from the prokaryotes and from the simple bacteriophages which can infect them and grow in them. This has been largely due to the comparative ease with which the prokaryotes can be grown and infected under laboratory conditions and the fast generation time which has made the isolation of genetic mutants a comparatively simple exercise. In addition, the DNA content of prokaryote cells is about one hundred-fold smaller than that of eukaryote cells so that the analytical task is relatively simple. In the 1940s laboratory techniques for growing eukaryote cells in culture were developed and the much larger task of analysing the arrangement and location of specific gene functions in eukaryote cells was initiated. While these complex systems are as yet imperfectly understood, one major advance that resulted from the cell culture methodology, was the ability to grow and replicate the small animal viruses in the laboratory and the function and sequence of these has been an area of intense research activity over the last decade and has yielded much information about the nature of eukaryote gene organization and expression.

4.2 BACTERIA

The fundamental differences between prokaryotic (bacterial) and eukaryotic (animal) cell structure are described in detail in Chapter 5. By far the majority of genetic experiments have been performed in varying strains of the bacterium *Escherichia coli*, with a much smaller number being in *Bacillus subtilis*, *Micrococcus luteus* and other bacterial species. There are a wide variety of strains of *E. coli* itself, each possessing its own properties with respect to susceptibility of bacteriophage infection and to antibiotic resistance. In addition, many bacteria possess sex pili which are thought to play a major functional role in genetic recombination (Section 4.3.1). While bacteria normally divide by binary fission, the transfer of genetic information between two types of bacteria is possible. Strains of bacteria which are able to donate their chromosome are known as high frequency recombinant or HFr strains and have been widely used in the study of bacterial gene function (Section 4.3.1). Mutants are readily isolated by a technique known as *replica plating* whereby colonies of bacteria are grown on a plate of nutrient agar which contains all the essentials for growth. A velvet pad is then used to transfer small portions of these colonies to a second agar plate which contains minimal medium in which certain mutants will not grow. The colonies which are present on the first, but not the second, plate can then be selected. The process of mutant selection is greatly enhanced if the cells are first grown in minimal medium in the presence of an inhibitor which will kill all the growing bacteria but not the mutants which are unable to replicate in that medium.

Bacteria which possess the full complement of bacterial genes are usually referred to as 'wild type' bacteria or prototrophs and mutants are classified according to their missing functions. Thus a thr^- mutant is a bacterium which requires threonine in its growth medium as it lacks the correct information to make the enzymes involved in threonine synthesis in their fully functional state. Such a nutritional mutant is called an *auxotroph*.

While most bacteria carry all their genetic information on a single, circular chromosome, some possess in addition small extrachromosomal elements known as plasmids which carry genes coding for functions such as drug resistance. These molecules are usually comparatively small and circular and can be present in one or several copies per cell (Section 4.3). These plasmids have become widely used as tools in genetic engineering (Chapter 14). For a more detailed description of bacterial growth and physiology, there are several specialized texts on the subject which can be consulted [1,2,3].

4.2.1 The bacterial chromosome

The chromosome of *E. coli* is a single circular molecule of 4.5 million kilobase pairs which effectively has a circumference of 1 mm but must be contained in a bacterial cell with a diameter in the order of 1 μm. A complex packaging mechanism is therefore necessary in order to ensure that all the DNA is folded within the bacterial cell in a manner which will not inhibit transcription, nor allow entanglement of the two daughter strands to occur during replication. This is done by two main mechanisms. Firstly, the DNA is folded into between 40 and 100 loops, and secondly, each of these quasi circles is itself supercoiled independently of the others [4,5]. Each loop is maintained as an independent physical area by DNA : RNA interactions since the loops are abolished by ribonuclease (Chapter 6), but not by proteolytic enzymes. Deoxyribonuclease abolishes the superhelical nature of the loops and limited treatment with this enzyme will therefore lead to the abolition of only some of the supercoiling since only a small number of loops will be attacked [6,7].

The intact folded *E. coli* chromosome is isolated by non-ionic detergents in the presence of high molarities of salt in two possible forms, free and membrane associated. The free form which sediments at 1600–1700 *s* consists of about 60% DNA, 30% RNA and 10% protein, the bulk of which is the enzyme RNA polymerase (Chapter 10). The membrane-bound form which sediments contains an additional 20% of membrane-attached protein. The exact nature of the membrane attachment sites is not yet known.

4.2.2 The bacterial division cycle

As in eukaryotes (see below) the general rule in bacteria is that chromosome replication (DNA synthesis) and cell division occur alternately. Under slow growth conditions (i.e. doubling times greater than 60 min) there is a gap between completion of a round of DNA synthesis and cell division. Cell division takes place at a fixed time (generally about 60 min) after initiation of a round of DNA synthesis [8,9]. However, bacteria, under suitable conditions, can grow much more quickly than one division every 60 min and in these situations DNA synthesis becomes continuous, i.e. it no longer occupies a restricted part of the cell cycle [10,11]. Although cell division still occurs 60 min after initiation of a round of DNA synthesis the cell does not wait until cell division has occurred before initiating a second (or even a third) round of DNA synthesis. This results in some cells, immediately prior to division having two copies of part of the DNA (the last to be duplicated) and up to eight copies of other regions (those duplicated first) (see Fig. 4.1).

Protein synthesis is required to initiate a round of chromosome replication and hence an amino acid auxotroph, when starved of essential amino acids, will complete ongoing rounds of replication and come to rest with one unreplicated chromosome per cell. In contrast a thymine-requiring mutant when deprived of thymine cannot make DNA but the

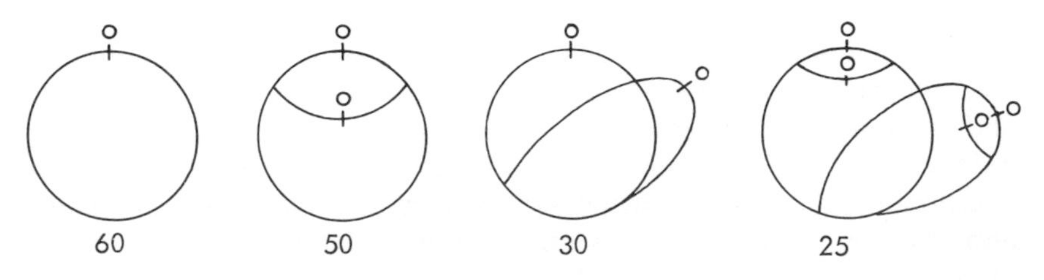

Fig. 4.1 Diagrammatic representation of the structure of one chromosome (of the two present in the cell) immediately prior to cell division in *E. coli* growing at different rates [11]. ○ is the origin of chromosome replication.

chromosome is brought into such a condition that when thymine is restored a second round of replication is initiated.

4.2.3 Bacterial transformation

Pneumococci *(Diplococcus pneumoniae)* may be classified into a number of different types each characterized by the ability to synthesize a specific serologically distinct and chemically distinct capsular polysaccharide. In 1928 Griffith observed that a particular strain of pneumococci cultivated *in vitro* under specific conditions lost the ability to form the appropriate polysaccharide and consequently grew on solid media in so-called 'rough' colonies in contrast to the 'smooth' glistening colonies formed by encapsulated cells. If a living culture of such unencapsulated cells was injected into mice together with killed encapsulated pneumococci of type III, the organisms subsequently recovered from the animals were live virulent pneumococci of the encapsulated type III. It appeared therefore that some material present in the dead type III organisms had endowed the unencapsulated pneumococci with the capacity to synthesize the characteristic type III polysaccharide.

During the next five years it was shown that such pneumococcal transformation could be produced *in vitro,* that a cell-free extract could replace killed cells as the transforming agent, and that organisms which had undergone transformation did not spontaneously revert to their original type.

The chemical nature of the active principle remained obscure until 1944 when Avery, McLeod and McCarty [12], at the Rockefeller Institute in New York, showed that DNA extracted from encapsulated smooth strains of pneumococcus type III could, on addition to the culture medium, transform unencapsulated 'rough' cells into the fully encapsulated smooth type III. The smooth cells so developed could propagate indefinitely in the same form, producing more DNA with the same capabilities. The pneumococcal DNA had therefore initiated its own reduplication as well as inducing the specific inheritable property of capsule synthesis. In other words, it had executed two functions usually associated with the gene.

These observations stimulated further research into bacterial transformation, from which it emerged that the reaction was not limited to pneumococci but could be produced in a wide range of bacteria, e.g. *Haemophilus influenzae* [13,14], *E. coli* [15], and the meningococcus. Nor were these transformations limited to changes in serological type since they could also be used to endow bacteria with resistance to specific drugs or antibiotics or the ability to utilize particular nutrients. Transformation can be demonstrated for any characteristic whose acquisition can be measured in the recipient. The two stages required are (a) transfer and (b) utilization of the transferred material.

While the proportion of treated cells which may develop a new characteristic after exposure to appropriate DNA is usually small, figures as high as 17% have been recorded by Hotchkiss [16]. Among the factors influencing the yield is the capacity of the recipient strain to be transformed, since some strains are much more susceptible than others and *E. coli* requires treatment with $CaCl_2$ in order to act as a recipient at all [17]. However, even in a single population not all cells are competent to take up DNA and this may be a result of variable cell wall permeability. The most appropriate time for transformation to occur is just after cell division. When pneumococci are cooled to a temperature at which growth is arrested and then rewarmed so that they start to divide synchronously, transformations are exceptionally numerous.

The number of transformation events is proportional to the DNA concentration up to a plateau level and hence the frequency of transformation has been used as a measure of gene frequency (see Chapter 8).

Duplex DNA enters the recipient better than does single-stranded DNA and the saturating DNA concentration is reached when about 150 DNA molecules of molecular weight about 10^7 have been taken up by each bacterium [18].

Just how the transforming DNA enters the cell to be transformed is not fully understood, but it is known that the acquisition of the new characteristic induced by the DNA requires a period up to 1 hour and that after acquiring the new DNA a cell multiplies more slowly for some time than do its unchanged neighbours. The establishment of the mechanism for duplicating the new DNA requires still longer. The mechanism of the transformation reaction has been recently reviewed [19,20].

Although duplex DNA is better for transformation, on entry into the recipient cell one strand of the duplex is degraded. The major endonuclease of pneumococcus is associated with the cell membrane and it is responsible for degrading one of the DNA strands following action of an enzyme which introduces single strand breaks every 6000 bases [21]. In *E. coli* the recipient bacterium must be rec BC^- otherwise the rec BC nuclease degrades the incoming DNA.

For expression, the incoming single strand of DNA finds a complement in the recipient DNA and displaces one of the original duplex strands in a recombination-type mechanism (see Chapter 9). This integration event occurs at widely different efficiencies for different markers.

It is possible to transfer more than one inheritable characteristic to susceptible bacteria in a single DNA preparation, e.g. one specimen of DNA may carry the three characteristics of resistance to penicillin, resistance to streptomycin, and the ability to form a capsule in pneumococci. Such a specimen of DNA might bring about transformation in 5% of the recipient cells. Of the cells transformed, 98% would acquire only one of the three characteristics, 2% would acquire two of the characteristics, and only 0.01% would acquire all three. Clearly, therefore, the DNA preparation cannot convey a complete set of the donor's characteristics to the recipient, although certain characteristics appear to be linked. For example, the DNA factors responsible for streptomycin resistance tend to be coupled with those responsible for the ability of pneumococci to use mannitol as a source of energy.

Presumably all 'transformations' are essentially processes by which bacteria are endowed with enzyme-synthesizing capacities which they did not previously possess. The first direct proof of the presence of such a new enzyme in an organism after treatment with a DNA transforming factor was made by Marmur and Hotchkiss [22] who demonstrated that the ability to oxidize mannitol could be transferred to a non-utilizing strain of pneumococcus by culturing it in the presence of DNA prepared from a strain which possessed the power of utilizing this sugar. The organisms so transformed differ from the parents in possessing the new enzyme mannitol phosphate dehydrogenase.

An analogous somatic transformation has been reported by Szybalski for mammalian cells [23]. He isolated DNA from cultures of human cells of the strain D98S which contain the enzyme IMP pyrophosphorylase (EC2.4.2.8) responsible for the reaction:

$$\text{hypoxanthine} + \text{PRPP} \rightleftharpoons \text{IMP} + \text{PPi}$$

which is discussed further in Chapter 7. The addition of this DNA to cultures of cells of the strain D98/AH-2 which are deficient in the enzyme resulted in the appearance of IMP pyrophosphorylase-positive genetically transformed cells detected under highly selective conditions. The transforming activity was abolished by deoxyribonuclease but not by ribonuclease.

One problem with eukaryote DNA for use in transformation is that the required gene represents only a minute fraction of the total DNA and so its concentration is very low. One way of increasing the concentration of a particular gene within a DNA sample is by *transduction* using a bacteriophage (see below) or by using cloned DNA and such methods are considered in Chapter 14.

4.3 PLASMIDS AND BACTERIAL CONJUGATION [20,24]

Plasmids are duplex supercoiled DNA molecules which range in size from 2.5×10^6 to 1.5×10^8 daltons. They are stable elements which exist in bacteria in an extrachromosomal state. The large ones are present in only 1 or 2 copies per cell whereas there may be 20 or more copies per cell of the smaller ones [25,26]. They enjoy an autonomous, self-replicating status without lowering host viability [27–29].

There are three types of plasmid (1) F or sex factors which promote bacterial conjugation, (2) R factors which, in addition, confer resistance to certain drugs and (3) colicinogenic factors which produce colicins. A colicin is a protein which is toxic to related bacterial cells but not to the host cell.

4.3.1 F-factors

A bacterium possessing an F-factor (F^+) is male and the F factor induces the bacterium to make a tube or pilus by which the male bacterium becomes attached to a female bacterium, i.e. one not carrying the F-factor. When the F-factor replicates (see Chapter 8) one linear single-stranded copy of plasmid DNA may pass through the pilus 5′-end first into the female bacterium. The complementary strand is synthesized in the recipient which thereby becomes male as it now carries the F-factor. Soon all the cells in a population become F^+. The presence of an F-factor in a bacterium excludes the entry of a second F-factor, but more than one *type* of plasmid may be stably maintained in a cell if the two are not closely related. Generally F^+ cells are rare but each usually contains several F-factors. As well as the plasmid, parts of the bacterial chromosome may pass through the pilus and this leads to a low frequency of exchange of genetic markers between bacteria, i.e. a male (F^+) prototroph bacterium may transfer the gene required for tyrosine synthesis to a female tyr^- auxotroph. This is known as transfer by bacterial conjugation.

In one in 10000 F^+ cells the F-factor becomes integrated into the bacterial chromosome by reciprocal crossing over. A plasmid which can exist either autonomously or in an integrated state is called an episome. For insertion the F-factor breaks at a unique site, but this linear piece of DNA integrates at random with regard to position and orientation. Like the λ-prophage, an integrated plasmid loses its ability for independent replication [30,31]. However, *E. coli* mutants defective in initiation of chromosomal replication can be rescued by integration of an F-factor which takes over control of initiation [30]. The integrated plasmid still causes the formation of pili and can be transferred by conjugation, but now, between the leading 5′-end of the plasmid DNA and the trailing 3′-end, there is the entire bacterial chromosome which gets dragged along the conjugation tube, into the female cell. Since transfer of the whole chromosome takes about 60 min, breakage of the conjugation tube normally occurs before transfer is complete. When this occurs the second half of the F-factor is not transferred and so the recipient remains female. As the fragments transferred cannot replicate autonomously they are either lost, or are integrated into the recipient's chromosome which thereby shows a high frequency of recombination of genetic markers. Hence cells with an integrated F-factor are called *Hfr*.

Because of the mode of integration of an F-factor into the bacterial chromosome, different Hfr strains transfer the genes of the bacterial chromosome in a different order and with different polarity. However, closer study shows the gene order to be circularly permuted, providing evidence for the linear arrangement of genes on a circular chromosome [32].

Sometimes an Hfr strain may revert to F^+, and when this occurs a section of the bacterial chromosome may also be excised and be found in the plasmid (e.g. Flac contains the *lac* operon integrated into an F-factor which is thereby more than doubled in size). Such plasmids are known as F′-factors. This gene is now transferred to a recipient bacterium at conjugation (cf. transduction below) to produce a partial diploid and recombination readily occurs (see Chapter 9). Thus in $Flac^+$ *lac* is transferred infectiously but other markers are only transferred following integration.

4.3.2 R-factors

R-factors are similar to F-factors in that they induce the host bacterium to produce a pilus but they are more complex and consist of two parts. One region carries genes conferring resistance to a number of drugs while the second region enables the plasmid to be transferred to other bacteria by conjugation. Thus R222 has a molecular weight of 68×10^6 in *E. coli* and is made up of two parts (a 54×10^6 molecular weight resistance-transfer factor, RTF, and a 12×10^6 molecular weight resistance-determining unit, r) which exist separately in *Proteus mirabilis*[33]. Resistance is conferred to a series of drugs, usually by enzymes which modify or degrade the drug, e.g. chloramphenicol resistance depends on an enzyme which acetylates chloramphenicol; ampicillin resistance by a β-lactamase. Resistance to tetracycline is different in two ways: firstly the tet^r gene is carried not on r but on RTF; and secondly resistance is acquired by virtue of an inability to take up the drug into the cell. Not all R-factors confer resistance to all drugs; many carry only one drug-resistance marker.

4.3.3 Colicinogenic factors

There are several types of colicinogenic factors classified with respect to the nature of the colicin produced. They fall into two size classes: the larger ones resemble the F- and R-factors in that they possess analogous transfer systems; the small Col E factors are not transmissible and pass between cells only when conjugation is brought about by some other factor.

Plasmids, especially the smaller ones existing at high copy numbers are used as cloning vehicles in genetic engineering (see Chapter 14). The presence of drug-resistance markers is also helpful in the selection regimens used. One of the early vectors used as pSC101 derived from a fragment of a larger RTF, but derivatives of Col El are now in common use. Thus pBR322 has been engineered by removing some regions and inserting others, e.g. the tetracycline-resistance gene from pSC101, to give an autonomous plasmid of molecular weight 2.7×10^6 [116].

4.4. TRANSPOSABLE GENETIC ELEMENTS [34] (see also Chapter 13)

When a piece of DNA coding say for a tryptophan synthetic enzyme is transferred to a recipient bacterium (by conjugation, transformation or transduction, see p. 75) it becomes integrated into the recipient DNA at the site coding for that tryptophan enzyme, replacing the original piece of SNA. Thus the DNA becomes integrated at a *complementary* site by a process of legitimate recombination (see Chapter 9). Recently, however, it has been found that some recombination events occur between DNA regions which show little or no sequence complementarity. Such illegitimate recombination involves so-called transposable genetic elements or transposons which are discrete segments of DNA which have the ability to move around among the chromosomes and extra-chromosomal elements, i.e. plasmids or viruses. Thus the ampicillin-resistance gene referred to above rapidly moves from one plasmid to another and from one location to another on the plasmid in the form of a 4800 nucleotide pair fragment known as transposon 3 (Tn 3).

Sometimes, when the plasmid carrying a transposon is denatured and the separated DNA strands allowed to reanneal a characteristic structure is often seen in electron micrographs. The two ends of the transposon have the same base sequence but in the opposite orientation so that they can form an intrastrand base-paired region on reannealing. The central part of the transposon and the rest of the plasmid remain as single-stranded loops. These large repeated elements (700–1400 nucleotide pairs long) which exist at both ends of some transposons can also exist singly when they are known as insertion sequences (IS). An IS is similar to a transposon but does not carry any DNA coding for a protein. Both IS and transposons have inverted terminal repeats of about 30 nucleotide pairs. An IS is also able to insert itself into a chromosome almost at random. Insertion of an IS or a transposon is associated with the duplication of a smaller number (5, 9 or 11 depending on the IS) of base pairs originally present in the recipient DNA [35]. The presence of an IS in a plasmid may allow it to act as a transposon [235].

In the R-factor plasmids discussed above the resistance-determining factor consists of a series of transposons clustered together; each carrying the gene conferring resistance to one or more antibiotics. The RTF also has one transposon (Tn 10) carrying the gene for tetracycline resistance and may have an isolated insertion sequence [36].

The DNA of bacteriophage mu is a giant transposon which must be inserted into the bacterial chromosome for replication but then codes for proteins allowing it to infect other bacteria.

Transposons also occur in eukaryotes and circumstantial evidence for their existence was first obtained in studies with maize [37] (see also Chapter 13).

4.5 BACTERIAL VIRUSES OR BACTERIOPHAGES [38]

4.5.1 Structure

Bacteriophages contain single-stranded RNA (e.g. MS2), single-stranded DNA (e.g. ØX174 [38]), or double-stranded DNA (e.g. the

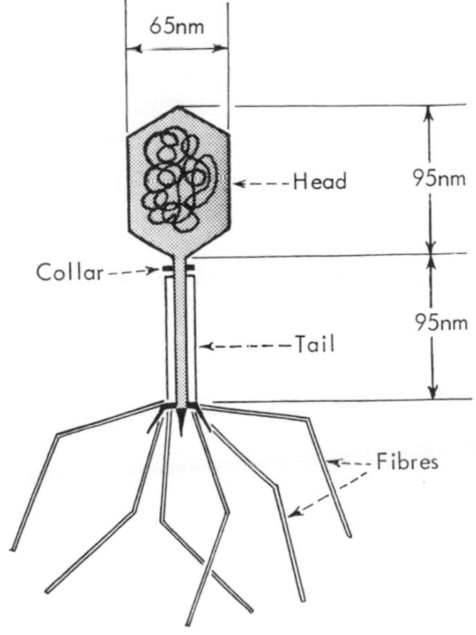

Fig. 4.2 The structure of a T-even phage particle.

T-phages). The single-stranded DNA bacteriophages are either spherical (e.g. ØX174) or filamentous (e.g. fd or M13 [40]) where the circular DNA molecule is wrapped in a protein coat and then formed into a filament of two nucleoprotein strands by bringing the opposite sides of the circle together. All the RNA bacteriophages are spherical. MS2 consists of 180 protein subunits all identical except for one (the maturation protein).

Many of the double-stranded DNA bacteriophages have a more intricate structure. The T-even bacteriophages (Fig. 4.2) have a head (which contains the DNA), a tail (through which the DNA is injected into the host cell), and a base plate with 6 tail fibres which recognize, and attach to, sites on the surface of the host cell [41].

The protein coat protects the nucleic acid from damage and also confers a specific host range on the protential infectivity of the particle.

Because of their simple structure a virus cannot multiply on its own. It must infect a bacterial, plant or animal cell and use the host cell's machinery for protein and nucleic acid synthesis.

4.5.2 Bacteriophage DNA (Table 4.1)

Viral DNA varies in mol. wt from a little over 10^6 to more than 10^8 (cf. the value of 2.2×10^9 for *E. coli* DNA) and unlike the DNA of bacteria and eukaryotes it can often be extracted from the virus without degradation. Such intact DNA molecules have revealed a striking variety of tertiary structure. The DNA of bacteriophage ØX174 is single stranded and in the form of a continuous cyclic chain containing 5386 nucleotides. DNA from bacteriophage G4 and M13 is similar. One of the first things that happen on infection of a bacterium with such single-stranded DNA is that it is converted into the duplex replicative form (RF) (see Chapter 8).

(a) Cohesive ends

When DNA extracted from bacteriophage λ(mol. wt 30×10^6) is heated to 65° and cooled slowly, its sedimentation coefficient is 37S, but when it is quick-cooled its sedimentation coefficient is only 32S [42]. This behaviour is a result of the 5′-ends, of the DNA projecting as single strands beyond the 3′-ends, the two single-stranded regions being complementary. These cohesive ends can base-pair and convert the DNA into a cyclic molecule which is disrupted at 65°. The complementary sequences of the cohesive ends, which are 12 nucleotides long, have been determined [43]. The hydrogen-bonded cyclic form can be converted with polynucleotide ligase (p. 248) into a cyclic form with both strands continuous. *E. coli* DNA polymerase, which adds on nucleotides to the 3′-ends, abolishes the ability to form cyclic molecules, but the ability is regained after treatment with *E. coli* exonuclease III which removes the

Table 4.1 Properties of some viral nucleic acids.

	Host cell	Mol. wt. $\times 10^{-6}$	Single- or double-stranded	Shape	Terminal repetition	Circular permutation	Single-strand 'nicks'
DNA Bacteriophages							
T2	*E. coli*	130	Double	Linear	Yes	Permuted	No
T5	*E. coli*	85	Double	Linear	—	Unique	Yes
T7	*E. coli*	25	Double	Linear	Yes	Unique	No
λ	*E. coli*	32	Double	Linear	Yes (exposed)	Unique	No
ΦX174	*E. coli*	1.7	Single	Cyclic	—	—	No
P22	*Salmonella*	26	Double	Linear	Yes	Permuted	No
RNA Bacteriophage							
MS2	*E. coli* (male)	1.1	Single	Linear	—	—	—
DNA Animal viruses							
polyoma	Mammals	3	Double	Cyclic	—	—	No
herpes	Man	68	Double	—	—	—	—
RNA Animal viruses							
poliovirus	Man	2.2	Single	Linear	—	—	—
reovirus	Mammals	12	Double	Linear (in several pieces)	—	—	—
Plant virus							
TMV	Tobacco plant	2	Single	Linear	—	—	—

newly added bases from the 3′-ends [44]. Electron microscopy has shown the 37S form to be cyclic and the 32S form to be linear. Some other lysogenic bacteriophages contain DNA with a similar structure (e.g. φ80 [45].

(b) Terminal repetition

The DNA of bacteriophage T7 (mol. wt 25×10^6) is double-stranded and linear, and the sequence (about 0.7% of the total) at the beginning is repeated at the end of each molecule [46]. Treatment of such molecules with *E. coli* exonuclease III results in the formation of cohesive ends which cause circle formation under suitable conditions. Terminal repetition has been detected in the DNA from several bacteriophages (e.g. T2, T4, T3, P22) and is believed to play a role in replication by enabling multiple length concatamers to be formed (see p. 262).

(c) Circular permutation

If the linear double-stranded DNA of bacteriophage T2 (mol. wt 130×10^6) is denatured and allowed to reanneal slowly, circular molecules are formed which can be detected with the electron microscope [47]. These are formed because the bacteriophage DNA molecules do not have a unique sequence, but the population is a collection of molecules with sequences which are circular permutations of each other (Fig. 4.3). The molecules also show terminal repetition of the sequences at the beginning. The DNAs from bacteriophages T4 and P22 also exhibit circular permutation and terminal repetition.

(d) Nicks

Three specific breaks have been found by electron microscopy in one strand of the double-stranded linear DNA extracted from bacteriophage T5 [48]. Accordingly, when the DNA is denatured, five single-stranded pieces are produced (instead of two) from each T5 DNA molecule. Bacteriophages SP8 and SP50 may also contain specific breaks.

(e) Modification of viral DNA

The DNAs of the T-even bacteriophages contain 5-hydroxymethyl-cytosine in place of cytosine [50] and the hydroxyl group of this base can be glucosylated (Tables 4.2 and 4.3) [51]. Growth of the bacteriophage in a UDP-glucose-deficient host produces bateriophages with DNA which is not glucosylated, and such DNA is degraded when the bacteriophages infect the normal host [52].

The DNAs of many bacteriophages are modified by host specific mechanisms when grown in certain strains of *E. coli (E. coli* K12 and *E. coli* B), but not when grown in other strains (e.g. *E. coli* C) [53,54].

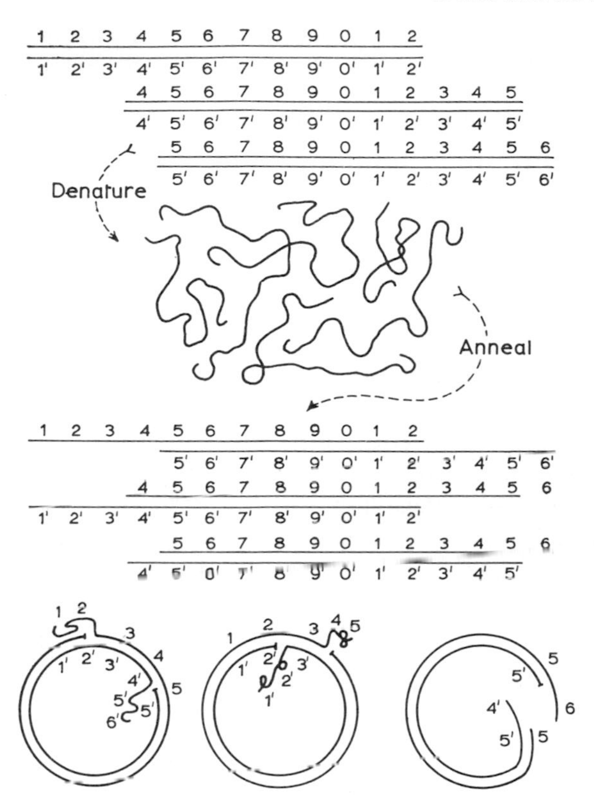

Fig. 4.3 Formation of cyclic DNA by denaturing and annealing a permutated collection of duplexes [49]. Notice that each permutation is also terminally repetitious. One repetitious terminal from each strand cannot find a complementary partner and is left out of the circular duplex. Their separation depends on the relative permutation of the partner chains. (Reproduced by permission of Academic Press Inc. from [49].)

Table 4.2 Molar proportions of bases in the DNA from certain strains of coliphage.

Strain	Adenine	Guanine	Thymine	Cytosine	Hydroxy-methyl-cytosine
T2	32.4	18.3	32.4	—	17.0
T4	32.4	18.3	32.4	—	17.0
T6	32.5	18.3	32.5	—	16.7
T5	30.3	19.5	30.8	19.5	—
T3	23.7	26.2	23.5	27.7	—

Table 4.3 Per cent glucosylation of hydroxymethyl-cytosine residues in the DNA of T-even bacteriophages [51].

	T2	T4	T6
Unglucosylated	25	0	25
α-Glucosyl	70	70	3
β-Glucosyl	0	30	0
β-Glucosyl-α-glucosyl (Diglucosyl)	5	0	72

Bacteriophage λ modified by growth in K12 or B will only grow efficiently in the same strain, K12 or B respectively or in C. Unmodified bacteriophage will only grow efficiently in C.
When the infecting bacteriophage fails to grow (i.e. is restricted), the bacteriophage DNA is degraded. The restriction/modification system is considered in detail in Chapter 6.

4.5.3 Life cycle

The replication of a virus can be considered in stages: (1) adsorption of the virus on to the host cell; (2) penetration of the viral nucleic acid into the cell; (3) development of virus specific functions, alteration of cell functions, replication of the nucleic acid, and synthesis of other virus constituents; (4) assembly of the progeny virus particles; (5) release of virus particles from the cell [25].

(1) Viruses will only infect certain specific cells, i.e. they have a limited host-range because the coat (or tail) will only recognize and adsorb to specific sites on the appropriate cell walls. The host range of the T-phages is a property of their tail fibres. Some T-even bacteriophages are free to adsorb to the bacterial cell wall site only in the presence of tryptophan; in the absence of this amino acid the tail fibres are folded back and attached to the tail sheath [5]. The initial reversible interaction of the tail fibres with the cell wall is followed by the formation of a permanent attachment. The small male-specific bacteriophages (e.g. MS2, R17) attach only to the f-pili of male *E. coli* cells [56] (see p. 62).

(2) The penetration of the viral nucleic acid into the cell involves a phage mechanism (e.g. T-phage), host-cell mechanism (e.g. MS2), or possibly the simple removal of the coat once the virus is in the cell.

After attachment to the cell, the lysozyme present in the base of the bacteriophage T4 tail probably hydrolyses part of the cell wall. This allows the tail core to penetrate into the cell as the contractile tail sheath contracts and the small amount of ATP present in the phage tail is hydrolysed to ADP [57]. How the DNA passes from the head through the tail and into the cell (a process equivalent to passing a 10-metre long piece of string down a straw) is not understood.

The injection of bacteriophage T5 DNA takes place in two stages. Eight per cent of the DNA (the first step transfer DNA) enters the cell and directs mRNA and protein synthesis. One of these proteins is required to complete the injection of the remaining DNA [58].

After attachment of bacteriophage MS2 to the f-pilus the RNA leaves the phage and is then transported inside the length of the pilus to the cell. This last step requires cellular energy [59,60]. An alternative mechanism proposed for entry of M13 into male *E. coli* involves retraction of the pilus with the filamentous bacteriophage attached [61]. Replication to the duplex form is necessary for the bacteriophage DNA to be drawn into the host, and when entry is effected a considerable fraction of the capsid protein is deposited in the inner cell membrane [25].

(3) Once uncoated, isolated viral DNA will not infect cells as well as intact virus. Thus following penetration of viral DNA into a cell the infective titre drops. With bacteriophage it remains low for maybe 20 min (the lag phase) and then suddenly rises up to 100-fold, indicating the presence of many new mature virus particles. During this lag phase the metabolic processes of the cell are being modified after virus infection [58,62]. Viral mRNA directs the synthesis of specific enzymes, and the rates of host cell DNA, RNA, and protein synthesis are altered. The viral nucleic acid replicates and virus constituents are synthesized. These different viral functions are divided into two groups, the early and late functions, which appear to be controlled either at the level of transcription or translation. Early functions include the biosynthesis of enzymes required for the replication of the nucleic acid, and late functions include the formation of the virus coat and other constituents. T-even bacteriophages turn off the synthesis of host cell DNA and redirect the synthesis of DNA precursors to fit their particular requirements (e.g. hydroxymethylcytosine triphosphate is produced; see Chapter 8).

The replication of virus nucleic acids is described in Chapter 8.

(4) The assembly of virus particles appears to be either spontaneous [63] or to involve a series of virus-directed steps [64].

More complex viruses are pieced together step by step. This process has been partially characterized for T4 and λ [53, 65]. Wood and Edgar [64] have shown that about 45 viral genes are involved in T4 assembly, and many of the steps have been characterized by *in vitro* complementation using the partially finished pieces (e.g. heads, tails, fibres) found in cells infected with different mutants under non-permissive conditions. Eight genes have been assigned to the component parts of the head, and a further 8 implicated in the assembly of these components. The head, which then appears to be morphologically complete, requires the action of two more gene products before it can interact spontaneously with assembled tails. The base plate components (12 genes) are assembled in 2 steps, then the core (3 genes) and the sheath (1 gene) are added by progressive polymerization from the base plate. Two gene products are required to finish the tails before they can be joined to the heads. The assembled heads and tails are modified (1 gene) before the tail fibres are added. This last step requires complete tail fibres (2 genes for components and 3 for assembly) and a labile factor (L) which has many properties of an enzyme. Phage particles assembled *in vitro* are active and possess characteristics which vary with the source of the parts (e.g. the genotype of the head and the host range of the tail fibres).

(5) Interference with the normal metabolic processes may lead to the eventual death of the infected cell, followed by natural lysis, but in some instances it has been shown that the virus actively causes cell lysis. Bacteriophage T4 codes for a lysozyme [66] which digests the host cell wall, causing the release of the progeny virus particles. Bacteriophage M13 and related filamentous viruses do not kill the host bacterium but pass out through the cell membrane picking up coat protein on the way.

4.5.4 The Hershey–Chase experiment

The classic experiment of Hershey and Chase [67] showed that on infection of *E. coli* with T2 bacteriophage only the DNA enters the bacterium. Thus only the DNA replicates and carries the information to specify new virus. To demonstrate this they grew T2 bacteriophage in the presence of ^{32}P-labelled phosphate and ^{35}S-labelled amino acids. This resulted in phage containing DNA labelled with ^{32}P (there is no sulphur in DNA) and protein labelled wth ^{35}S. The labelled virus was then allowed to infect unlabelled bacteria and the progeny isolated. It was found that much of the ^{32}P (i.e. the DNA) was present in the progeny virus, but none of the ^{35}S-labelled protein. Vigorous shaking of the culture within a few minutes of infection was found to dislodge the empty ^{35}S-labelled protein coat of the virus from the bacteria. On centrifugation this coat remained in the supernatant fraction when the bacteria sedimented. The bacteria carried the ^{32}P-labelled viral DNA which replicated and produced new phage particles. This shows that the viral DNA carries the genetic material and that the coat protein is not essential once the DNA is inside the bacterium.

4.5.5 The information content of bacteriophage DNA

The large bacteriophages have many genes. Thus T4 is known to have 135 genes of which 53 are concerned with bacteriophage assembly; 34 of these code for structural proteins. About 60 genes in T4 are non-essential but enable the virus to cope with sub-optimal conditions and allow it to extend the host range [68].

The small single-stranded bacteriophages, many of whose DNA has been sequenced, possess very few genes. ØX174 and the related bacteriophage G4 have eleven genes, six of which are concerned with coat-protein synthesis or assembly, one with cell lysis and the remainder with DNA synthesis (see Chapter 8). The combined molecular weight of these eleven proteins adds up to 262 000 which is greater than can be coded for by

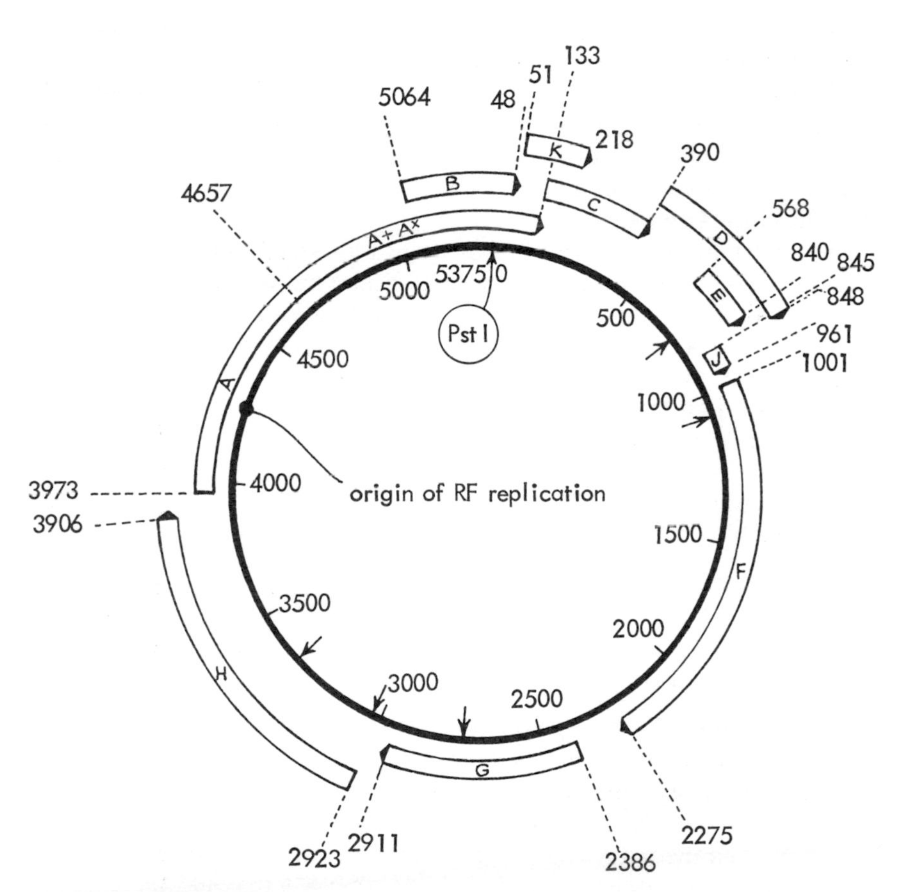

Fig. 4.4 Physical map and coding potential of ØX174 DNA. The numbers refer to nucleotides relative to the single PstI site. The other arrows in the inner circle show the sites of action of HpaII. The open bars indicate the regions coding for the different proteins (A–K) and bars of different radii demonstrate the use of all three coding frames. The thick arrows indicate the C-terminus of the proteins.

the small genome if traditional concepts apply. However in ØX174 the nucleotide sequence [69] shows that gene B is encoded with gene A but in a different reading frame (see Chapter 12) and gene E is encoded within gene D. Gene K is read in a second reading frame and overlaps the gene A/C junction (see Fig. 4.4). In fact gene C starts 5 nucleotides before the end of gene A so in this region all three reading frames are used. Gene A′ produces a protein identical to the carboxy terminus of the gene A protein: in this case there are two sites for initiation of translation. It is clear that the small bacteriophages make full use of their DNA and detailed comparisons of the nucleotide sequence of the genomes of ØX174 and G4 suggest the possibility of the existence of several other genes whose products have not yet been identified [70,71].

The small RNA bacteriophages (e.g. R17) contain all their genetic information in a single strand of RNA about 3300 nucleotides long (which can code for about 1100 amino acids or about 3–4 proteins of average size). Three complementation groups corresponding to three viral coded proteins have been identified (p. 369), and these may represent the total genetic potential of the virus [72]. In order, they are: the coat protein (129 amino acids, mol wt 14000), the maturation protein (mol. wt about 37000 [73], and the RNA-dependent RNA polymerase (p. 369) mol. wt about 50000). It seems likely that there is only one maturation protein molecule per particle and that this protein is concerned with the assembly of the RNA into the bacteriophage. In the infected cell, the synthetase is an early protein made about 10 minutes after infection, while the coat and maturation proteins are late proteins made (in appropriate amounts) about 20 minutes later. The mechanism of this control is considered in Chapter 13.

4.5.6 Virus mutants

The viruses provide a unique opportunity to characterize completely the information content of a functional genome.

A mutation in a viral genome can cause the inactivation of a gene function. If the missing function can be characterized (e.g. a missing enzyme) then the nature of the gene, and the mutant, is defined. However, if the function is essential, the mutation is lethal and the mutant cannot be propagated and studied. The discovery of conditional lethal mutants [74–76], therefore, which lack a gene function under one set of conditions but regain it under another set of conditions, has revolutionized the study of virus, and in particular, phage genetics. Stocks of conditional lethal mutants can be grown under permissive conditions and then the mutant gene function studied and identified under non-permissive conditions.

Two classes of conditional lethal mutation are particularly useful. Temperature-sensitive mutants [75] are not viable at the non-permissive temperature (e.g. 42°) but grow at the permissive temperature (e.g. 30°). This is due to a single base change in the DNA causing the introduction of the wrong amino acid into the protein, and reducing the stability of its configuration at the higher temperature.

Amber mutants [74] of a bacteriophage will only grow in a permissive host (which contains a suppressor). This is caused by a single base change altering an amino acid coding triplet to UAG (p. 391), which is read as stop under the normal, non-permissive conditions, and the protein is terminated at that point. Permissive host bacteria, which suppress the mutation, have a species of tRNA which translates UAG as an amino acid and allows the protein to be completed. Different amino acids are added by different classes of permissive host.

When two virus mutants, with mutations in different genes, infect the same cell under non-permissive conditions, the missing function of each can be supplied by the other, i.e. complementation takes place, and progeny viruses are formed. If the two mutants have mutations in the same gene, complementation cannot occur and no progeny are formed. Thus complementation between mutants is used to determine the number of complementation groups, i.e. the number of different genes in which the mutations occur. If sufficient mutants have been isolated for each gene of the virus to be represented, then the total number of essential viral genes can be estimated.

The joint growth of two bacteriophages in the same cell can also result in the formation of a few progeny phages, recombinants, which carry genetic characters of both parental bacteriophages (see Chapter 9). A study of the frequency of formation of recombinants from parental bacteriophages carrying known mutations allows the construction of a genetic map; the frequency of recombinants with both genetic characters is related to the distance between the two mutation points on the genome. Such a map shows the relative positions of the mutations and therefore of the genes in which these mutations occur, and it can be linear or circular [76].

4.5.7 Lysogeny and transduction [22]

When a virulent bacteriophage infects a cell, the virus replicates and the cell is killed. The temperate bacteriophages, however, can either kill the cell or lysogenize it [77]. A lysogenic cell usually carries the DNA of the bacteriophage integrated in the cell genome (as a prophage). The bacteriophage genes are transcribed (in a controlled way), replicated, and inherited along with the cell genes. Most of the bacteriophage functions are repressed (e.g. those involved in the lytic development), but others are not (e.g. those involved in the maintenance of lysogeny). One gene function which is not repressed in the lysogenic cell confers immunity against further infection of the cell by the same bacteriophage. This immunity is quite specific.

The survival of virulent bacteriophages depends on a continuous supply of susceptible bacteria (e.g. in sewage, a rich source of bacteriophages) while the temperate bacteriophages can survive and replicate with a limited population of cells which are protected from further infection. Prophages are either integrated at a specific site on the bacterial chromosome (e.g. λ which normally attaches near the *gal* locus of *E. coli* [78], (see Chapter 9) or at random (e.g. phage mu which causes inactivation of a gene into which it integrates).

The stable lysogenic cell can be induced by various agents, most of which interfere with normal DNA synthesis [79]. This results in the bacteriophage DNA being cut out of the cell genome, probably by a single recombination event [80]. The DNA can then replicate and function in a virulent manner, resulting in cell lysis and liberation of a burst of progeny bacteriophage.

Occasionally, on induction, mistakes are made in the excision of the bacteriophage DNA and small pieces may be left behind or sections of bacterial DNA may be excised along with the bacteriophage DNA. This bacterial DNA may be replicated along with the bacteriophage DNA and packaged into the virus particle. On reinfection of new bacterium these genes may become integrated into the new genome. *This process is called transduction.* Thus bacteriophage λ, which integrates next to the *gal* gene, may pick up this gene on induction to form the composite bacteriophage λ*gal*. The *gal* gene is then replicated along with and is introduced into a new bacterium. If this recipient is gal^- it could thus acquire the ability to metabolize galactose. Some prophages (e.g. P1 [81]) are not integrated but remain in a stable extrachromosomal state like plasmids. Only very occasionally are they packed and virions released. This packaging is very non-specific in that any DNA of the right length, including fragments of host chromosome, can be packed (i.e. a 'headfull' mechanism). On infecting a new bacterium genes of the original host are introduced at random and P1 is known as a generalized transducing bacteriophage.

Along with transformation and conjugation, transduction is one of the three ways in which DNA can be transferred between cells. Transfection is the rather inefficient process whereby cells may be infected with naked DNA rather than with intact virus. It involves uptake of the entire viral genome.

4.6 ANIMAL AND PLANT VIRUSES

4.6.1 Structure

Most plant viruses contain single-stranded RNA as in TMV (tobacco mosaic virus), but a few contain double-stranded RNA as in wound tumour virus [82] or double-stranded DNA as in cauliflower mosaic virus [83]. The virus particles can be rod-shaped (e.g. TMV) or spherical (e.g. cowpea chlorotic mottle virus).

An extensive study of TMV (for reviews see [84–86]), which has a particle weight of 4×10^7 and is rod-shaped measuring 15×300 nm, has led to a detailed picture of its structure (Fig. 4.5). A helical array of about 2100 identical protein subunits of mol. wt. 17400 surround a single-stranded RNA molecule of mol. wt. 2×10^6. Cowpea chlorotic mottle virus is spherical and is composed of 180 identical subunits of mol. wt. about 20000 arranged on the surface of an icosahedron in 32 morphological units of 20 hexamers (on the faces) and 12 pentamers (on the vertices) [63, 87].

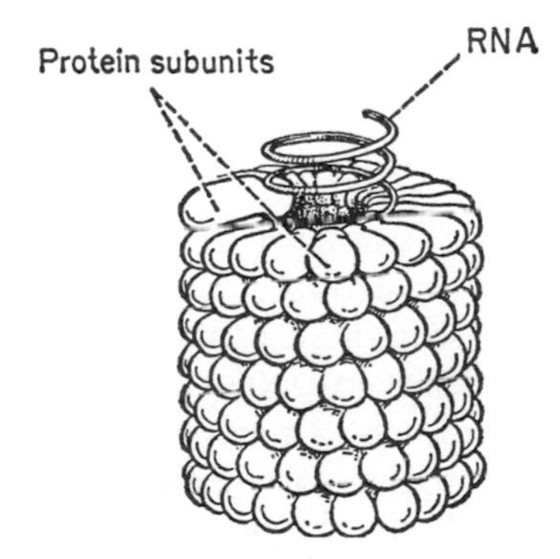

Fig. 4.5 Segment of the tobacco mosaic virus particle showing the protein subunits forming a helical array [84]. The RNA lies in a helical groove in the protein subunits, some of which have been omitted to show the top two turns of the RNA helix.

Animal viruses contain single-stranded RNA as in poliovirus, double-stranded RNA as in reovirus [82], single-stranded DNA as in minute virus of the mouse [88], or double-stranded DNA as in polyoma virus. Some viruses have both a virus-specified protein coat and a lipoprotein envelope similar to the cell cytoplasmic membrane (e.g. influenza virus, herpes virus [89]).

The small animal RNA viruses such as poliomyelitis virus, resemble plant viruses and contain one molecule of single-stranded RNA (mol. wt.

$1–3 \times 10^6$) per particle [90]. The replication of these viral RNA molecules appears to involve a double-stranded intermediate (see p. 371).

A second type of single-stranded RNA viruses includes the retroviruses, oncornaviruses or RNA tumour viruses such as the leukaemia viruses. These viruses contain RNA of molecular weight around 10^7 daltons (60–70S) but this can be dissociated to subunits of about 3×10^6 molecular weight. Replication of this RNA takes place by way of a DNA intermediate and involves the enzyme known as reverse transcriptase (see p. 266) [91].

Each reovirus particle contains about 12×10^6 daltons of RNA which, when extracted, is in pieces of three sizes (mol. wt. 2.3×10^6, 1.3×10^6, and 8×10^5), plus 50–100 single-stranded oligonucleotides rich in adenine [92].

The genome of herpes simplex virus (HSV) (mol. wt. about 10^8) is remarkable in being made up of two parts joined in tandem. Each part has a region which is repeated at the two ends in an inverted orientation. As the two parts can be joined in either orientation there are four possible sequences present in the HSV genome [93] i.e.

↦ ⟼
↤ ⟼
↦ ⟻
↤ ⟻

The genome of adenoviruses (mol. wt. $20–25 \times 10^6$) has a sequence of about 100 nucleotide pairs repeated in an inverted fashion at the two ends [94] and the 5′ end of both strands is covalently attached to a protein of molecular weight 55 000 (see Chapter 8).

The smallest animal viruses (the parvoviruses) have a linear single stranded DNA genome of molecular weight $1.2–2.2 \times 10^6$. Some (e.g. minute virus of mouse–MVM) have a hairpin duplex at both ends of the DNA while others (e.g. adenoassociated virus–AAV) have a hairpin at one end only [95].

The small animal tumour viruses SV40 and polyoma (papovaviruses) exist in the cell in the form of nucleoprotein particles where the cylic duplex DNA molecule (5224 and 5292 nucleotide pairs respectively) is wrapped around histone octomers to form a nucleosomal structure—the minichromosome (see p. 91). There are 21 nucleosomes associated with with the cyclic chromosome [46–98] which, because of the presence of topoisomerases (see p. 253) is present in a form of minimum energy. In the cell or when isolated in low-salt conditions the minichromosome is highly compact. When the histones are removed and the DNA isolated it is present in a supercoiled form (form I) with about 26 superturns per molecule (i.e. there is an average of 1.25 superhelical turns per nucleosome). A single nick in one of the DNA strands will convert the duplex into a relaxed open circular form (form II).

Three forms of double-stranded DNA can be isolated from purified polyoma virus [99, 100]. The supercoiled form (21S, component I), the open cyclic form (16S, component II), and a linear form (14.5S, component III). Viral component III forms a minor fraction of the linear virion DNA which is mainly cellular DNA that has been encapsulated in virus particles [101]. Electron microscopy (Plate I) has revealed structures which can be clearly identified with the supercoiled, open cyclic, and linear forms and the three forms are readily separated by electrophoresis on agarose gels.

The DNA of the larger animal viruses (e.g. adenovirus or herpes simplex virus) although not naked in the cell does not appear to have a nucleosomal structure similar to that of the host cell DNA. It is for this reason that the smaller SV40 and polyoma virus have been used as models for animal cell chromatin structure.

4.6.2 Life cycle

The life cycle of animal viruses is in many ways similar to that of bacteriophage, the main difference being that a cycle takes 20–60 h rather than 20–60 min. Uptake mechanisms are simpler. Thus polyoma virus attaches to neuraminidase-sensitive sites on mouse cells [102] and is then taken up entirely by pinocytosis and is carried to the nucleus where the DNA is uncoated. An eclipse period follows which may involve disruption of host cell metabolism. On the other hand the small tumour viruses, SV40 and polyoma, stimulate the synthesis of host cell DNA [103, 104] particularly if the cells are in a resting state before infection [105].

The reconstitution of the rod-shaped TMV has been studied extensively [106–109]. When the coat protein and viral RNA are mixed in the correct ionic environment, virus particles are formed which possess up to 80 per cent of the original infectivity. Attempts to reconstitute small spherical viruses have proved more difficult. MS2 RNA and coat protein form morphologically complete particles which are not infective, possibly because they lack the maturation protein [110]. Cowpea chlorotic mottle virus can be partially degraded and reassembled to form particles indistinguishable from the original virus, and separated protein and nucleic acid have been mixed under conditions such that infectious particles are formed which have the same appearance, serological properties, and sedimentation coefficient as intact virus [111]. Such experiments are consistent with the suggestion that the nature of the virus coat subunits alone directs the size and shape (spherical or rod) of the completed virus particles [112].

4.6.3 The information content of viral DNA

The large animal viruses such as herpes simplex virus (HSV) can code for more than 100 proteins, and, as with the large bacteriophage (e.g. T4) many of these proteins duplicate the functions of host proteins. Unlike the smaller animal viruses the HSV genome has a random amount of the dinucleotide CpG which is deficient in the genome of the host and the smaller viruses [111–115].

The papovavirus genome codes for two (SV40) or three (polyoma) early proteins known as tumour antigens as they are expressed in transformed cells (see below). Later in infection three virion coat proteins (VP1, VP2 and VP3) are synthesized. The two (or three) tumour antigens

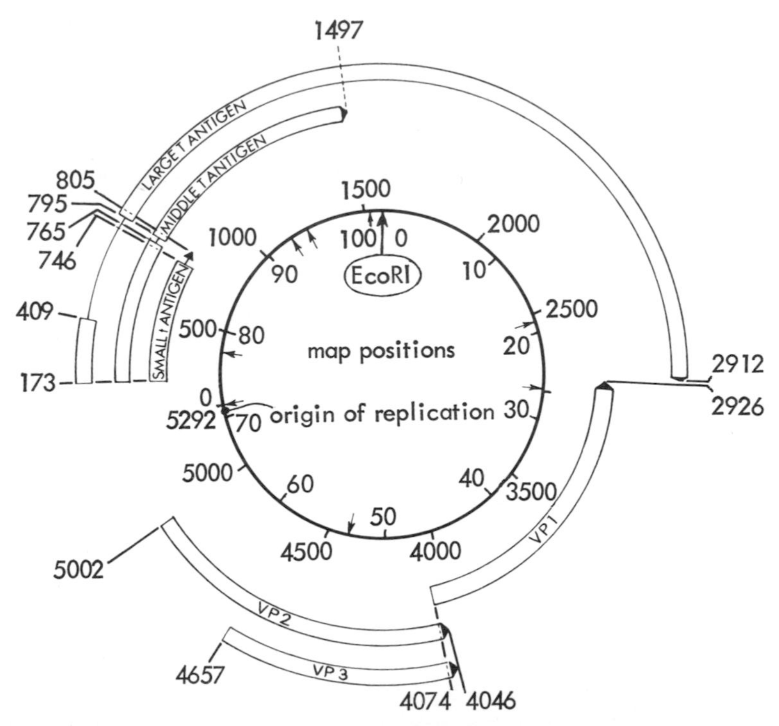

Fig. 4.6 Physical map and coding potential of polyoma DNA. The numbers inside the inner circle show the polyoma DNA divided into 100 map units starting at the single EcoR1 site. Other arrows show the sites of action of HpaII. All other numbers refer to nucleotides measured from the HpaII 3/5 junction. The open bars indicate the regions coding for the various proteins —the thick arrows indicate the C-terminus. The single line joining the open bars for the antigens indicate these proteins are coded for by non contiguous regions of DNA (Section 11.3).

N.B. Bars of different radii do not imply different reading frames.

are translated from messenger RNA molecules derived by processing a single transcript from the early half of the genome starting near the origin of replication. This region of the DNA is 2700 nucleotide pairs long and in polyoma codes for large T (molecular weight about 100 000), middle T (molecular weight 55 000) and small t (molecular weight 17 000) antigens. In order to achieve this more than one reading frame is used (cf. ØX174). Although all three proteins have the same N-terminus showing they are coded for by the same reading frame to start with this has a termination signal after about 600 nucleotides leading to production of t antigen. RNA splicing (see Section 11.3) changes the reading frame in some of the transcripts to frame 2 (large T antigen) or frame 3 (middle T antigen).

The late region of polyoma virus (coding for the virion proteins) also extends from near the replication origin but in the opposite direction to the early region. It is about 2100 nucleotide pairs long and codes for proteins requiring two reading frames. All three messenger RNAs have the same leader but splicing leads to synthesis of VP2 and VP3 in one reading frame (VP3 is identical to the C-terminus of VP2) and VP1 in a different partly overlapping frame (see Fig. 4.6) [113–115].

Thus, like ØX174, polyoma virus and SV40 have more coding information than a co-linear relationship between DNA, RNA and protein would suggest but this is achieved in a somewhat different manner than with the small bacteriophage.

4.6.4 Tumour viruses and animal cell transformation

Some animal viruses (tumor viruses) can alter (transform) the infected cell, without killing it, so that it has new properties which are typically neoplastic [117–119]. Uninfected hamster cells do not form tumours when injected into new-born hamsters, and do not grow when suspended in nutrient agar, but polyoma-transformed hamster cells form tumours and grow in agar [117].

Several DNA-containing tumour viruses have been identified, including polyoma, SV40, rabbit papilloma, human papilloma, and adenoviruses 7, 12 and 18.

Some of these viruses can interact with different cells in different ways. Polyoma virus will transform hamster cells but will replicate in, and kill, mouse cells. SV40, on the other hand, transforms mouse cells and kills green monkey cells. A temperature-sensitive mutant of polyoma [121] will transform mouse cells at the non-permissive temperature, 38·5°C, and then replicate when the temperature of the transformed cell is reduced to 31°C [122]. This suggests that at least one virus function necessary for replication is not required for transformation. It is not understood why the different interactions occur with the different cell types.

Several lines of evidence show that the viral genome is present in the transformed cells, in a stable inheritable integrated state (cf. the lysogenic state of some bacteria). mRNA isolated from polyoma transformed cells will hybridize with polyoma virus DNA [123]. SV40-transformed mouse cells can be fused with green monkey cells using inactivated Sendai virus, and the hybrid cells liberate active SV40 particles [118, 124].

Although some SV40-transformed cells appear to contain only one copy of the viral genome per cell [125], others contain up to nine copies [126]. Integration of viral DNA into the host genome suppresses all but the early viral functions. How the T antigens of SV40 and polyoma virus bring about cell transformation is not clear. The large T antigens and polyoma middle T antigen may have protein kinase activity [127, 128] and middle T antigen is also found on the cell surface where it may mediate cell/cell interactions. The middle T antigen may be the (tumour specific) transplantation antigen (TSTA) which causes rejection of transplantable tumours by animals previously immunized with the virus [120]. The possible roles of T antigen in transformed cells have recently been reviewed [129, 130].

The leukaemia viruses and the sarcoma viruses are closely related RNA tumour viruses. The leukaemia viruses are a natural cause of leukaemia and the sarcoma viruses transform cells in tissue culture. Avian, murine, and feline RNA tumour viruses have been isolated; the feline sarcoma virus will transform human cells in tissue culture, but there is no evidence which relates human leukaemia to such viruses.

The RNA tumour viruses contain single-stranded RNA of molecular weight about 10^7 (70S) [131] along with a considerable amount of 4S material and smaller amounts of 28S, 18S, and 7S RNA [132] and DNA [133] which may be of cellular origin. The virus particles, which are enveloped, also contain an RNA-dependent DNA polymerase [134, 135] which, in the presence of the four deoxyribonucleoside triphosphates and an RNA template, catalyses the formation of first an RNA–DNA hybrid and then double-stranded DNA [136] (see Chapter 8). In cells transformed by RNA tumour viruses the DNA copy of the viral RNA is integrated into the host genome but only one gene is expressed. This *src* gene product is a protein kinase located in the cell membrane [120, 137].

4.7 EUKARYOTE DNA

4.7.1 The eukaryote cell cycle

Bacterial cells in rapid growth synthesize DNA almost continuously. However in eukaryote cells, DNA synthesis takes place in a restricted time period during the *cell cycle* (Fig. 4.7). In mitosis the cells divide and

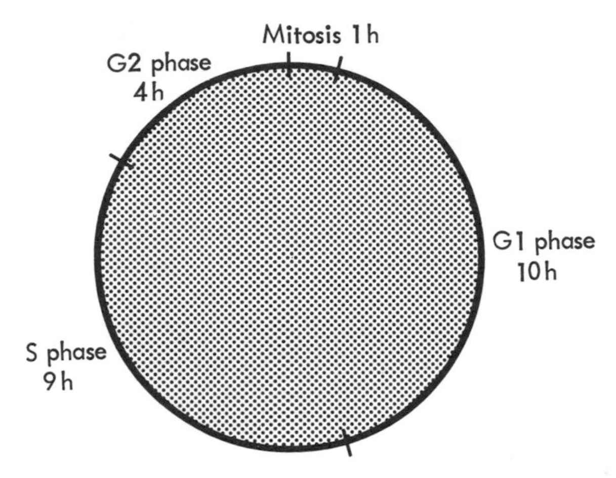

Fig. 4.7 The eukaryote cell cycle G1, S and G2 phases are together known as interphase. The actual times involved vary with cell type and culture conditions.

form two daughter cells. This is then followed by a gap period designated G1 during which no DNA synthesis occurs. The length of the various gap periods and indeed the length of the entire cell cycle depends very much upon the cell type but Fig. 4.7 shows a typical example of a cell in culture with a replication time of 24 h. G1 is followed by S phase in which DNA synthesis occurs and then by a second shorter gap period known as G2. The cells then go into mitosis again. The combination of G1,S and G2 phases is known as *interphase* and under these conditions the cell nucleus is clearly visible in the light microscope [138].

The actual process of mitosis can be subdivided into several discrete stages. At the end of the interphase two *poles* are formed in the cell by the *centrioles* (Fig. 4.8). In prophase, the chromosomes condense to form two identical, linked chromatids joined at the *centromere* and the nuclear envelope disappears. Fine fibres of microtubules form between the two poles. In metaphase the chromosomes line up in the centre of the cell at the equatorial region to form the *metaphase plate.* In anaphase the two daughter chromosomes are pulled apart into the two poles and in telophase the two new nuclear envelopes form.

A population of growing cells will normally consist of cells at all stages of the cell cycle. A variety of methods may be used to *synchronize* the population so that all are dividing at the same time. Metaphase cells may be shaken from the glass or plastic of the tissue culture vessel and separated from cells at other stages of the cycle. Alternatively, chemical inhibitors such as thymidine may be used to stop all the cells in S phase from completing the cycle, since excess thymidine inhibits DNA

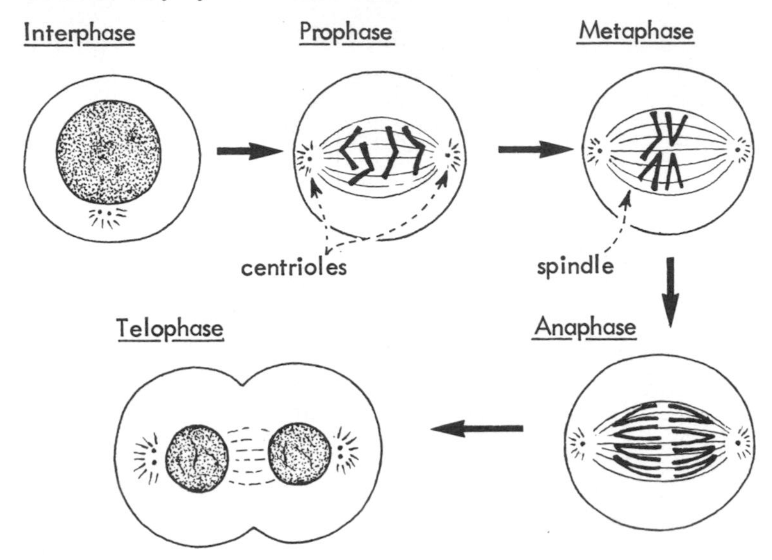

Fig. 4.8 The process of mitosis.

synthesis. When this method is used, it is usual to apply the thymidine for two periods of about 16 hours separated by a period of about 8–10 hours in the absence of thymidine. During the first exposure cells making DNA are stopped throughout S phase while cells in G1, G2 and mitosis continue to grow until they reach the beginning of S phase. On removal of the thymidine all the cells traverse S phase and so during the second exposure they now all accumulate at the G1/S border. Other methods are covered in more specialist texts [141, 204].

4.7.2 Chromosomes

Eukaryote DNA is contained in a relatively small number of chromosomes which varies according to the species (Table 4.4). No direct correlation can be made between the amount of DNA in the nucleus and the number of chromosomes in which it is contained [139]. Somatic cells of each species have two copies or homologues of each chromosome with the exception of the sex chromosomes for which the female carries two X chromosomes and the male an X and a Y. Germ cells contain only one copy of each chromosome and it is to these cells that the term *haploid chromosome content* or *haploid DNA content* refer. While most somatic cells are *diploid, tetraploid* cells with four and *octaploid* cells with eight copies of each chromosome can also be found, particularly in cells in culture. *Aneuploid* cells have an abnormal chromosome complement, which is not necessarily an increase on the diploid condition for each

Table 4.4 Haploid DNA content and chromosome number of a variety of eukaryotes.

	Haploid DNA content		Haploid chromosome number
	Picograms	Kilobases (kb)	
Saccharomyces cerevisiae (yeast)	0.0045	7×10^3	17
Drosophila melanogaster (fruit fly)	0.06	1×10^5	4
Xenopus laevis (toad)	14	22×10^6	18
Gallus domesticus (chicken)	0.6	1×10^6	39
Mus musculus (mouse)	1.5	2.3×10^6	20
Homo sapiens (human)	1.8	2.8×10^6	23
Zea mays (corn)	10	15×10^6	10
HeLa cells (human cell culture, aneuploid)	15.8	23.7×10^6	70–164 (mean 82)

chromosome type, so that some may be present in greater numbers than others. The characteristic number and morphology of the chromosomes in any particular cell type is known as the *karyotype* of that cell and is usually determined in metaphase when the chromosomes are highly condensed and readily stained by basic dyes [140]. In interphase, the chromosomes are spread out in the nucleus and cannot be individually distinguished.

Metaphase chromosomes consist of two DNA duplexes, highly compacted by proteins and joined in the middle at the *centromere*. Chromosomes with a centromere towards one end are known as *acrocentric* and those with a more centrally located centromere are *metacentric*.

Additional secondary constrictions are usually the site of rRNA genes which form the nucleolus in an interphase nucleus and are known as *nucleolar organisers*. These are only found on a few chromosomes. Detailed cytogenetic studies of chromosomes involve the use of fluorescent dyes such as quinacrine mustard which form a characteristic band pattern (known in this case as Q bands) visible on fluorescence microscopy [149]. In this way chromosome deletions or rearrangements may be more closely observed. A large variety of banding techniques is now available for cytogenetic studies [150].

4.7.3 Haploid DNA content (C value)

One of the most striking features of eukaryote DNA is the great quantity of it which is present in each cell (Table 4.4). The amount is about an order of magnitude in excess of that required for the known gene-coding capacities of the cells and the logical conclusion is that the bulk of the DNA is not expressed but is essentially redundant or has a function which is as yet imperfectly understood [142–146]. For each class of eukaryote, the minimum size of genome increases with the stage of evolutionary development of the animal [147]. However, certain amphibia have a C value one hundred-fold in excess of man and even more in excess of other amphibia and the reason for this is not clear. The phenomenon is known as the *C value paradox* [148]. Logically, the species with the greater amount of DNA should have the advantage of greater coding potential and the disadvantage of the requirement to replicate very large amounts of DNA during cell division.

In additon to the C value paradox, different cell types within the same species can vary in their haploid DNA content. Amphibian oocytes have a very large amount of cytoplasm to provide with essential components of protein synthesis such as ribosomes and consequently have greatly amplified numbers of genes coding for ribosomal RNA. Eukaryote cells in culture tend to become aneuploid with the passage of time, apparently adapting to their culture conditions. In addition, they have the capacity to amplify certain genes several hundred-fold in response to external stimuli. Thus the gene coding for the enzyme dihydrofolate reductase is amplified in cells which are resistant to methotrexate a drug which interferes with one carbon unit metabolism [151].

4.7.4 Gene frequency

Eukaryote DNA is by convention divided into three frequency classes of unique DNA, moderately repetitive DNA and highly repeated, or satellite DNA. There is, in fact, considerable overlap between the three categories which are probably better classified by their Cot values (Chapter 3) since this is the manner in which the classification is achieved experimentally.

Unique DNA in human cells is generally classified as that DNA which has a Cot value of 1000 moles of nucleotide seconds/litre and over. It comprises about half of the total haploid DNA content and is thought to consist of the sequences coding for most enzyme functions for which there is thought to be only one or a small number of genes in each haploid cell. The genes coding for the various chains of haemoglobin or the enzyme glucose 6-phosphatase fall into this category.

Moderately repetitive DNA in human cells is usually taken to be that fraction of the DNA which reanneals with a Cot value of between 100 and 1000 moles of nucleotide seconds/litre. It comprises 30–40% of the cellular DNA. The sequences represented in this group are generally thought to be those coding for proteins which form major structural components of the cell such as the histones. The genes for rRNA and tRNA also fall into this category.

Highly repetitive DNA is also frequently referred to as *satellite DNA* or simple sequence DNA and is that fraction of human DNA which reanneals with a Cot value of under 100 moles of nucleotide seconds/litre. The origin of the name satellite relates to the method of its isolation on caesium chloride buoyant-density gradients of sheared DNA where it will sometimes form a separate satellite band due to its differing content of adenine and thymine residues. The simplest known satellite DNA is poly dAT which occurs in certain crabs. Other satellites can have any number from four to several hundred base pairs which are repeated in tandem fashion along the genome (Fig. 4.9).

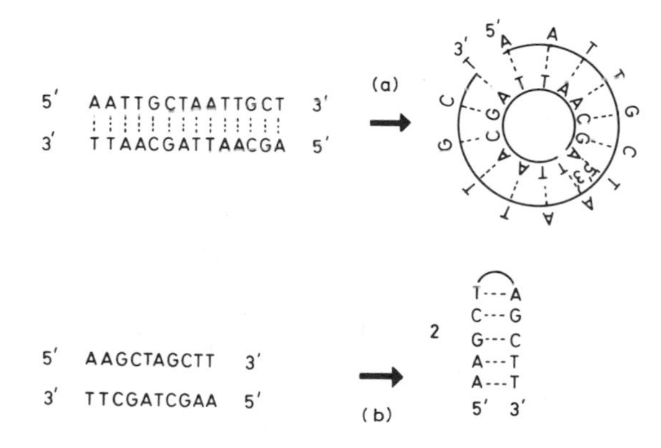

Fig. 4.9 Two examples of unusual structures formed by eukaryotic chromosomal DNA on denaturation and reannealing. (a) Repeated DNAs such as satellite DNA can form circular molecules. (b) DNAs with inverted repetition form hairpin loops or palindromes.

Human DNA has been shown by density-gradient centrifugation to have four main satellites [152, 143] and by dye binding [154] and restriction endonuclease cleavage [155] to have two additional satellites. The distribution among chromosomes varies with some chromosomes appearing to have virtually none of the satellite sequences which have been isolated so far and some (noteably the Y chromosome) to be heavily covered in satellite sequences [156]. In general, satellite DNA appears to be concentrated near the centromere of the chromosomes in the heterochromatin fraction (Section 4.9.1). Originally, it was thought that satellite DNA was not transcribed since RNA of corresponding sequence was seldom isolated but occasional cases of satellite transcription have been reported [157].

The arrangement of unique and moderately repetitive genes differs from that of the clustered satellite DNA. Unique sequences of about 1000 base pairs are interspersed with moderately repetitive sequences of about 300 base pairs for about 50% of the genome and unique sequences of several thousand base pairs are interspersed with repetitive sequences for a further 30–40% [158–161]. The 300 nucleotide interspersed sequences are closely related to one another and in particular contain a highly conserved 30 nucleotide segment present in both human and hamster DNA and also near the replication origin of papovavirus DNA [242]. While this pattern appears to hold for most eukaryotes, *Drosophila melanogaster* is atypical in having a longer interspersion ratio, the unique sequences being closer to 13 000 base pairs in length and the repeated regions of 5500 base pairs [162].

The sequences which flank specific genes whose coding function is known are only described as yet in a few cases. Human haemoglobin genes of the beta group comprise the epsilon, delta, gamma and beta chains which are produced at various stages during development and are clustered within the same 60kb fragment of DNA. A non-globin repeat sequence of about 400 base pairs in length is interspersed between the genes coding for all these globins, sometimes in more than one copy [163]. In the other large-cloned fragment which is well analysed, that of the chicken ovalbumin gene, two similar genes with considerable homology can be shown to be located on the same 46kb fragment, though the nature of the intervening DNA is not known [164].

A special class of *foldback* or *palindromic* DNA sequences comprising 3–6% of eukaryote DNA has also been characterized. The size range is from 300 to 1200 base pairs and the molecules are all below a Cot value of 10^{-5} moles of nucleotide seconds/litre. This palindromic DNA is represented in all frequency classes and widely distributed throughout metaphase chromosomes. Its function is not yet known [175–177] (Fig. 4.9).

4.7.5 Eukaryote gene structure

In recent years a fundamental difference between prokaryote and eukaryote genes has been uncovered by experimental analysis. Whereas prokaryote genes occupy a single uninterrupted sequence of DNA, the majority of eukaryote structural genes so far analysed have shown intervening sequences in the middle of the gene. These intervening sequences which have also been named 'introns' may be small and single as in the case of the gene for tyrosine suppressor tRNA [165], or large and multiple as is the case for the ovalbumin gene which has seven inserted sequences so that the entire gene occupies a length of 7.7kb with only a small fraction of this DNA being actually used for coding purposes [166, 167, 171]. Current evidence suggests that these intervening regions are transcribed into RNA (Chapter 11) and then removed by internal processing events in the cell nucleus until the RNA transcript is the size of the final mRNA product [168, 172, 173]. Intervening sequences appear to have a wider evolutionary freedom for mutation than the coding sequences of the DNA [169]. Where genes such as the rDNA genes in *Drosophila* occur in more than one copy in the cell, there is a preference for the transcription of the gene which has no intervening sequences [170]. In general intervening sequences are less frequent in the lower eukaryotes [171]. While the majority of structural genes display this phenomenon, it is not universal even in higher eukaryotes and is not found in the majority of tRNA genes or in the genes coding for the histones [172, 173]. The full functional significance of intervening sequences remains improperly understood and has been suggested to reflect the continual process of evolution within the eukaryote chromosome [145, 146, 174].

4.8 THE STRUCTURE OF THE EUKARYOTE CHROMOSOME

4.8.1 Chromatin

Eukaryote chromosomes in metaphase are generally referred to as chromosomes, but in interphase the term chromatin is more generally used to describe the nucleoprotein fibres in the cell nucleus. Originally chromatin was loosely defined and subdivided into only two main classes, euchromatin and heterochromatin. Heterochromatin comprised the dense, readily stained areas of the nucleus or chromosome and was thought to represent inactive chromatin which was not undergoing transcription whereas the more loosely packed euchromatin was thought to represent the transcriptionally active material. These concepts have been largely superseded by recent research in the area but are in occasional use.

Chromatin consists of DNA, RNA and proteins. The actual weight ratio between the three varies greatly with the tissue or cell of origin of the material, but in general the amount of protein is equal to or greater than the amount of DNA while the amount of RNA is comparatively small [175]. The protein content of chromatin can be further subdivided into the histone and non-histone proteins. The former group consists of a few types of molecule which are present in very large amounts and the latter of a diverse range of protein molecules, each present in much smaller quantity. Historically the bulk of the non-histone proteins were referred to as the acidic proteins but this classification is no longer appropriate.

4.8.2 The histones

There are five major types of histone molecule in the eukaryote cell nucleus. These are now classified as histones H1, H2A, H2B, H3 and H4 since a CIBA Foundation Symposium of 1975 [178]. Prior to this a variety of differing nomenclatures were adopted by differing laboratories. The histones are basic proteins of low molecular weight [179, 180] which show remarkably small sequence variation within and among species. They are readily isolated by salt or acid extraction of chromatin and can be separately purified on the basis of size [181] or charge [182]. In 2M-NaCl all the histones are dissociated from DNA. However, in 0.5M-NaCl histone 1 alone is dissociated so that the functions of the residual or 'core' histones may be investigated [183, 184]. In the absence of DNA the 'core' histones will associate with each other, the predominant species being a tetramer in the case of histones H3 and H4 and a dimer in the case of H2A and H2B. At high ionic strength, all four histones form complexes which have variously been described as heterotypic tetramers each with one histone molecule or octamers with two of each type of core histone molecule [185, 186]. They also readily form long, linear aggregates whose functional significance is not clear.

Modification of histones occurs readily in the cell nucleus. They are found in methylated [187], acetylated [188], phosphorylated [189, 190] and ADP-ribosylated [243] states under varying metabolic conditions. These modifications have been suggested to affect their basic function of DNA packaging [191, 192].

A special type of histone known as histone 5 is found in the nucleated erythrocyte of fish, amphibians and birds. It bears many similarities to histone 1 and is thought to maintain the highly repressed state of the chromatin these cell types [193, 194]. In sperm cells, histones are replaced by other small basic proteins known as *protamines* [195, 196].

4.8.3 Eukaryote chromatin structure

The DNA from a human cell is of the order of 1 metre in length and must be condensed into a cell nucleus whose diameter is of the order of 10μm in such a manner as to maintain accessibility and prevent tangling during replication. The packaging problem presented is therefore even greater than that for bacteria. Eukaryote cells achieve this condensation by a series of packaging mechanisms involving the histones and some other chromosomal proteins. While the higher order mechanisms of packaging are as yet imperfectly understood, the initial coiling of DNA into nucleosomes and polynucleosomes was the subject of intensive research in the late 1970s and is now substantially elucidated.

4.8.4 The nucleosome

The nucleosome, or core particle of chromatin, is a small nucleoprotein particle which is variously likened in shape to a hatbox, a pill, or a clamlike platysome. Its dimensions are 10nm × 10nm × 6nm (Fig. 4.10). Within each nucleosome are eight histone molecules two of each of histones 2A, 2B, 3 and 4, the latter two being believed to be near the centre of the particle. Wound round the histones are 138–140 base pairs of DNA forming 1.75 turns of negative toroidal superhelix so that there are 80 base pairs in each turn. While the DNA is essentially on the outside, certain regions are protected from nuclease digestion to a greater extent than others and are presumably more extensively covered by proteins. It is possible to label the termini of the nucleosomal DNA with polynucleotide kinase and ^{32}P-labelled ATP and subsequently digest the DNA with DNase I (Chapter 6). Electrophoresis of the digest on polyacrylamide gels followed by autoradiography shows a ladder of fragments differing in size by an integral ten base pairs, but certain size categories occur with greatly reduced frequencies indicating reduced nuclease accessibility at these points. While it is not yet finally established whether the DNA is smoothly curved or kinked [197, 198, 207] there is no firm evidence to support the latter hypothesis. Much of the details of the structure of the nucleosome have been determined by Klug and his colleagues using X-ray crystallography [199] though other techniques have also been employed [200, 201]. Many of the details of nucleosome structure have been described in a recent symposium [202] and extensively reviewed [201, 190].

There is little sequence specificity in the DNA requirement for nucleosome formation though a slight preference for AT rich DNA can be demonstrated. This may simply reflect the fact that AT rich DNA is more readily bent into a curve. Nucleosome formation *in vitro* requires the slow dialysis of a mixture of DNA and histones from high to low salt conditions [203]. While nucleosomes can readily be formed with modified histones such as acetylated histone 4 together with the other three, such particles show an increased lability in physicochemical terms and an increase susceptibility to DNase I suggesting that they are more unstable. Under biological conditions, the nucleosome appears to be remarkably stable and to have little tendency either to travel along a single length of DNA or to move among preformed chromosomes [205, 206].

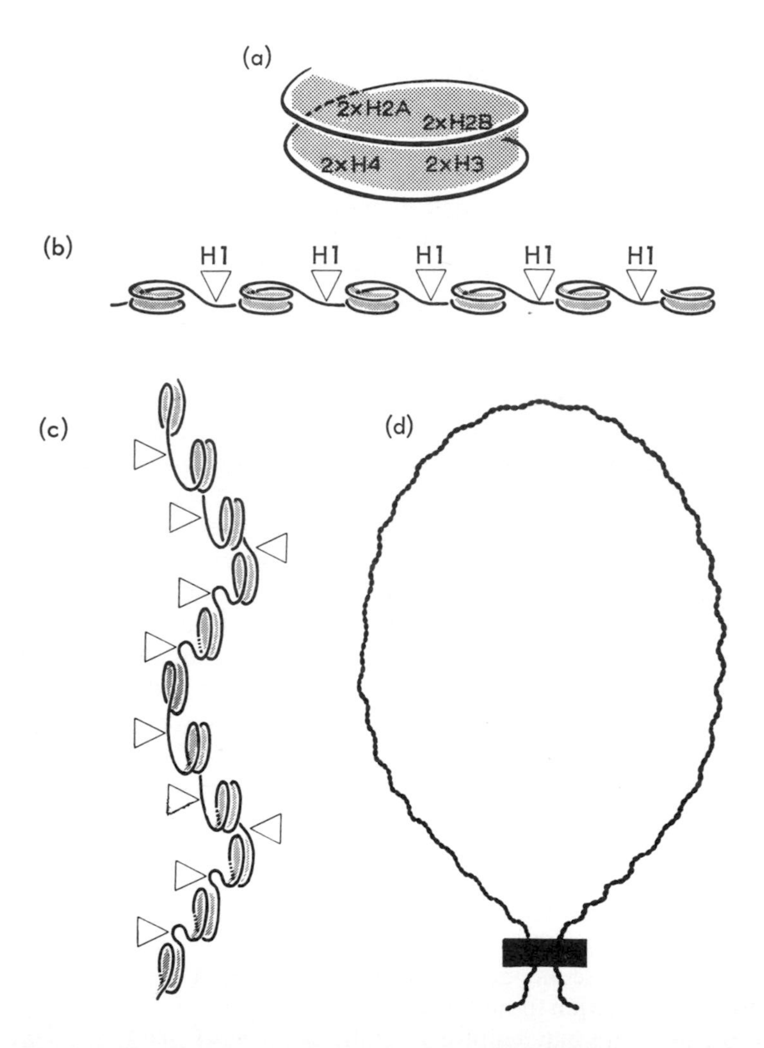

Fig. 4.10 The structure of chromatin. (a) A diagrammatic structure of a single nucleosome with eight histone molecules in the centre and 1.75 turns of DNA on the outside. (b) Several nucleosomes in an oligonucleosome showing the internucleosomal position of histone 1. (c) The coiling of an H1-containing polynucleosome into a helix whose pitch decreases with increasing ionic strength. (d) A polynucleosome loop restrained so that removal of the protein leads to a loop of superhelical DNA. Loop sizes are in the region of 400–1000 nucleosomes.

4.8.5 Oligonucleosomes and polynucleosomes

Whereas the core nucleosome is a sufficiently discrete entity to be isolated and crystallized, oligonucleosomes are more heterogeneous with respect to their DNA content. They are readily isolated by digestion of chromatin with microccocal nuclease from *Staphyloccocus aureus* (Chapter 6) followed by sucrose-gradient sedimentation [208]. Dinucleosomes comprise two nucleosomes joined by a single piece of linking DNA and trinucleosomes contain three nucleosomes joined by two linking segments of DNA. The variability in 'linker' DNA size contributes to the size heterogeneity of the oligomers within any broad size class. All cell types appear to have their own specific size range of linker and their chromatin can be characterized by its *nucleosome repeat frequency,* by gentle digestion with microccocal nuclease followed by electrophoresis of the DNA products on polyacrylamide gels beside markers of known size. The gels show a characteristic banding pattern and the size difference between the middle of any two DNA bands in number of base pairs is taken as the nucleosome repeat frequency. This parameter has the rather unusual characteristic in that it appears to be larger for chromatin which is totally inactive, such as that from chick erythrocyte which has a repeat frequency of 212 base pairs [196, 209], and smaller for chromatin from active cells, such as yeast which has a repeat frequency of 165 base pairs [210]. It would be logical to anticipate that active chromatin might be less closely packed than inactive chromatin but this is not the case. Table 4.5 gives the repeat frequencies of chromatin from some eukaryote tissues. It is interesting to note that two cell types which are from the same organ can differ in their repeat distance [211]. Current evidence suggests that histone H1 is attached to both the nucleosome and the linker DNA.

The 10 nm fibre of chromatin which can be seen in the electron microscope, is composed of chains of single nucleosomes (Plate II). This chain is further compacted into a 30 nm fibre at higher salt concentrations if histone 1 is present. The nature of the 30 nm fibre has been much disputed and both a superbead [212] and helical solenoid of coiled beads have been suggested [213]. Current experimental evidence favours a helical structure whose pitch decreases with increase in ionic strength [214] and which at physiological ionic strength values must be close to that of a contact helix.

In the absence of histone 1 nucleosomes and polynucleosomes are soluble over a wide range of ionic strength. However, when histone 1 is present they precipitate above NaCl concentrations of 80 mM and $MgCl_2$ concentrations of 2 mM. This provides a useful mechanism for the separation of nucleosomes with or without histone 1 [215, 216].

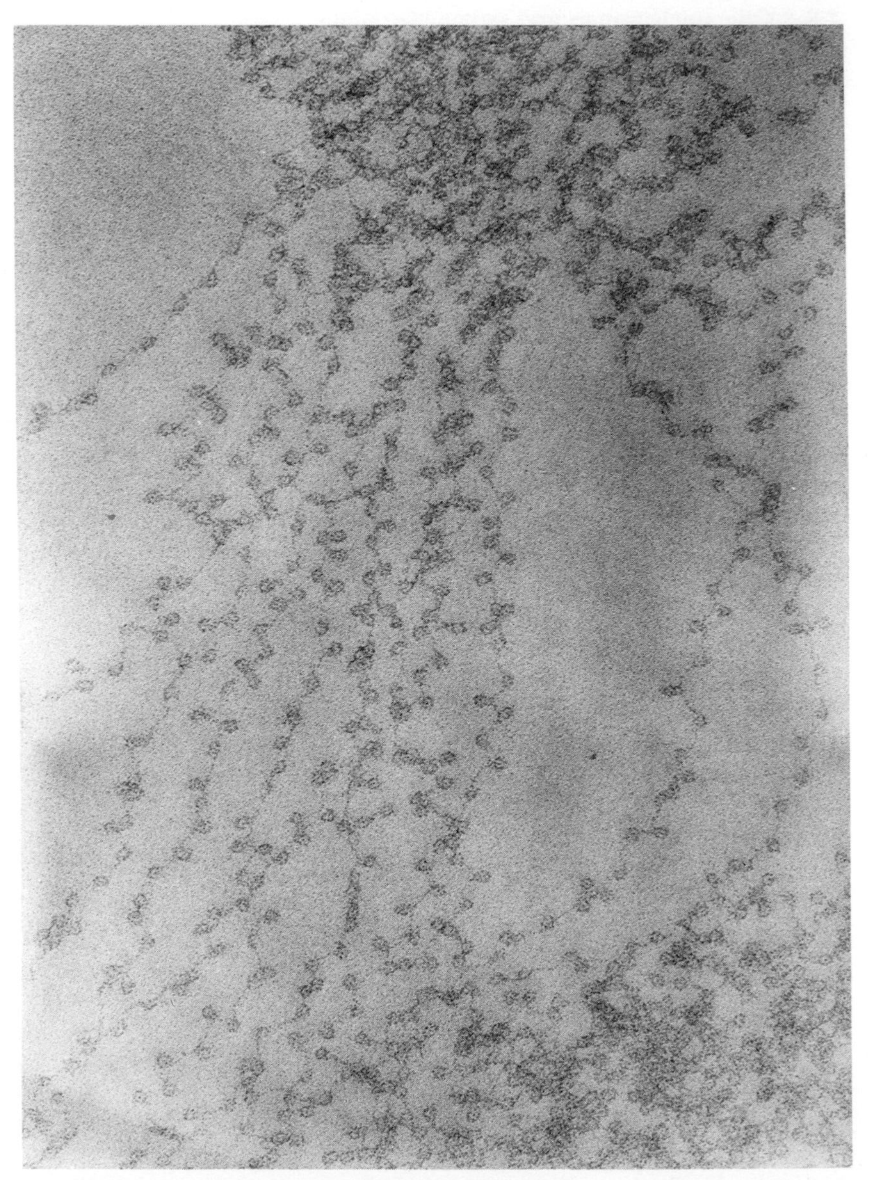

Plate II Chromatin fibres streaming out of a chicken erythrocyte nucleus. The bead-like structures now termed nucleosomes or nu-bodies are about 7 nm in diam. The connecting strand is around 14 nm in length. The sample was negatively strained with 5mM-uranyl acetate in water and the magnification 285000. (By courtesy of Drs D.E. Olins and A.L. Olins.)

Table 4.5

Cell type	Nucleosome repeat frequency
Rabbit cortical neurone	162
Yeast	165
HeLa	188
Rat foetal liver	193
Rat liver	196
Rabbit cerebellar neuron	200
Chicken erythrocyte	212

Whereas polynucleosomes with up to 100 beads are readily isolated from chromatin, by gentle microccocal nuclease digestion, polynucleosomes in the size range above 400 beads have not been isolated and it is clear that another superstructural restraint operates at this level of organization.

4.8.6 Chromosome superstructure

The higher order of packaging of both chromatin and metaphase chromosomes resembles that found in bacterial chromosomes. When cells at any stage of the cell cycle are lysed in the presence of non-ionic detergents which disrupt the membranes, high levels of EDTA (ethylene diamine tetra-acetic acid) to chelate the bivalent metal ions which activate nucleases, and high molarities of sodium chloride which dissociates proteins, an equivalent to the bacterial nucleoid can be isolated (Plate III). This material displays all the dye-binding and sedimentation characteristics of supercoiled DNA in loops of 100 kb in size. Unlike the bacterial chromosomal loops, these are stabilized by proteins rather than RNA since proteases but not ribonucleases abolish the supercoiling [217,218]. Since each nucleosome contributes 1.75 superhelical turns to the DNA which surrounds it, the packaging may differ from that in bacterial systems in that the DNA is essentially in the relaxed state in the chromosome while packaged by the histones, and only displays superhelical characteristics when the protein is dissociated. However, it is possible that, even in the presence of the histones, some superhelical twist remains on the DNA and until the helical sense of the solenoidal superstructure has been established the amount of supercoiling in the intact chromosome is difficult to calculate.

Metaphase chromosomes which have been deproteinized are readily visualized in the electron microscope as looped structures with the thread of DNA in each loop entering and leaving the chromosome at the same

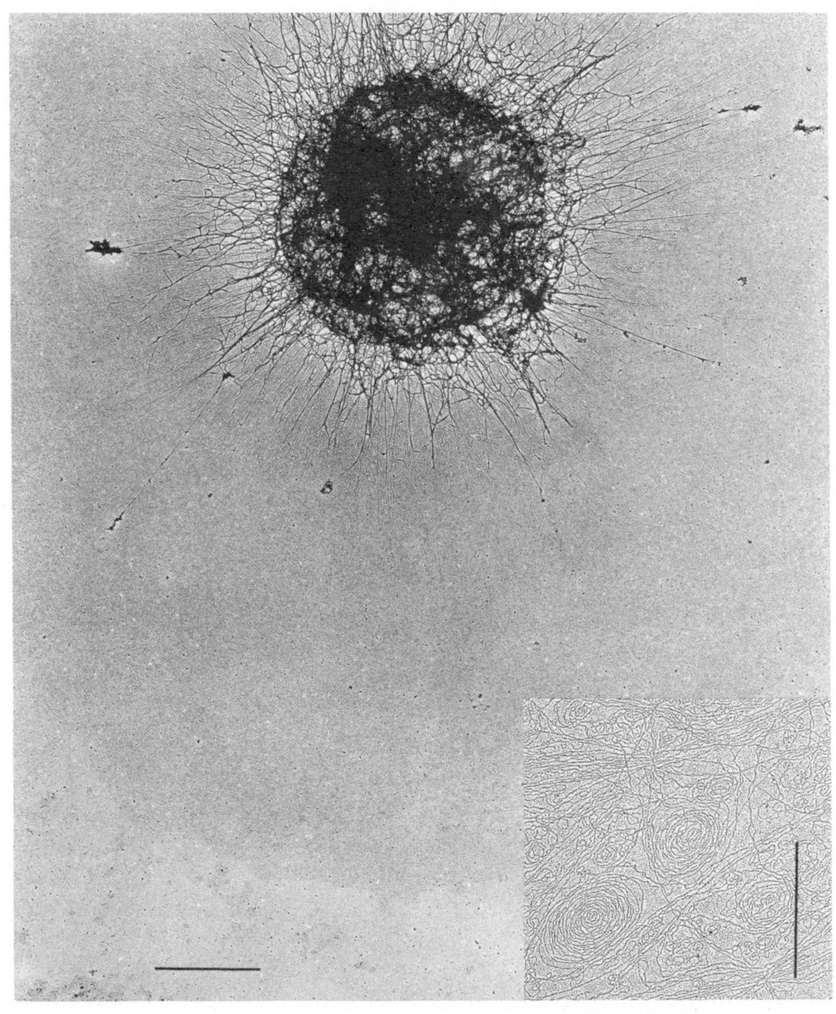

Plate III An electromicrograph of some of the DNA in a human (HeLa) cell. The nuclear DNA has been spread throughout most of the field to form a skirt which surrounds the collapsed skeleton of the nucleus. A tangled network of DNA fibres radiates from the nuclear region. The bar represents 5μm.

Inset: Only at the very edge of the skirt can individual duplexes be resolved. Most appear as collapsed toroidal or interwound superhelices, indicating that the linear DNA must be unbroken and looped, probably by attachment to the nuclear skeleton. The bar represents 1 μm. (By courtesy of Dr P. R. Cook and Dr S. J. McReady.)

central point. It is possible to digest 99% of the DNA from these metaphase preparations leaving behind a morphologically intact central chromosome 'scaffold'. The remaining DNA falls into the moderately repetitive frequency class. The proteins of the 'scaffold' are a heterogeneous group forming seven or eight major bands and many minor bands on polyacrylamide-gel electrophoresis [219, 220].

4.8.7 Polytene chromosomes

In certain tissues, notably the salivary glands of the fruitfly *Drosophila* and some other species, the DNA molecules do not segregate after replication but remain together during several rounds of the cell cycle. By geometric progression, after ten rounds of replication there are more than one thousand DNA strands lying alongside each other with their specific sequences and specific bound proteins matched. This forms an excellent experimental system for analysis. Firstly, *Drosophila* has a simple chromosome complement of four, the centromere remains under replicated and a highly characteristic star-like structure with the spokes radiating from the centromere can readily be identified by light microscopy. Unlike metaphase chromosomes, these chromosomes are in the stretched out interphase condition and sites can be clearly identified where 'puffing' due to transcriptional activity occurs. Puffs can even be induced in certain sites by the application of specific insect hormones so that the process of transcription can be effectively visualized with respect to time [221, 222].

Polytene chromosomes have a highly characteristic banding pattern with dark and light bands of differing densities and dispositions at various points in the arms. In addition, the ease with which *Drosophila* mutants can be isolated means that a lack of a specific gene function can be correlated with the absence of a specific band. Although it is not possible to identify unique genes on ordinary metaphase chromosomes because of the low specific radioactivity of the small amount of mRNA or cDNA involved leading to weak autoradiography, the large number of binding sites make it possible to perform *in situ* hybridization successfully on polytene chromosomes even in cases where a gene has a low overall frequency with respect to haploid DNA content. As well as their use for the identification of specific RNA hybridization sites, these chromosomes are extensively used in the analysis of the non-histone chromosomal proteins by immunofluorescence since an antibody can be raised against a specific non-histone protein and used together with a fluorescent second antibody which binds to the first, to identify the binding site on the DNA of the original antigenic protein [223, 224].

4.8.8 Lampbrush chromosomes

In certain amphibian oocytes such as those of *Triturus* or *Xenopus*, the chromosomes are to be seen stretched out while undergoing transcription. It is possible to see the two matching loops of DNA emerging from the central stem of the paired DNA duplexes and on each loop the newly synthesized RNA chains can be observed undergoing packaging into ribonucleoprotein particles. This type of chromosome provides a second excellent system for direct visualization of the transcriptional process and it is particularly used for the study of proteins which package both the DNA undergoing transcription, and the newly synthesized RNA. In addition, the various types of polymerizing enzymes may be identified. The lampbrush chromosomes cannot be used to identify sites of specific genes as can the polytene chromosomes but genes actively transcribing rRNA and tRNA can be identified [225, 226]. While they form a valuable system for analysis, they are less widely used than polytene systems because of their limited scope.

4.8.9 Non-histone chromosomal proteins

The non-histone chromosomal proteins include those such as RNA polymerase which have enzyme activities, packaging and processing proteins, the contracile proteins which are presumed to be associated with mitotic division, scaffolding proteins and nucleoid proteins and proteins presumed to be associated with the nuclear pore complexes. Most can be selectively purified from chromatin under various extraction procedures [227]. In additon, the non-histone chromosomal proteins are assumed to contain those proteins which are responsible for the control of gene expression in eukaryotes. However, the expectation is that the amount of any protein responsible for the activation of a specific structural gene would be very small indeed and consequently it would be extremely difficult to purify such proteins by conventional techniques of protein fractionation. Attempts to identify such proteins therefore usually involve the use of cell-specific antisera [228, 229] or the preferential binding to sequences of cloned DNA containing the genes in question (Chapter 14).

One group of non-histone proteins which is present in sufficient amounts for purification and sequence analysis is the high-mobility group or HMG proteins [230]. The proteins can be extracted from chromatin by 0.35 M-NaCl and are soluble in 2% (w/v) trichloracetic acid. The most abundant, HMG proteins 1, 2 and 17, have been extensively studied and the latter in particular is believed to be preferentially associated with transcriptionally active chromatin (Chapter 10) [216].

4.9 EXTRANUCLEAR DNA

4.9.1 Mitochondrial DNA

Mitochondrial DNA is usually found in circular, double-stranded, supercoiled molecules the exceptions being the linear mitochondrial DNA molecules from *Tetrahymena* and *Paramecium*. Most eukaryote mitochondrial DNA molecules fall in the molecular-weight range of 10 million (15 kb) and can therefore code for 15 to 20 genes. However, some yeast ones are considerably larger and *Saccharomyces* has a genome of molecular weight 50 million (75 kb). Yeast mitochondrial DNA molecules have been widely studied since they are readily amenable to genetic analysis. When yeast cells are grown in the presence of mutagens such as ethidium bromide 'petite' mutants are formed in which large sections of the mitochondrial genome are deleted with the remaining segments being amplified so that the DNA molecules remain the same size. The mutants are readily propagated by the yeast cells and can be isolated and analysed for their coding potential.

Clearly mitochondrial DNA can only code for a small percentage of the mitochondrial proteins. However, the information carried in the yeast mitochondrial genome is vital for the proper sequential assembly of the cytochrome chains on the membranes since it codes for major subunits of the cytochromes which, together with nuclear coded proteins, form the respiratory chain. In addition, mitochondrial DNA carries ATPase genes and genes for mitochondrial tRNA and rRNA molecules. In evolutionary terms it represents an interesting mixture of prokaryote and eukaryote since the rRNA genes are of the bacterial type and yet have inserted sequences which are characteristic of eukaryotes. There is no evidence to suggest that mitochondrial DNA is packaged in nucleosomes. The subject of mitochondria and their genetics has been extensively reviewed [231–234, 244].

4.9.2 Chloroplast DNA

Chloroplast DNA is in general much larger than mitochondrial DNA being in the molecular weight range 100 million (150 kb). In contrast to mitochondria which have from one to ten molecules per organelle [244] chloroplasts tend to have a very large number of copies of the DNA molecule in each organelle, in some cases greater than one hundred. Like mitochondrial DNA it carries the coding information for essential membrane components, tRNA and rRNA [236–238, 244]. Mutants similar to the petite mutants may be isolated but show a tendency to revert, possibly because of the multiplicity of copies of the master gene in the chloroplast. All known chloroplast DNA molecules are circular and supercoiled.

4.9.3 Kinetoplast DNA

The kinetoplast is part of a highly specialized mitochrondrion found in certain groups of flagellated protozoa such as the trypanosomes. Its DNA (kDNA) consists of an interlinked series of many thousands of circular DNA molecules which vary in size from 0.6 kb to 2.4 kb depending on the type of trypanosome from which they are obtained. These components, which are known as minicircles, are further interlinked with a much smaller number of larger circular DNA molecules of above 30 kb in size known as maxi circles in the majority of systems analysed. While maxi circles appear to perform the conventional functions of mitochondrial DNA in trypanosomes, minicircles are microheterogeneous in sequence and size and there is no evidence to suggest that they are ever transcribed so that it is possible that they may fulfil some structural rather than coding role [239–241, 244].

REFERENCES

1 Mandelstam, J. and McQuillen, K. (1973), *Biochemistry of Bacterial Growth,* Blackwell Scientific Publications, Oxford.
2 Brock, T.D. (1970), *Biology of Microorganisms,* Prentice Hall, New Jersey.
3 Dawes, I.W. and Sutherland, I.W. (1976), *Microbial Physiology,* Basic Microbiology, **4,** Blackwell Scientific Publications, Oxford.
4 Stonington, O. and Pettijohn, D. (1971), *Proc. Natl. Acad. Sci. USA,* **68,** 6.
5 Worcel, A. and Burgi, L. (1972), *J. Mol. Biol.,* **71,** 127.
6 Pettijohn, D. and Hecht, R. (1973), *Cold Spring Harbor Symp. Quant. Biol.,* **38,** 31.
7 Worcel, A., Burgi, E., Robinson, J. and Carlson, C.L. (1973), *Cold Spring Harbor Symp. Quant. Biol.,* **38,** 43.
8 Pardee, A.B. and Rosengurt, E. (1975), *MTP International Review of Science: Biochemistry Series One,* Vol. 2 (ed. C.F. Fox), p. 155.
9 Cooper S. (1979), *Nature,* **280,** 17.
10 Abbo, F.E. and Pardee, A.B. (1960), *Biochim. Biophys. Acta,* **39,** 478.
11 Cooper, S. and Helmstetter, C.E. (1968), *J. Mol. Biol.,* **31,** 519.
12 Avery, O.T., McLeod, C.M. and McCarty, M. (1944), *J. Exp. Med.,* **79,** 137.
13 Alexander, H.E. and Leidy, G. (1951), *J. Exp. Med.,* **93,** 345.
14 Zamenhof, S., Leidy, G., Alexander, H.E., Fitzgerald, P.L. and Chargaff, E. (1952), *Arch. Biochem. Biophys.,* **40,** 50.
15 Boivin, A. (1947), *Cold Spring Harbor Symp. Quant. Biol.,* **12,** 7: (1948), *C.R. Soc. Biol. Paris,* **142,** 1258.
16 Hotchkiss, R.D. and Weiss, E. (1956), *Sci. Amer.,* **195,** 48.

17 Mandel, M. and Higa, A. (1970), *J. Mol. Biol.,* **53,** 159.
18 Hayes, W. (1968), *The Genetics of Bacteria and their Viruses,* 2nd Edn, Blackwell Scientific Publications, Oxford.
19 Notani, N.K. and Setlow, J.K. (1974), *Progr. Nucleic Acid. Res. Mol. Biol.,* **14,** 39.
20 Lewin, B. (1977), *Gene Expression*-3, John Wiley and Sons, Inc., New York.
21 Lacks, S. and Neuberger, M. (1975), *J. Bacteriol.,* **124,** 1321.
22 Marmur, J. and Hotchkiss, R.D. (1953), *J. Biol. Chem.,* **214,** 383.
23 Szybalska, E.H. and Szybalski, W. (1962), *Proc. Nat. Acad. Sci. USA,* **48,** 2026.
24 Bukhari, A.I., Shapiro, J.A. and Adhya, S.L. (1977), *DNA Insertion Elements, Plasmids and Episomes,* Cold Spring Harbor.
25 Kornberg, A. (1974), *DNA Synthesis,* Freeman, San Francisco.
26 Goebel, W. and Schremph, H. (1972), *Biochim. Biophys. Acta,* **262,** 32.
27 Meynell, G.G. (1972), *Bacterial Plasmids,* Macmillan, London.
28 Sherratt, D.J. (1974), *Cell,* **3,** 189.
29 Campbell, A. (1969), *Episomes,* Harper and Row, New York.
30 Nishimura, Y., Caro, L., Berg, C.M. and Hirota, Y. (1971), *J. Mol. Biol.,* **55,** 441.
31 Caro, L.G. and Berg, C.M. (1969), *J. Mol. Biol.,* 45, 325.
32 Jacob, F. and Wollman, E.L. (1961), *Sexuality and the Genetics of Bacteria,* Academic Press, New York.
33 Nishimura, Y., Ishibashi, M., Meynell, E. and Hirota, Y. (1967), *J. Gen. Microbiol.,* **49,** 89.
34 Cohen, S.N. and Shapiro, J.A. (1980), *Sci. Amer.,* **242** (2), 36.
35 Johnsrud, L., Calos, M.P. and Miller, J.H. (1978), *Cell,* **15,** 1209.
36 Shapiro, J.A. (1977), *Proc. Natl. Acad. Sci. USA,* **76,** 1933.
37 McClintock, B. (1965), *Brookhaven Symp. in Biol.,* No. 18, p. 162.
38 Luria, S.E., Darnell, J.E. Baltimore, D. and Campbell, A. (1978), *General Virology,* 3rd Ed, John Wiley & Sons, New York.
39 Sinsheimer, R.L. (1959), *J. Mol. Biol.,* **1,** 43.
40 Marvin, D.A. and Hoffman-Berling, H. (1963), *Z. Naturforsch,* **186,** 884.
41 Kellenberger, E. (1962), *Adv. Virus Res.,* **8,** 1.
42 Hershey, A.D., Burgi, E. and Ingraham, L. (1963), *Proc. Nat. Acad. Sci. USA,* **49,** 748.
43 Wu, R. and Taylor, E. (1971), *J. Mol. Biol.,* **57,** 491.
44 Strack, H.B. and Kaiser, A.D. (1965), *J. Mol. Biol.,* **12,** 36.
45 Yamagishi, H., Nakamura, K. and Ozeki, H. (1965), *Biochem. Biophys. Res. Commun.,* **20,** 727.
46 Ritchie, D.A., Thomas, C.A., MacHattie, L.A. and Wensinv, P.C. (1967), *J. Mol. Biol.,* **23,** 365.
47 Thomas, C.A. and MacHattie, L.A. (1964), *Proc. Natl. Acad. Sci. USA,* **52,** 1297.
48 Bujard, H. (1969), *Proc. Natl. Acad. Sci. USA,* **6,** 1167.
49 Colter, J.S. and Parenchych, W. (Eds) (1967), *The Molecular Biology of Viruses,* Academic Press, New York.
50 Wyatt, G.R. and Cohen, S.S. (1950), *Biochem. J.,* **55,** 774.

51 Lehman, I.R. and Pratt, E.A. (1960), *J. Biol. Chem.*, **235,** 3254.
52 Hattman, S. and Fukasawa, T. (1963), *Proc. Natl. Acad. Sci. USA,* **50,** 297.
53 Hershey, A.D. (ed.) (1971), *The Bacteriophage Lambda,* Cold Spring Harbor, New York.
54 Arber, W. (1968), *Symposium No.* 18, *Society for General Microbiology,* p. 295, University Press, Cambridge.
55 Stent, G.S. and Wollman, E.L. (1950), *Biochim. Biophys. Acta,* **6,** 307.
56 Crawford, E.M. and Gesteland, R.F. (1964), *Virology,* **22,** 165.
57 Kozloff, L.M. and Lute, M. (1959), *J. Biol. Chem.*, **234,** 534.
58 McCorquodale, D.J., Oleson, A.E. and Buchanan, J.M. (1967), in *The Molecular Biology of Viruses* (eds J. S. Colter and W. Paranchych), Academic Press, New York, p. 31.
59 Paranchych, W. (1966), *Virology,* **28,** 90.
60 Brinton, C.C. and Beer, H. (1967), *The Molecular Biology of Viruses,* (eds J. S. Colter and W. Paranchych), Academic Press, New York, p. 251.
61 Marvin, D.A. and Hohn, B. (1969), *Bact. Rev.*, **33,** 172.
62 Keir, H.M. (1968), *Symposium No.* 18, *Society for General Microbiology,* p. 67, University Press, Cambridge.
63 Leberman, R. (1968), *Symposium No.* 18, *Society for General Microbiology,* p. 183, University Press, Cambridge.
64 Wood, W.B. and Edgar, R.S. (1967), *Sci. Amer.*, **217,** (1), 60.
65 Weigle, J.J. (1966), *Proc. Natl. Acad. Sci. USA,* **55,** 1462.
66 Streisinger, G., Mukai, F., Dreyer, W.J., Miller, B. and Horiuchi, S. (1961), *Cold Spring Harbor Symp. Quant. Biol.*, **26,** 25.
67 Hershey, A.D. and Chase, M. (1952), *J. Gen. Physiol.*, **26,** 36.
68 Kornberg, A. (1980), *DNA Replication,* W. & H. Freeman & Co., San Francisco.
69 Sanger, F., Air, G.M., Barrell, B.G., Brown, N.L., Coulson, A.R., Fiddes, J.C., Hutchinson, C.A., Slocombe, P.M. and Smith, M. (1977), *Nature,* **265,** 687.
70 Godson, G.N., Barrell, B.G., Staden, R. and Fiddes, J.C. (1978), *Nature,* **276,** 236.
71 Fiddes, J.C. and Godson, C.W. (1979), *J. Mol. Biol.*, **133,** 19.
72 Tooze, J. and Weber, K. (1967), *J. Mol. Biol.*, **28,** 311.
73 Steitz, J.A. (1968), *J. Mol. Biol.*, **33,** 923.
74 Epstein, R.H., Bolle, A., Steinberg, C.M., Kellenberger, E., Boy De La Tour, E., Chevalley, R., Edgar, R.S. Sussman, R.S. Denhardt, G.H. and Lielausis, A. (1963), *Cold Spring Harbor Symp. Quant. Biol.*, **28,** 375.
75 Edgar, R.S. and Lielausis, I. (1964), *Genetics,* **49,** 649.
76 Edgar, R.S. and Epstein, R.H. (1965), *Sci. Amer.*, **212** (2), 71.
77 Thomas, R. (1968), *Symposium No.* 18, *Society for General Microbiology,* University Press, Cambridge, p. 315.
78 Lederberg, E.M. and Lederberg, J. (1953), *Genetics,* **38,** 51.
79 Tomizawa, J.I. and Owaga, T. (1967), *J. Mol. Biol.*, **23,** 247.
80 Campbell, A. (1962), *Adv. Genetics,* **11,** 101.
81 Jacob, F. and Wollman, E.L. (1957), in *The Chemical Basis of Heredity* (eds W. D. McElroy and B. Glass), John Hopkins, Baltimore, p. 468.

82 Gomatos, P.J. and Tamm, I. (1963), *Proc. Natl. Acad. Sci. USA*, **50,** 878.
83 Russell, G.J., Follett, E.A.C. and Subak-Sharpe, J.H. (1971), *J. Gen. Virol.*, **11,** 129.
84 Klug, A. and Caspar, D.L.D. (1960), *Adv. Virus Res.*, **7,** 225.
85 Knight, C.A. (1963), *Chemistry of Viruses, Protoplasmatologia,* Vol. 5, No. 2.
86 Markham, R. (1963), *Progress in Nucleic Acid Research,* (ed. J. N. Davidson and W. E. Cohn), Academic Press, New York, Vol. 2, p. 61.
87 Bancroft, J., Hills, G. and Markham, R. (1967), *Virology,* **31,** 354.
88 Crawford, L.V., Follett, E.A., Burdon, M.G. and McGeoch, D.J. (1969), *J. Gen. Virol.*, **4,** 37.
89 Watson, D.H. (1968), *Symposium No.* 18, *Society for General Microbiology,* University Press, Cambridge, p. 208.
90 Crawford, L.V. (1968), *Symposium No.* 18, *Society for General Microbiology,* University Press, Cambridge, p. 163.
91 Green, M. and Gerard, G.F. (1974), *Prog. Nucleic Acid Res. Mol. Biol.*, **15,** 1 (ed. W. E. Cohn).
92 Bellamy, A.R. and Joklik, W.K. (1967), *Proc. Natl. Acad. Sci. USA,* **58,** 1389.
93 Subak-Sharpe, J.H. and Timbury, M.C. (1977), *Comp. Virol.*, **9,** 89.
94 Tolun, A., Alestrom, P. and Pettersson, V. (1979), *Cell,* **17,** 705.
95 Astell, G.R., Smith, M., Chow, M.B. and Ward, D.C. (1979), *Cell,* **17,** 691.
96 Griffith, J. (1975), *Science,* **187,** 1202.
97 Germond, J.E., Hirst, B., Oudet, P., Gross-Bellard, M. and Chambon, P. (1975), *Proc. Natl. Acad. Sci. USA,* **72,** 1843.
98 Shure, M. and Vinograd, J. (1976), *Cell,* **8,** 215.
99 Dulbecco, R. and Vogt, M. (1963), *Proc. Natl. Acad. Sci. USA,* **50,** 236.
100 Weil, R., and Vinograd, J. (1963), *Proc. Natl. Acad. Sci. USA,* **50,** 730.
101 Winocour, E. (1967), *The Molecular Biology of Viruses* (ed. J.S. Colter and W. Paranchych), Academic Press, New York, p. 577.
102 Crawford, L.V. (1962), *Virology,* **18,** 117.
103 Dulbecco, R., Hartwell, L.H. and Vogt, M. (1965), *Proc. Natl. Acad. Sci. USA,* **53,** 403.
104 Winocour, E., Kaye, A.N. and Stollar, V. (1965), *Virology,* **27,** 156.
105 Fried, M. and Pitts, J.D. (1968), *Virology,* **34,** 761.
106 Fraenkel-Conrat, H. and Williams, R.C. (1955), *Proc. Natl. Acad. Sci. USA,* **41,** 690.
107 Fraenkel-Conrat, H. and Singer, B. (1957), *Biochim. Biophys. Acta,* **24,** 540.
108 Fraenkel-Conrat, H. and Singer, B. (1964), *Virology,* **23,** 354.
109 Butler, P.J.G. and Klug, A. (1978), *Sci. Amer.*, **239** (5), 52.
110 Hohn, T. (1967), *Eur. J. Biochem.*, **2,** 152.
111 Bancroft, J.B. and Hiebert, E. (1967), *Virology,* **32,** 354.
112 Crick, F.H.C. and Watson, J.D. (1956), *Nature,* **177,** 473.
113 Reddy, V.B., Thimmappaya, B., Dhar, R., Subramanian, K.N., Zain, B.S., Pan, J., Ghosh, P.K., Celma, M.L. and Weissman, S.M. (1978), *Science,* **200,** 494.
114 Fiers, W., Contreras, R., Haegeman, G., Rogiers, R., Van de Vorde, A., van Heuvrswyn, H., van Hevreweghe, J., Volckaert, G. and Ysebaert, M. (1978), *Nature,* **273,** 113.

115 Soeda, E., Arrand, J.R., Smoler, N., Walsh, J.E. and Griffin, B.E. (1980), *Nature,* **283,** 445.
116 Bolivar, F. (1979), *Life, Sci.,* **25,** 807.
117 Tooze, J. (1973), *The Molecular Biology of Tumour Viruses,* Cold Spring Harbor Lab.
118 Dulbecco, R. (1967), *Sci. Amer.,* **216** (4), 28.
119 MacPherson, I. (1967), *Br. Med. Bull.,* **23,** No. 2, 144.
120 Erikson, R.L., Collett, M.S., Erikson, E. and Purchio, A.F. (1979), *Proc. Natl. Acad. Sci. USA,* **76,** 6260.
121 Fried, M. (1965), *Proc. Natl. Acad. Sci. USA,* **53,** 486.
122 Cuzin, F., Vogt, M., Dieckmann, M. and Berg, P. (1970), *J. Mol. Biol.,* **47,** 317.
123 Benjamin, T.L. (1966), *J. Mol. Biol.,* **16,** 359.
124 Watkins, J.F. and Dulbecco, R. (1967), *Proc. Natl. Acad. Sci. USA,* **58,** 1396.
125 Gelb, L.D., Kohne, D.E. and Martin, M.A. (1971), *J. Mol. Biol.,* **57,** 129.
126 Ozanne, B., Vogel, A., Sharp, P., Keller, W. and Sambrook, J. (1973), *Lepetit Colloq. Biol. Med.,* **4.**
127 Smith, A.E., Smith, R., Griffin, B. and Fried, M. (1979), *Cell,* **18,** 915.
128 Griffin, J.D., Spangler, G. and Livingston, D.M. (1979), *Proc. Natl. Acad. Sci. USA,* **76,** 2610.
129 Crawford, L.V. (1980), *TIBS* (Feb.), **39.**
130 Rigby, P. (1979), *Nature,* **282,** 781.
131 Duesberg, P.H. (1968), *Proc. Natl. Acad. Sci. USA,* **60,** 1511.
132 Duesberg, P.H. (1970), *Current Topics Microbiol. Immunol.,* **51,** 74.
133 Levinson, W.E., Varmus, H.E., Garapin, A.C. and Bishop, J.M. (1972), *Science,* **175,** 76.
134 Temin, H.M. and Mitzutani, S. (1970), *Nature,* **266,** 1211.
135 Baltimore, D. (1970), *Nature,* **226,** 1209.
136 Spiegelman, S., Burny, A., Das, M.R., Keydar, J., Schlom, J., Travnicek, M. and Watson, V. (1970), *Nature,* **227,** 563.
137 Willingham, M.C., Jay, G. and Pastan, I. (1979), *Cell,* **18,** 125.
138 Mitchison, J.M. (1971), *The Biology of the Cell Cycle,* University Press, Cambridge.
139 Yunis, J.J. (1976), *Science,* **191,** 1268–1270.
140 Yunis, J.J. (1977), *Molecular Structure of Human Chromosomes,* Academic Press, New York.
141 Lewin, B. (1974), *Gene Expression,* Vol. 2, Wiley, London.
142 Thomas, C.A. (1971), *Ann. Rev. Genetics,* **5,** 237–256.
143 Callan, H.G. (1967), *J. Cell. Sci.,* **2,** 1–7.
144 Hinegardner, R. (1976), in *Molecular Evolution* (ed. F.J. Ayala), Sinauer Sunderland, p. 179–199.
145 Doolittle, W.F. and Spienza, C. (1980), *Nature,* **284,** 601–603.
146 Orgel, L.E. and Crick, F.H.C. (1980), *Nature,* **284,** 604–607.
147 Britten, R.J. and Davidson, E.H. (1968), *Science,* **165,** 349–357.
148 Ohno, S. (1971), *Nature,* **234,** 134–137.

149 Caspersson, T., Hulten, M., Linsten, J. and Zech, L. (1971), *Hereditas,* **67,** 147–149.
150 Dutrillaux, B. (1977), in *The Molecular Structure of Human Chromosomes* (ed. J. Yunis), Academic Press, New York, p. 233–262.
151 Flintoff, W.F., Davidson, S.V. and Siminovitch, L. (1976), *Somat. Cell Genet.,* **2,** 245–262.
152 Jones, K.W. and Corneo, G. (1971), *Nature New Biol.,* **233,** 268—271.
153 Evans, H.J., Gosden, J.R., Mitchell, A.R. and Buckland, R.A. (1974), *Nature,* **251,** 346–347.
154 Manuelidids, L. (1978), *Chromosoma,* **66,** 1–21.
155 Maio, J.J., Brown, F.L. and Musich, P.R. (1977), *J. Mol. Biol.,* **117,** 637–655.
156 Miklos, G.L.G. and John, B. (1979), *Amer. J. Hum. Genet.,* **31,** 264–280.
157 Varley, J.M., Macgregor, H.C. and Erba, H.P. (1980), *Nature,* **283,** 686–688.
158 Davidson, E.H., Hough, B.R., Amenson, C.S. and Britten, R.J. (1973), *J. Mol. Biol.,* **77,** 1–23.
159 Schmid, C. and Deininger, P.L. (1975), *Cell,* **6,** 345–358.
160 Firtel, R.A. and Kindle, K. (1975), *Cell,* **5,** 401–411.
161 Angerer, R.C., Davidson, E.H. and Britten, R.J. (1975), *Cell,* **6,** 29–39.
162 Manning, J.E., Schmid, C.W. and Davidson, N. (1975), *Cell,* **4,** 141–155.
163 Fritsch, E.F., Lawn, R.M. and Maniatis, T. (1980), *Cell,* **19,** 959–972.
164 Royal, A., Garapin, A., Cami, B., Perrin, F., Mandel, J.L., LeMeur, M., Bregegere, F., Gannon, F., Le Pennec, J.P., Chambon, P. and Kurilsky, P. (1979), *Nature,* **279,** 125–132.
165 Goodman, H.M., Olson, M.V. and Hall, B.D. (1977), *Proc. Natl. Acad. Sci. USA,* **00,** 5454–5457.
166 Dugaiczyk, A., Woo, S.L., Lai, E.C., Mace, M.L., McReynolds, C. and O'Malley, B.W. (1978), *Nature,* **274,** 328–333.
167 Mandel, J.L., Breathnach, R., Gerlinger, P., LeMeur, M. Gannon, F. and Chambon, P. (1978), *Cell,* **14,** 641–653.
168 Abelson, J. (1979), *Ann. Rev. Biochem.,* **48,** 1035–1069.
169 Lomedico, P., Rosenthal, N., Efstratiadis, A., Gilbert, W., Kolodner, R. and Tizard, R. (1979), *Cell,* **18,** 545–558.
170 Wellauer, P.K. and David, I.B. (1977), *Cell,* **10,** 193–212.
171 Doel, M.T., Houghton, M., Cook, E.A., and Carey, N.H. (1977), *Nucl. Acids Res.,* **4,** 3701–3713.
172 Tilghman, S.M., Curtis, P.J., Tiemcier, D.C., Leder, P. and Weissman, C. (1978), *Proc. Natl. Acad. Sci., USA,* **75,** 1309–1313.
173 Knapp, G., Beckman, J.S., Johnson, P.F., Fuhrman, S.A. and Abelson, J. (1978), *Cell,* **14,** 221–236.
174 Crick, F.H.C. (1979), *Science,* **204,** 264–271.
175 Wilson, B. and Thomas, C. (1974), *J. Mol. Biol.,* **84,** 115–144.
176 Schmid, C.W. and Deininger, P.L. (1975), *Cell,* **6,** 345–358.
177 Dott, P.J., Chuang, C.R. and Saunders, G.F. (1976), *Biochemistry,* **15,** 4120–4125.
178 Fitzsimons, D.W. and Wolstenholme, G.E.W. (1975), *Ciba Found. Symp. Struct. Funct. Chromatin,* **28.**

179 Philips, D.M.P. (ed.) (1971), *The Histones and Nucleohistones,* Plenum, New York.
180 Hnilica, L.S. (1972), *The Structure and Biological Function of Histones,* Chemical Rubber Publ. Co., Cleveland, Ohio.
181 Von Holt, C. and Brandt, W.F. (1977), in *Method in Cell Biology,* XVI (ed. D. M. Prescott), Academic Press, New York, p. 205–225.
182 Luck, J.M., Rasmussen, P.S., Satake, K. and Tsvetikov, A.N. (1958), *J. Biol. Chem.,* **193,** 265–271.
183 Frederiq, E. (1971), in *Histones and Nucleohistones* (ed. D.M.P. Philips), p. 136–159.
184 Christiansen, G. and Griffith, J. (1977), *Nucleic Acids Res.,* **4,** 1837–1848.
185 Campbell, A.M. and Cotter, R. (1976), *FEBS Lett.,* **70,** 209–213.
186 Thomas, J.O. and Butler, P.J.G. (1977), *J. Mol. Biol.,* **116,** 769–781.
187 Borun, T.W., Pearson, D. and Paik, W.K. (1972), *J. Biol. Chem.,* **247,** 4288–4293.
188 Riggs, M.G., Whittaker, J.R., Neuman, J.R. and Ingram, V.M. (1977), *Nature,* **268,** 462–464.
189 Rubin, C.S. and Rosen, O.M. (1975), *Ann. Rev. Biochem.,* **44,** 831–887.
190 Elgin, S.C.R. and Weintraub, H. (1975), *Ann. Rev. Biochem.,* **44,** 725–774.
191 Vidali, G., Boffa, L.C., Bradbury, E.M. and Allfrey, V.G. (1978), *Proc. Natl. Acad. Sci. USA,* **75,** 2239–2244.
192 Mathis, D.J., Oudet, P., Wasylyk, B. and Chambon, P. (1978), *Nucleic Acids Res.,* **5,** 3523–3532.
193 Schiffman, S. and Lee, P. (1974), *Br. J. Haematol.,* **27,** 101–114.
194 Schiffman, S., Rapaport, S.I. and Patch, M.J. (1963), *Blood,* **22,** 733–741.
195 Somer, J.B. and Castaldi, P.A. (1970), *Br. J. Haematol.,* **18,** 147–156.
196 Swart, A.C.W. and Hemker, H.C. (1970), *Biochim. Biophys. Acta,* **222,** 692–695.
197 Noll, M. (1974), *Nucleic Acids Res.,* **1,** 1573.
198 Crick, F.H.C. and Klug, A. (1975), *Nature,* **255,** 530.
199 Finch, J.T., Lutter, L.C., Rhodes, D., Brown, R.S., Rushton, B., Levitt, M. and Klug, A. (1977), *Nature,* **269,** 29.
200 Pardon, J.F., Cotter, R.I., Lilley, D.M.J., Worcester, D.L., Campbell, A.M., Wooley, J.C. and Richards, B.M. (1978), *Cold Spring Harbor Symp. Quant. Biol.,* **42,** 11–22.
201 Kornberg, R.D. (1977), *Ann. Rev. Biochem.,* **46,** 931–954.
202 *Cold Spring Harbor. Symp. Quant. Biol.* (1978), Vol.**42.**
203 Germond, J.E., Hirt, P., Oudet, M. Gross Bellard, M. and Chambon, P. (1975), *Proc. Natl. Acad. Sci. USA,* **72,** 1843.
204 Adams, R.L.P. (1980), *Cell Culture for Biochemists,* Elsevier/North Holland.
205 Beard, P. (1978), *Cell,* **15,** 955–967.
206 Manser, T., Thacher, T. and Rechsteiner, M. (1980), *Cell,* **19,** 993.
207 Lutter, L.C. (1978), *Cold Spring Harbor Symp. Quant. Biol.,* **42,** 137.
208 Sahasrubuddhe, C.G. and Van Holde, K.E. (1974), *J. Biol. Chem.,* **249,** 152.
209 Morris, N.R. (1976), *Cell,* **9,** 627.
210 Thomas, J.E. and Furber, V. (1976), *FEBS Lett.,* **66,** 274–280.

211 Thomas, J.E. and Thomson, R.J. (1977), *Cell,* **10,** 633.
212 Renz, M. Nehls, P. and Hozier, J. (1977), *Proc. Natl. Acad. Sci. USA,* **74,** 1879.
213 Finch, J.T. and Klug, A. (1976), *Proc. Natl. Acad. Sci. USA,* **73,** 1897.
214 Campbell, A.M., Cotter, R.I. and Pardon, J.F. (1978), *Nucleic Acids Res.,* **5,** 1571.
215 Campbell, A.M. and Cotter, R.I. (1977), *Nucleic Acids Res.,* **4,** 3877.
216 Goodwin, G.H., Mathew, C.G.P., Wright, C.A., Venkov, C.D. and Johns, E.W. (1979), *Nucleic Acids Res.,* **7,** 1815.
217 Cook, P.R. and Brazell, I.A. (1978), *Eur. J. Biochem.,* **84,** 465.
218 Benyajati, C. and Worcel, A. (1976), *Cell,* **9,** 393.
219 Adolph, K.W., Cheng, S.M. and Laemmli, U.K. (1977), *Cell,* **12,** 805.
220 Adolph, K.W., Cheng, S.M., Paulson, J.R. and Laemmli, U.K. (1977), *Proc. Natl. Acad. Sci. USA,* **74,** 4937.
221 Lewis, M., Helmsing, P.J. and Ashburner, M. (1975), *Proc. Natl. Acad. Sci. USA,* **72,** 3604.
222 Korge, G. (1975), *Proc. Natl. Acad. Sci. USA,* **72,** 4550.
223 Jamrich, M. Greenleaf, A.L. and Bautz, E.K.F. (1977), *Proc. Natl. Acad. Sci. USA,* **74,** 2079.
224 Silver, L.M. and Elgin, S.C.R. (1977), *Cell,* **11,** 971.
225 Franke, W.W., Scheer, V., Spring, H., Trendelenburg, M.F. and Krohne, G. (1976), *Exp. Cell. Res.,* **100,** 233.
226 Sommerville, J. (1979), *J. Cell. Sci.,* **40,** 1.
227 *Methods in Cell Biology,* Vol. 16 (Gen. ed. D.M. Prescott; Vol. eds G. Stein, J. Stein and L. J. Kleinsmith) (1977), Section D.
228 Campbell, A.M., Briggs, R.C., Bird, R.E. and Hnilica, L.S. (1978), *Nucleic Acids Res.,* **6,** 205.
229 Dunn, J.H.J., Lyall, R.M., Briggs, R.C., Campbell, A.M. and Hnilica, L.S. (1980), *Biochem. J.,* **185,** 277.
230 Walker, J.M., Goodwin, G.H. and Johns, E.W. (1976), *Eur. J. Biochem.,* **62,** 461.
231 Borst, P. and Griwell, L.S. (1978), *Cell,* **15,** 705.
232 Bacilla, M. Horecker, B.L. and Stoppani, A.I.M. (eds) (1978), *Biochemistry and Genetics of Yeast,* Academic Press, New York.
233 Bandlow, W., Schweyen, R.J., Wolf, K. and Kaudewitz, F. (1977), *Genetics and Biogenesis of Mitochondria,* De Gruyter, Berlin.
234 Saccone, C. and Kroon, A.M. (1976), *The Genetic Function of Mitochondrial DNA,* North Holland, Amsterdam.
235 Ohtsubo, E., Zenilman, M. and Ohtsubo, H. (1980), *Proc. Natl. Acad. Sci, USA,* **77,** 750.
236 Klein, A. and Bonhoeffer, F. (1972), *Ann. Rev. Biochem.,* **41,** 301.
237 Smith, H. (1975), *Nature,* **254,** 13.
238 Ohta, N., Sager, R. and Inouye, M. (1975), *J. Biol. Chem.,* **250,** 3655.
239 Borst, P. and Hoeijmakers, J.H.J. (1979), *Plasmid,* **2,** 20.
240 Kleisen, C.M., Borst, P. and Weijers, P.J. (1976), *Eur. J. Biochem.,* **64,** 141–151.

241 Kleisen, C.M., Weislogel, P.O., Fonck, K. and Borst, P. (1976), *Eur. J. Biochem.*, **64,** 153.
242 Jelinek, W.R., Toomey, T.P., Leinwand, L., Duncan, C.H., Biro, P.A., Choidary, A.V., Weissman, S.M., Rubin, C.M., Houck, C.M., Deininger, W.F. and Schmid, C.W. (1980), *Proc. Natl. Acad. Sci. USA,* **77,** 1398.
243 Smulson, M. (1979), *Trends in Biochemical Sciences,* **4,** 225.
244 Gillham, N.W. (1978), *Organelle Heredity,* Raven Press, New York.

The cellular RNAs: their location and structure

5

Developments in the biochemistry of RNA have been profoundly influenced by simultaneous advances in cytology. Consequently consideration of the place of RNA in the life of the cell must be prefaced by a brief outline of the fine structure of the cell.

5.1 THE ANIMAL CELL

A schematic diagram of a typical animal cell is shown in Fig. 5.1. Inside the cell membrane is the cytoplasm in which are suspended numerous inclusions, the largest of which is the more or less centrally placed nucleus bounded by a double membrane pierced by a number of pores. In tissue sections fixed and stained by the usual methods inclusions of basophilic material can often be seen in the cytoplasm. These represent endoplasmic reticulum and ribosomes coagulated by the fixative. In electron micrographs the mitochondria (dimensions 0.5–5 μm × 0.3–0.7 μm) appear as oval profiles. Each mitochondrion is bounded by an outer and an inner membrane about 5 nm in thickness, the inner membrane being connected with a series of incomplete partitions, the cristae mitochondriales, which project into the interior of the organelle dividing it into a series of interconnecting compartments.

One of the most interesting cytoplasmic components as revealed by electron microscopy is a complex mixture of strands and vesicles which was termed the endoplasmic reticulum. It is limited by a membrane about 5 nm thick separating the content of the tubules and vesicles from the general matrix of the cytoplasm and giving the whole component the character of a finely divided vacuolar system. A dense area of the endoplasmic reticulum is associated with the Golgi body.

The structure of the endoplasmic reticulum in the liver cell has been

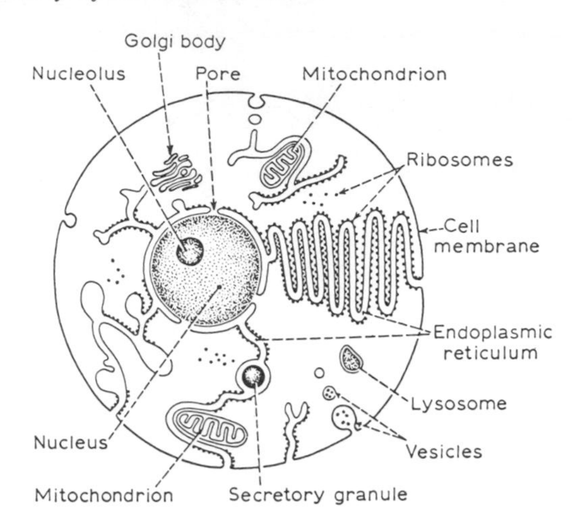

Fig. 5.1 Schematic representation of a typical animal cell.

intensively studied in very thin sections in the electron microscope. It can be represented by numerous profiles of circular, oval, or elongated shape with smaller diameter measuring 40–150 nm. They are bounded by a very fine membrane and have an apparently homogeneous content. Two types of profile can be distinguished, smooth and rough surfaced. The smooth-surfaced profiles (40–100 nm in diameter) are circular, oval or irregular in shape and correspond to vesicles and contorted tubules linked together in a tightly meshed reticulum. The rough-surfaced profiles are more numerous and are of length 50 nm to 5 μm with a fairly constant diameter of about 50 nm. They frequently occur in more or less parallel arrays separated from each other at fairly regular intervals. The rough surface is due to the presence of small round electron-dense particles (10–20 nm diameter) attached to the outside surface of the limiting membrane. These particles are known as *ribosomes*, and are also found free in the cytoplasm and in the mitochondria. They are specially abundant in rapidly proliferating cells and in secretory cells. They are present in bacterial and plant cells (and in their chloroplasts) as well as in mammalian cells.

5.2 THE BACTERIAL CELL

Micro-organisms are of such diverse complexity and varied morphology that it is not possible to give more than the briefest outline of their structure, but the following dscription of the common rod-shaped bacterial cell may serve to indicate the principal features (Fig. 5.2). The

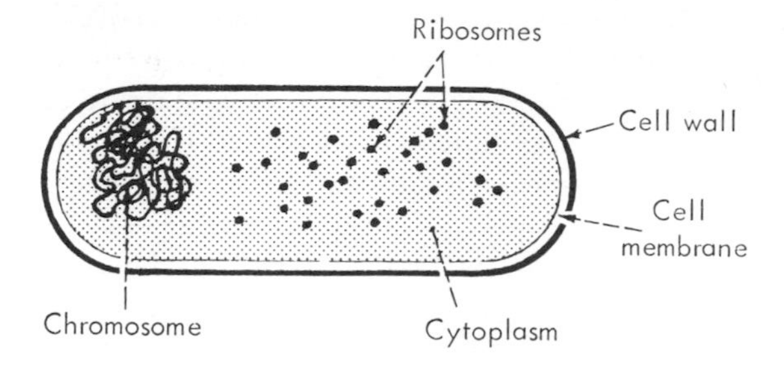

Fig. 5.2 Diagrammatic representation of a bacterial cell.

bacterial cell consists of a protoplast including nuclear body and cytoplasm bounded by a delicate cytoplasmic membrane. The membrane itself is in close contact with the rigid cell wall of characteristic shape. In many bacteria the cell wall is surrounded by a much wider capsule which usually consists of complex polysaccharides. The cytoplasm contains granules of various types, some of which are reserve food materials, but the most important and abundant cytoplasmic particles are the bacterial ribosomes. The bacterial nuclear body which has not the easily definable structure of the animal cell nucleus, contains DNA as the basic genetic material, or chromosome.

When the cell wall is eliminated by digestion with the enzyme lysozyme, the membrane and its contents are released as the osmotically sensitive protoplast. Gram-positive organisms yield protoplasts which are free of wall constituents, but gram-negative organisms yield osmotically sensitive spheres known as spheroplasts which retain fragments of the wall.

5.3 THE SEPARATION OF SUBCELLULAR COMPONENTS

From the chemical point of view most information has been obtained by the study of cell components separated by the process of differential centrifugation from cells disrupted in bulk in a suitable medium. Nuclei are sometimes prepared by procedures involving treatment of the finely divided tissue with a weak acid, such as citric acid, followed by differential centrifugation and washing with very dilute acid. The citric acid method has been developed and improved by Dounce [1, 2] and by Mirsky and Pollister [3], in whose papers full details are given. Dounce [2] has isolated nuclei by the citric acid method at different pH values and has pointed out that nuclei prepared at pH values much below 3.0 undoubtedly lose much of their histone content and so give high analytical values for nucleic acid and lipid when analysed in bulk. On the other hand, nuclei isolated at pH 6.0–6.2 appear to lose some RNA and probably also some protein. Since most methods for the preparation of clean nuclei free from contamination with cytoplasmic residues involve repeated washing either with dilute sodium chloride solution or with dilute citric acid, it is not surprising that, as Mirsky and his colleagues [4] point out, the protein content of such isolated nuclei, as determined by gross chemical analysis, is considerably below the values found for similar nuclei isolated from non-aqueous media. Such nuclei were originally prepared by Behrens [5] by a method in which powdered freeze-dried tissue was allowed to sediment out in columns of organic solvents of graded density. This method has subsequently been modified and improved [3, 6, 7] and the nuclei so obtained have the advantage of retaining all their acid-soluble constituents as well as all the nuclear proteins. More recently glycerol has replaced the petroleum ether originally used as the homogenization medium [8].

Useful methods for isolating nuclei in sucrose solutions have also been described [9, 10]. A commonly employed procedure is that of Schnieder and Peterman [11] in which 0.25M-sucrose containing 1.8mM-$CaCl_2$ is used. Philpot and Stanier [12] have employed for the isolation of rat liver nuclei a medium containing 0.3M-sucrose, 65mM-potassium glycerophosphate, 1mM-$MgCl_2$ and 40% glycerol. Widnell and Tata [14] have prepared metabolically active nuclei by homogenizing the tissue in 0.32M-sucrose–3mM-$MgCl_2$ with subsequent purification in 2.2M-sucrose–1mM-$MgCl_2$.

The criteria for assessing the quality of preparations of nuclei are: (i) their morphological appearance in the ordinary light microscope, in the phase-contrast microscope, and in the electron microscope; (ii) the presence of enzymes such as NAD pyrophosphorylase (EC2.7.7.1) which are known to be exclusively of nuclear origin; (iii) the absence of cytoplasmic enzymes such as cytochrome oxidase (EC1.9.3.1) or glucose-6-phosphatase (EC3.1.3.9).

Such criteria are met in preparations of nuclei made by quick and simple methods involving the use of detergents such as Triton X-100 (a member of the octylphenoxyethanol series) [14], Tween 80 (polyoxethylene sorbitan monoleate) [15], or sodium deoxycholate [16].

Metabolically active nuclei can also be prepared by gradient centrifugation in sucrose and dextran [17] or Ficoll (a high polymer of sucrose) [18].

Methods of isolation have been discussed in several reviews [19–24]. Even the cleanest preparations, as judged by the standards of conventional microscopy, may show some adhering cytoplasmic debris when examined in the electron microscope [25]. The major problem, as with the isolation of all subcellular fractions, is that in the attempt to obtain microscopically clean nuclei damage may occur with consequent loss of normal intranuclear components. It must be emphasized that the cell nucleus is not a homogeneous structure. Therefore, while gross chemical analysis on bulk material can give in general terms the nature and relative amounts of the various constituents, it tells us nothing about their distribution in the various regions of the nucleus which are studied by the cytologist [26, 27].

Much effort has gone into the subfractionation of nuclei and methods are available for the isolation of purified nucleoli [28–34].

Nuclei are sedimented from a cell homogenate by low-speed centrifugation, and this leaves the mitochondria as the largest particles remaining in the supernatant fraction. In order to prevent aggregation of the smaller particles it is essential to include sucrose in the homogenization medium. In 0.88M-sucrose the mitochondria retain their rod-like morphological characteristics and their ability to stain supravitally with the dye Janus Green B. At this concentration of sucrose, however, on account of the high viscosity and density of the medium, very high centrifugal speeds have to be employed to sediment the subcellular fractions, and a compromise is to use 0.25M-sucrose as a medium in which aggregation of granules does not occur and in which mitochondria can readily be prepared with the same biochemical properties as those obtained in 0.88M-sucrose although they no longer stain with Janus Geen B and are spherical rather than elongated in shape. For the separation by differential centrifugation of the subcellular fractions from a homogenate in 0.25M-sucrose prepared in a Potter–Elvehjem homogenizer, a force of 700 **g** is employed to remove nuclei and general cell debris, including unbroken cells. After removal of the nuclear fraction the extract is centrifuged at 8500 **g** for 10 minutes to bring down the mitochondria and at 100000 **g** for 60 minutes to bring down the microsomes. The supernatant fraction is said to be derived from the cell sap or *cytosol* and contains no easily sedimentable material. Many variations of this scheme of differential centrifugation are, of course, possible and the different schemes available have been reviewed [35, 36]. The microsomes (diameter 16–150 nm) isolated by such procedures [37] are too small to be resolved by the light microscope and were at first 'cytochemical concepts without any known morphological counterpart in the intact cell'. In an electron microscope study of sections of the microsome pellet from liver tissue the predominant structural element is represented by membrane-bound profiles recalling those found in the endoplasmic reticulum in sections in the intact liver cell. These profiles appear to correspond in three dimensions to tubules or cisternae and may be smooth surfaced, though the majority carry on their surface small dense particles similar to those observed in electron micrographs of whole cells. The microsomes therefore are not artefacts introduced by homogenization of the tissue but, as fragments of the endoplasmic reticulum, they represent cytoplasmic structures known to pre-exist in the intact cell.

When suspensions of microsomes are treated with sodium deoxycholate they are disrupted into an unsedimentable portion derived from the membranous component and containing most of the protein and nearly all of the phospholipid, pigment, and enzymes, and a particulate portion sedimentable at 100 000 **g** containing almost 20 per cent of the protein and nearly all of the RNA of the microsomes which is itself the bulk of the RNA of the cytoplasm. These small particles, which must be clearly distinguished from the microsomes themselves, contain approximately equal amounts of RNA and protein and are in fact isolated ribosomes [38, 39]. While this description refers essentially to animal cells, it should be kept in mind that bacterial cells also contain ribosomes which can be isolated by differential centrifugation [40–42].

5.4 ISOLATION OF RNA

The method of choice may vary according to the type of tissue employed and the particular RNA species to be isolated. Probably the most commonly employed method for the preparation of undegraded RNA in good yields is based on treatment at elevated temperatures (63°C) with a detergent (e.g. sodium dodecyl sulphate) to release and denature protein together with a solvent such as phenol for the denatured protein. The aqueous layer obtained upon centrifugation contains RNA and polysaccharides. Both are precipitated by ethanol but RNA may be extracted from the precipitate with 2-methoxyethanol from phosphate buffer. After dialysis the RNA is precipitated with ethanol. There are several modifications of the phenol method [43]; they have been discussed in detail by Kirby [44–46] and by Georgiev [47]. Traces of DNA can be removed by treatment with pancreatic DNase (purified free of RNase activity).

5.5 THE RNA OF THE CYTOSOL

The *cytosol* or *cell sap* contains the soluble proteins of the cytoplasm and also a collection of small RNA molecules (23 000–28 000 daltons) known as transfer RNAs (tRNAs). These account for some 10 to 15 per cent of the total cellular RNA.

In the case of rapidly dividing bacterial cells there are about 4×10^5 tRNA molecules of perhaps fifty or so different varieties. The precise number of varieties is not yet known but there is one or more for every type of amino acid. In a mammalian cell the total number per cell can be as high as 10^8. tRNAs can readily be extracted from the cytosol of most cells with buffered aqueous phenol and sediment in the 4S region on zonal ultracentrifugation (see p. 119). The precise chain length of the different varieties appears to vary over the rather narrow range of 75–85 nucleotides.

In 1965 Holley and his colleagues [48, 49] worked out the complete sequence of nucleotides in the alanine transfer RNA (tRNA^{Ala}) of yeast (Fig. 5.3) This work which took several years to complete, was recognized by the award of a Nobel Prize three years later. This was followed shortly by the primary structures for yeast tRNA^{Ser} which exists in two forms (1 and 2) differing only in three nucleotides (Fig. 5.4) [50].

The primary structures of around 130 cytosolic tRNAs are now known [51]. Just about all these sequences can be fitted to the same hydrogen-bonded secondary structure of loops and short helical region as shown in Fig. 5.5, the general features of which are as follows. (1) An amino acid arm, a helix of 7 base pairs, terminating at the 3′-end in an unpaired-C—C—A_{OH} sequence to which an amino acid becomes attached under the influence of appropriate enzymes as described in Chapter 12. (2) A dihydrouracil loop (I) containing 8–11 nucleotides at the end of a helical

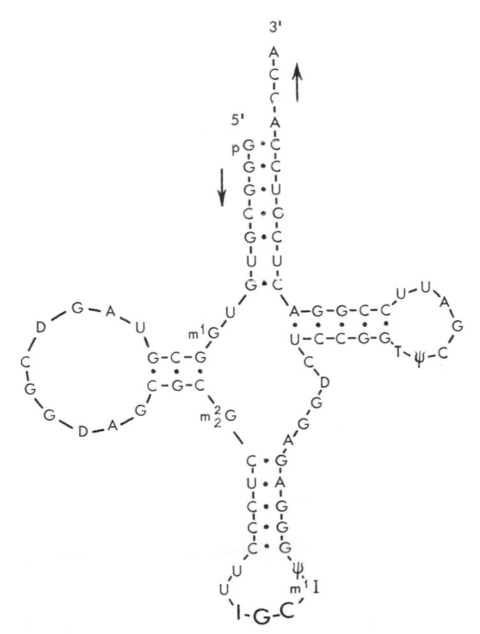

Fig. 5.3 The structure of tRNA^{Ser} from yeast [48]. The anticodon is shown in heavy type. For an explanation of the symbols denoting the minor bases, the list of abbreviations at the beginning of the book should be consulted.

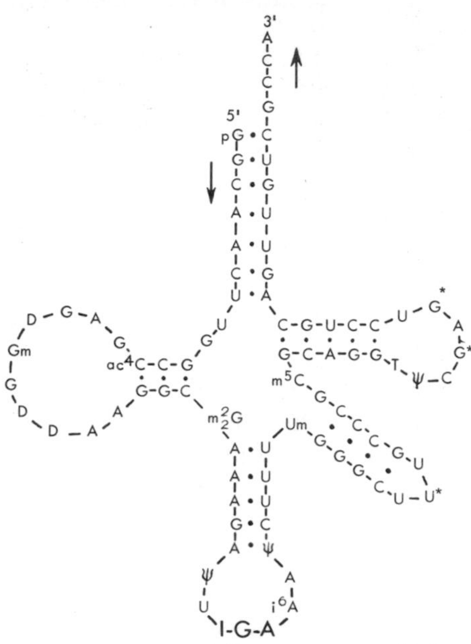

Fig. 5.4 The structure of $tRNA_2^{Ser}$ from yeast [50]. The anticodon is shown in heavy type. In the isoacceptor $tRNA_1^{Ser}$ the three nucleotides indicated with asterisks are different, U being replaced by C and G by A.

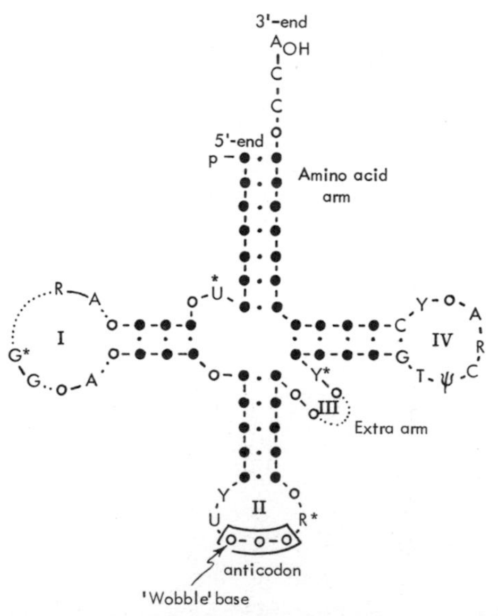

Fig. 5.5 Generalized cloverleaf secondary structure of tRNA. The solid circles represent bases in helical regions, paired by hydrogen bonds (centre dots), while the open circles represent unpaired bases. R is a purine nucleoside, Y is a pyrimidine nucleoside, T is ribothymidine, and ψ is pseudouridine. An asterisk indicates that the nucleoside may be modified.

stem of three or four base pairs. (3) An anticodon loop (II) containing 7 nucleotides at the end of a helical stem of five base pairs. The anticodon itself, a group of three nucleotides at the centre of the loop, is discussed in Chapter 12. (4) An 'extra arm' (III) which varies widely from species to species, containing 3–18 nucleotides. (5) A pseudouridine-ribothymidine loop (IV) containing seven nucleotides, including the sequence -T-ψ-C- at the end of a helical stem of five base pairs (eukaryotic initiator tRNAs, however, do not have this particular sequence). Whilst there are considerable data to support this 'clover-leaf' arrangement in two dimensions, only recently have there been any unambiguous data regarding tertiary structure. The results of a systematic crystallization study of yeast $tRNA^{Phe}$ were crystals suitable for X-ray analysis at 3 Å resolution, the results of which have already been discussed in Chapter 3. As also mentioned there, an additional bonus has been a greater appreciation of the rules that appear to govern RNA secondary structure [52]. Only G·C, A·U and *sometimes* G·U pairs are allowed. All helices consist of at least three base pairs without G·U pairs or four base pairs when G·U pairs are allowed. Looped out bases are non-existent or very rare and 'hairpin' loops contain at least three nucleotides [52].

Certain bacterial tRNAs are involved in cell-wall peptidoglycan synthesis. These have lost a number of the features common to those involved in the synthesis of proteins on ribosomes.

5.6 THE RNA OF RIBOSOMES AND POLYSOMES

The name 'ribosome' was introduced in 1957 to distinguish the particulate material of the microsomes from the membrane material. The ribosomes are electron-dense particles of diameter 20 nm containing about 40 per cent protein and 60 per cent RNA which are found in all types of living cell, both free and attached to the endoplasmic reticulum. They play a vital part in the process of protein synthesis (Chapter 12) during which they become attached to a strand of messenger RNA to form complexes known as polysomes.

It is customary to characterize macromolecules and small particles (e.g. ribosomes) by their sedimentation coefficients expressed in Svedberg units (S). The sedimentation coefficient of a particle or molecule depends on both its molecular weight and its shape and is proportional to its rate of sedimentation in a centrifugal field.

The *E. coli* ribosome (sedimentation value 70S, mass 2.7×10^6 daltons) is composed of a 30S subunit (mass 0.9×10^6 daltons) and a 50S subunit (mass 1.8×10^6 daltons) (Fig. 5.6). When the magnesium concentration is reduced below 0.5mM the ribosomal particle dissociates into the subunits;

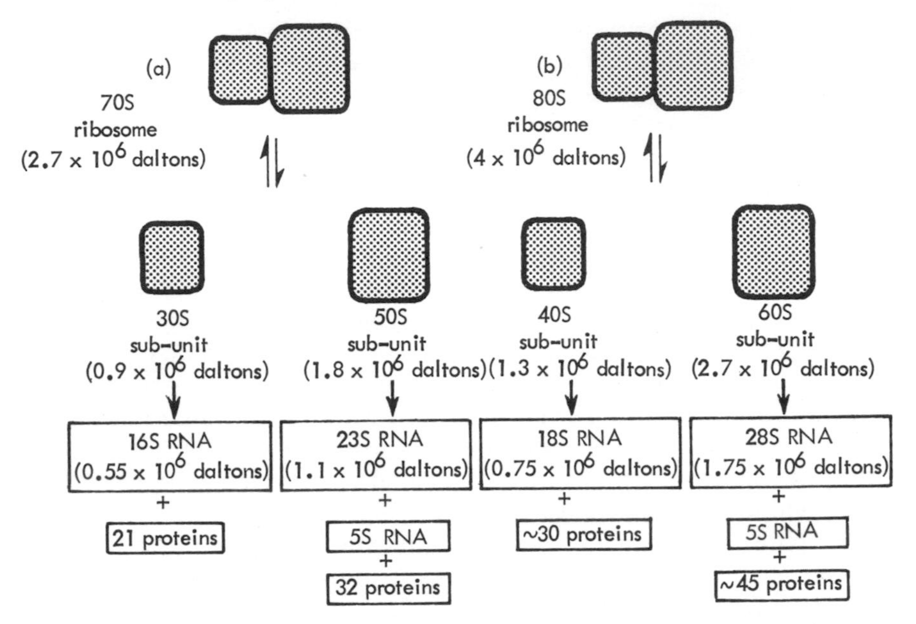

Fig. 5.6 S-values and molecular weights of the components of (a) a ribosome from *E. coli* and (b) a mammalian ribosome.

they reunite when the magnesium concentration is raised again. In the mammalian cells the ribosome has a sedimentation value of about 80S with subunits of about 40S and 60S (Fig. 5.6). Ribosomal RNA comprises about 80% of the total RNA of the cell. In *E. coli* the 50S and 30S ribsomal subunits yield RNAs of molecular weights 1.1×10^6 (23S RNA) and 0.55×10^6 (16S RNA) respectively [53]. The two RNAs contain different base ratios, differ in base sequences [54, 55], and can hybridize with different sites on the bacterial genome (p. 334). The 80S ribosomes from mammalian cells also yield two rRNA components corresponding to those from the 40S and 60S ribosomal subunits. The smaller component generally has a molecular weight of 0.75×10^6 (18S) while the larger has a molecular weight of 1.76×10^6 (28S).

The 18S component is common to all animals but the larger (28S) component has evolved with each major step in animal evolution from 1.4×10^6 daltons in sea urchins to 1.75×10^6 daltons in mammals. In higher plants, ferns, algae, fungi and some protozoa the main ribosomal RNAs are 1.3×10^6 and 0.7×10^6 daltons. Whilst the ribosomal RNA from most organisms possess broadly similar base compositions with guanine plus cytosine contents of between 50–60% there are some exceptions to this e.g. *Drosophila* (40%) and *Tetrahymena* (43%).

Associated with the ribosomes are also two RNAs of low molecular weight: (i) 5S RNA containing about 120 nucleotides, which is associated with the larger ribosomal subunit and has been obtained from bacterial

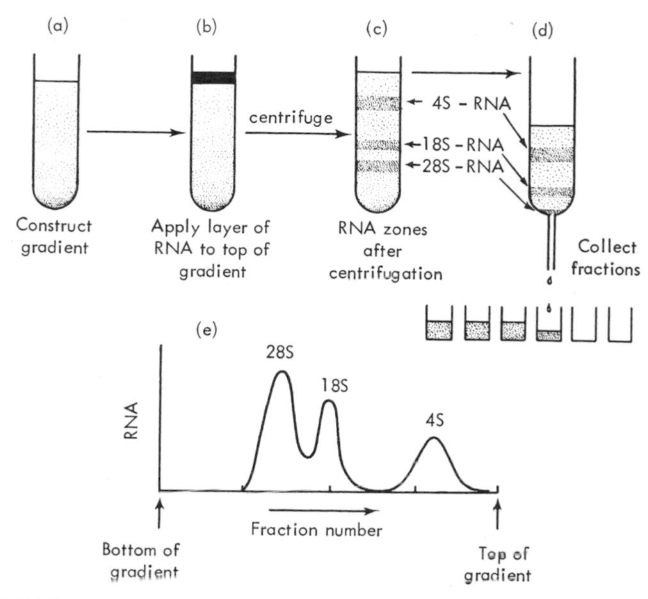

Fig. 5.7 The process of gradient centrifugation. A sucrose gradient is constructed in a centrifuge tube at (a), the RNA solution is then applied carefully as a layer on top (b), and, on centrifuging, the RNA separates out into its main components according to molecular weight and shape of the molecules (c). When the tube is punctured (d), the gradient is collected a few drops at a time in tubes to give sequential fractions. The amount of RNA in each fraction is then determined by ultraviolet absorption measurements so as to give the pattern shown at (e).

[56, 57], animal [58–62] and plant cells [63, 64]. It, however, does not occur on mitochondrial ribosomes, (ii) 5.8S RNA, of chain length about 130 nucleotides, which is associated with the 28S RNA [65, 66] in the large subunits of animal and plant ribosomes (p. 339).

These main RNA fractions can also be readily separated in the process of zone centrifugation through sucrose density gradients [67, 68]. In this process an ultracentrifuge tube is prepared containing sucrose solution increasing in concentration from 5 per cent at the top to 25 per cent at the bottom. The solution of RNAs is carefully layered on the top and the tube is centrifuged at high speed for several hours (Fig. 5.7). The bottom of the tube is then punctured with a hypodermic needle and a series of samples of a few drops each is collected. The nucleic acid content of each sample is estimated by measurement of ultraviolet adsorption. An example of the results obtained by this method is shown in Fig. 5.7. For large-scale work centrifugation in zonal rotors may be employed.

One of the most delicate methods for fractionating small amounts of ribonucleic acids is zone electrophoresis through polyacrylamide gels [69–77]. The fractions separate as discrete bands which may be located by scanning in ultraviolet light. The type of separation obtained by this method is illustrated in Fig. 5.8. It can be used on a preparative scale for the fractionation of RNA [78], and also in two dimensions to effect the separation of low-molecular-weight RNAs (e.g. tRNAs) [79, 80]. Normally the mobility of RNAs through the gels depends not only on molecular size but also on secondary structure. However, the technique can be modified using formamide as electrophoretic solvent to determine RNA molecular weight [81]. The formamide destroys secondary structure and renders the RNAs conformationally homogeneous. Prior treatment of the RNA with 2.2M-formaldehyde [82], 10mM-methyl mercury [82] or 1M-glyoxal (see Plate v) and 50% dimethyl sulphoxide have also been used to eliminate secondary structure [83]. Allied methods of separation include zone electrophoresis in starch gels [84], agarose [85], or in composite gels of agarose–polyacrylamide [86, 89].

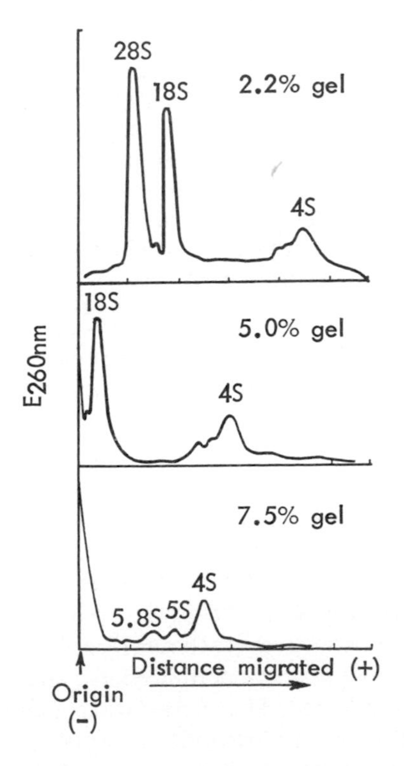

Fig. 5.8 Schematic diagram illustrating he electrophoretic separation of RNA components (isolated from tumour cell cytoplasm using hot phenol and detergent) which can be achieved with the aid of polyacrylamide gels of various concentrations.

5.6.1 5S ribosomal RNA

E. coli 5S ribosomal RNA was one of the first nucleic acid molecules to be sequenced [88], however the secondary structure has been a matter of some controversy for more than a decade. Numerous attempts have been made to deduce a secondary structure for this RNA by applying the rules deduced from studies on tRNA and the 'diagonal' procedure of Tinoco and his colleagues [89]. A possible model [90] which is consistent with the physical data is shown in Fig. 5.9.

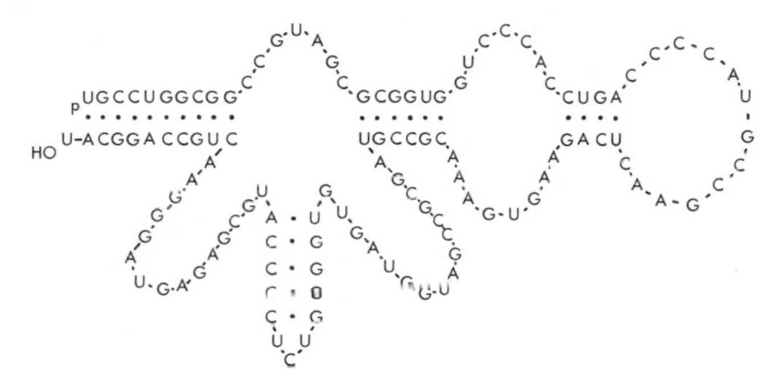

Fig. 5.9 A possible structure of *E. coli* 5S rRNA [90].

The sequence of some 16 prokaryotic and 25 eukaryotic 5S RNA species is now known [91]. All contain 116–121 nucleotides. Amongst the bacterial species there is considerable conservation of sequence. For example the 5S RNA from *E. coli* and *Ps. fluorescens* have two thirds of their sequence in common. In eukaryotes no difference is detected between human KB cell 5S RNA and that from two mouse cell lines whereas 90–95% of the *Xenopus* 5S RNA is the same as that of the human and mouse species.

5.6.2 16S and 23S ribosomal RNAs of prokaryotes

The pioneering work of Ebel and his colleagues has led to the evaulation of the complete sequence of the 16S ribosomal RNA species in *E. coli* [92] mainly using rapid RNA sequencing methods. The proposed sequence contains 1541 nucleotides and agrees completely with the sequence of one (rrnB) of the seven ribosomal RNA cistrons [180] (there is some minor heterogeneity between the various cistrons). Various approaches have been used to examine the topography of this molecule e.g. electron microscopy, nuclease digestion, single-strand specific chemical modification using agents such as kethoxal [186]. It appears that the RNA chain contains approximately 50 helical elements organised roughly into five structural domains by virtue of long-range base pairing interactions [186, 187, 204, 205]. It is an RNA chain folded in space much like a typical protein chain [188]. The complete sequence (2904 nucleotides) for the *E. coli* rrnB cistron 23S RNA has also been achieved using the Maxam-Gilbert DNA sequencing techniques (see Chapter 3) on cloned restriction fragments [93].

As is the case for tRNA a small number of specific nucleotides of ribosomal RNAs are modified during transcription. *E. coli* 16S and 23S RNAs possess 22 and 27 methyl groups respectively. Most of these groups are on various base moieties only a few being on ribose [94]. The 16S rRNA has been examined from a wide range of prokaryotes, and it appears that there are some highly conserved regions most of which contain the methylated nucleoside residues, and are clustered in the 3′-half of the molecule [95]. Closely related bacteria have similar 5′-terminal sequences whereas these sequences differ in distantly related species. 3′-terminal sequence analysis of various bacterial rRNAs showed that the 3′-terminus of 16S and 23S RNA differs between different bacteria [96–98]. However the 3′-terminus of all 16S RNAs contains a pyrimidine rich sequence that may be involved in specific interactions with mRNA (see Chapter 12).

5.6.3 The large ribosomal RNAs of eukaryotes

The large ribosomal RNAs of mammlian origin are also methylated. There are 46 methylated nucleosides in the mammalian 18S component and 71 in the larger mammalian species (28S). However, unlike the bacterial situation most (95%) of the methyl groups are on ribose moieties and the rest on bases [99]. Again it appears that the regions of eukaryotic ribosomal RNA molecules that contain these methylated nucleosides are remarkably well conserved amongst the vertebrates and some also occur in non-vertebrates such as yeast [99]. Another difference between eukaryotic and bacterial ribosomal RNAs is that eukaryotic rRNA contains considerable numbers of pseudouridine residues. 18S has approximately 37 and 28S around 60 [99].

DNA-sequencing methods have been used to examine the 3′-terminii of 18S rRNA from mouse, silkworm, wheat embryo and slime mold. The 3′-terminal 20 nucleotides of these RNAs are highly conserved and also exhibit a strong homology with the 3′-end of *E. coli* 16S rRNA. However, the 16S sequence CCUCC implicated in mRNA binding by bacterial ribosomes (see Chapter 12) is absent from each eukaryotic sequence [100]. The use of treatments that specifically disrupt hydrogen bonding (e.g. heat, dimethyl sulphoxide, formamide etc.) has revealed hidden breaks in the large ribosomal RNAs of amoeba, sea urchins, chloroplasts and a variety of insects. In insects, the break occurs close to the centre of the large (26S) rRNA [101]. While the large ribosomal RNA molecules of bacteria and vertebrates are covalently continuous, denaturation of the large vertebrate 28S rRNA releases a small 5.8S RNA species which is normally hydrogen bonded to the parent 28S RNA molecule [99].

5.6.4 5.8S ribosomal RNA

The presence of 5.8S rRNA associated with the large rRNA has been demonstrated in all eukaryotic cells so far examined. The primary sequence of 5.8S molecules from nine different sources is now known [91]. They vary from 161 to 167 nucleotides in length and like the larger ribosomal RNAs they can obtain methylated nucleosides and pseudouridine [99].

5.7 MESSENGER RNA

Most of the messenger RNA in living cells is found associated with ribosomes in the *polyribosomes* or polysomes [102]. Polysomes are very delicate structures and the greatest care must be taken during the process of preparation and isolation [103–105] in order to prevent mechanical breakage or degradation by RNase. Sedimentation on a sucrose gradient separates polysomes containing different numbers of ribosomes. Messenger RNA (mRNA) molecules are now known to account for 3–5 per cent of the total cellular RNA. They are quite heterogeneous with regard to size. The typical prokaryotic mRNA is known to carry information for more than one protein, and are termed *polycistronic*. The mRNA for the *E. coli lac* operon codes for three different polypeptides (see p. 438) and that for the *trp* operon for five. Some *monocistronic* mRNAs also exist in bacteria such as a lipoprotein mRNA of *E. coli*. Eukaryotes on the other hand seem to lack *polycistronic* mRNAs. Due to the large size and metabolic instability, specific bacterial mRNAs have only been isolated in isotopically detectable amounts, generally by hybridization to the DNA of a transducing phage carrying the appropriate genes. The best characterized polycistronic mRNAs are the RNAs of the small RNA bacteriophages (MS2, Qβ, R17, f2). In 1976 Fiers and his colleagues [106] determined the entire 3569 nucleotides in MS2 RNA and the arrangement of the cistrons is shown in Fig. 5.10 (a). Noteworthy are non-coding sequences in both terminal and internal positions. A few bacterial mRNAs such as the relatively small *monocistronic* mRNA for a 78 amino acid precursor to the lipoprotein in the outer membrane of *E. coli* are sufficiently stable and present in sufficiently large amounts to allow isolation and sequence characterization [107]. The general layout of this mRNA is shown in Fig. 5.10 (b) and again the presence of non-coding regions at both the 5′- and 3′-terminii should be noted. Nuclease digestion data indicate a very stable hairpin stem and loop at the 3′-end which may be important in the termination of transcription of this mRNA [107] (see Chapter 10). In eukaryotes, since highly differentiated cells synthesize a fedw proteins in great predominance, they provide a favourable source of specific mRNAs, for in these cells mRNAs turnover relatively slowly so that reasonable amounts can be isolated. Besides judicious choice of cell or tissue, several general technical advances have facilitated the isolation of specific eukaryotic mRNAs [108]: (a) The control of ribonuclease activity during isolation and deproteinization with inhibitors such as diethylpyrocarbonate, or naturally occurring inhibitors such as that from rat liver. (b) Poly (A) absorption methods which capitalize on the fact that a high proportion of eukaryotic messenger RNAs contain a covalently attached 3′-terminal poly (A) segment ranging from 20 to 250 nucleotides in length. Thus poly (A)-containing mRNA can be separated from other cellular RNAs by affinity chromatography on oligo (dT)-cellulose or poly (U)-agarose. (c) Sensitive electrophoretic separation techniques using agarose or polyacrylamide gels under denaturing conditions which was already mentioned on page 120 permits the fine resolution of different species on the basis of size. (d) The establishment of cell-free protein synthesizing systems capable of translating exogenous mRNAs, thus allowing the

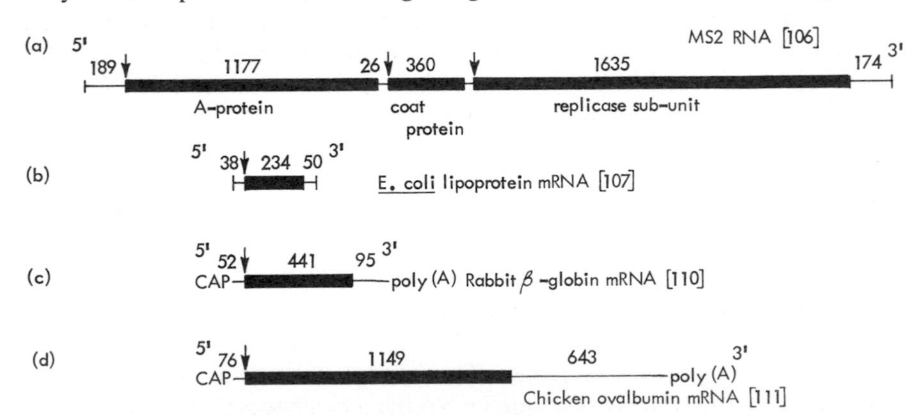

Fig. 5.10 The general layout of various mRNAs (a) a polycistronic prokaryotic mRNA (b) a monocistronic prokaryotic mRNA (c) and (d) monocistronic eukaryotic mRNAs. Thin line represents non-coding regions, thicker line represents coding regions. ↓ represents the location of an initiation codon AUG or GUG (see Chapter 12). The numbers refer to nucleotide lengths.

mRNA purification to be monitored. These include systems derived from mouse ascites tumour cell extracts [181], wheat germ extracts [182], and lysed reticulocytes pretreated with micrococcal nuclease to remove endogenous haemoglobin mRNA [183]. (e) Polysome immunoprecipitation in which antibodies to a specific protein are used to precipitate selectively only those polysomes which are synthesizing a particular antigen thus allowing an enrichment in a particular mRNA. (f) Sequence determination in which complementary DNAs (cDNAs) can be made for various mRNAs using RNA-dependent DNA polymerase (reverse transcriptase) (see Chapter 9). This enzyme requires a *primer* such as oligo (dT) which binds to the 3′-poly (A) segment of the mRNA which can then serve as *template* for the synthesis of a complementary strand of DNA (see Fig. 5.11). Generally the cDNA is synthesized in the presence of the drug actinomycin D which prevents the synthesis of a second strand of DNA [178]. Single-stranded cDNA can fold back upon itself and provide a 3′-OH end for the synthesis of a second strand (see Chapter 14). The size of the cDNA depends on the nature of the template and the substrate concentration. Full length copies of mRNAs require the addition of high concentrations of the deoxyribonucleoside 5′-triphosphates but it appears that this problem may be due to low levels of phosphatase activity in the reverse transcriptase preparations and can be alleviated by the addition of sodium fluoride or pyrophosphate [178]. In principle reverse transcriptase will transcribe any RNA template or part of an RNA molecule as long as a suitable short primer of known sequence is provided complementary to a specific sequence in the RNA to initiate cDNA synthesis.

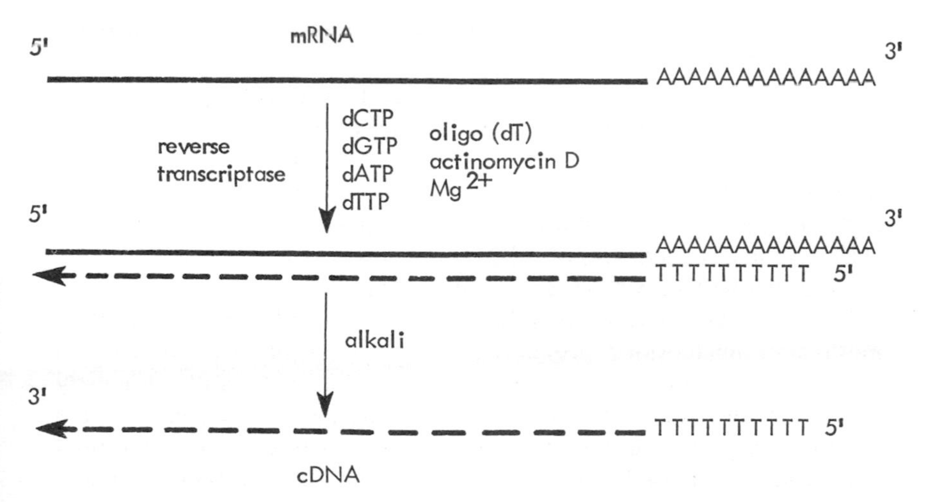

Fig. 5.11 A schematic diagram to illustrate the synthesis of a cDNA to an mRNA.

However, the size of the cDNA produced in such cases will depend where DNA synthesis was initiated. Another approach is to add on a stretch of poly (A) (or any other nucleotide) as the 3′-OH of a given RNA [179]. As is detailed in Chapter 14 cDNAs can be 'cloned' in appropriate plasmids and subsequently sequenced by the methods devised for DNA (see Chapter 3). cDNAs can also be used to detect and quantitate mRNAs by DNA.RNA hybridization as well as to facilitate their isolation e.g. cDNA- cellulose affinity columns [see 108].

These procedures and others have permitted the isolation and characterization of a growing number of eukaryotic mRNAs e.g. α and β globin mRNAs (9S), lens crystallin mRNAs (10S and 14S), protamine mRNA (6S), histone mRNAs (9S), ovalbumin mRNA (15S), procollagen mRNA (27S), silk fibroin mRNA (32S), vitellogenin mRNA (29S), myosin (26S) to mention only a few.

Like the *E. coli* lipoprotein mRNA, eukaryotic mRNAs have regions at their 5′- and 3′-terminii which do not specify any amino acid sequence [see Fig. 5.10 (c and d)]. As already mentioned a conspicious feature of *some* eukaryotic mRNAs is a long uninterrupted sequence of adenosine nucleotides at their 3′-end. Additionally a short sequence AAUAAA some 20 or so nucleotides from the poly (A) tract is common to all known eukaryotic mRNA sequences [109].

Although poly (A) tracts have been detected on mRNAs from diverse organisms such as vertebrates, plants, slime moulds and yeast [112], it is now clear that not all eukaryotic mRNAs have this structural feature. In particular histone mRNAs were shown to lack poly (A) [113, 114] (except H5 mRNA [185]) and more recently other non-polyadenylated messengers have been identified in a number of eukaryotic cell types including HeLa cells (where it amounted to 30% of the total mRNA) [115], sea urchin embryos [116], BHK (hamster) cells [117], slime molds [118], L-cells [119] and plant cells [120]. Since the mRNA fraction lacking poly (A) in HeLa cells does not hybridize extensively to cDNA made from poly (A) containing mRNA it appears that at least some is a separate class of mRNA although no functional differences are apparent [115]. A small proportion of mRNA sequences may be bimorphic, i.e. occur either with or without a poly (A)-tail [121].

Like tRNA and ribosomal RNA eukaryotic mRNAs also contain methylated nucleosides, probably 3 to 5 per molecule depending on its size [122]. One or two of these are located internally and are mostly N^6-methyladenosine. The other modified nucleosides are clustered in a complex 'cap' structure at the 5′-end of the mRNAs. The 5′-terminal nucleoside is 7-methylguanosine linked through its 5′-carbon by 3 phosphates to the 5′-carbon of a 2′-*O*-methyl nucleoside sometimes followed by a second 2′-*O*-methyl nucleoside [123, 124] (see Fig. 5.12). The capped sequences are of the form $m^7G(5')ppp(5')NmpNp$ . . . (Type 1) or $m^7G(5')ppp(5')NmpNmp$ . . . (Type 2). Many different modified 5′-terminal sequences are present in mRNA of mammalian cells [123, 124]. The first 2′-*O*-methyl can be any of the normal four or N^6-methyl-2′-*O*-methyladenosine, and the second again any of normal four. Some mRNAs from yeast, slime molds, and plant and animal viruses have a Type O structure which is $m^7G(5')ppp(5')Np$. . . . Although cap structures are found on mRNAs that lack poly (A) (e.g. histone mRNA) they are missing from some viral RNAs that function as messengers (e.g. polio virus RNA, EMC virus RNA and satellite tobacco necrosis virus RNA [125–127]). On the other hand polio virus RNA does have a polypeptide (4000–6000 daltons) covalently linked to its 5′-terminal uridine residue, the bond being O^4-(5′-uridylyl)tyrosine [128].

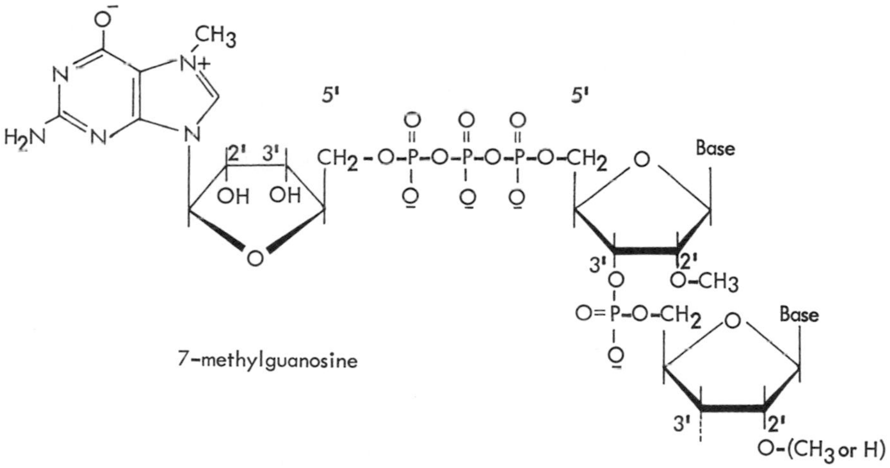

Fig. 5.12 The 5′-terminal structure of eukaryotic mRNAs.

Relatively little is known about the precise secondary and tertiary structure of mRNA. Sedimentation analysis and electrophoresis studies have indicated myosin and ovalbumin mRNAs to have more extended chains than is the case for ribosomal RNA [112]. Around 40–60% of the nucleotides of globin mRNA may be in double-helical regions [129, 130]. There is certainly considerable physical evidence for secondary structure in RNA of MS2 or Qβ bacteriophages [106, 131] (see Fig. 5.13). Tentative structures involving base pairing between distant sequences have been proposed for these phage RNAs [106, 131].

Before leaving the subject of messenger RNA it should be pointed out that eukaryotic mRNA is usually complexed with a number of proteins. For instance, treatment of isolated polysomes with EDTA dissociates the

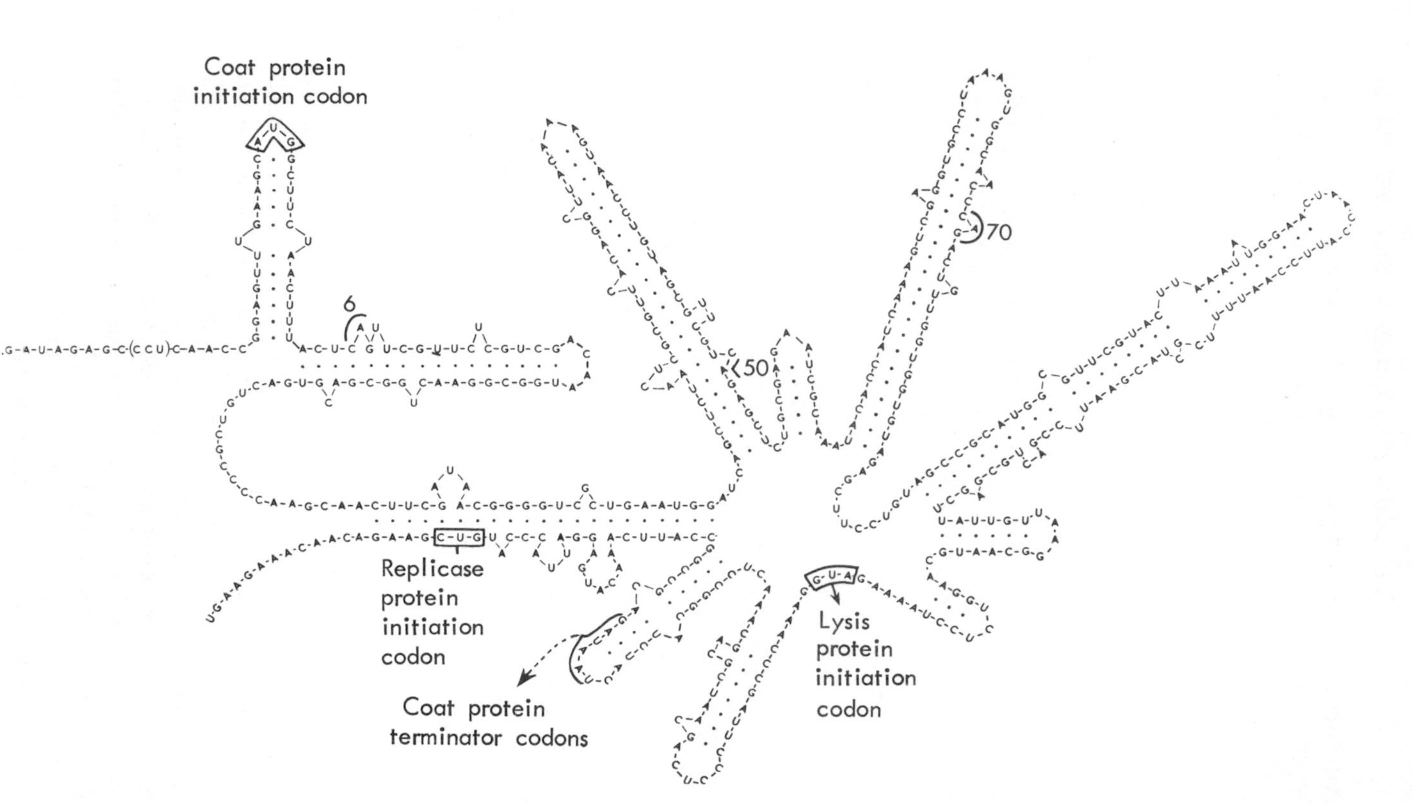

Fig. 5.13 The nucleotide sequence in MS2 RNA before the coat protein cistron (folded in the proposed 'flower' conformation), the intercistronic region preceding the replicase cistron, and the beginning of the replicase cistron [131]. The numbers indicate amino acid positions at which the amber mutations discussed in Section 13.3 are located.

ribosomal subunits and releases mRNA molecules in the form of messenger ribonucleoprotein (mRNP) complexes [132]. The complex of globin mRNA and protein (1:4) sediments at around 15S [133, 134]. The protein component is made up of 9–10 polypeptides ranging in molecular weight from 40000 to 120000. A major polypeptide of 73000 daltons is found which corresponds to the 'poly (A)-binding protein' and which appears to be characteristic of all polyribosomal mRNP [135]. Electron microscopy indicates extensive coverage of the mRNA and the proteins probably interact at precisely defined sites along the messenger molecule. Not all the cytoplasmic mRNA is, however, located on polysomes. 10–35 per cent of globin mRNA in avian erythroblasts [134] is found free in the cytosol as a ribonucleoprotein complex sedimenting around 20S. In this non-polysomal mRNA–protein complex the polypeptides comprise 12 major species in the range 20–120000 daltons and differ substantially from those found in the 15S polysomal mRNP described above [136] and the poly (A)-binding protein is absent. Whereas the 15S mRNP can be translated in cell-free systems to yield globin, the free 20S mRNP cannot be translated. Perhaps there is a role for the free mRNP proteins in the translational regulation of the associated mRNA [137]. There seem to be two components in the free mRNP population, one which may be a transport form of mRNA prior to translation on the polysomes, and another that is not translated in a given cell (see Chapter 13).

5.8 MITOCHONDRIAL RNAs

Not only do mitochondria contain their own DNA (see Chapter 11) they also contain their own ribosomes, tRNAs and mRNA [138–141].

Mammalian mitochondria contain ribosomes that sediment at 55S (fungal ribosomes are bigger, 73S). The 55S ribosome dissociates into 40S and 30S subunits and contains RNAs of 0.56×10^6 daltons (16S) and 0.36×10^6 daltons (12S) respectively, which contain modified nucleosides (yeast mitochondrial ribosomal RNA sediments at 21S and 15S). A notable difference, however, between mitochondrial ribosomes and cytoplasmic ribosomes is their apparent lack of the small 5S rRNA.

Mitochondria contain tRNAs but these differ from the corresponding tRNAs of the cell sap (see also Chapter 12) although they are still of the same size range and contain modified nucleotides. Another notable difference is the presence of fMet-$tRNA_f^{Met}$, which is absent from the cell sap.

With regard to the occurrence of mRNA species in mitochondria there are data indicating a number of discrete species terminating in 80 nucleotide-long poly (A) tails in mitochondria of mammals and insects (yeast mitochondrial mRNA do not have poly (A) tails).

5.9 CHLOROPLAST RNAs

Like mitochondria, chloroplasts also contain cyclic double-helical DNA molecules as well as ribosomes and some specific tRNAs and mRNAs [142] (see Chapter 11). The two large ribosomal RNAs from ribosomes of higher plant chloroplasts sediment at around 70S and are 1.1×10^6 and 0.56×10^6 daltons respectively [144]. Additionally there is a detectable 5S rRNA associated with the large subunit. Although plant mRNA is known to have poly (A), the mRNA specified by chloroplast DNA for the large subunit of ribulose 1,5-diphosphate carboxylase in *C. reinhardii* lacks poly (A) [143].

An interesting feature of the 16S ribosomal RNA from *Zea mays* is its extensive sequence homology with *E. coli* 16S ribosomal RNA.

5.10 RNA OF THE CELL NUCLEUS

Nuclei usually only contain a small fraction (4 per cent) of the total cellular RNA. In the resting nucleus a large part of this nuclear RNA is concentrated in the nucleolus which is particularly prominent in those cells in which a strongly basophilic cytoplasm indicates the presence of a high concentration of RNA. The nucleolus usually gives a negative Feulgen reaction, except perhaps in the peripheral regions adjacent to the nucleolus-associated chromatin. The central regions show a strong affinity for basic dyes, which is removed by ribonuclease, and a strong absorption of ultraviolet light at 257 nm, which is likewise abolished by ribonuclease [145]. This suggests that the central portion of the nucleolus, at least, is composed of ribonucleoprotein.

The DNA associated with the nucleolus is from a region of the chromosome known as the nucleolar organizer which codes for the synthesis of ribosomal RNA.

The biological significance of the nucleolus has been extensively studied and reviewed [146–154]. Its role in ribosome formation has been deduced from several lines of evidence. (i) Its intense basophilia, indicating a high concentration of RNA, is particularly obvious in cells in which protein synthesis is very active. (ii) Accumulation of RNA in the nucleolus can be selectively prevented by low doses of actinomycin D. (iii) The accumulation of newly formed ribosomal RNA in the cytoplasm of growing cells can be prevented by irradiation of the nucleolus with a microbeam of ultraviolet light [155]. (iv) Mutants of the toad *Xenopus* in which the nucleolus is lacking do not survive beyond the gastrula stage and are unable to produce ribosomal RNA [156]. They lack virtually all the DNA complementary to 28S and 18S RNA. (v) Kinetic studies in cells labelled with radioactive uridine have indicated that the nucleolus is the site of synthesis of 45S RNA which is the precursor of ribosomal RNA. This matter is discussed further in Chapter 11.

High-molecular-weight nuclear RNA can be found at sites other than the nucleolus. This RNA is very heterogeneous in sedimentation behaviour (20–100S), hence the term heterogeneous nuclear RNA (hnRNA) [157–159]. Characteristically these molecules ($1–15 \times 10^6$ daltons) which account for about one per cent of the cellular RNA exhibit a marked tendency to form intermolecular aggregates [160] so that size determination using electrophoresis or gradient analyses [161] must be carried out under strictly denaturing conditions (see p. 120). Whereas low levels of actinomycin D will selectively inhibit the synthesis of nucleolar RNA (see above) hnRNA synthesis occurs in the euchromatic regions of the nucleus. However, metabolic decay of hnRNA within the nucleus is notably rapid ($t_{1/2}$ 20–60 min) but a proportion (a quarter to a third) have poly (A) tracts of about 200 nucleotides long at their 3′-ends. 5′-cap structures are also detectable in a high proportion of hnRNA molecules [159]. Recent data (see Chapter 11) indicate that among the hnRNA population are mRNA precursors in various stages of the post-transcriptional processes that yield cytoplasmic mRNA. These are usually observed as ribonucleoprotein fibril structures [167] sometimes in association with chromatin (see Chapter 11 and Plate V).

Whilst some of the low-molecular-weight RNA detectable in the nucleus is the short-lived precursor to cytoplasmic tRNA (see Chapter 11) there is an important group of low-molecular-weight or small nuclear RNAs (snRNAs) which are metabolically quite stable [158]. The snRNAs account for about 0.5 per cent of the total cellular RNA and can usually be resolved into 11–14 well-defined bands on gel electrophoresis [161–163] depending on the source of nuclei. They range in size from 93 (4.5S RNA III) to about 30S (species K) nucleotides and have a varied nucleotide composition and sequence [164]. The nomenclature used is complex but some of these in order of increasing size are 4.5S RNA, U-1 (or D), U-2 (or C), U-3 (or A) L and K. The primary sequences of 4.5S RNA I, U-1, U-2 and U-3 B of Novikoff cells for example are known [189–192]. These sequences show no similarity with each other, but U-1, U-2 and U-3 have 5′-termini similar to the 'cap' structure of the mRNAs. In addition they contain a large number of pseudouridine residues and methylated bases and/or ribose residues. Their secondary structure seems more open than that of tRNA and they have no amino acid acceptor activity. Whilst, types U-1 and U-2 are mainly nucleoplasmic and type U-3 is mainly nucleolar, they can also be detected in the cytoplasm. Type L is actually preferentially cytoplasmic [158, 161, 165]. Relatively little is known about the structural similarities between the snRNAs in different species. Within a given species the snRNA pattern is independent of the cell type from which the RNA is derived. As the different low-molecular-weight RNAs can have different subcellular locations it seems reasonable to assume that they may have different functions. They are only partly associated with chromatin, thus a regulatory role as modulators of transcription seems unlikely, although nuclear 4.5S RNA has been shown to affect the availability of DNA in mammalian chromatin to act as template for RNA synthesis [200]. On the other hand it has been known that certain RNAs shuttle non-randomly between cytoplasm and nucleus and a role for small nuclear RNAs in programming chromosomal information has been proposed [201, 202]. As will be discussed more fully in Chapter 11 recent interest in the snRNAs concern their possible role in RNA 'splicing' mechanisms. It appears that certain of the snRNAs (e.g. U-1 and U-2) are specifically associated with proteins [193] which in turn may be associated with the ribonucleoprotein structures containing the mRNA precursors undergoing post-transcriptional processing [166, 194–196]. Some low-molecular-weight RNAs are found *in vivo* hydrogen-bonded to nuclear and cytoplasmic poly (A)-terminated RNAs from various mammalian cells [197–199]. However, these small RNAs may also regulate mRNA translation (U-1 can bind to ribosomes and inhibit translation [203]).

5.11 RNA SEQUENCE DETERMINATION

Early methods employed in RNA sequence determination consisted essentially of controlled degradation of RNA with various enzymes and separation of the oligonucleotide products by chromatography or electrophoresis [168]. By examining the products of hydrolysis at various stages by using appropriate combinations of enzymes it is possible to determine the composition of each fragment and to work out how fragments may be pieced together to establish the sequence in the complete RNA chain [169].

Initially the oligonucleotides that were liberated by enzymic digestion were detected by their absorption of ultraviolet light. This limited both the scale and the rate at which sequence determination could progress. To reduce the scale, Sanger and his colleagues [170, 171] prepared ^{32}P-labelled RNAs by *in vivo* labelling and studied enzymic digests of these, detecting and estimating the ^{32}P-labelled oligonucleotides after two-dimensional electrophoretic separation by autoradiographic and radioactive counting techniques. Fractionation in this 'fingerprinting' technique (see Fig. 5.14) in the first dimension is by high-voltage ionophoresis on cellulose acetate at pH 3.5 and in the second dimension by ionophoresis with 7 per cent formic acid on DEAE-paper, and ion-exchange paper of opposite charge to that of the nucleotide. Alternarively thin-layer chromatography on DEAE-cellulose (diethylamino-ethylcellulose) may be employed, and in the method known as 'homo-chromatography' a mixture of non-radioactive nucleotides in solution is used as displacing ions. The probable composition of an oligonucleotide in the 'fingerprint' can often be predicted from its relative position.

Amongst the enzymes which have been used to generate partial or complete digestion products are T_1 ribonuclease (T_1 RNase) (from *Aspergillus oryzae*) which specifically breaks the internucleotide bonds between guanosine 3′-monophosphate and the 5′-hydroxyl groups of

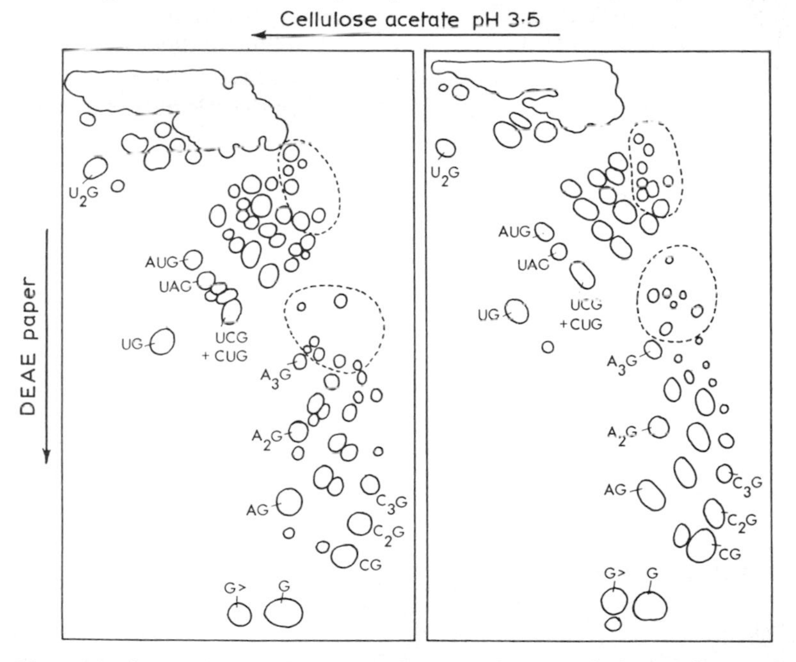

Fig. 5.14 Diagram illustrating autoradiogram of 'fingerprinting' of digests of ribosomal RNA labelled with ^{32}P from HeLa cells. Left, 18S RNA. Right, 28S RNA. First run, right to left, electrophoresis on cellulose acetate at pH 3.5. Second run, down, electrophoresis on DEAE paper with 7% formic acid. A few of the oligonucleotides shown have been named on the diagram. There are several qualitative or quantitative differences between the 18S and 28S RNA patterns. Regions showing conspicuous differences are indicated by circles. (By courtesy of M. Salim and Dr B.E.H. Maden.)

adjacent nucleotides; pancreatic RNase (from bovine pancreas) which breaks between 3′-phosphoryl primidine nucleotides and adjacent nucleotides, and Phy 1-RNase from *Physarum polycephalum* which pref-shows a specificity for the internucleotide bonds adjacent to purine nucleotides, and Phy 1-Rnase from *Physarum polycephalum* which preferentially digests at guanine, adenine and uracil residues. For a more detailed coverage of these earlier approaches the reader is referred to [168].

Several methods have been developed recently for rapid RNA sequencing. The general principle [172, 173] of these newer methods depends on the generation of a set of radioactively labelled oligonucleotides terminated with a unique type of RNA nucleotide, followed by their electrophoretic separation and autoradiographic detection on polyacrylamide gels as described for the 'rapid read-off' gel methods already described in Chapter 3 for DNA sequencing.

For example partial specific endonuclease hydrolysis of 5′-^{32}P-labelled RNA can be used [173]. The required end labelling at the 5′-OH groups, with a ^{32}P-ester group is achieved using polynucleotide kinase (induced in *E. coli* by T4 bacteriophage) and γ-^{32}P-ATP as the ^{32}P-phosphate donor

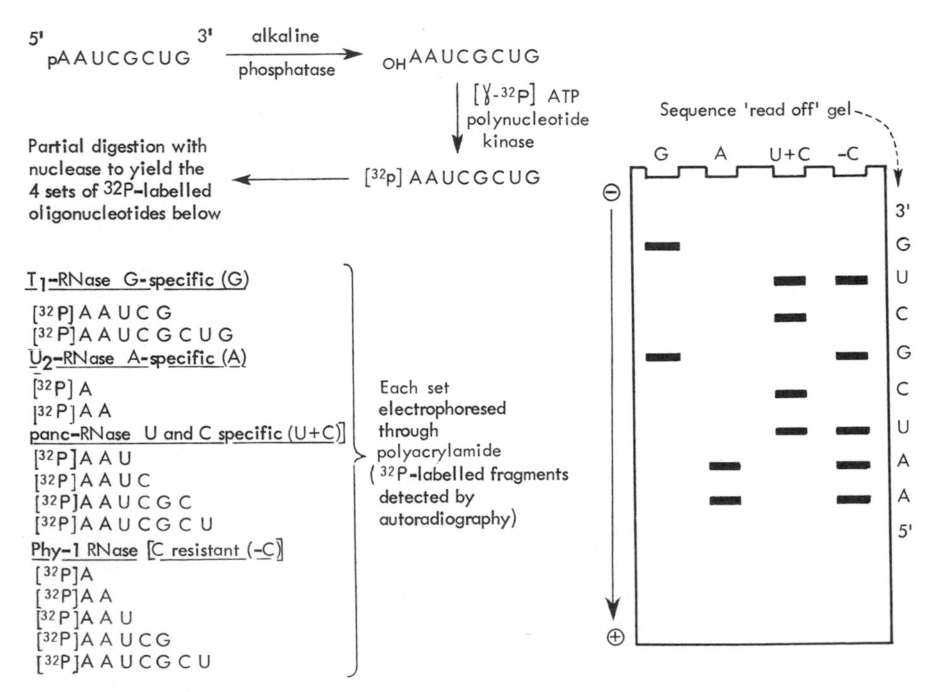

Fig. 5.15 A schematic diagram illustrating the principle of a type of rapid RNA sequencing technique.

(see also Chapter 6). Partial digestion of this 5′-end-labelled RNA is then carried out separately with T_1 RNase (to yield a series of ^{32}P-labelled fragments ending with a guanine residue), with U_2-RNase (which will yield a series of ^{32}P-labelled fragments ending with adenine residues *if conditions of partial digestion are used*), with pancreatic RNase (to yield a series of fragments ending with cytosine *or* uracil residues) and with Phy 1-RNase which will cleave at residues *other than cytosine*. Each set of 5′-^{32}P-labelled oligonucleotide fragments is then electrophoresed separately through polyacrylamide gels under denaturing conditions and detected subsequently by autoradiography. Like the approaches now used for DNA sequencing it is possible in principle (see Fig. 5.15) to 'read-off' the RNA nucleotide sequence from the autoradiograph bearing in mind the different specificities of the nucleases used and the ability to separate oligonucleotides of differing length by polyacrylamide-gel electrophoresis.

Modification of cytosine residues by methoxyamine–sodium bisulphite mixtures leads to the unfolding of RNA secondary structure and helps to generate a more uniform set of fragments after partial enzyme hydrolysis [179]. Moreover the position of these modified cytosines in an RNA sequence can be determined by virtue of their resistance to enzymic cleavage [174].

A recent improved direct RNA sequencing method involves chemical hydrolysis with hot formamide to induce one cleavage per molecule. The 5′-OH ends of the fragments produced are then labelled with γ-^{32}P-ATP and polynucleotide kinase and the fragments separated in the first dimension on a polyacrylamide gel. The fragments are then blotted from the gel onto a DEAE-cellulose thin layer, and there digested with RNase T_1, RNase T_2 and pancreatic RNase which liberates the 5′-terminal nucleotide as a ^{32}P-labelled 5′,3′-nucleoside diphosphate which can then be identified by electrophoresis in the second dimension [184].

Other methods of generating sets of radioactively labelled fragments involve the specific termination with nucleoside triphosphate analogues (3′-deoxyuridine 5′-triphosphate [175] or 2′-3′-dideoxy ribonucleotides [176]) in complementary RNA [175] synthesis catalysed by Qβ replicase [175] or complementary DNA synthesis using reverse transcriptase [177] using the RNA to be sequenced as template [175, 176].

REFERENCES

1 Dounce, A.L. (1943), *J. Biol. Chem.*, **147**, 685; (1943), *ibid.*, **151**, 221, 235.
2 Dounce, A.L. (1955), in *The Nucleic Acids*, (ed. E. Chargaff and J.N. Davidson), Academic Press, New York.

3 Mirsky, A.E. and Pollister, A.W. (1943), *J. Gen. Physiol.*, **30,** 117.
4 Allfrey, V., Stern, H., Mirsky, A.E. and Saetren, H. (1952), *J. Gen. Physiol.*, **35,** 529.
5 Behrens, M. (1938), *Aberhalden's Handbuch der biologische Arbeitsmethoden,* Sect. 5, Part 10, p. 1363.
6 Dounce, A.L., Tishkoff, G.H., Barnett, S.R. and Freer, R.M. (1950), *J. Gen. Physiol.,* **33,** 629.
7 Kay, E.R.M., Smellie, R.M.S., Humphrey, G.F. and Davidson, J.N. (1956), *Biochem. J.,* **62,** 160.
8 Kirsch, W.M., Leitner, J.W., Gainey, M., Schutz, D., Lasher, R. and Nakane, P. (1970), *Science,* **168,** 1592
9 Busch, H. (1967), *Methods Enzymol.,* **12A,** 421, (ed. L. Grossman and K. Moldave), Academic Press, New York.
10 Wang, T.Y. (1967), *Methods Enzymol.* **12A,** 417, (ed. L. Grossman and K. Moldave), Academic Press, New York.
11 Schneider, R.M. and Peterman, R. (1950), *Cancer Res.,* **10,** 751.
12 Philpot, J. St. L. and Stanier, J.E. (1956), *Biochem. J.,* **63,** 214.
13 Widnell, C.C. and Tata, J.R. (1964), *Biochem. J.,* **92,** 313.
14 Hymer, W.C. and Kuff, E.L. (1964), *J. Histochem. Cytochem.,* **12,** 359.
15 Fisher, H.W. and Harris, H. (1962), *Proc. R. Soc., B,* **156,** 521.
16 Penman, S. (1966), *J. Mol. Biol.,* **17,** 117.
17 Fischer, W.D. and Cline, G.B. (1963), *Biochim, Biophys, Acta,* **68,** 640.
18 Allfrey, V.G., Littau, V.C. and Mirsky, A.E. (1964), *J. Cell Biol.,* **21,** 213.
19 Methods of Separation of Subcellular Structural Components (1963), *Biochem. Soc. Symp.* No. 23.
20 Allfrey, V. (1959), *The Cell,* Vol. 1, p. 193, (ed. A.E. Mirsky and J. Brachet), Academic Press, New York.
21 Siebert, G. and Smellie, R.M.S. (1957), *Int. Rev. Cytol.,* **6,** 383.
22 Siebert, G. (1967), in *Methods in Cancer Research,* (ed. H. Busch), Academic Press, New York, **2,** 287.
23 Siebert, G. (1967), in *Methods in Cancer Research,* (ed. H. Busch, Academic Press, New York, **3,** 47.
24 Roodyn, D.B. (1972), in *Subcellular Components,* p. 15 (ed. G.D. Birnie), Butterworths, London.
25 Davison, P.F. and Mercer, E.H. (1956), *Exp. Cell Res.,* **11,** 237.
26 Bouteille, M., Laval, M. and Dupuy-Coin, A.M. (1974), in *The Cell Nucleus,* Vol. 1, p. 3 (ed. H. Busch), Academic Press, New York.
27 Dupraw, E.J. (1970), *DNA and Chromosomes.* Holt, Rinehart and Winston, New York.
28 Smetana, K. and Busch, H. (1974), in *The Cell Nucleus,* Vol. 1, p. 73 (ed. H. Busch), Academic Press, New York.
29 Muramatsu, M. and Busch, H. (1967), in *Methods in Cancer Research,* (ed. H. Busch), **2,** 303.
30 Busch, H., Hodneth, J.L., Morris, H.P., Neogy, R. and Unuma, T. (1968), in *Methods in Cancer Research,* (ed. H. Busch), Academic Press, New York, **4,** 179.
31 Vincent, W.S. (1952), *Proc. Natl. Acad. Sci., USA,* **38,** 139.

32 Monty, K.M., Litt, M., Kay, E.R.M. and Dounce, A.L. (1956), *J. Biophys. Biochem. Cyto.*, **2,** 127.

33 Maggio, R., Siekevitz, P. and Palade, G.E. (1963), *J. Cell Biol.*, **18,** 293.

34 Busch, H. (1967) *Methods Enzymol.*, **12A,** 448.

35 Birnie, G.D. and Fox, S.M. (1969), *Subcellular Components: Preparation and Fractionation,* Butterworths, London.

36 G.D. Birnie (ed.) (1972), *Subcellular Components,* Butterworths, London.

37 Tata, J.R. (1972), in *Subcellular components* (ed. G.D. Birnie), Butterworths, London, p. 185

38 Bonanou-Tzedaki, S.A. And Arnstein, H.R.V. (1972), in *Subcellular Components* (ed. G. D. Birnie), Butterworths, London, p. 215.

39 Birnie, G.D., Fox, S.M. and Harvey, D.R. (1972), in *Subcellular Components* (ed. G. D. Birnie), Butterworths, London, p. 235.

40 Gillchriest, W.C. and Boch, R.M. (1958), *Microsomal Particles and Protein Synthesis* (ed. R. B. Roberts), p. 1.

41 Kurland, C.G. (1971), *Methods Enzymol.*, **20,** 379.

42 Takanami, M. (1967), *Methods Enzymol.*, **12A,** 491.

43 Ralph, R.K. and Bellamy, A.R. (1964), *Biochim. Biophys. Acta,* **87,** 9.

44 Kirby, K.S. (1964) *Progress in Nucleic Acid Research*, Vol. 3, p. 1 (eds J. N. Davidson and W. E. Cohn), Academic Press, New York.

45 Kirby, K.S. (1965), *Biochem. J.*, **96,** 266.

46 Kirby, K.S. (1968), *Methods Enymol.*, **12B,** 87.

47 Georgiev, G.P. (1967), *Prog. Nucleic Acid Res. Mol. Biol.*, **6,** 259.

48 Holley, R.W. (1966), *Sci. Amer.*, **214** (2), 30.

49 Holley, R.W., Apgar, J., Everett, G.A., Madison, J.T., Marquisee, M. Merrill, S.H., Penswick, J.R. and Zamir, A. (1965), *Science,* **147,** 1462.

50 Zachau, H.G., Dutting, D., Feldman, H., Melchers, F. and Karau, W. (1966), *Cold Spring Harbor Symp. Quant. Biol.*, **31,** 417.

51 Sprinzl, M. Greuter, F., Spelzhaus, A. and Gauss, D.H. (1980), *Nucleic Acids Research,* **8,** 1.

52 Tinoco, J., Uhlenbeck, O.C. and Levine, M.D. (1971), *Nature,* **230,** 362.

53 Kurland, C.G. (1960), *J. Mol. Biol.*, **2,** 83.

54 Aronson, A.I. (1962), *J. Mol. Biol.*, **5,** 453.

55 Aronson, A.I. (1963), *Biochim. Biophys. Acta,* **72,** 176.

56 Spirin, A.S. (1963), in *Progress in Nucleic Acid Research,* Vol. 1, p. 301 (ed J.N. Davidson and W.E. Cohn), Academic Press, New York.

57 Brownlee, G.G., Sanger, F. and Barrell, B.G. (1968), *J. Mol. Biol.*, **34,** 379.

58 Morrell, P. and Marmur, J. (1968), *Biochemistry,* **7,** 1141.

59 Knight, E.J.R. and Darnell, J.E. (1967), *J. Mol. Biol.*, **28,** 491.

60 Forget, B.G. and Weissman, S.M. (1969), *J. Biol. Chem.*, **244,** 3148.

61 Watson, J.D. and Ralph, R.K. (1967), *J. Mol. Biol.*, **26,** 451.

62 Ford, P.S. (1973), *Biochem. Soc. Symp.*, **37,** 69.

63 Paynes, P.I. and Dyer, T.A. (1971), *Biochem. J.*, **124,** 87.

64 Soave, C., Galante, E. and Torti, G. (1970), *Bull. Soc. Chim. Biol.*, **52,** 857.

65 Rubin, G.M. (1974), *Eur. J. Biochem.*, **41,** 197.

66 Pene, J.J., Knight, E., Jr. and Darnell, J.E. (1968), *J. Mol. Biol.*, **33,** 609.

67 McConkey, E.H. (1967), *Methods Enzymol.*, **12A,** 620.

68 Vinograd, J. and Hearst, J.E. (1962), *Progress in Chemistry of Organic Natural Products,* **20,** 372.
69 Loening, U.E. (1967), *Biochem. J.,* **102,** 251.
70 Grossbach, U. and Weinstein, I.B. (1968), *Anal. Biochem.,* **22** (2), 311.
71 Burdon, R.H. and Clason, A.E. (1969), *J. Mol. Biol.,* **39,** 113.
72 Caton, J.E. and Goldstein, G. (1971), *Anal. Biochem.,* **42,** 14.
73 Loening, U.E. (1968), *Chromatographic and Electrophoretic Techniques,* 2nd Edn (ed. I. Smith), Heinemann, London, Vol. 2, p. 437.
74 Richards, E.G. and Gratzer, W.B. (1968), *Chromatographic and Electrophoretic Techniques* (ed. I. Smith), Heinemann, London, Vol. 2, p. 419.
75 Richards, E.G. and Lecanidou, R. (1971), *Anal. Biochem.,* **41,** 43.
76 Richards, E.G. and Temple, C.J. (1971), *Nature,* **230,** 92.
77 De Wachter, R. and Fiers, W. (1971), *Methods Enzymol.,* **21,** 167.
78 Lanyon, W.G., Paul, J. and Williamson, R. (1968), *FEBS Lett.,* **1,** 279.
79 Stein, M. and Varrichio, F. (1974), *Anal. Biochem.,* **61,** 112.
80 Reddy, R. Sitz, T.O., Ro-Choi, T.S. and Busch, H. (1974), *Biochem. Biophys. Res. Commun.,* **56,** 1017.
81 Pinder, J.C., Staynor, D.Z. and Gratzer, W.B. (1974), *Biochemistry,* **13,** 5373.
82 Lehrach, H., Diamond, D., Wozney, J.M. and Boedtker, H. (1977), *Biochemistry,* **16,** 4743.
83 McMaster, G.K. and Carmichael, G.G. (1977), *Proc. Natl. Acad. Sci. USA,* **74,** 4835.
84 Goldthwait, D.A. (1959), *J. Biol. Chem.,* **234,** 3245.
85 McIndoe, W. and Munro, H.N. (1967), *Biochem. Biophys. Acta,* **134,** 458.
86 Ringborg, U., Egyhezi, E., Daneholt, B. and Lambert, B. (1968), *Nature,* **220,** 1037.
87 Floyd, R.W., Stone, M.P. and Joklik, W.F. (1974), *Anal. Biochem.,* **59,** 599.
88 Brownlee, G.G., Sanger, F. and Barrell, B.G. (1968), *J. Mol. Biol.,* **34,** 379.
89 Tinoco, J., Uhlenbeck, O.C. and Levine, M.D. (1971), *Nature,* **230,** 362.
90 Fox, G.E. and Woese, C.R. (1975), *Nature,* **256,** 505.
91 Erdmann, V.A. (1980), *Nucleic Acids Res.,* **8,** 31.
92 Carbon, P., Ehresmann, C., Ehresmann, B. and Ebel, J-P. (1979), *Eur. J. Biochem.,* **100,** 399.
93 Brosius, J., Dull, T.J. and Noller, H.F. (1980), *Proc. Natl. Acad. Sci. USA,* **77,** 201.
94 Kurland, C.G. (1972), *Annu. Rev. Biochem.,* **41,** 377.
95 Woese, C.R., Fox, G.E., Zablen, L., Uchida, T., Bonen, L., Pechman, K., Lewis, B.J. and Stahl, D. (1975), *Nature,* **254,** 83.
96 Shine, J. and Dalgarno, L. (1974), *Proc. Natl. Acad. Sci. USA,* **71,** 1342.
97 Shine, J. and Dalgarno, L. (1975), *Nature,* **254,** 34.
98 Shine, J. and Dalgarno, L. (1975), *Eur. J. Biochem.,* **57,** 221.
99 Maden, B.E.H., Khan, M.S.N., Hughes, D.G. and Goddard, J.P. (1977), *Biochem. Soc. Symp.,* **42,** 165.
100 Hagenbuchle, O., Santer, M., Steitz, J.A. and Mans, R.J. (1978), *Cell,* **13,** 551.

101 Shine, J., Hunt, J.A. and Dalgarno, L. (1974), *Biochem. J.*, **141**, 617.
102 Rich, A. (1963), *Sci. Amer.*, **209** (6), 44.
103 Rich, A. (1967), *Methods Enzymol.*, **12A**, 481.
104 Haschemeyer, A.E.V. and Gross, J. (1967), *Biochim. Biophys. Acta*, **145**, 76.
105 Wettstein, F.O., Staehelin, T. and Noll, H. (1963), *Nature*, **197**, 430.
106 Fiers, W., Contreras, R., Duerinck, F., Haegeman, G., Iserentant, D., Merregaert, J., Min Jou, W., Molemans, F., Raeymackers, A., Vanden Berghe, A., Vokkaert, G. and Ysebaert, J. (1976), *Nature*, **260**, 500.
107 Nakamura, K., Pirtle, R.M., Pirtle, I.M., Takeishi, K. and Inouye, M. (1980), *J. Biol. Chem.*, **255**, 210.
108 Taylor, J.M. (1979), *Annu. Rev. Biochem.*, **48**, 681.
109 Proudfoot, N. (1967), *J. Mol. Biol.*, **107**, 491.
110 Efstratiadis, A., Kafatos, F.C. and Maniatis, T. (1977), *Cell*, **10**, 571.
111 Breathnach, R., Benoist, C., O'Hare, K., Gannon, F. and Chambon, P. (1978), *Proc. Natl. Acad. Sci. USA*, **75**, 4853.
112 Brawerman, G. (1976), *Prog. Nucleic Acids Res. Mol. Biol.*, **17**, 118.
113 Adesnick, M. and Darnell, J.E. (1972), *J. Mol. Biol.*, **67**, 397.
114 Greenberg, J.R. and Perry, R.P. (1972), *J. Mol. Biol.*, **72**, 91.
115 Milcarek, C., Prince, R. and Penman, S. (1974), *J. Mol. Biol.*, **89**, 435.
116 Nemer, M., Graham, M. and Dubroff, L.M. (1974), *J. Mol. Biol.*, **89**, 735.
117 Burdon, R.H., Shenkin, A., Douglas, J.T. and Smillie, E.J. (1976), *Biochim. Biophys. Acta*, **474**, 254–267.
118 Lodish, H.G., Jacobson, A., Firtel, R., Alton, T. and Tuchmon, J. (1974), *Proc. Natl. Acad. Sci. USA*, **71**, 5103.
119 Greenberg, J.R. (1976), *Biochemistry*, **15**, 3516.
120 Gray, R.F. and Cashmore, A.R. (1976), *J. Mol. Biol.*, **108**, 5.
121 Milcarek, C. (1979), *Eur. J. Biochem.*, **102**, 467.
122 Perry, R.P. and Kelley, D.E. (1974), *Cell*, **1**, 37.
123 Adams, J.M. and Cory, S. (1975), *Nature*, **255**, 28.
124 Shatkin, A. (1976), *Cell*, **9**, 645.
125 Frisby, D., Eaton, M. and Fellner, P. (1976), *Nucleic Acids Res.*, **3**, 2771.
126 Wimmer, E., Chang, A.Y., Clark, J.M. and Reichmann, M.E. (1968), *J. Mol. Biol.*, **38**, 59.
127 Hewlett, M.J., Rose, J.K. and Baltimore, D. (1976), *Proc. Natl. Acad. Sci. USA*, **73**, 327.
128 Rothberg, P.G., Harris, T.J.R., Nomoto, A. and Wimmer, E. (1978), *Proc. Natl. Acad. Sci. USA*, **75**, 4868.
129 Favre, A., Morel, C. and Scherrer, K. (1975), *Eur. J. Biochem.*, **57**, 147.
130 Holder, J.E. and Lingrel, J.B. (1975), *Biochemistry*, **14**, 4206.
131 Weissmann, C., Billeter, M.A., Goodman, H.M., Hindley, J. and Weber, H. (1973), *Annu. Rev. Biochem.*, **42**, 303.
132 Perry, R. and Kelley, D. (1968), *J. Mol. Biol.*, **35**, 37.
133 Morel, C., Gander, E., Herzberg, M., Dubochet, J. and Scherrer, K. (1973), *Eur. J. Biochem.*, **38**, 443.
134 Gander, E., Stewart, A., Morel, C. and Scherrer, K. (1973), *Eur. J. Biochem.*, **36**, 455.

135 Jeffrey, W. (1977), *J. Biol. Chem.*, **252,** 3525.
136 Vincent, A., Civelli, O., Buri, J. and Scherrer, K. (1977), *FEBS Lett.*, **77,** 281.
137 Civelli, O., Vincent, A., Buri, J. and Scherrer, K. (1976), *FEBS Lett.*, **72,** 71.
138 Borst, P. (1977), *Trends in Biochemical Sciences,* **2,** 31.
139 Borst, P. and Grivell, L.A. (1978), *Cell,* **15,** 705.
140 Kroon, A.M. and Saccone, C. (1974), *The biogenesis of mitochondria,* Academic Press, New York.
141 Tzagoloff, A., Macino, G. and Sebald, W. (1979), *Annu. Rev. Biochem.*, **48,** 419.
142 Ciferri, O. (1978), *Trends in Biochemical Sciences,* **3,** 256.
143 Malnoe, P., Rochaix, J.D., Chua, N.H. and Spahr, P.F. (1979), *J. Mol. Biol.,* **133,** 417.
144 Loening, U.E. (1968), *J. Mol. Biol.,* **38,** 355.
145 Schwarz, Z. and Kossel, H. (1980), *Nature,* **283,** 739.
146 Davidson, J.N. and Waymouth, C. (1946), *J. Physiol.,* **105,** 191.
147 Vincent, W.S. (1955), *Int. Rev. Cytol.,* **4,** 269.
148 Stich, H. (1956), *Experientia,* **12,** 7.
149 Sirlin, J.L. (1962), *Prog. Biophys. Biophys. Chem.,* **12,** 27.
150 Sirlin, J.L. and Jacob, J. (1962), *Nature,* **195,** 114.
151 Chipchase, M.I.H. and Birnstiel, M.L. (1963), *Proc. Natl. Acad. Sci, USA,* **50,** 1101.
152 Busch, H., Byvoet, F. and Smetana, K. (1963), *Cancer Res.,* **23,** 313.
153 Miller, Jr., O.L. and Beatty, B.R. (1969), in *Handbook of Molecular Cytology* (ed. A. Lima-de-Faria), p. 605.
154 Busch, H. and Smetana, K. (1970), *The Nucleolus,* Academic Press, New York.
155 Perry, R.P., Hall, A. and Errera, M. (1961), *Biochim. Biophys. Acta,* **49,** 47.
156 Brown, D.D. and Gurdon, J.P. (1964), *Proc. Natl. Acad. Sci. USA,* **51,** 139.
157 Darnell, J.E. (1968), *Bact. Rev.,* **32,** 262.
158 Weinberg, R.A. (1973), *Annu. Rev. Biochem.,* **42,** 329.
159 Perry, R.P. (1976), *Annu. Rev. Biochem.,* **45,** 605.
160 Federoff, N., Wellauer, P.K. and Wall, R. (1977), *Cell,* **10,** 597.
161 Weinberg, R.A. and Penman, S. (1968), *J. Mol. Biol.,* **38,** 289.
162 Busch, H., Ro-Choi, T.S., Prestayko, A.W., Shibata, H. Croke, S.T., El-Khatib, S.M., Choi, Y.C. and Mauritzen, C.M. (1971), *Perspectives Biol. Med.,* **15,** 117.
163 Hellung-Larsen, P., Tyrsted, G. and Frederiksen, S. (1973), *Expt. Cell. Res.,* **80,** 393.
164 Ro-Choi, T.S., Reddy, R., Henning, D., Takano, T., Taylor, C.W. and Busch, H. (1972), *J. Biol. Chem.,* **247,** 3205.
165 Frederiksen, S. and Hellung-Larsen, P. (1975), *FEBS Lett.,* **58,** 374.
166 Lerner, M.R., Boyle, J.A., Mount, S.M., Wolin, S.L. and Steitz, J.A. (1980), *Nature,* **283,** 220.
167 Samarina, O.P., Lukanidin, E.M., Molnar, J. and Georgiev, G.P. (1968), *J. Mol. Biol.,* **33,** 251.

168 Brownlee, G.G. (1972), *Determination of sequences in RNA*, Elsevier/North Holland, Amsterdam, New York.
169 Holley, R.W., Apgar, J., Everett, G.A., Madison, J.T., Marquisee, M., Merrill, S.H., Penswick, J.R. and Zamir, A. (1965), *Science*, **147**, 1462.
170 Sanger, F. (167), *Methods Enzymol.*, **12**, 361.
171 Sanger, F. and Brownlee, G.G. (1970), *Biochem. Soc. Symp.*, **30**, 183.
172 Donis-Keller, H., Maxam, A.M. and Gilbert, W. (1977), *Nucleic Acids Res.*, **4**, 2557.
173 Simoncsits, A., Brownlee, G.G., Brown, R.S., Rubin, J.R. and Guilley, H. (1977), *Nature*, **269**, 833.
174 Mazo, A.M., Mashkova, T.D., Audonina, T.A., Ambartsumyan, N.S. and Kisselev, L.L. (1979), *Nucleic Acids Res.*, **7**, 2469.
175 Kramer, F.R. and Mills, D.R. (1978), *Proc. Natl. Acad. Sci. USA*, **75**, 5334.
176 Sanger, F., Nicholson, S. and Coulson, A.R. (1977), *Proc. Natl Acad. Sci. USA*, **74**, 5463.
177 Hamlyn, P.H., Brownlee, G.G., Chang, C.C., Gait, J.M. and Milstein, C. (1978), *Cell*, **15**, 1067.
178 Verma, I.M. (1977), *Biochim. Biophys. Acta*, **473**, 1.
179 Hell, A., Young, B.D. and Birnie, G.D. (1976), *Biochim. Biophys. Acta*, **442**, 37.
180 Brosius, J., Palmer, M.L., Kennedy, P.T. and Noller, H.F. (1978), *Proc. Natl. Acad. Sci. USA*, **75**, 4801.
181 Mathews, M.B. and Korner, A. (1970), *Eur. J. Biochem.*, **17**, 328.
182 Marcus, K. and Dudock, B. (1974), *Nucleic Acids Res.*, **1**, 1385.
183 Pelham, H.R.B. and Jackson, R.S. (1976), *Eur. J. Biochem.*, **67**, 247.
184 Tanaka, Y., Dyer, T.A. and Brownlee, G.G. (1980), *Nucleic Acids Res.*, **8**, 1259.
185 Molgaard, H.U., Perucho, M. and Ruiz-Carrillo, A.)1980), *Nature*, **283**, 502.
186 Noller, H. (1980), in *Ribosomes: Structure, Function and Genetics*, p. 3 (eds G. Chambliss, G.R. Craven, J. Davies, L. Kahan, M. Nomura), Baltimore University Park Press.
187 Wollenzien, P., Hearst, J.E., Thammana, P. and Cantor, C.R. (1979), *J. Mol. Biol.*, **135**, 255.
188 Cantor, C.R., Wollenzien, P.L. and Hearst, J.E. (1980), *Nucleic Acids Res.* **8**, 1855.
189 Ro-Choi, T.S. and Busch, H. (1974), in *The Molecular Biology of Cancer* (ed. H. Busch), Academic Press, New York, p. 241.
190 Reddy, R., Ro-Choi, T.S., Henning, D. and Busch, H. (1974), *J. Biol. Chem.*, **249**, 6486.
191 Shibata, A., Ro-Choi, T.S., Reddy, R., Choi, Y.C., Henning, D. and Busch, H. (1975), *J. Biol. Chem.*, **250**, 3909.
192 Reddy, R., Henning, D. and Busch, H. (1979), *J. Biol. Chem.*, **254**, 11097.
193 Raj, N.B.K., Ro-Choi, T.S. and Busch, H. (1975), *Biochemistry*, **14**, 4380.
194 Flytzanis, C., Alonso, A., Lous, C., Kreig, L. and Sekeris, C.E. (1978), *FEBS Lett.*, **96**, 201.
195 Lerner, M.R. and Steitz, J.A. (1979), *Proc. Natl. Acad. Sci. USA*, **76**, 5495.

196 Howard, E.F. (1978), *Biochemistry,* **17,** 3229.
197 Jelinek, W. and Leirwand, L. (1978), *Cell,* **15,** 205.
198 Harada, F. and Kato, N. (1980), *Nucleic Acids Res.,* **8,** 1273.
199 Harada, F., Kato, N. and Hoshino, H. (1979), *Nucleic Acids Res.,* **7,** 909.
200 Kanehisa, T., Kitazune, Y., Ikata, K. and Tanaka, Y. (1977), *Biochim. Biophys. Acta,* **475,** 501.
201 Goldstein, L., Wise, G.E. and Beeson, M. (1973), *Expt. Cell Res.,* **76,** 281.
202 Goldstein, L. (1976), *Nature,* **261,** 519.
203 Rao, M.S., Blackstone, M. and Busch, H. (1977), *Biochemistry,* **16,** 2756.
204 Woese, C., Magrum, L.J., Gupta, R., Siegel, R.B., Stahl, D.A., Kop, J., Crawford, N. Brosius, J., Gutell, R., Hogan, J.J. and Noller, H.F. (1980), *Nucleic Acids Res.,* **8,** 2275.
205 Glotz, C. and Brimacombe, R. (1980), *Nucleic Acids Res.,* **8,** 2377.

Nucleases and related enzymes

Enzymes which catalyse the breakdown of nucleic acids by hydrolysis of phosphodiester bonds have been found in almost all biological systems [1–6]. Some, the *ribonucleases,* are quite specific for RNA, others, the *deoxyribonucleases,* act only on DNA, while a third group of non-specific *nucleases* is active against either nucleic acid.

The *phosphorylases,* polynucleotide phosphorylase and pyrophosphorylase, are also capable of depolymerizing RNA, but their degradative role *in vivo* is uncertain and they are dealt with elsewhere in this volume (Chapter 11).

The *phosphomonoesterases* act on polynucleotides or oligonucleotides with a terminal phosphate group or on a mononucleotide to liberate inorganic phosphate. Their substrates will often be products of nuclease action.

In all nucleolytic enzymes tested, the P–O bond is cleaved, as shown by ^{18}O incorporation [7].

6.1 CLASSIFICATION OF NUCLEASES

Classification schemes for the nucleases have been discussed by Laskowski [1] and by Barnard [6]. Three main features of nuclease action can be used as a basis for classification.

The first of these is *substrate specificity,* i.e. action on RNA, DNA, or both, as discussed above. The second is *mode of attack*; polynucleotides can be attacked at points within the polymer chain *endolytically* or stepwise from one end of the chain *exolytically.* Thus we may have *endonucleases* which produce oligonucleotides and cause rapid changes in physical properties (e.g. in viscosity of DNA), and *exonucleases* which produce mononucleotides but with rather less drastic effects on nucleic acid physical properties. A few enzymes appear to act as both endo- and exo-nuclease, e.g. *micrococcal nuclease* [8]. The third feature is *mode of phosphodiester bond cleavage.* Most biological polymers can, like proteins and carbohydrates, be split in only one way; polynucleotides can be cleaved in two ways to give products bearing (i) *5′-phosphoryl end groups* by hydrolysis of the bond between the 3′-OH and the phosphate group or (ii) *3′-phosphoryl end groups* by hydrolysis of the bond between the 5′-OH and the phosphate group.

Additional criteria may be used to define further the action of a nuclease. These include *specificity towards secondary structure of substrate, direction of attack by exonuclease* (3′→5′ or 5′→3′), and *preferential endonucleolytic bond cleavage,* e.g. GpX→Gp, by ribonuclease T_1. However, in only a few cases (e.g. the restriction endonucleases) are base specificities absolute, and relative differences in reaction rates with different bases are more common.

Experimental details for the preparation and handling of several of these enzymes are to be found in the handbooks edited by Cantoni and Davies [3] and by Grossman and Moldave [9].

6.2 RIBONUCLEASES (RNases)

6.2.1. Endonucleases forming 3′-phosphate groups

(a) *Pancreatic ribonuclease* (EC 3.1.4.22)* (for reviews see [1, 10–12]). In 1920 Jones [13] described a heat-stable enzyme present in the pancreas which was capable of digesting yeast RNA. The enzyme was purified by Dubos and Thompson [15] and was crystallized in 1940 by Kunitz [16] who named it *ribonuclease.*

Crystalline pancreatic RNase prepared by the method of Kunitz tends to be contaminated with traces of proteolytic enzymes which have on occasion given rise to misleading results. The crystallization of pancreatic RNase absolutely free from proteolytic contaminants has been described by McDonald [17].

Pancreatic RNase is a very small protein, mol. wt 13 700, is stable over a wide pH range, and is remarkably resistant to heat in slightly acid solution, although it is readily inactivated by alkali. It has no action on DNA and is strongly antigenic. Its maximum activity is in the range pH 7.0–8.2, with the optimum at pH 7.7. Its optimum temperature is 65°C.

As the result of the work of Moore, Stein, and their collaborators [18, 19], the sequence of amino acids in the pancreatic RNase molecule has

* These index numbers refer to the classification system in the Report of the Enzyme Commission [14].

been fully worked out and the active site and mechanism of action determined [20–23]. Recently, the complete structure of the protein was obtained by X-ray crystallography [24, 25] and the total chemical synthesis of pancreatic RNase A has been achieved [26–28].

Pancreatic RNase is a highly specific endonuclease which splits the bond between the phosphate residue at C-3′ in a *pyrimidine* nucleotide to C-5′ in the next nucleotide in sequence. The basic feature of its action is an intramolecular attack on the phosphodiester bond using the 2′-OH group to form an obligatory 2′:3′-cyclic phosphate intermediate which is then hydrolysed by the enzyme to give pyrimidine 3′-phosphates either as free nucleotides or as terminal nucleotide residue in an oligonucleotide (Fig. 6.1). The products of short periods of pancreatic RNase action on RNA are the cyclic 2′:3′-phosphates of cytidine and uridine together with oligonucleotides terminating in a pyrimidine nucleotide carrying a cyclic phosphate group [29–31]. Pancreatic RNase and other RNases whose mode of action is similar have been classified as *cyclizing* by the

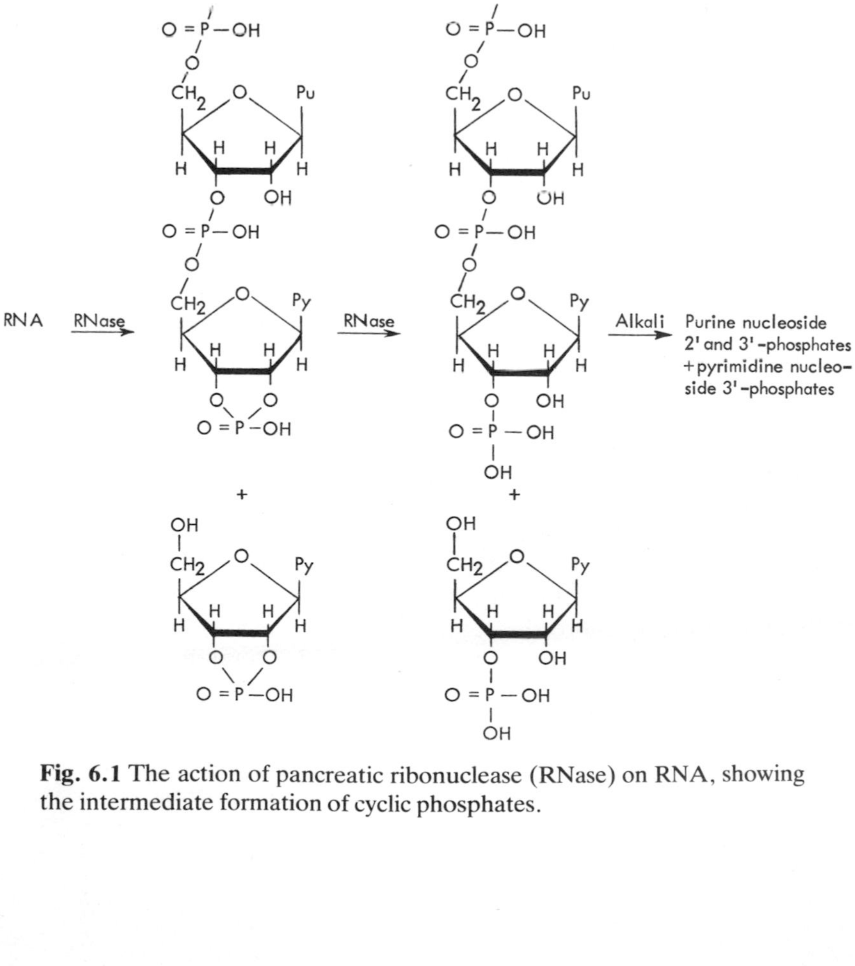

RNA → RNase → RNase → Alkali → Purine nucleoside 2′ and 3′-phosphates + pyrimidine nucleoside 3′-phosphates

Fig. 6.1 The action of pancreatic ribonuclease (RNase) on RNA, showing the intermediate formation of cyclic phosphates.

Standing Committee on Enzymes of the International Union of Biochemistry [14]. The cyclizing RNases have the possibility of forming both the 2′- and 3′-monoester by hydrolysis of the cyclic phosphate (Fig. 6.1); most, if not all, form the 3′-ester exclusively.

Subsequent treatment of oligonucleotides resulting from pancreatic ribonuclease digestion with alkali yields purine 2′- and 3′-nucleotide by cyclization and fission as described earlier, together with the 3′-isomers of the terminal nucleotides. These pyrimidine nucleotides are, of course, stable in alkaline solution; in acid they quickly yield an equilibrium mixture of the 2′- and 3′-phosphates.

Thus, pancreatic RNase may be regarded as a highly specific phosphodiesterase which will hydrolyse only secondary phosphate esters of pyrimidine nucleoside 3′-phosphates. It will therefore also hydrolyse the cyclic 2′:3′-secondary phosphates of the pyrimidine nucleosides.

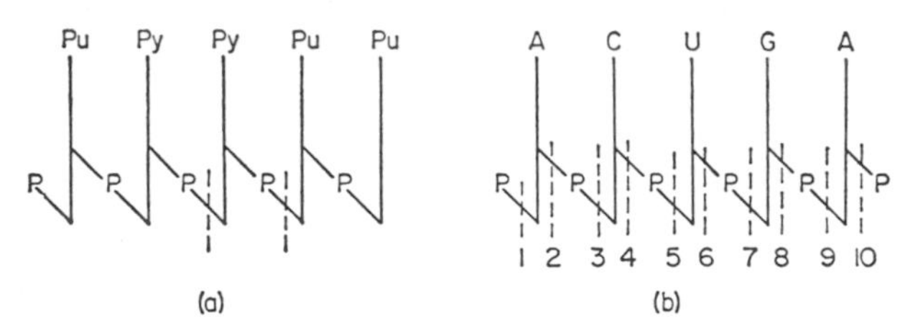

Fig. 6.2 (a) The pentanucleotide containing 3 purine and 2 pyrimidine nucleotide units is split by ribonuclease at the broken lines. (b) A pentanucleotide containing 2 adenine nucleotide residues and one residue each of cytosine, uracil, and guanine nucleotides with monoesterified phosphate residues at each end is split by pancreatic ribonuclease at positions 5 and 7, by ribonuclease T_1 at position 9, by 5′-monoesterase at 1, by 3′-monesterase at 10, by venom diesterase at 2, 4, 6, and 8, and by spleen diesterase at 3, 5, 7, and 9.

The action of pancreatic RNase may be illustrated as follows. The pentanucleotide shown in Fig. 6.2(a), in which Pu and Py represent purine and pyrimidine residues respectively, will be hydrolysed at the points shown by the broken lines, while the ribopolynucleotide chain shown in Fig. 6.2(b), which may also be expressed as pApCpUpGpAp, will be broken at positions 5 and 7 to yield pApCp + Up + GpAp.

To take a slightly more elaborate case, the polynucleotide A-C-C-C-C-A-G-G-G-U-U-U-A-G-U-Cp would be split by RNase thus:

A-C-/C-/C-/C-/-A-G-G-G-U-/U-/U-/A-G-U-/Cp

to yield

A-Cp + A-G-G-G-Up + A-G-Up + 4Cp + 2Up

i.e. 6 pyrimidine nucleotides, a dinucleotide, a trinucleotide, and pentanucleotide each with a terminal pyrimidine nucleotide residue linked to C-3′ of the preceding purine nucleotide but composed otherwise of purine nucleotide residues.

Pancreatic RNase also digests certain of the polyribonucleotides produced by the action of the polynucleotide phosphorylase described in Chapter 11. Thus poly(A) and poly(I) are not split by the enzyme whereas poly(C) and poly(U) yield the 3′-mononucleotides. However, the specificity of pancreatic RNase for pyrimidines is not absolute, since Ap diester bonds in a polynucleotide are also attacked, albeit considerably less readily than Up diester bonds [32].

The action of the enzyme may be demonstrated by making use of the fact that a solution of uranyl acetate in dilute trichloroacetic acid completely precipitates RNA but not its split products [33]. Pancreatic RNase renders about half the phosphorus of the nucleic acid non-precipitable by the uranyl reagent. Enzyme activity may also be determined spectrophotometrically [34–36], manometrically [37], or by the action on RNA labelled with ^{32}P [38]. The methods available have been reviewed [23].

(b) *Ribonuclease T_1* (EC 3.1.4.8) [39] has been the subject of intensive study, and our knowledge of its chemistry approaches that of pancreatic RNase. It is obtained from *Aspergillus oryzae* and specifically hydrolyses the internucleotide bonds of RNA between 3′-GMP and the 5′-OH groups of adjacent nucleotides (Fig. 6.2). Ribonuclease T_1 is a small heat-stable and acid-stable endonuclease. A second enzyme from the same source, ribonuclease T_2 preferentially attacks Ap residues and will digest tRNA almost totally to 3′-monophosphates [40]. Enzymes similar to ribonuclease T_1 are common in fungi and bacteria, e.g. ribonuclease U_1 from *Ustilago sphaerogena* [41] and ribonuclease N_1 from *Neurospora crassa* [237].

One important outcome of the purification and characterization of RNA endonucleases with particular specificities has been their use in the production of defined fragments of RNA. This has led to the complete sequence analysis of several RNA species. Among the RNases employed have been pancreatic and T_1 although T_2, U_2 (from *U. sphaerogena)* and Phy 1 (from *Physarum polycephalum*) are also used. This topic is illustrated in Chapter 5.

Many tissues and organisms contain RNA endonucleases which produce 3′-monophosphates, although most of these have not been so closely studied as the two enzymes just described. Some of the more important of these enzymes deserve brief mention.

(c) *Rat liver* contains several ribonucleases [42–44]. Two are endonucleases forming 3′-monophosphates. One has a pH optimum at about 6.0 and hydrolyses all phosphodiester bonds in RNA with the production of nucleoside 2′:3′-cyclic phosphates. The other has a pH optimum at about 8.0. It hydrolyses phosphodiester bonds only between adjacent pyrimidine nucleotides leaving purine-rich oligonucleotide tracts.

(d) *Mouse L cell nucleoli* [45] and probably also the nucleoli from *Novikoff hepatoma* [46] and *HeLa cells* [47] contain an endonucleolytic activity for single-stranded RNA with possible relevance to ribosomal precursor RNA processing. It is inhibited by Mg^{2+} and has a marked preference for cytidylate residues.

(e) *Nuclei* from rat mammary glands contain a calcium-activated pyrimidine specific endonuclease yielding oligonucleotide 3′-phosphates [247], whilst nuclei from bovine brain have an activity with an acidic optimum again yielding oligonucleotide 3′-phosphates [248].

(f) *Ribonuclease C* from human plasma yields oligonucleotide 2′:3′-cyclic phosphates and shows a high preference for phosphodiester linkages adjacent to cytidylic residues [264].

(g) *Ribonuclease NU from human KB cells*. This enzyme exhibits a high degree of specificity in that it can cleave the precursor to tyrosine tRNA in two places and the single-stranded RNA of phage Φ80 in four specific places [48].

(h) *Ribonuclease I* from *E. coli* normally occurs in the free state but on conversion of the cells into spheroplasts it is released into the medium. In broken cell preparations it is found in association with 30S ribosomal subunits [49]. Its intracellular role is unclear since strains lacking this enzyme are perfectly viable [50].

(i) *Ribonuclease III* from *E. coli* is also found in ribosome fractions [51–54]. It attacks only double-stranded RNA and requires Mg^{2+} or Mn^{2+} and K^+, Na^+, or NH_4^+. Considerable biochemical and genetical data implicate this nuclease in the post-transcriptional processing of ribosomal RNA precursors and of messenger RNA precursors (see Chapter 11). Analogous enzyme activities are found in animal cells [55–57].

6.2.2 Endonucleases which form 5′-monophosphates

The mechanism by which these nucleases cleave the phosphodiester bond is noncyclizing, i.e. a direct attack of water on a 3′,5′-phosphodiester is catalysed and thus a 2′-OH group is not required. For this reason, many 5′-monophosphate-forming endonucleases will attack RNA and DNA.

(a) *Ribonuclease IV* from *E. coli* [53, 58, 59]. This appears to be fairly specific in that it will cleave phage R17 RNA into a 15S fragment carrying the 5′ terminus of the original molecule and a 21S fragment lacking the 5′-end.

(b) *Ribonuclease P* of *E. coli*[60, 61]. This highly specific endonuclease is found in association with ribosomes and is responsible for the removal, as a single fragment, of the extra nucleotides from the 5′-ends of tRNA precursor molecules as shown in Chapter 11 (Fig. 11.4). Mutants deficient in this enzyme accumulate transfer RNA precursor molecules [75]. Its ionic requirements are very similar to those of ribonuclease III mentioned above. A similar activity can be detected in mammalian cells [62, 63].

(c) *Ribonuclease H.* This endonucleolytic activity was originally discovered in calf thymus tissue [64, 65] and specifically degrades the RNA strand of DNA–RNA hybrid to acid-soluble products with 5′-monophosphate ends. It does not degrade single-stranded RNA or DNA, double-stranded RNA or DNA, or the DNA strand of the DNA–RNA duplex. So far it has been purified from calf thymus tissue [71, 72], rat liver [66], chick embryo cells [67], KB cells [67], *Ustilago* [68], and *E. coli* [69, 70]. Cellular ribonuclease H has no associated DNA polymerase activity, and the enzyme from eukaryotic cells has a molecular weight of 70–90 000 [67, 72].

(d) *Rat liver alkaline ribonuclease I.* This enzyme cleaves RNA non-specifically to give products with a 5′-monophosphate terminus. The activity requires Mg^{2+} and operates maximally at pH 7.5 [49]. Similar activities have been reported in nuclei from pig heart, mouse ascites cells [78, 246], and HeLa cell nucleoli [265].

6.2.3 RNA exonucleases

(a) *Ribonuclease II* from *E. coli.* RNase II is only loosely bound to the ribosomes and acts on single-stranded RNAs as an exonuclease in the 3′→5′ direction. It requires both K^+ and Mg^{2+} ions [73, 74, 76, 77], and the products of its action are 5′-mononucleotides and a residual oligonucleotide fraction. Regarding a possible intracellular role, it may be involved in 'trimming' of tRNA, mRNA, and rRNA precursors after endonucleolytic cleavages by RNase P or RNase III [75].

(b) 3′-OH *specific exonucleases of mammalian nuclei.* Exonucleases specific for 3′-OH ends of single-stranded RNAs have been detected in ascites cell nuclei [78], rat liver nuclei [78], mouse L-cell nuclei [79] and HeLa cell nucleoplasm [266]. They all yield 5′-mononucleotides.

(c) *Oligoribonuclease* from *E. coli.* This enzyme activity shows a marked preference for oligoribonucleotides. In fact the reaction rate is inversely proportional to the chain length of the substrate. For full activity Mn^{2+} is required and the exonucleolytic cleavage is in the 3′→5′ direction yielding 5′-mononucleotides [80, 81]. Its cellular role may be to complete the digestion of oligoribonucleotides for example resistant to RNase II as described above.

(d) *Ribonuclease H of RNA tumour virus particles.* RNase H also turns out to be a ubiquitous activity of oncornavirions [82]. However, at least in the case of avian myeloblastosis virions [69], the nuclease activity appears to be an intregal component of the virion RNA-dependent DNA polymerase (reverse transcriptase) [83]. Additionally the polymerase-associated RNase H is unlike the previously mentioned RNase H species

Table 6.1 Some ribonucleases of *E. coli* (for references see text).

	Type of activity	Substrate	Ionic requirement	Cleavage product(s)
RNase I	endo-(2′-, 3′-nucleotides as intermediates)	Single-stranded RNAs	—	3′-Mononucleotides
RNase II	exo (3′→5′ direction)	Single-stranded RNAs	K^+, Mg^{2+}	5′-Mononucleotides plus some resistant oligonucleotides
RNase III	endo-	Double-stranded RNAs	K^+, Mg^{2+}	3′-Phosphorylated oligonucleotides (10–25 units)
RNase IV	endo- (specific position)	R17 RNA		Two specific fragments (see text)
RNase P	endo- (specific) position)	tRNA precursor	Mg^{2+}, Mn^{2+}, K^+, NH_4^+	tRNA plus fragment (see text)
RNase H	endo-	RNA strand of DNA–RNA hybrid	K^+, Mg^{2+}	5′-Phosphorylated oligonucleotides
OligoRNase	exo-(3′→5′ direction)	Short oligoribonucleotides	Mn^{2+}	5′-Mononucleotides
Polynucleotide phosphorylase	exo-(3′→5′ direction)	Single-stranded RNAs	Mg^{2+}	5′-Nucleoside diphosphates (see text)

of prokaryotic and eukaryotic origin in that it acts exonucleolytically in *both* 5′→3′ and 3′→5′ directions [69].

(e) One enzyme which acts, in a sense, like a limited RNA exonuclease is *CCA pyrophosphorylase*. This has been purified from *E. coli* [84] and, in the presence of inorganic phosphate, forms CTP and ATP by removal of the CpCpA specifically from the 3′-OH terminus of tRNA. The reaction is freely reversible and the action towards both bases appears to be catalysed by one protein [85].

(f) *Polynucleotide phosphorylase* can be regarded as depolymerase of RNA. It catalyses the reversible reaction whereby a polyribonucleotide reacts with inorganic phosphate to yield ribonucleoside diphosphates. It is discussed in greater detail in Chapter 11. Most of the well-characterized exonucleases which degrade RNA are also active against DNA, and these enzymes will be dealt with as non-specific nucleases.

The properties of some of the ribonucleases known to occur in *E. coli* are summarized in Table 6.1.

Other ribonucleases of *E. coli* which are believed to play a role in the post-transcriptional processing of primary RNA transcripts are dealt with more fully in Chapter 11.

6.2.4 Ribonuclease inhibitors

Rat liver contains a protein which acts as a powerful inhibitor of pancreatic RNase but does not affect RNase T_1 or plant RNases [86, 87]. Heparin also inhibits pancreatic RNase. The clay *bentonite* is a powerful inhibitor of RNase and is commonly employed to prevent degradation of RNA during isolation [88]. Polyvinyl sulphate and *Macaloid* have often been employed in a similar way, but, at least for sea urchin ribonuclease activity, they are less effective inhibitors than bentonite [89].

Diethyl pyrocarbonate *(Baycovin)* has been used as an RNase inhibitor [90], particularly in the extraction of nucleic acids. It has an advantage over several other inhibitors in that it is water-soluble, but there is evidence [91] that it reacts with RNA.

Ribonuclease from human skin [92] may seriously contaminate glassware, dialysis tubing, and other laboratory materials. A similar problem exists for deoxyribonuclease.

6.3 NON-SPECIFIC NUCLEASES

6.3.1 Endonucleases

(a) *Micrococcal nuclease* (EC 3.1,4,7). This enzyme is found in cultures of *Staphylococcus* and degrades DNA to a mixture of nucleoside 3′-

monophosphates and oligonucleotides with 3′-phosphate termini [93]. It attacks RNA and, preferentially, heat-denatured DNA. It requires Ca^{2+} for maximum activity. Its structure and chemical properties have been recently reviewed [94, 95]. 3′-Phosphate-forming endonucleases have been reported to be present in a variety of snake venoms [96].

(b) *Neurospora crassa nuclease* [97]. This enzyme has been considerably purified from conidia of *Neurospora* and attacks DNA or RNA to give oligonucleotides with a 5′-phosphate terminus. It exhibits a preference for guanosine or deoxyguanosine residues, but its most interesting property is an absolute requirement for denatured polynucleotide. It is active under a wide range of conditions and requires Ca^{2+} or Mg^{2+}.

(c) *Nuclease S1 from Aspergillus oryzae.* This enzyme is very similar to the *N. crassa* nuclease in that it hydrolyses phosphodiester bonds in single-stranded DNA or RNA [98, 99]. It shows a requirement for Zn^{2+}.

(d) *Nuclease P_1* from *Pencillium citrinum* splits 3′-5′ phosphodiester bonds in RNA and DNA as well as 3′-phosphomonoester bonds in mononucleotides and oligonucleotides without specificity [238].

(e) Several 5′-phosphate-forming endonucleases have been reported from a variety of mammalian cells, in invertebrates, plants, and bacteria. A common feature is a requirement for Mg^{2+} [100]. One of these, the *mung bean nuclease* of Laskowski [101] has been extensively purified.

6.3.2 Non-specific exonucleases

(a) *Venom phosphodiesterase* [102]. The venom of several species of snakes contains a phosphodiesterase which is commonly employed in the

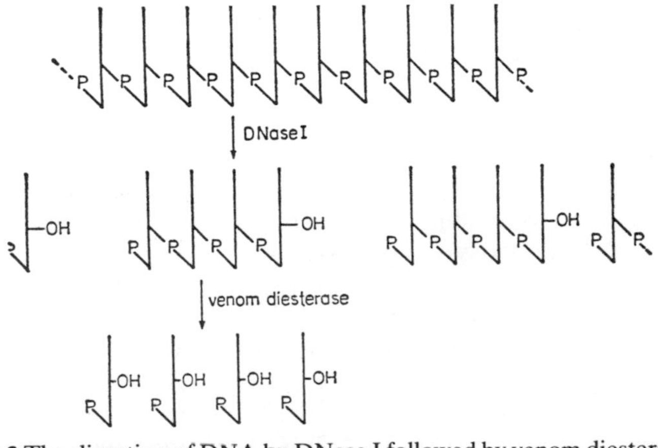

Fig. 6.3 The digestion of DNA by DNase I followed by venom diesterase to yield deoxyribonucleoside 5′-monophosphates.

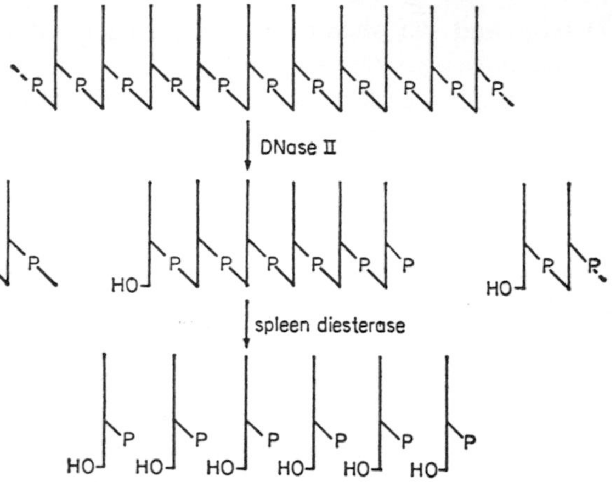

Fig. 6.4 The digestion of DNA by DNase II followed by spleen diesterase to yield deoxyribonucleoside 3′-monophosphates.

preparation of nucleoside 5′-phosphates. The enzyme occurs naturally in association with a high concentration of phosphomonoesterase from which it can be freed by chromatography and acetone fractionation [103, 104].

Venom diesterase hydrolyses RNA to nucleoside 5′-monophosphates (Fig. 6.2) starting at the 3′-hydroxyl end of the chain, and is also active in hydrolysing the oligonucleotides produced by the action of deoxyribonuclease I on DNA to deoxyribonucleoside 5′-phospates (Fig. 6.3). The presence of a 3′-phosphoryl terminal group confers resistance on the substrate.

(b) *Spleen phosphodiesterase* [105]. This enzyme hydrolyses RNA to nucleoside 3′-monophosphates (Fig. 6.2) starting at the 5′-hydroxyl end, and also acts on the mixture of oligonucleotides produced from DNA by spleen deoxyribonuclease II to yield deoxyribonucleoside 3′-phosphates (Fig. 6.4) [106]. It is inactive with oligonucleotides carrying a 5′-phosphomonoester end-group.

6.4 DEOXYRIBONUCLEASES (DNases)

6.4.1. Endonucleases

The two deoxyribonucleases which were the first to be purified and characterized are both endonucleases. The first type exemplified by *pancreatic deoxyribonuclease (DNase I)* is a 5′-phosphomonoester former. The second type (*DNase II*) which is found in spleen and thymus is a 3′-phosphomonoester former (Figs. 6.3 and 6.4). It has become increasingly difficult to classify DNA endonucleases as DNase I-type or

DNase II-type and we should probably regard these as extremes between which most activities will fall [1].

The activity of DNA endonucleases is generally measured by estimating the release of acid-soluble products from DNA, either as ultraviolet-absorbing material or as radioactive label. These methods are useful in the presence of extensive endonuclease action; where extreme sensitivity has been required, supercoiled circular viral DNA is used as a substrate. Only one phosphodiester bond cleavage is required to alter the physical properties of such molecules and allow separation of intact and cleaved molecules.

(a) *Pancreatic deoxyribonuclease* (*DNase I*) (EC 3.1.4.5) (for reviews see [10, 93, 107]). This enzyme breaks down DNA into oligonucleotides of average chain length 4 units with a free hydroxyl group on position 3′ and a phosphate group on position 5′ (Fig. 6.3). It requires magnesium ions and has an alkaline optimum pH in the range 6·8–8·2.

The method of purification was described in 1946 by McCarty [108], who used a 0·25N-sulphuric acid extract of ox pancreas from which the enzyme could be prepared by fractionation with ammonium sulphate between 0·17 and 0·3 saturation. The method has been modified by Kunitz [109] so as to yield a crystalline preparation which has also been described by McDonald [107]. A valuable additional method is that of Polson [110]. The enzyme has a molecular weight of 31 000 and an isoelectric point of 4·7. It has two disulphide bonds which are very readily reduced by mercaptoethanol with concomitant inactivation. Ca^{2+} ions prevent this inactivation.

The enzyme is activated by magnesium ions (optimum concentration 3mM) or manganese ions, and the nature of the divalent cation qualitatively affects specificity [111]. It is inhibited by fluoride [112], and citrate at 0·01M inhibits completely the magnesium-activated but not the manganese-activated enzyme.

Citrate, borate, and fluoride exert their inhibitory action by removing the activating magnesium ions while other inhibitors such as sodium sulphide and thioglycollic acid appear to react with the functional groups of the enzyme protein [113].

Pancreatic DNase hydrolyses native DNA more rapidly than denatured DNA. As the early products of the reaction are worse substrates than the initial DNA the enzyme is autoretarding. In the early stages of the reaction single-stranded nicks are producd towards the centre of the DNA molecule [114] but later on the Pu-p-Py bond is preferentially cleaved leading to a final product of di- and oligo-nucleotides. Small oligodeoxyribonucleotides, apurinic acid [115], or deaminated single-stranded DNA [116] are not hydrolysed. The biosynthetic polymers poly(dA)·poly(dT), poly(dI)·poly(dC), poly(dG)·poly(dC) are degraded in part by pancreatic DNase. The resistance of the poly(dC) chain in the latter two co-polymers to hydrolysis by the enzyme is overcome by adding Ca^{2+} to the Mg^{2+} or by replacing Mg^{2+} by Mn^{2+} [111]. The enzyme can be freed of ribonuclease contamination by electrophoresis [117] or by ion-exchange chromatography [118], and has been shown to exist in multiple forms [101].

(b) *Deoxyribonuclease II* (*DNase II*) (EC 3.1.4.6) (for reviews see [10, 119]). A deoxyribonuclease of molecular weight 40000, pH optimum in the range 4·5–5·5, and no requirement for magnesium ions has been isolated from spleen and thymus.

Double-stranded DNA is degraded by splenic DNase II, in part by a 'one-hit' process that hydrolyses both strands of the double helix at the same point [114, 120]. This initial phase of the reaction is followed by the slower release of oligonucleotides of chain length from 14 to 100 nucleotides. The final stage produces oligonucleotides of average chain length 6 units, which have a free 5′-hydroxyl group and a phosphate residue on position 3′ (Fig. 6.4).

The properties of the two main types of DNase are summarized in Table 6.2

Table 6.2 The properties DNase I and DNase II.

	DNase I	DNase II
Substrate	DNA	DNA
pH optimum	7–8	4–5
Activators	Mg^{2+}, Mn^{2+}, Ca^{2+}	0·3M-Na^{+}
Inhibitors	Citrate, EDTA	Mg^{2+}
Product	5′-Phosphoryl terminated oligonucleotides	3′-Phosphoryl terminated oligonucleotides

(c) *Streptococcal deoxyribonuclease.* Streptococcal deoxyribonuclease (streptodornase) is a deoxyribonuclease of the endonuclease type, cleaving the 3′-phosphate bond, and producing 5′-phosphoryl-terminated fragments of various lengths [121]. Only traces of mononucleotides are produced, together with small amounts of dinucleotides, but the majority of the fragments are larger than dinucleotides. The preferential cleavage involves the pY-R bonds. The optimal pH is 7, and divalent cations are required. The optimum Mg^{2+} ion concentration is below 20 mM [122].

At least two other distinct DNases have been characterized from group A *Streptococci* [123].

(d) *Endonuclease I from* E. coli [121, 124–126]. This enzyme is an endonuclease which attacks DNA producing scissions at many points along the DNA chain. At each scission an exonucleolytic activity removes about 400 nucleotides [127] which are released as a mixture of oligonucleotides of average chain length 7 units terminated by a 5′-phosphoryl group. It is highly specific for DNA, attacking native DNA seven times more readily than denatured DNA to give random double-stranded breaks [128]. RNA is an inhibitor of the action of this enzyme.

(e) *Endonuclease II from* E. coli [129, 130] produces single-strand breaks in double-stranded DNA which has been alkylated with monofunctional alkylating agents (e.g. methyl methanesulphonate). To do this it acts in conjunction with an *N*-glycosidase which first removes the alkylated base. The role in repair of endonuclease II (and endonuclease V which specifically cleaves uracil containing DNA) is considered in Chapter 8. Endonuclease III which cleaves duplex DNA containing thymine dimers is also considered in Chapter 9. Endonuclease II has a pH optimum at pH 8–9 and is stimulated by Mg^{2+} or Mn^{2+}. It produces single-stranded nicks with 5′-phosphoryl end groups. (A similar enzyme which recognizes thymine dimers in duplex DNA is also discussed in Chapter 9). It has been shown recently that endonuclease II and exonuclease III activities are present in the same enzyme molecule [131, 132].

(f) *ATP-dependent endonucleases.* A number of enzymes are known which can hydrolyse or unwind duplex DNA while hydrolysing ATP. Their role in DNA replication is considered in Chapter 8. The enzyme from *M. luteus* acts preferentially on native double-stranded DNA to give fragments with 5′-monophosphate termini [134]. Three molecules of ATP are hydrolysed for every phosphodiester bond cleaved [135]. Associated with purified preparations is an exonuclease activity towards native DNA, also ATP-dependent. These activities have been implicated in the recombination process.

(g) *Phage-induced endonucleases.* Nucleases are induced following infection of bacteria with a variety of bacteriophages. Two of the best studied are endonuclease II and endonuclease IV induced after infection of *E. coli* with phage T4 [137–139] (see p. 265).

(1) *Endonuclease II* makes single-strand breaks in double-stranded DNA other than that of T4 to give products (at least from phage λ) of about 10^3 nucleotides. These have 5′-phosphoryl and 3′-OH termini. It differs from *E. coli* endonuclease II in its inability to attack T4 DNA (glucosylated or not).

(2) *Endonuclease IV* hydrolyses single-stranded DNA to give considerably smaller products than endonuclease II, again with 5′-phosphoryl termini, but with dCMP exclusively in that position. DNA containing hydroxymethylcytosine (i.e. T4 DNA) is inactive as a substrate.

A mechanism has been suggested [139] whereby these enzymes, with the help of a bacteriophage-induced exonuclease (see p. 265) may be involved in the degradation of host DNA after T4 infection (Fig. 6.5).

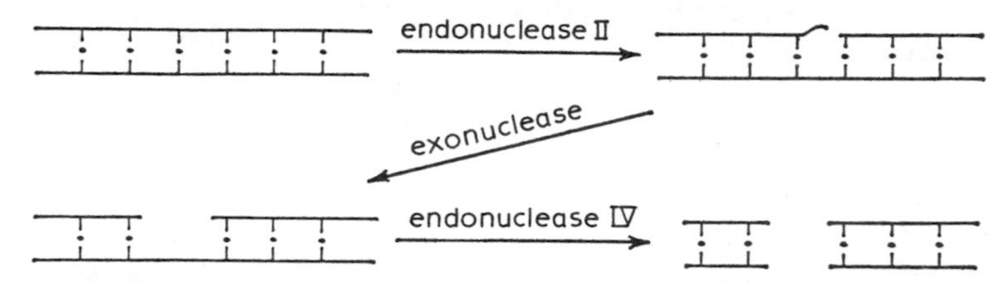

Fig. 6.5 Hypothetical scheme for degradation of cell DNA after infection by bacteriophage T4.

(3) Another endonuclease which plays a specific role in the replication of bacteriophage DNA is that coded for by gene A of ØX174. This breaks a single phosphodiester bond to initiate phage DNA synthesis [140] (see p. 259).

(h) *Mammalian virus endonucleases*. Much less is known about DNases induced by mammalian viruses than about phage-induced activities. However, deoxyribonuclease activities are increased after infection with several mammalian viruses. Some of these form part of the virus structure, and an example is the *endonuclease* associated with the penton protein of *adenovirus 2*. This enzyme preferentially attacks native DNA, cleaves both strands to give large fragments, and is inhibited by tRNA. It is active against all DNAs tested except glucosylated T4 DNA [141]. An endonuclease is also present in the core of *vaccinia virus* particles [142].

6.4.2 DNA exonucleases

The bacterial DNA-specific exonucleases (phosphodiesterases) from *E. coli* are of considerable general interest and their mode of action has been worked out in some detail [124, 143–147]. Their properties are summarized in Table 6.3 and [268].

(a) E. coli *exonuclease I* [143, 244, 148]. This enzyme hydrolyses heat-denatured single-stranded DNA and has hardly any effect on native double-stranded DNA. It is an exonuclease hydrolysing the DNA chain stepwise beginning at the 3′-hydroxyl end, and releasing deoxyribonucleoside 5′-monophosphates until only a dinucleotide is left.

The enzyme does not cleave free dinucleotides or the 5′-terminal dinucleotide portion of a polydeoxyribonucleotide chain, but it can degrade bacteriophage DNAs containing glucosylated hydroxymethylcytosine (p. 264) quantitatively to their constituent mononucleotides. It has no effect on polyribonucleotides.

Table 6.3 Properties of DNA exonucleases from *E. coli* [268].

Enzyme	Reqd. DNA structure	End group attacked	Extent of action	Products
Exonuclease I	Single-stranded	3′-OH	Up to terminal dinucleotide	Mono- and dinucleoside 5′-monophosphates
Exonuclease II (3′→5′ activity associated with DNA polymerase I)	Prefers single-stranded	3′-OH	Complete	Nucleoside 5′-monophosphates
Exonuclease III	Double-stranded	3′-OH or 3′-OP (initial attack removes terminal P_i)	To 40 per cent degradation	P_i nucleoside 5′-monophosphates and larger, single-stranded oligonucleotides
Exonuclease IV	Prefers oligonucleotides	3′-OH	Complete (cleaves dinucleotides)	Nucleoside 5′monophosphates
Exonuclease V	Single- and double-stranded	3′OH or 5′-OH		Oligonucleotides
Exonuclease VI (5′→3′ activity associated with DNA polymerase I)	Double-stranded	5′-OH or 3′-OP	Excises mismatched regions	Mostly mono- and dinucleotides with 20–25% longer
Exonuclease VII	Single-stranded	3′-OH or 5′-OH		Oligonucleotides

Other enzymes which preferentially attack single-stranded DNA are found in liver [149, 150] and in lamb brain [151].

(b) E. coli *exonuclease III* (*DNA phosphatase-exonuclease*) [146, 147]. This enzyme is found in small amounts in close association with the DNA polymerase of *E. coli* but can be separated from it by chromatography. Its exonuclease action is very similar to that of *E. coli* exonuclease II but, in addition, it acts as a phosphatase highly specific for a phosphate residue esterified to the 3′-hydroxyl terminus of a DNA chain (Fig. 6.6). It does not release inorganic phosphate from deoxyribonucleoside 3′- or 5′-monophosphates from oligodeoxyribonucleotides of short chain length or from 3′-phosphoryl-terminated RNA, but it does attack DNA with a phosphoribonucleotide terminus.

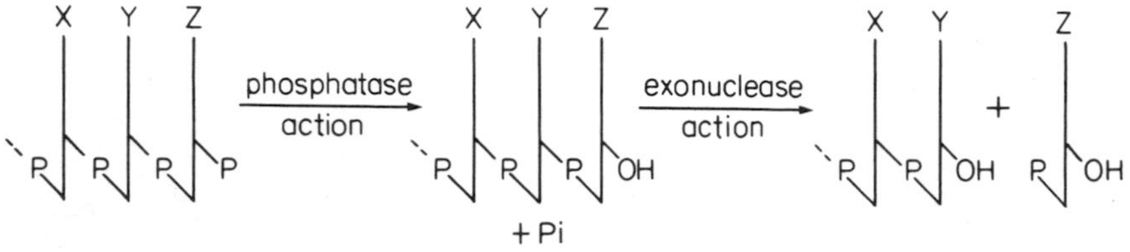

X Y Z — P P P P —(phosphatase action)→ X Y Z — P P P OH + Pi —(exonuclease action)→ X Y — P P OH + Z — P OH

Fig. 6.6 Sequential action of *E. coli* exonuclease III (DNA phosphatase–exonuclease) on a DNA chain terminated by a nucleotide carrying a 3′-phosphate group.

As an exonuclease it carries out a stepwise attack on the 3′-hydroxyl end of the DNA chain releasing mononucleotides (Fig. 6.5) but it acts only on double-stranded DNA, degrading it until 35–45 per cent has been digested. If the enzyme begins its attack from both 3′-hydroxyl ends of the double-stranded molecule (Fig. 6.7), when nearly half has been degraded, the residual acid-insoluble DNA will be single-stranded and resistant to further attack although it is still susceptible to the action of exonuclease I.

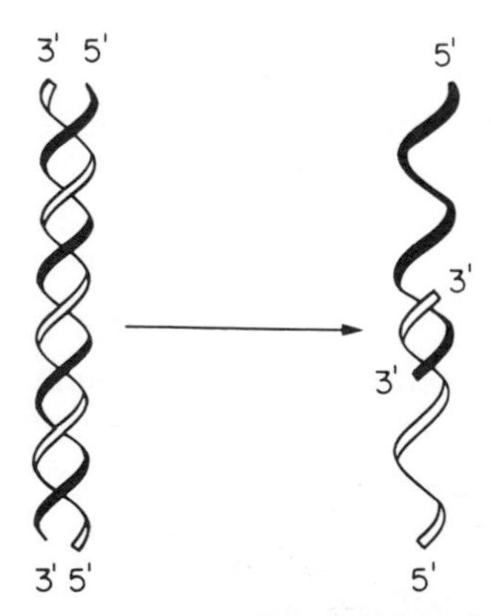

Fig. 6.7 Mechanism of action of stepwise attack of *E. coli* exonuclease III on native DNA beginning at the 3′-hydroxyl terminus.

Similar enzymes are found in *Diplococcus pneumonia* [152] and *B. subtilis* [153]. An enzyme of opposite polarity has been found in bacterial cells infected with bacteriophage λ which produces deoxyribonucleoside 5′-monophosphates stepwise from the 5′-end of DNA chains [142]. This enzyme is involved in λ-recombination [154].

(c) E. coli *exonuclease IV*. This exonuclease shows little activity towards single- or double-stranded DNA, and exhibits a considerable preference (twenty-fold) for DNA predigested with pancreatic deoxyribonuclease. In this sense it could be termed *oligonucleotide diesterase*. It can be separated by DEAE-cellulose chromatography into two fractions (IVA and IVB) [155].

(d) E. coli *exonuclease V*. The role of this enzyme in repair and recombination is discussed on p. 295. It is made of two subunits and one of them (β) is defective in mutants with altered *rec B* or *rec C* genes. The enzyme shows DNA-dependent ATPase activity as well as ATP-dependent DNase activity and may act as both an endo- and an exo-nuclease [156].

(e) *Exonucleases associated with* E. coli *DNA polymersae I*. The terms exonuclease II and VI have been used to define the 3′→5′ and the 5′→3′ DNA exonuclease activities which form part of the protein of *E. coli* DNA polymerase I (Chapter 8) [157].

3′→5′ *activity* [145]. This activity resides in the same protein molecule that possesses DNA polymerase activity (p. 237). Like exonuclease I it commences attack at the 3′-hydroxyl terminus of a polydeoxyribonucleotide chain, with the stepwise release of deoxyribonucleoside 5′-monophosphates, but unlike exonuclease I it also attacks dinucleotides. It will, for example, hydrolyse the oligonucleotide pT-T-T-T-T to 5pT. The enzyme attacks denatured DNA in preference to native DNA [158] and has no effect on oligonucleotides bearing a 3′-phosphomonoester group or on RNA. Most bacterial and phage DNA polymerases so far investigated (with the possible exception of those from *B. subtilis*) have an associated 3′→5′ exonuclease [157].

Evidence that the enzyme is in fact an exonuclease with these properties comes from several sources.

(1) Exhaustive digestion of ^{32}P-labelled d(A-T) copolymer results in the conversion of 99 per cent of the ^{32}P into an acid-soluble form which can be accounted for in terms of 5′-monophosphates.

(2) Partial digestion results in the release of a proportion of radioactivity which is the same as the proportion of monophosphates formed.

(3) When d(A-T) copolymer specifically labelled with ^{32}P-dTMP at the 3′-hydroxyl end is used as substrate, 90 per cent of the ^{32}P-labelled material is made acid-soluble when less than 10 per cent of the unlabelled nucleotides from the interior of the chain have been released. This indicates attack from the 3′-hydroxyl end of the chain.

(4) Treatment of DNA having transforming activity (p. 59) from *B. subtilis* results in a 36 per cent drop in viscosity with 46 per cent of the initial transforming activity still present. With an endonuclease (DNase I) a 36 per cent drop in viscosity is accompanied by a drop in transforming

activity to 0·1 per cent of the original value owing to breakage of the chains at critical regions.

(5) With ^{32}P-labelled native DNA as substrate the decrease in viscosity is more rapid than the release of ^{32}P-mononucleotides. This suggests an exonucleolytic attack on a double helical polynucleotide of opposite polarity with initiation of the hydrolysis at the two 3′-hydroxyl groups at opposite ends leaving the opposing strand in single-stranded form of low viscosity.

5′→3′ *activity*. This activity is specific towards native DNA and will function in the presence and absence of a 5′-phosphate group on the substrate. The products are mostly mononucleotides, with 20–25 per cent dinucleotides or longer [159].

The DNA polymerase I molecule is susceptible to protease action such that it specifically splits into two fragments. One of these (76 000 mol. wt.) has DNA polymerase I and 3′→5′ exonuclease activity; the other (34 000 mol. wt.) shows 5′→3′ exonuclease activity but only if the cleavage is carried out in the presence of DNA [160] (p. 238). *B. subtilis* DNA polymerase contains no 5′→3′ exonuclease activity [153] but a separate enzyme exists with this function. 5′→3′ exonuclease activity is also present in *E. coli* DNA polymerase III, but is absent from *E. coli* DNA polymerase II.

(f) *Exonuclease VII* degrades single-stranded DNA from either end, releasing oligonucleotides. It does not require ATP [271].

6.4.3 Virus-induced DNA exonucleases

Phage T4 induced DNA polymerase has a 3′→5′ exonuclease activity similar to that of the host polymerase [158]. In addition Koerner [161] has described an *exonuclease A* from bacteriophage T4-infected cells which is distinct from the polymerase-associated activity. This is an *oligonucleotide diesterase* (as *E. coli* exonuclease IV) which liberates 5′-monophosphates quantitatively from the 3′-terminus [162]. It is possible that this exonuclease is involved in degradation of cell DNA (see Fig. 6.5).

Phage SP3 induces in *B. subtilis* SB19 an exonuclease which attacks the 5′-terminus of single-stranded DNA, but releases only dinucleotides (90 per cent) and trinucleotides [163]. At no time during digestion can larger fragments be detected.

Poxvirus particles contain an exonuclease activity and a second is induced in infected cells [164]. Exonuclease is induced after *herpes virus* infection [165]. The enzyme is a 5′-phosphate-former and acts on native or denatured DNA.

6.4.4. Mammalian DNA exonucleases

Activities corresponding to two exonucleases are found in *rabbit liver* [166, 167]. *DNase III* preferentially attacks denatured DNA from the 3′-terminus to give dinucleotides in addition to monomers. *DNase IV* is specific for native DNA and attacks from the 5′-terminus.

6.4.5 Deoxyribonuclease inhibitors

Most animal tissues contain enzymes similar to DNase I and DNase II. Many tissues also contain inhibitors of both, of which the best known are the protein inhibitors of DNase I, which are particularly abundant in the crop gland of pigeons [93] and in calf spleen [168].

6.5 DNA METHYLATION

6.5.1 Cellular DNA

Methylated bases occurring in DNA fall into two classes. The first class contains thymine and hydroxymethylcytosine (in T4 DNA) which are incorporated into DNA from dTTP or HMdCTP by DNA polymerase. The second class contains bases arising from methylation of a preformed polydeoxynucleotide, i.e. methylation occurs *after* DNA synthesis. This second class of methylated bases contains only 5-methylcytosine, first discovered by Hotchkiss in calf thymus DNA [85] and 6-methyladenine which appears to be restricted to the DNA from lower organisms. 6-Methyladenine (where the methyl group is a substituent on the 6-amino group) may be present as the sole product of polynucleotide methylation, e.g. in *E. coli* $15T^-$ which contains 1 mole per cent base as 6-methyladenine [169].

5-Methylcytosine, which is widespread in nature, may also occur alone and to greatly differing extents, from an almost negligible level in insect DNA [170, 171, 216, 217] through 1·0–1·5 per cent in mammalian DNA [170] to 5–6 per cent in plant DNA [170, 172]. Variation in DNA methylation occurs between organelles in certain cell types. Main band DNA from *Euglena gracilis* is methylated but satellite DNA is not [173]; in tobacco leaves nuclear DNA is methylated but not chloroplast DNA [172]. On the other hand, the level of methylation of mouse nuclear satellite DNA is twice that of the main band [174]. The chromosomal genes for ribosomal RNA are methylated in *Xenopus* but the amplified ribosomal RNA genes present in the oocyte are not [175]. In the sea urchin some chromosomal genes are methylated and others not [218].

6.5.2 Viral DNA

5-Methylcytosine and 6-methyladenine also occur in bacteriophage DNA, in addition to many other modifications, e.g. glucosylated 5-hydroxymethylcytosine [176]. Bacteriophage T2, T4, T7, and P1 [177, 178] contain 6-methyladenine while T3 and T5 have a complete lack of methylated bases other than thymine [179]. Polyoma, herpes simplex type I, and pseudorabies virus DNAs are not methylated at the polynucleotide level [180–182].

6.5.3 The methylation reaction

Methylation of DNA takes place at the polynucleotide level [183], and the reactions are catalysed by specific enzymes, the DNA methylases (methyl-transferases). The source of methyl groups is methionine, in the form of an intermediate with a high free energy of hydrolysis, *S*-adenosyl-L-methionine [184]. The mechanism is illustrated in Fig. 6.8.

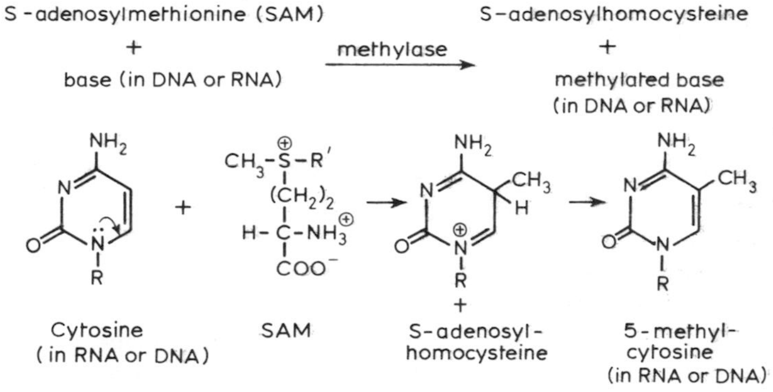

S-adenosylmethionine (SAM) + base (in DNA or RNA) —methylase→ S-adenosylhomocysteine + methylated base (in DNA or RNA)

Cytosine (in RNA or DNA) + SAM → S-adenosyl-homocysteine → 5-methyl-cytosine (in RNA or DNA)

Fig. 6.8 The nucleic acid methylase reaction.

At least one, and possibly two, DNA methylases have been purified from *E. coli W* [185] and four DNA methylases have been purified from *Haemophilus influenzae* Rd [186]. These enzymes methylate specific sequences corresponding to those cleaved by the restriction endonucleases (see below and Table 6.4).

Certain bacteriophages induce DNA methylase activities which are distinct from those of their host cells and which may therefore play a role in the infective process. Among these are T1, T2, and T4, while T3 produces an enzyme which cleaves *S*-adenosylmethionine and thereby inhibits DNA methylation [187, 188].

6.5.4 Mammalian DNA methylase activity

This is normally found in the chromatin fraction of the cell [189–193] and the majority of methyl groups are added shortly after the DNA is made [194, 195]. However, there is a delay before methylation is complete and homologous DNA will therefore act as substrate for mammalian DNA methylase [192, 194, 196].

One interesting feature of methylation of DNA concerns the doublet CpG. This doublet occurs with low frequency in mammalian DNAs (see Chapter 8) and is present as 5-methyl CpG primarily; this accounts for most of the 5-methylcytosine in animal DNA [197, 198].

6.5.5 Function of DNA methylation

Methylation of specific sites in DNA by a host-specific methylase now seems to form the modification role in modification-restriction phenomena observed in bacteria [176] (see below). However, the major methylase in *E. coli* is the dam methylase (DNA adenine methylase) and this appears not to function in restriction modification [200, 219]. Neither does introduction into *E. coli* B of the cytosine methylase from *E. coli* K12 affect restriction modification [219]. These results point to an additional major role for DNA methylation which is unconnected with restriction modification and a role in recombination has been postulated [220]. In *E. coli*, continuing DNA synthesis *in vivo* appears to require the formation of a normally methylated DNA template [199]. However, mutants of *E. coli* K12 with highly undermethylated DNA grow normally [200] and removal of 6-methyladenine from T2 and T4 DNA has no effect on several biological properties [201].

No function has been found for the high levels of cytosine methylation which occur in most eukaryotic cells. Some sites which are methylated in the DNA from certain tissues are unmethylated in DNA from other tissues and the suggestion has been made that genes which are being transcribed have certain sites unmethylated [221, 222]. However, in the sea urchin the genes for histone and ribosomal RNA are always unmethylated whether or not they are being transcribed [218] and in insects methylation is almost totally absent [216, 217].

The DNA of animal viruses (SV40, polyoma, adeno, herpes) is not normally methylated despite the fact that these viruses grow in cells containing methylated DNA. However, when adenovirus or herpes viruses integrate into the host-cell genome to bring about cell transformation (see p. 81) methylation of the viral DNA occurs. In adenovirus transformed cells those regions of the adenovirus DNA which are expressed (i.e. those coding for the early genes) are not as highly methylated as those regions which are not expressed [249]. In contrast, in cells transformed with herpesvirus viral DNA is more than 80 per cent methylated except in virus producing lymphoid cells [250].

In bull sperm DNA certain satellites are not methylated. These satellites are methylated in somatic cell DNA and, as the proportion of satellite in bovine DNA is high, this leads to a much lower proportion of cytosine methylation in bovine sperm DNA which is not apparent in sperm DNA from other species [251, 252]. Variable levels of methylation in satellite DNAs and the absence of methylation from amplified (but not chromosomal) ribosomal DNA in *Xenopus* oocytes [253] suggests a role for methylation in reactions involving the breakage and joining of DNA molecules e.g. in recombination.

Another postulated function for DNA methylation is in *mismatch repair* [254]. If an incorrect (but normal) base is inserted into DNA during replication it is usually removed by proof reading (see p. 239). If proof reading fails DNA is synthesized with a mismatch. No error-correction mechanism known has the means of knowing which base is correct and which incorrect. However, since the daughter strand does not become methylated for a finite time (2 min to several hours in eukaryotes [255, 256]) after synthesis correction mechanisms may exist which recognize unmethylated DNA and replace the mismatched bases in the daughter strand prior to methylation.

DNA methylation has also been postulated to have a function in *differentiation, cell ageing* and *biological clocks* [257, 258, 261]. In eukaryotes most methylcytosines occur in the dinucleotide m^5CpG. As the complement of this is also m^5CpG methyl groups occur in pairs along the DNA duplex [259]. However, not all CpGs are methylated. It is believed that a signal for methylation of a CpG in the daughter strand following DNA replication normally involves a methylated CpG in the parental strand. However, if this restriction were to be relaxed at certain stages of development, then normally unmethylated pairs of CpG would become methylated; possibly resulting in a modified pattern of gene transcription. These additional methylations may occur in a highly programmed manner at each replication cycle [257, 258] and thereby provide a means of counting cell divisions. To counter such ageing processes some mechanism must exist in gametogenesis or early embryogenesis to remove these extra modifications from the genome.

In *E. coli* certain spontaneous base substitutions occur as a result of deamination of 5-methylcytosine to thymine [260]. Although this reaction was postulated many years ago as a mechanism underlying eukaryotic cell differentiation [261] there is little evidence for such a function, though a role in cell ageing remains possible.

A review of eukaryotic DNA methylation has appeared recently [267].

6.6 DNA RESTRICTION AND MODIFICATION

6.6.1 The phenomenon

When a bacteriophage is transferred from growth on one host to a different host, its efficiency is frequently impaired several thousand-fold. However, when those bacteriophages which do survive and multiply are used to reinfect the second host they now grow normally.

The initial poor growth is caused by the action on the bacteriophage DNA of highly specific bacterial endonucleases (known as restriction endonucleases). Following restriction, the invading bacteriophage DNA is rapidly degraded to nucleotides by exonuclease action, although certain regions may be rescued by recombination [202].

Host (i.e. bacterial) DNA is not degraded because the nucleotide sequence which is recognized by the restriction endonucleases has been modified by methylation (see above). The few molecules of bacteriophage DNA which survive the initial infection do so because they are themselves modified by the host methylase before the restriction enzyme has time to act. Similarly, methylation of progeny DNA renders it resistant in the second infection [203].

A related phenomenon is the degradation of cytosine containing DNA in *E. coli* infected with T-even bacteriophage whose own DNA contains hydroxymethylcytosine (see p. 68).

6.6.2 Restriction endonucleases

There are two classes of restriction enzymes. Those responsible for restriction in *E. coli* strains B and K are representative of class I and most of the remainder so far investigated are in Class II [204–206].

(a) *Class I* (e.g. EcoB, EcoK, EcoPI) are complex multifunctional proteins which cleave unmodified DNA in the presence of S-adenosyl-L-methionine (SAM or AdoMet), ATP, and Mg^{2+} [29]. These enzymes, which are also methylases and ATPases, are multisubunit enzymes. They have two α subunits (mol. wt. 135 000), two β subunits (mol. wt. 60 000) and γ subunit (mol. wt. 55 000) [205]. Before interacting with DNA the *E. coli* K restriction enzyme binds to SAM [223, 208]. Binding to SAM takes place rapidly and is followed by a slower allosteric modification of the enzyme to an activated form which then interacts with DNA at a non-specific site. The enzyme moves to the recognition site on the DNA (SK) and its subsequent reaction depends on the state of this recognition site.

The recognition sites for the EcoK and EcoB type I restriction enzymes are shown in Fig. 6.9. The adenines with an asterisk in the EcoB sequence are methylated by the EcoB enzyme and the corresponding adenines in the EcoK sequences are probably the site of methylation by the EcoK restriction methylase [224–226].

The recognition sequence both consist of a group of three bases and a group of four bases separated by six (EcoK) or eight (EcoB) unspecified bases. Four of the seven specified bases are conserved and the methylated adenines in both cases are separated by eight bases.

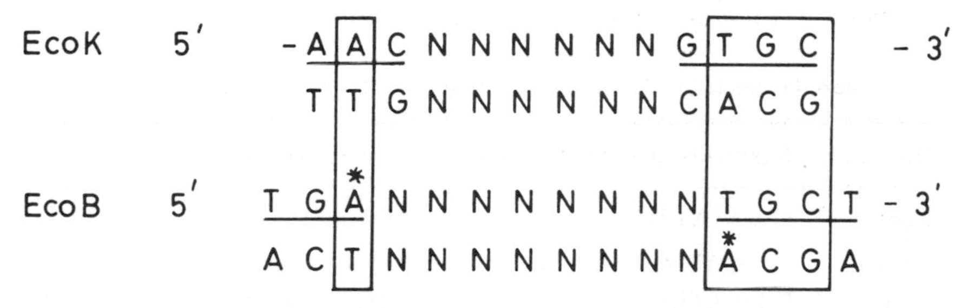

Fig. 6.9 The recognition sequences for the EcoK and EcoB restriction modification enzymes [226].

If the recognition site is methylated in both strands of the DNA the enzyme does not recognize it. If the site is methylated on one strand only (as would be the case with DNA immediately following synthesis) the enzyme binds to this site and methylates the second strand in a reaction stimulated by ATP [227]. Methylation can also occur at completely unmethylated sites but only at a rate about 0.2% of that at half-methylated sites [227]. Normally if the site is unmodified in both strands the enzyme is triggered into its restriction mode, but restriction does not occur at the binding site.

In the presence of ATP the DNA is made to loop past the enzyme which remains bound to the recognition site [228]. This translocation leads to formation of supercoiled loops which become relaxed on subsequent cleavage [269]. Cleavage of the looped out DNA occurs at non-random sites. Cleavage occurs in two stages (first one strand, followed some time later by a break on the opposite strand) and at this point the enzyme ceases to be a nuclease (i.e. it performs only one restriction event) and becomes a vigorous ATPase. About 10^5 ATP molecules are hydrolysed for each restriction event [202]. (EcoPI differs from EcoK in that it does not require SAM for cleavage and does not become an ATPase). The requirement for SAM in the initial binding reaction is perhaps a safety device to prevent the breakdown of host DNA formed when conditions of methionine deprivation have limited methylation [202].

(b) *Class II* (e.g. EcoRI, HapII) are simpler enzymes which require only Mg^{2+} for activity. They have molecular weights ranging from 20000 to 100000 and EcoRI has two identical subunits. These enzymes, which are neither methylases nor ATPases, recognize a specific site on the DNA, and if this is unmodified cleavage occurs at this site. DNA modified on one strand is not a substrate for restriction, but is a substrate for a complementary methylase which recognizes the same specific nucleotide sequence. All sites are 4–6 nucleotide pairs long and have a two-fold

Table 6.4 Class II DNA restriction and modification enzymes.

Enzyme	Restriction and modification site	Bacterial strain
EcoRI	G ↓ A*ATTC	*E. coli* RY13
EcoRII	↓ C*C$^{A}_{T}$GG	*E. coli* R245
Ava I	C ↓ Y°CGRG	*Anabaena variabilis*
Bam HI	G ↓ GATCC	*Bacillus amyloliquifaciens* H
Bgl II	A ↓ GATCT	*Bacillus globiggi*
Hha I	G°CG ↓ C	*Haemophilus haemolyticus*
Hae III	GG ↓ CC	*Haemophilus aegyptius*
Hind III	*A ↓ AGCTT	*Haemophilus influenzae* Rd
Hinf I	G ↓ ANTC	*Haemophilus influenzae* Rf
Hpa II	C ↓ *CGG	*Haemophilus parainfluenzae*
Hga I	GACGCNNNNN ↓	*Haemophilus gallinarum*
Sma I	CCC ↓ GGG	*Serratia marcescens* Sb
Dpn	GATC	*Diplococcus pneumoniae*

↓ *shows the site of cleavage and* * *shows the site of methylation of the corresponding methylase where known.* ° *shows the site of action of a presumed methylase, i.e. methylation at this site blocks restriction but the methylase has not yet been isolated. More complete lists are found in references* [232–234, 262, 263].

rotational symmetry (see Table 6.4) which suggests that the two enzyme subunits may be arranged with two-fold symmetry [206]. Cleavage at the site may be staggered by up to five nucleotides (e.g. EcoRI) when identical self complementary, cohesive termini are produced. BsuI and HindII produce even breaks without single-stranded termini (see Table 6.4). The recognition site for the HinfI restriction site has one unspecified base [229] and frequently bases are only partially specified, i.e. they must be a purine or a pyrimidine (see Table 6.4). Several enzymes do not cleave within the recognition sequence thus, enzyme HphI cleaves DNA eight bases before the sequence TCACC [230]. In these cases neither the recognition nor the cleavage sequence is symmetrical. With EcoRII the cleavage site is a sequence of five nucleotide pairs and hence not all the ends are mutually cohesive. In this case the complementary methylase modifies two cytosines in different sequences.

At least one restriction endonuclease (DnpI) is specific for methylated DNA but its function is unclear as *Diplococcus pneumoniae* also contains an enzyme of opposite specificity (DpnII) [231].

Comprehensive lists of Class II restriction enzymes can be found in several reviews [232–234, 262, 263].

A restriction enzyme is found in crown gall tumours [236] and an enzyme which cuts a monkey satellite DNA into discrete size fragments has been isolated from the testes of the African Green monkeys but whether this is a restriction enzyme needs yet to be proven [235].

6.6.3 Nomenclature

The restriction endonucleases are named according to the system described by Smith and Nathans [209]. The first three letters give the name of the bacterium (e.g. Eco for *E. coli*) and the fourth indicates the strain (e.g. EcoR, Hind). Where more than one restriction enzyme is found in a particular strain these are indicated by Roman numerals (e.g. HinaI, HinaII).

Enzymes recognizing identical nucleotide sequences are called isoschizomers. However, isoschizomers do not necessarily cleave in the same position and may respond to methyl groups on different bases in the recognition sequence.

6.6.4 Applications

Since their discovery the class II restriction enzymes have become increasingly used as tools for the biochemist. Because of their ability to make relatively few specific cuts they can be used as a first step in the sequencing of DNA (see Chapter 3) or in the isolation of specific genes. The use of a series of restriction enzymes allows a physical map to be made of genes or small chromosomes. Thus the small chromosome of the tumour virus SV40 (simian virus 40) is cleaved by each restriction enzyme into a small number of discrete pieces which are readily separated by electrophoresis in gels of polyacrylamide or agarose [206, 210, 211]. The size of the fragments can be determined from their speed of migration in the gel or by direct length measurements using the electron microscope. The order of the fragments in the genome can be determined by a variety of techniques such as analysis of partial digests or successive cleavage by multiple restriction endonucleases [206, 212]. The knowledge of the chromosome map thus gained may be used to compare closely related organisms and to localize various functions to discrete regions of the chromosome. Thus it has been shown that early viral functions (see Chapter 4) are coded for by the continuous ragments A, H, I, and B on the HindII/HindIII restriction map of SV40 (see Fig. 6.10) [213]. Similarly the origin and direction of replication of several viral and plasmid chromosomes have been discovered in particular by using EcoRI [214, 215] (see Chapter 8).

As cleavage by many of these restriction enzymes is staggered, this leads to production of fragments with 'sticky' ends (i.e. termini with overlapping self-complementary sequences) which can be rejoined by DNA ligase (see Chapter 8). Moreover, fragments from different genomes can be joined together to form hybrid genomes, and this is the basis for genetic engineering considered in detail in Chapter 14.

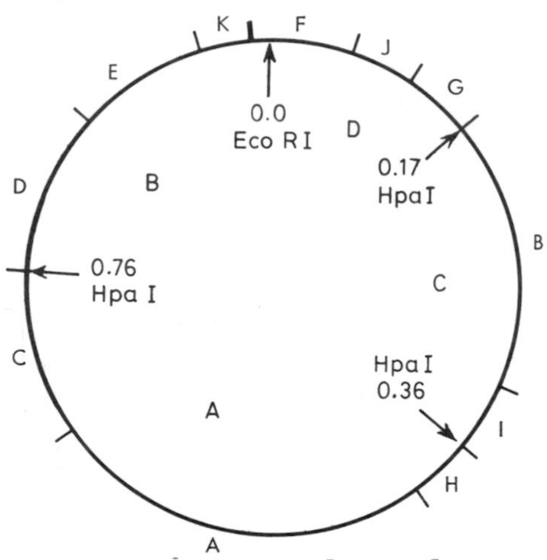

Fig. 6.10 Map of the SV40 chromosome [211, 212]. The fragments enumerated outside the circle in order of decreasing size are produced by the joint action of HindII and HindIII. Those enumerated within the circle are produced by the HpaI and EcoRI.

The use of certain pairs of restriction enzymes has enabled the extent of methylation of sites in individual eukaryotic genes to be investigated. Thus HpaII will only cleave the sequence CCGG if it lacks a methyl group on the internal cytosines of both strands of the duplex DNA whereas MspI will cleave the same sequence irrespective of whether the internal cytosine is methylated [222, 217].

6.7 PHOSPHATASES (PHOSPHOMONOESTERASES) AND KINASES

Phosphomonoesterases remove, as inorganic orthophosphate, the terminal monoesterified phosphate group from mononucleotides or oligonucleotides. The 5′-nucleotidases which have been prepared from seminal plasma and snake venom remove the phosphate group from nucleoside 5′-phosphates [2]. Rye-grass contains a 3′-nucleotidase [239]. The aklaline phosphatase from *E. coli* hydrolyses a wide range of compounds containing monoesterified phosphate. *E. coli* contains a second 3′-nucleotidase (in addition to the exonuclease III) which may remove the 2′- or 3′- but not the 5′-terminus of DNA or RNA [240] and acts on all the natural mononucleotides. A similar activity is induced after T-even bacteriophage infection; it attacks 3′-monophosphates or 3′-phosphoryl termini of DNA and may be involved in preparing a priming site for the polymerization step in DNA repair (see Chapter 9).

Another type of enzyme which may be involved in DNA repair is *polynucleotide (5-hydroxyl) kinase* which has been detected in T2 and T4 infected *E. coli* [241, 242] and also in rat liver nuclei [243]. This enzyme specifically catalyses the incorporation of ^{32}P from [γ-^{32}P]ATP into 5′-hydroxyl groups of DNA [242]. Such a phosphorylation of 5′ends may be a prelude to DNA ligase action in a repair process (see Chapter 8). RNA can also act as acceptor of phosphate groups, and the enzyme together with ^{32}P-ATP of very high specific radioactivity has been used to label the digestion products of non-radioactive RNA prior to sequence analysis [244, 245] (see also Chapter 5).

REFERENCES

1 Laskowski, M., Sr. (1967), *Adv. Enzymology,* **29,** 165.

2 Colowick, S.P. and Kaplan, N.O. (1955), *Methods Enzymol.,* Vol. 2, pp. 427–450 and 561–570: (1963) Vol. 6, pp. 40–55. Academic Press, New York.

3 Cantoni, G.L. and Davies, D.R. (eds.), (1966) *Procedures in Nucleic Acid Research.* Harper and Row, New York.

4 Shugar, D. and Sierakowska, H. (1967), *Prog. in Nucleic Acid Res. Mol. Biol.,* **7,** 369 (J.N. Davidson and W.E. Cohn, eds.). Academic Press, New York.

5 Lehman, I.R. (1967), *Annu. Rev. Biochem.,* **36,** 645.

6 Barnard, E.A. (1965), *Annu. Rev. Biochem.,* **38,** 677.

7 Hilmoe, R.J., Heppel, L.A., Springhorn, S.S. and Koshland, D.E., Jr. (1961), *Biochim. Biophys. Acta,* **53,** 214.

8 de Meuron-Landolt, M. and Privat de Garilhe, M. (1964), *J. Amer. Chem. Soc.,* **78,** 4642.

9 Grossman, R. and Moldave, K. (eds.) (1967–68), *Methods in Enzymology,* Vols. 12A and 12B; (1971) Vols. 20 and 21. Academic Press, New York.

10 Khorana, H.G. (1961), *The Enzymes,* 2nd Ed, Vol. 5, p. 79 (P.D. Boyer, H. Lardy and K. Myrbäck, eds.). Academic Press, New York.

11 Anfinsen, C.B. and White, F.H. (1961), *The Enzymes* 2nd Ed, Vol. 5, p. 95 (P.D. Boyer, H. Lardy and K. Myrbäck, eds.). Academic Press, New York.

12 Witzel, H. (1963), *Progress in Nucleic Acid Research,* Vol. 2, p. 221 (J.N. Davidson and W.E. Cohn, eds.). Academic Press, New York.

13 Jones, W. (1920), *Amer. J. Physiol.,* **52,** 203.

14 Enzyme Nomenclature (1972), Recommendations of the International Union of Biochemistry. Elsevier, Amsterdam.
15 Dubos, R.J. and Thompson, R.H.S. (1938), *J. Biol. Chem.*, **124,** 501.
16 Kunitz, M. (1940), *J. Gen. Physiol.*, **24,** 15.
17 McDonald, M.R. (1955), *Methods in Enzymology,* Vol. 2, p. 427 (S.P. Colowick and N.O. Kaplan, eds.). Academic Press, New York.
18 Smyth, D.G., Stein, W.H. and Moore, S. (1963), *J. Biol., Chem.*, **238,** 227.
19 Stein, W.H. (1964), *Fed. Proc.*, **23,** 599.
20 Findlay, D., Herries, D.G., Mathias, A.P., Rabin, B.R. and Ross, C.A. (1962), *Biochem. J.*, **85,** 152.
21 Bernhard, S. (1968), *The Structure and Function of Enzymes.* Benjamin, New York.
22 Gutfruend, H. (1965), *An Introduction to the Study of Enzymes.* Blackwell Scientific Publications, Oxford.
23 Scheraga, H.A. and Rupley, J.A. (1962), *Adv. Enzymology,* **24,** 161.
24 Kartha, G., Bello, J. and Harker, D. (1967), *Nature,* **213,** 862.
25 Wyckoff, H.W., Hardman, K.D., Allewell, N.M., Inagami, T., Johnson, L.N. and Richards, F.M. (1967), *J. Biol. Chem.*, **242,** 3984.
26 Gutte, B. and Merrifield, R.B. (1969), *J. Amer. Chem. Soc.*, **91,** 501.
27 Hirschmann, R., Nutt, R.F., Veber, D.F., Vitali, R.A., Varga, S.L., Jacob, T.A. Holly, F.W. and Denkewalter, R.G. (1969), *J. Amer. Chem. Soc.*, **91,** 507.
28 Bernd, G. and Merrifield, R.B. (1971), *J. Biol. Chem.*, **246,** 1922.
29 Markham, R. and Smith, J.D. (1952), *Biochem. J.*, **52,** 552.
30 Cohn, W.E. and Volkin, E. (1953), *J. Biol. Chem.*, **203,** 319.
31 Brown, D.M., Dekker, C.A. and Todd, A.R. (1952), *J. Chem. Soc.*, **2715.**
32 Beers, R.F. (1960), *J. Biol. Chem.*, **235,** 2393.
33 MacFayden, D.A. (1934), *J. Biol. Chem.*, **107,** 297.
34 Crook, E.M., Mathias, A.P. and Rabin, B.R. (1960), *Biochem. J.*, **74,** 234.
35 Kunitz, M. (1946), *J. Biol. Chem.*, **164,** 563.
36 Dickman, S.R., Aroskar, J.P. and Kropf, R.B. (1956), *Biochim. Biophys. Acta,* **21,** 539.
37 Zittle, C.A. and Reading, E.H. (1945), *J. Biol. Chem.*, **160,** 519.
38 Roth, J.S. and Milstein, S.W. (1952), *J. Biol. Chem.*, **196,** 489.
39 Egami, F., Takahashi, K. and Uchida, T. (1964), *Progress in Nucleic Acid Research,* Vol. 3, p. 59 (J.N. Davidson and W.E. Cohn, eds.). Academic Press, New York.
40 Uchida, T. and Egami, F. (1967), *J. Biochem.*, **61,** 44.
41 Glitz, D.G. and Dekker, C.A. (1964), *Biochem.*, **3,** 1391.
42 De Lamirande, G., Allard, C., DaCosta, H.C. and Cantero, A. (1954), *Science,* **119,** 351.
43 Roth, J.S. (1954), *J. Biol. Chem.*, **208,** 180.
44 Reid, E. and Nodes, J.T. (1959), *Ann. N.Y. Acad. Sci.*, **81,** 618.
45 Winicov, I. and Perry, R.P. (1974), *Biochem.*, **13,** 2908.
46 Prestayko, A.W., Lewis, B.C. and Busch, H. (1975), *Biochim. Biophys. Acta,* **319,** 323.
47 Mirault, M.E. and Scherrer, K. (1972), *Eur. J. Biochem.*, **28,** 197.

48 Bothwell, A.L.M. and Altman, S. (1975), *J. Biol. Chem.*, **250,** 1460.
49 Neu, H.C. and Heppel, L.A. (1964), *J. Biol. Chem.*, **239,** 3893.
50 Gesteland, R. (1966), *J. Mol. Biol.*, **16,** 67.
51 Robertson, H.D. Webster, R.E. and Zinder, N.D. (1967), *Virology*, **32,** 718.
52 Robertson, H.D., Webster, R.E. and Zinder, N.D. (1968), *J. Biol. Chem.*, **243,** 82.
53 Schweitz, H. and Ebel, J.P. (1971), *Biochemie*, **53,** 582.
54 Robertson, H.D. (1971), *Nature New Biol.*, **229,** 169.
55 Stern, R. (1970), *Biochem. Biophys. Res. Commun.*, **41,** 608.
56 Robertson, H.D. and Matthews, M.B. (1973), *Proc. Natl. Acad. Sci. USA*, **70,** 225.
57 Stern, R. and Wilczeck, J. (1973), *Fed. Proc.*, **32,** 620.
58 Spahr, P.F. and Gesteland, R.F. (1968), *Proc. Natl Acad. Sci. USA*, **58,** 876.
59 Gesteland, R.F. and Spahr, P.F. (1969), *Cold Spring Harbor Symp. Quant. Biol.*, **34,** 707.
60 Altman, S. and Smith, J.D. (1971), *Nature New Biol.*, **233,** 35.
61 Robertson, H.D., Altman, S. and Smith, J.D. (1972), *J. Biol. Chem.*, **247,** 5243.
62 Altman, S., Bothwell, A.L.M. and Stark, B.C. (1975), *Brookhaven Symp. Biol.*, **26,** 12.
63 Burdon, R.H. (1975), *Brookhaven Symp. Biol.*, **26,** 138.
64 Stein, H. and Hausen, P. (1969), *Science*, **166,** 393.
65 Hausen, P. and Stein, H. (1970), *Eur. J. Biochem.*, **14,** 278.
66 Roewekamp, W. and Sekeris, C.E. (1974), *Eur. J. Biochem.*, **43,** 405.
67 Keller, W. and Crouch, R. (1972), *Proc. Natl. Acad. Sci. USA*, **69,** 3360.
68 Banks, G.R. (1974), *Eur. J. Biochem.*, **47,** 499.
69 Leis, J.P., Berkower, I. and Hurwitz, J. (1973), *Proc. Natl. Acad. Sci. USA*, **70,** 466.
70 Weatherford, S.C. Weisberg, L.S., Achord, D.T. and Apirion, D. (1973), *Biochem. Biophys. Res. Commun.*, **69,** 1307.
71 Stavrinopoulos, J.S. and Chargaff, E. (1973), *Proc. Natl. Acad. Sci. USA*, **70,** 1959.
72 Haberkern, R.C. and Cantoni, G.L. (1973), *Biochem.*, **12,** 2389.
73 Spahr, P.F. and Schlessinger, D. (1963), *J. Biol. Chem.*, **238,** PC2251.
74 Singer, M.F. and Tolbert, G. (1964), *Science*, **145,** 593.
75 Schedl, P., Primakoff, P. and Roberts, J. (1975), *Brookhaven Symp. Biol.*, **26,** 53.
76 Castles, J.J. and Singer, M. (1969), *J. Mol. Biol.*, **40,** 1.
77 Venkov, P., Schlessinger, D. and Longo, D. (1971), *J. Biol.*, **108,** 601.
78 Lazarus, H.M. and Sporn, M.B. (1967), *Proc. Natl. Acad. Sci. USA*, **57,** 1386.
79 Perry, R.P. and Kelley, D.E. (1972), *J. Mol. Biol.*, **70,** 265.
80 Niyogi, S.K. and Datta, A.K. (1975), *J. Biol. Chem.*, **250,** 7307.
81 Datta, A.K. and Niyogi, S.K. (1975), *J. Biol. Chem.*, **250,** 7313.
82 Grandgenett, D., Gerard, G. and Green, M. (1972), *J. Virol.*, **10,** 1136.
83 Grandgenett, D., Gerard, G. and Green, M. (1973), *Proc. Natl. Acad. Sci. USA*, **70,** 230.
84 Deutscher, M.P. (1970), *J. Biol. Chem.*, **245,** 4225.

85 Hotchkiss, R.D. (1948), *J. Biol. Chem.*, **175**, 315.
86 Roth, J.S. (1959), *Ann. N.Y. Acad. Sci.*, **81**, 611.
87 Shortman, K. (1962), *Biochim. Biophys. Acta*, **55**, 88.
88 Singer, B., Fraenkel-Conrat, H. and Tsugita, A. (1961), *Virology*, **14**, 54.
89 Daigneauet, R., Bellemare, G. and Cousineau, G.H. (1971), *Lab. Practice*, **20**, 487.
90 Rosén, C.G. and Fedorscàk, I. (1966), *Biochim. Biophys. Acta*, **130**, 401.
91 Solymosy, F., Hüvös, P., Gulys, A., Kapovits, I., Gaal, O., Bagi, G. and Garkas, G.L. (1971), *Biochim. Biophys. Acta*, **238**, 406.
92 Holley, R.W., Apgar, J. and Merrill, S.H. (1961), *J. Biol. Chem.*, **236**, PC42.
93 Laskowski, M. (1971), *The Enzymes*, 3rd Edn, Vol. IV, p. 289 (P.D. Boyer, ed.), Academic Press, New York.
94 Cotton, F.A. and Hazen, E.E. (1971), *The Enzymes*, 3rd Edn, Vol. IV, p. 153
95 Anfinsen, C.B., Cuatrecasas, P. and Taniuchi, H. (1971), *The Enzymes*, Vol. IV, p. 177 (P.D. Boyer, ed.), Academic Press, New York.
96 Richards, G.M., du Vair, G. and Laskowski, M. (1965), *Biochemistry*, **4**, 501.
97 Linn, S.and Lehman, I.R. (1965), *J. Biol. Chem.*, **240**, 1287, 1294.
98 Lehman, I.R. (1960), *J. Biol. Chem.*, **235**, 1474.
99 Shisido, K. and Ando, T. (1972), *Biochim. Biophys. Acta*, **287**, 477.
100 Isaksson, L.A and Phillips, J.H. (1968), *Biochim. Biophys. Acta*, **155**, 63.
101 Mikulski, A.J. and Laskowski, M. (1970), *J. Biol. Chem.*, **245**, 5026.
102 Laskowski, M. (1971), *The Enzymes*, 3rd End, Vol. IV, p. 313 (P.D. Boyer, ed.), Academic Press, New York.
103 Cohn, W.E., Volkin, E. and Khym, J.X. (1957), *Biochem. Prep.*, **5**, 49.
104 Butler, G.C. (1955), *Methods in Enzymology*, Vol. 2, p. 561, (S.P. Colowick and N.O. Kaplan, eds.), Academic Press, New York.
105 Bernardi, R. and Bernardi, G. (1971), *The Enzymes*, 3rd Edn, Vol. IV, p. 329 (P.D. Boyer, ed.), Academic Press, New York.
106 Heppel, L.A. and Hilmoe, R.J. (1955), *Methods in Enzymology*, Vol. 2, p. 565 (S.P. Colowick and N.O. Kaplan, eds.), Academic Press, New York.
107 McDonald, M.R. (1955), *Methods in Enzymology*, Vol. 2, p. 437 (S.P. Colowick and N.O. Kaplan, eds.), Academic Press, New York.
108 McCarty, M. (1946), *J. Gen. Physiol.*, **29**, 123.
109 Kunitz, M. (1948), *Science*, **108**, 19.
110 Polson, A. (1956), *Biochim. Biophys. Acta*, **22**, 61.
111 Bollum, F.J. (1965), *J. Biol. Chem.*, **240**, 2599.
112 Tamm, C. and Chargaff, E. (1951), *Nature*, **168**, 916.
113 Gilbert, L.M., Overend, E.G. and Webb, M. (1951), *Exp. Cell Res.*, **2**, 349.
114 Young, E.T. and Sinsheimer, R.L. (1965), *J. Biol. Chem.*, **240**, 1274.
115 Tamm, C., Shapiro, H.S. and Chargaff, E. (1952), *J. Biol. Chem.*, **199**, 313.
116 Matsuda, M. and Ogoshi, H. (1966), *Biochim. Biophys. Acta*, **119**, 210.
117 Worthington Biochemical Corporation, Catalogue.
118 Sandeen, G. and Zimmerman, S.B. (1966), *Anal. Biochem.*, **14**, 269.
119 Bernardi, G. (1971), *The Enzymes*, 3rd Edn, Vol. IV, p. 271 (P.D. Boyer, ed.), Academic Press, Yew York.

120 Bernardi, G. (1965), *J. Mol. Biol.,* **13,** 603.
121 Lehman, I.R. (1971), *The Enzymes,* 3rd Edn, Vol. IV, p. 251 (P.D. Boyer, ed.), Academic Press, New York.
122 Potter, J.L. and Laskowski, M. (1959), *J. Biol. Chem.,* **234,** 1263.
123 Yasmineh, W.G. and Gray, E.D. (1968), *Biochem.,* **7,** 105.
124 Lehman, I.R. (1963), *Progress in Nucleic Acid Research,* Vol. 2, p. 84 (J.N. Davidson and W.E. Cohn, eds.), Academic Press, New York.
125 Lehman, I.R., Roussos, G.G. and Pratt, E.A. (1962), *J. Biol. Chem.,* **237,** 819.
126 Lehman, I.R. (1963), *Methods in Enzymology,* Vol. 6, p. 44 (S.P. Colowick and N.O. Kaplan, eds.), Academic Press, New York.
127 Radloff, R., Bauer, W. and Vinograd, J. (1967), *Proc. Natl. Acad. Sci. USA,* **57,** 1514.
128 Studier, F.W. (1965), *J. Mol. Biol.,* **11,** 373.
129 Friedberg, E.C. and Goldthwaite, D.A. (1969), *Cold Spring Harbor Symp. Quant. Biol.,* **33,** 271.
130 Friedberg, E.C. and Goldthwaite, D.A. (1969), *Proc. Natl. Acad. Sci. USA,* **62,** 934.
131 Yajko, D.M. and Weiss, B. (1975), *Proc. Natl. Acad. Sci. USA,* **72,** 688.
132 Ljungquist, S., Nyberg, B. and Lindahl, T. (1975), *FEBS Lett.,* **57,** 169.
133 Buttin, G. and Wright, M.R. (1968), *Cold Spring Harbor Symp. Quant. Biol.,* **33,** 259.
134 Anai, M., Takakata, H. and Takagi, Y. (1970), *J. Biol. Chem.,* **245,** 767.
135 Anai, M., Hirahashi, T., Yamauaka, M. and Takagi, Y. (1970), *J. Biol. Chem.,* **245,** 775.
136 Goldmark, P.J. and Linn, S. (1970), *Proc. Natl. Acad. Sci. USA,* **67,** 434.
137 Hurwitz, J., Becker, A., Gefter, M. and Gold, M. (1967), *J. Cell. Comp. Physiol.,* Suppl. 1, **70,** 181.
138 Sadowski, P., Ginsberg, B., Yudelevitch, A., Fesnier, L. and Hurwitz, J. (1968), *Cold Spring Harbor Symp. Quant. Biol.,* **33,** 165.
139 Sadowski, P.D. and Hurwitz, J. (1969), *J. Biol. Chem.,* **244,** 6182, 6192.
140 Henry, J.J. and Knippers, R. (1974), *Proc. Natl. Acad. Sci. USA,* **71,** 1549.
141 Burlingham, B.T., Doerfler, W., Pettersson, U. and Philipson, L. (1971), *J. Mol. Biol.,* **60,** 45.
142 Pogo, B.G.T. and Dales, S. (1969), *Proc. Natl. Acad. Sci. USA,* **63,** 820.
143 Lehman, I.R. (1960), *Biol. Chem.,* **235,** 1479.
144 Lehman, I.R. (1963), *Methods in Enzymology,* Vol. 6, p. 40 (S.P. Colowick *and N.O. Kaplan, eds.), Academic Press, New York.*
145 Lehman, I.R. and Richardson, C.C. (1964), *J. Biol. Chem.,* **239,** 233.
146 Richardson, C.C. and Kornberg, A. (1964), *J. Biol. Chem.,* **239,** 242.
147 Richardson, C.C., Lehman, I.R. and Kornberg, A. (1964), *J. Biol. Chem.,* **239,** 251.
148 Lehman, I.R. and Nussbaum, A.L. (1964), *J. Biol. Chem.,* **239,** 2628.
149 Kellock, M.G., Smellie, R.M.S. and Davidson, J.N. (1962), *Biochem. J.,* **84,** 112P.
150 Burdon, M.G., Smellie, R.M.S. and Davidson, J.N. (1964), *Biochim. Biophys. Acta,* **91,** 46.

151 Healy, J.W., Stollar, D. and Levine, L. (1963), *Methods in Enzymology,* Vol. 6, p. 49 (S.P. Colowick and N.O. Kaplan, eds.), Academic Press, New York.
152 Lacks, S. and Greenberg, B. (1967), *J. Biol. Chem.,* **242,** 3108.
153 Okazaki, T. and Kornberg, A. (1964), *J. Biol. Chem.,* **239,** 259.
154 Radding, C.M., Szpirer, J. and Thomas, R. (1967), *Proc. Natl. Acad. Sci. USA,* **57,** 277.
155 Jorgensen, S.E. and Koerner, J.F. (1966), *J. Biol. Chem.,* **241,** 3090.
156 Lieberman, R.P. and Oishi, M. (1974), *Proc. Natl. Acad. Sci. USA,* **71,** 4816.
157 Kornberg, A. (1974), *DNA Synthesis.* Freeman, San Francisco.
158 Cozzarelli, N.R., Kelly, R.B. and Kornberg, A. (1969), *J. Mol. Biol.,* **45,** 513.
159 Kelly, R.B., Atkinson, M.R., Huberman, J.A. and Kornberg, A. (1969), *Nature,* **224,** 495.
160 Klenow, H. and Overgaard-Hansen, K. (1970), *FEBS Lett.,* **6,** 25.
161 Oleson, A.E. and Koerner, J.F. (1964), *J. Biol. Chem.,* **239,** 2935.
162 Short, E.C., Jr. and Koerner, J.F. (1969), *J. Biol. Chem.,* **244,** 1487.
163 Trilling, D.M. and Aposhian, H.V. (1968), *Proc. Natl. Acad. Sci. USA,* **60,** 214.
164 McAuslan, B.R. (1971), in *Strategy of the Viral Genome,* p. 25. Ciba Symposium Volume. Churchill Livingstone, Edinburgh, London.
165 Morrison, J.M. and Keir, H.M. (1968), *J. Gen. Virol.,* **3,** 337.
166 Lindahl, T., Gally, J.A. and Edelman, G.M. (1969), *J. Biol. Chem.,* **244,** 5014.
167 Lindahl, T., Gally, J.A. and Edelman, G.M. (1969), *Fed. Proc.,* **28,** 348.
168 Lindberg, U. (1964), *Biochim. Biophys. Acta,* **72,** 237.
169 Dunn, D.B. and Smith, J.D. (1958), *Biochem. J.,* **68,** 627.
170 Wyatt, G.R. (1951), *Biochem. J.,* **48,** 584.
171 Wyatt, G.K. and Linzen, B. (1965), *Biochim. Biophys. Acta,* **103,** 588.
172 Tewari, K.K. and Wildman, S.G. (1966), *Science,* **153,** 1269.
173 Ray, D.S. and Hanawalt, P.C. (1964), *J. Mol. Biol.,* **9,** 812.
174 Salomon, R., Kaye, A.M. and Herzberg, M. (1969), *J. Mol. Biol.,* **43,** 581.
175 Dawid, I.B., Brown, D.D. and Reader, R.H. (1970), *J. Mol. Biol.,* **51,** 341.
176 Arber, W. and Linn, S. (1969), *Annu. Rev. Biochem.,* **38,** 467.
177 Haussman, R. and Gold, M. (1966), *J. Biol. Chem.,* **241,** 1985.
178 Hudnik-Plevik, T.A. and Melechen, N.E. (1967), *J. Biol. Chem.,* **242,** 4118.
179 Gefter, M. Haussman, R., Gold, M. and Hurwitz, J. (1966), *J. Biol. Chem.,* **241,** 1995.
180 Kaye, A.M. and Winocour, E. (1967), *J. Mol. Biol.,* **24,** 475.
181 Low, M., Hay, J. and Keir, H.M. (1969), *J. Mol. Biol.,* **46,** 205.
182 Low, M., Mechie, M. and Hay, J. (1971), *Biochem. J.,* **124,** 63P.
183 Borek, E. and Srinivasan, P.R. (1966), *Annu. Rev. Biochem.,* **35,** 275.
184 Gold, M., Hurwitz, J. and Anders, M. (1964), *Proc. Natl. Acad. Sci. USA,* **50,** 164.
185 Gold, M. and Hurwitz, J. (1964), *J. Biol. Chem.,* **239,** 3858.
186 Roy, P.H. and Smith, H.O. (1973), *J. Mol. Biol.* **81,** 427.

187 Gold, M., Haussman, R., Maitra, U. and Hurwitz, J. (1964), *Proc. Natl. Acad. Sci. USA,* **52,** 292.

188 Krueger, D.H., Presber, W., Hansen, S. and Rosenthal, H.A. (1975), *J. Virol.,* **16,** 453.

189 Burdon, R.H., Martin, B.T. and Lal, B. (1967), *J. Mol. Biol.,* **38,** 357.

190 Shied, B., Srinivasan, P.R. and Borek, E. (1968), *Biochem.,* **7,** 280.

191 Kalousek, F. and Morris, N.R. (1969), *J. Biol. Chem.,* **244,** 1157.

192 Adams, R.L.P., Turnbull, J., Smillie, E.J. and Burdon, R.H. (1974), 9th F.E.B.S. Meeting, Symposium on Post Synthetic Modification of Macromolecules, p. 39.

193 Turnbull, J.F. and Adams, R.L.P. (1975), Abstracts of 10th F.E.B.S. Meeting, No. 178.

194 Burdon, R.H. and Adams, R.L.P. (1969), *Biochim. Biophys. Acta,* **174,** 322.

195 Kappler, J.W. (1970), *J. Cell Physiol.,* **75,** 21.

196 Adams, R.L.P. (1971), *Biochim. Biophys, Acta,* **254,** 205.

197 Doskocil, J. and Sormova, Z. (1965), *Coll. Czech. Chem. Commun.,* **30,** 38.

198 Grippo, P., Iaccarino, M., Parisi, E. and Scarano, E. (1968), *J. Mol. Biol.,* **36,** 195.

199 Lark, C. (1968), *J. Mol. Biol.,* **31,** 401.

200 Marinus, M.G. and Morris, N.R. (1974), *J. Mol. Biol.,* **85,** 309.

201 Novogrodsky, A., Gefter, M., Maitra, U., Gold, M. and Hurwitz, J. (1966), *J. Biol. Chem.,* **241,** 1977.

202 Arber, W. (1974), *Prog. Nucleic Acid Res. Mol. Biol.* (W.E. Cohn, ed.), **14,** 1.

203 Murray, K. and Old, R.W. (1974), *Prog. Nucleic Acid Res. Mol. Biol.* (W.E. Cohn, ed.), **14,** 117.

204 Boyer, H. (1971), *Annu. Rev. Microbiol.,* **25,** 153.

205 Meselson, M., Yuan, R. and Heywood, J. (1972), *Annu. Rev. Biochem.,* **41,** 447.

206 Nathans, D. and Smith, H.O. (1975), *Annu. Rev. Biochem.,* **44,** 273.

207 Meselson, M. and Yuan, R. (1968), *Nature,* **217,** 1110.

208 Yuan, R., Bickle, T.A. Ebbers, W. and Brack, C. (1975), *Nature,* **256,** 556.

209 Smith, H.O. and Nathans, D. (1973), *J. Mol. Biol.,* **81,** 419.

210 Sugisaki, H. and Takanami, M. (1973), *Nature New Biol.,* **246,** 138.

211 Sharp, P.A., Sugden, B. and Sambrook, J. (1973), *Biochemistry,* **12,** 3055.

212 Danna, K.J., Sack, G.H., Jr. and Nathans, D. (1973), *J. Mol. Biol.,* **78,** 363.

213 Khoury, G. Martin, M.A., Lee, T.N.H., Danna, K.J. and Nathans, D. (1973), *J. Mol. Biol.,* **78,** 377.

214 Fareed, G.C., Garon, C.F. and Salzman, N.P. (1972), *J. Virol.,* **10,** 484.

215 Lovett, M.A., Katz, L. and Helinski, D.R. (1974), *Nature,* **251,** 337.

216 Adams, R.L.P., McKay, E.L., Craig, L.M. and Burdon, R.H. (1979), *Biochim. Biophys. Acta,* **563,** 72.

217 Singer, J., Robert-Ems, J. and Riggs, A.D. (1979), *Science,* **203,** 1019.

218 Bird, A.P., Taggart, M.H. and Smith, B.A. (1979), *Cell,* **17,** 889.

219 Mamelak, L. and Boyer, H.W. (1970), *J. Bact.,* **104,** 57.

220 Chang, S. and Cohen, S.N. (1977), *Proc. Natl. Acad. Sci. USA,* **74,** 4811.

221 Mandel, J.L. and Chambon, P. (1979), *Nucl. Acids Res.,* **7,** 2081.

222 Waalwijk, C. and Flavell, R.A. (1978), *Nucl. Acids Res.*, **5**, 3231.
223 Hadi, S.M., Bickle, T.A. and Yuan, R. (1975), *J. Biol. Chem.*, **250**, 4159.
224 Ravetch, J.V., Horiuchi, K. and Zinder, N.D. (1978), *Proc. Natl. Acad. Sci. USA*, **75**, 226.
225 Lautenberger, J.A., Kan, N.C., Lackey, D., Linn, S., Edgell, M.H. and Hutchison, C.A. (1978), *Proc. Natl. Acad. Sci. USA*, **75**, 2271.
226 Kan, N.C., Lautenberger, J.A., Edgell, M.H. and Hutchison, C.A. (1979), *J. Mol. Biol.*, **130**, 191.
227 Vovis, G.F., Horiuchi, K. and Zinder, N.D. (1974), *Proc. Natl. Acad. Sci. USA*, **71**, 3810.
228 Bickle, T.A., Brack, C. and Yuan, R. (1978), *Proc. Natl. Acad. Sci. USA*, **75**, 3099.
229 Subramanian, K.M., Weissman, S.M., Zain, B.S. and Roberts, R.J. (1977), *J. Mol. Biol.*, **110**, 297.
230 Kleid, D., Humayun, Z., Jeffrey, A. and Ptashne, M. (1976), *Proc. Natl. Acad. Sci. USA*, **73**, 293.
231 Lacks, S. and Greenberg, B. (1975), *J. Biol. Chem.*, **250**, 4060.
232 Roberts, R.J. (1976), *CRC Crit. Rev. Biochem.*, **4**, 123.
233 Bingham, A.H.A. and Atkinson, T. (1978), *Trans. Biochem. Soc.*, **6**, 315.
234 Roberts, R.J. (1977), in *DNA Insertion Elements, Plasmids and Episomes*, (ed. Bukhari, A.I., Shapiro, J.A. and Adhya, S.L.), Cold Spring Harbor, p. 757.
235 Brown, F.L., Musich, P.R. and Maio, J.J. (1978), *Nucl. Acids Res.*, **5**, 1093.
236 Le Bon, J.M., Kado, C.I., Rosenthall, L.J. and Chirikjian, J.G. (1978), *Proc.*
237 Tokai, N., Uchida, T. and Egami, F. (1966), *Biochim. Biophys. Acta*, **128**, 218.
238 Fujimoto, M., Kuninaka, A. and Yoshina, H. (1974), *Agr. Biol. Chem.*, **38**, 785.
239 Schuster, L. and Kaplan, N.O. (1955), *Methods in Enzymology*, **2**, p. 551 (ed. S.P. Colowick and N.O. Kaplan). Academic Press, New York.
240 Becker, A. and Hurwitz, J. (1967), *J. Biol. Chem.*, **242**, 936.
241 Novagrodsky, A. and Hurwitz, J. (1965), *Fed. Proc. Fed. Am. Soc. Exp. Biol.*, **24**, 602.
242 Richardson, C.C. (1965), *Proc. Natl. Acad. Sci. USA*, **54**, 158.
243 Novagrodsky, A., Tal, M., Traub, A. and Hurwitz, J. (1966), *J. Biol. Chem.*, **241**, 2933.
244 Labrie, F. and Sanger, F. (1969), *Biochem. J.*, **114**, 29P.
245 Szekely, M. and Sanger, F. (1969), *J. Mol. Biol.*, **43**, 607.
246 Heppel, C.A. (1966), *Procedures in Nucleic Acid Res.*, **1**, 31,
247 Liu, D.K., Liao, W.S.-L., Fritz, P.J. (1977), *Biochemistry*, **16**, 3361.
248 Ittel, M.E., Niedergang, C., Munoz, D., Petek, F., Okazaki, H. and Mandel, P. (1975), *J. Neurochem.*, **25**, 171.
249 Sutter, D. and Doerfler, W. (1980), *Proc. Natl. Acad. Sci. USA*, **77**, 253.
250 Desrosiers, R.C., Mulder, C. and Fleckenstein, B. (1979), *Proc. Natl. Acad. Sci. USA*, **76**, 3839.
251 Kaput, J. and Sneider, T.W. (1979), *Nucl. Acids Res.*, **7**, 2303.

252 Adams, R.L.P. and Gibb, S. (1979), unpublished observations.

253 Dawid, I.B., Brown, D.D. and Reeder, R.H. (1970), *J. Mol. Biol.*, **51,** 953.

254 Radman, M. Villani, G., Boiteux, S., Kinsella, A.R., Glickman, B.W. and Spadari, S. (1978), *Cold Spring Harbor Symp. Quant. Biol.*, **43,** 937.

255 Kappler, J.W. (1970), *J. Cell Physiol.*, **75,** 21.

256 Burdon, R.H. and Adams, R.L.P. (1969), *Biochim. Biophys. Acta,* **174,** 322.

257 Riggs, A.D. (1975), *Cytogenet. Cell Genet.*, **14,** 9.

258 Holliday, R. and Pugh, J.E. (1975), *Science,* **187,** 226.

259 Bird, A.P. (1978), *J. Mol. Biol.*, **118,** 49.

260 Coulondre, C., Miller, J.H., Farabrough, P.J. and Gilbert, W. (1978), *Nature,* **274,** 775.

261 Scarano, E., Iaccarino, M., Grippo, P. and Parisi, E. (1967), *Proc. Natl. Acad. Sci. USA,* **57,** 1394.

262 Roberts, R.J. (1980), *Nucl. Acids Res.*, **8,** r 63.

263 Roberts, R.J. (1980), *Gene,* **8,** 329.

264 Schmuckler, M., Jewett, P.B. and Levy, C.C. (1975), *J. Biol. Chem.*, **250,** 2206.

265 Kwan, C.N. (1976), *J. Biol. Chem.*, **251,** 7132.

266 Kwan, C.N., Gotoh, S. and Schlessinger, D. (1974) *Biochim. Biophys. Acta,* **249,** 428.

267 Burdon, R.H. and Adams, R.L.P. (1980), *TIBS,* **5,** 294.

268 Chase, J.W. and Richardson, C.C. (1974), *J. Biol. Chem.*, **248,** 4545.

269 Yuan, R., Hamilton, D.L. and Burckhardt, J. (1980), *Cell,* **20,** 237.

7 The metabolism of nucleotides

7.1 ANABOLIC PATHWAYS

A supply of ribonucleoside and deoxyribonucleoside 5′-triphosphates is required for the biosynthesis of nucleic acids (see Chapters 8 and 10). The synthesis of these compounds occurs in two main stages, (1) the formation of the purine and pyrimidine ring systems and their conversion into the parent ribonucleoside monophosphates, inosine 5′-monophosphate (IMP) the uridine 5′-monophosphate (UMP); (2) a series of interconversions involving the reduction of ribonucleotides to deoxyribonucleotides and the phosphorylation of these nucleotides to form a balanced set of 5′-triphosphates (Fig. 7.8).

7.2 THE BIOSYNTHESIS OF THE PURINES

The pathway of biosynthesis of the purines leads directly to the production of their nucleoside 5′-monophosphates and this subject has been so extensively reviewed [1–8, 136] that only an outline need be given here.

It is known from experiments with isotopes that the sources of the atoms in the purine ring are as shown in Fig. 7.1.

The first step in the biosynthetic pathway is the phosphorylation of ribose 5′-phosphate by transfer of the pyrophosphoryl residue of ATP to carbon 1 of the ribose moiety to form 5-phosphoribosyl-1-pyrophosphate (PRPP) (Fig. 7.2).

The enzyme catalysing the reaction is *PRPP synthetase.* Assembly of the purine ring itself begins with the transfer of the γ-amino group of glutamine to PRPP forming 5-phosphoribosylamine (PRA) (Fig. 7.3) under the influence of the enzyme *phosphoribosyl pyrophosphate amidotransferase* (*amidophosphoribosyl transferase,* EC 2.4.2.14).

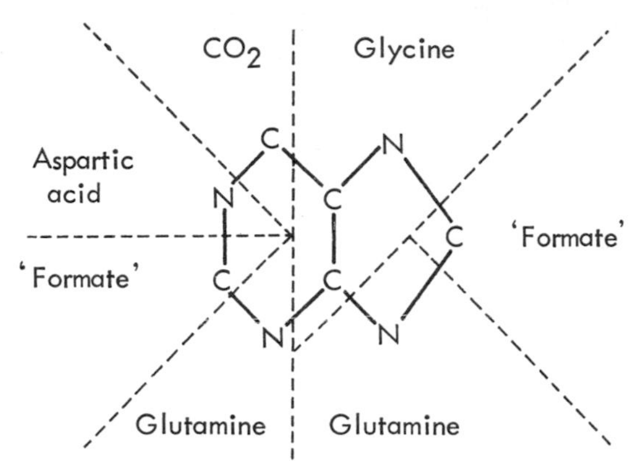

Fig. 7.1 Sources of the atoms in the purine ring.

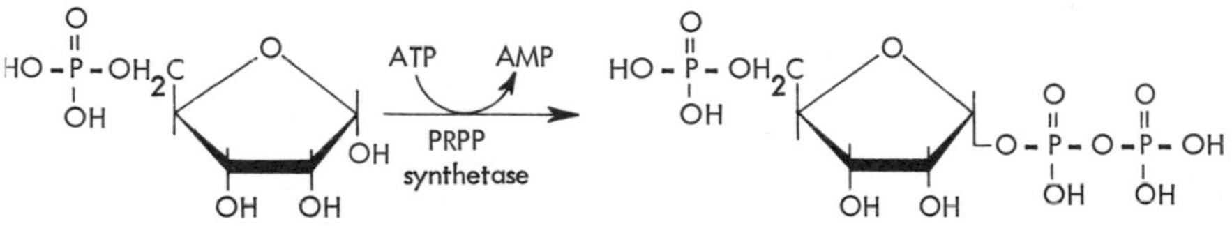

ATP → AMP, PRPP synthetase

Fig. 7.2 Formation of 5-phosphoribosyl-1-pyrophosphate (PRPP).

Glycine then reacts with PRA to give glycinamide ribonucleotide (GAR) a nucleotide-like compound in which the amide of glycine takes the place of the usual purine or pyrimidine base. The reaction sequence continues by formylation from N^5,N^{10}-methenyl-tetrahydrofolic acid to give formylglycinamidine ribonucleotide (formyl-GAM). Ring closure ensues, producing the imidazole ring compound 5-aminoimidazole ribonucleotide (AIR). Carboxylation of this compound gives 5-aminoimidazole-4-carboxylic acid ribonucleotide (carboxyl-AIR). The corresponding amide, 5-aminoimidazole-4-carboxamide ribonucleotide (AICAR) is produced in the subsequent two reactions via an intermediate compound, 5-aminoimidazole-4-succinocarboxamide ribonucleotide (succino-AICAR). The purine ring system is completed when N^{10}-formyltetrahydrofolic acid donates its formyl group to the 5-amino group of the imidazole carboxamide ribonucleotide. The complete parent ribonucleotide is inosinic acid (inosine 5′-monophosphate, IMP).

By virtue of the role of folic acid derivatives as donors of 1-carbon units at two different stages (B and D) in this sequence of reactions, analogues of folic acid such as aminopterin and amethopterin (methotrexate) are powerful inhibitors of purine biosynthesis. Because of this and because, as will be seen later, they also inhibit the formation of thymine nucleotides, these compounds interfere with the biosynthesis of nucleic acids and are widely used in the treatment of certain forms of

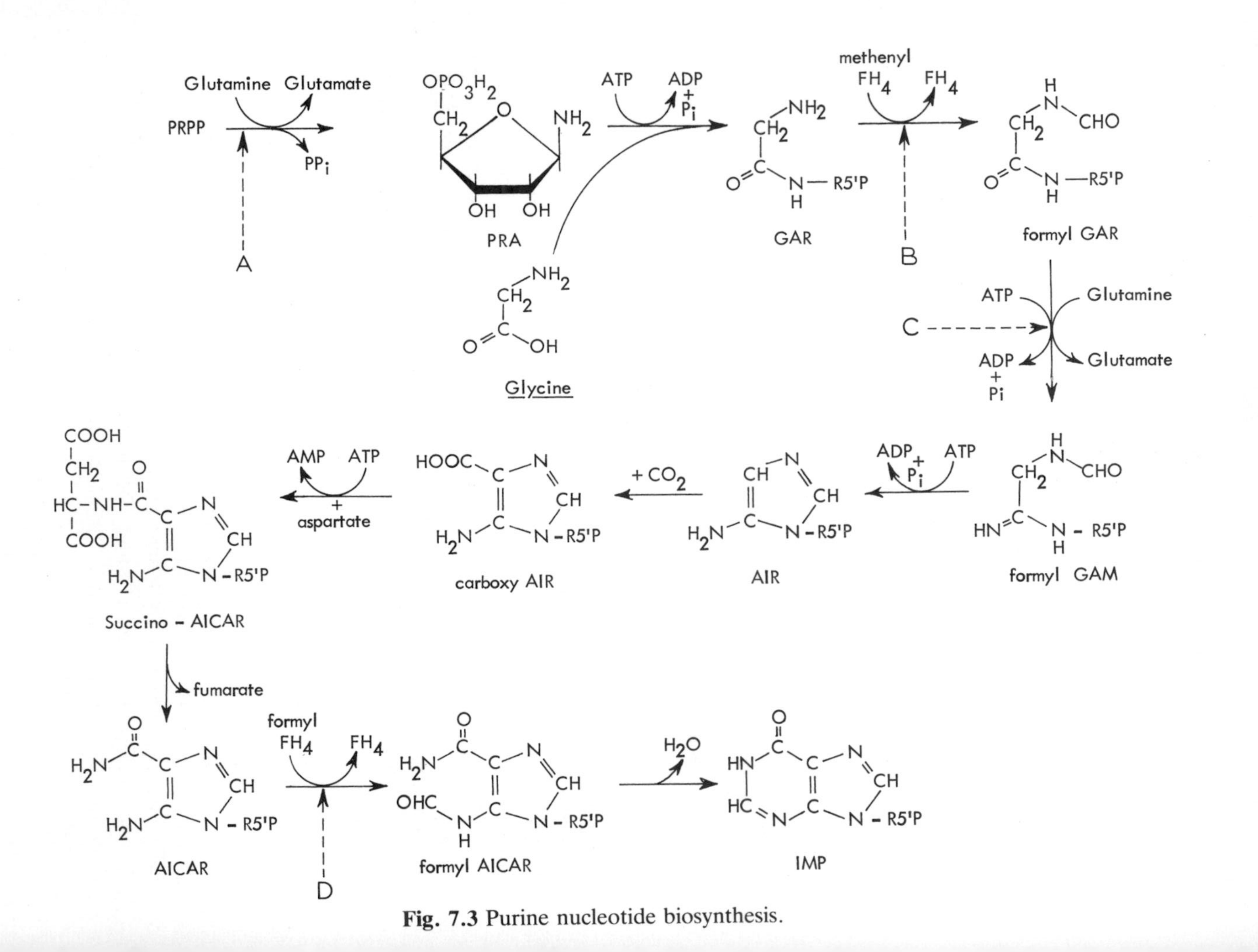

Fig. 7.3 Purine nucleotide biosynthesis.

cancer and related diseases [9–11]. In cell biology they are employed in studies on cell hybridization and transformation (see p. 192, 496).

Two other important inhibitors of purine nucleotide biosynthesis are azaserine:

$$\bar{N}{=}\overset{+}{N}{=}CH\text{-}CO\text{-}OCH_2CH(NH_2)\text{-}COOH$$

and 6-diazo-5-oxonorleucine (DON):

$$\bar{N}{=}\overset{+}{N}{=}CH\text{-}CO\text{—}CH_2\text{-}CH_2\text{-}CH(NH_2)\text{-}COOH$$

both of which are analogues of glutamine:

$$H_2N\text{-}CO\text{-}CH_2\text{-}CH_2\text{-}CH(NH_2)\text{-}COOH$$

and inhibit the sequence at the amination steps (A and C) from PRPP to PRA and from formyl-GAR to formyl-GAM [12, 133].

Inosine 5′-monophosphate is the common precursor of both adenosine and guanosine 5′-monophosphates. Amination of IMP to AMP proceeds in two stages with the intermediate formation of adenylosuccinic acid (Fig. 7.4).

This reaction, in which the amino group of aspartate is transferred to C-6 of IMP to give AMP, resembles the reaction above in which 5-aminoimidazole-4-carboxamide ribonucleotide is formed from 5-aminoimidazole carboxylic acid ribonucleotide (Fig. 7.3). One difference, however, is the requirement for GTP as coenzyme in the reaction forming adenylosuccinic acid from IMP.

The formation of GMP from IMP is also a two-stage reaction in which xanthosine 5′-monophosphate (XMP) is initially formed and then aminated to give GMP (Fig. 7.4). This amination reaction, like the earlier reactions in the sequence that utilize glutamine, is inhibited by azaserine and DON [134].

The two purine mononucleotides AMP and GMP are phosphorylated by kinases through the diphosphate stage to give ATP and GTP.

7.3 PREFORMED PURINES AS PRECURSORS

In 1947, Kalckar [13] demonstrated the interaction of purine bases and ribose-1-phosphate to yield nucleosides and inorganic phosphate. Such reactions, which are forms of transglycosidation, are reversible and are catalysed by enzymes termed *nucleoside phosphorylases* (EC 2.4.2.1):

$$\text{Hypoxanthine} + \text{ribose 1-phosphate} \rightleftharpoons \text{Inosine} + P_i$$

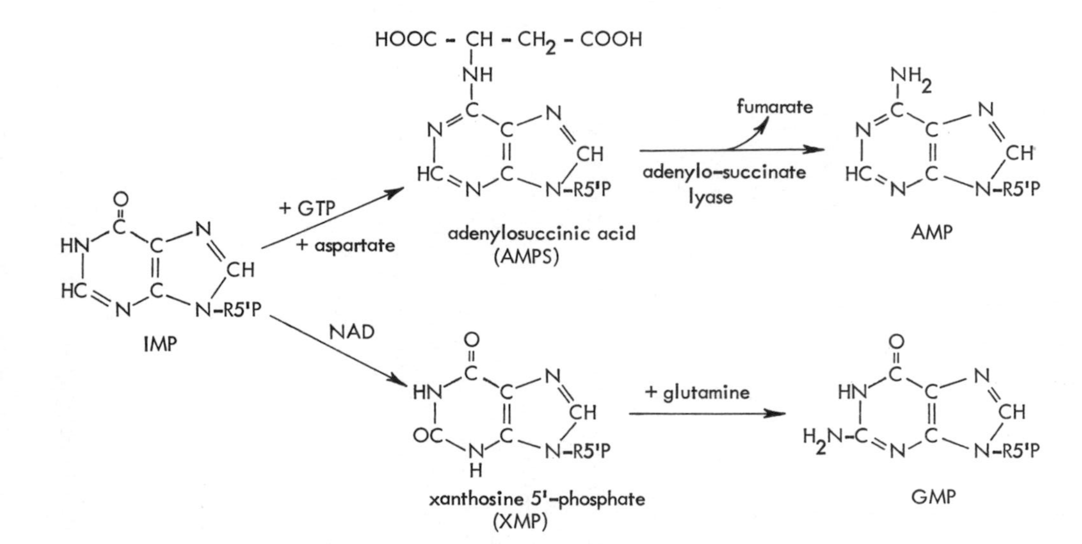

Fig. 7.4 Formation of AMP and GMP from IMP.

5-aminoimidazole-4-carboxamide can take the place of a purine base in this type of reaction.

Since most nucleosides can be phosphorylated by ATP under the influence of appropriate phosphokinases, a route exists for the biosynthesis of ribonucleotides from preformed purines.

A much more important mechanism for the conversion of bases into nucleotides involves PRPP which is also concerned in the *de novo* pathway described above. Under the influence of enzymes originally termed *nucleotide pyrophosphorylases* by Kornberg [14], but more correctly called *phosphoribosyltransferases,* bases can react with PRPP to form nucleotides and pyrophosphate [15].

The reactions catalysed by phosphoribosyl transferases are illustrated in Fig. 7.5.

One such enzyme, *adenine phosphoribosyltransferase* (EC 2.4.2.7), forms AMP from adenine and PRPP. A second enzyme, *hypoxanthine–guanine phosphoribosyltransferase* (EC 2.4.2.8) (HGPRT), converts hypoxanthine and guanine into IMP and GMP respectively in the presence of PRPP.

Sophisticated control mechanisms govern the pathway of purine nucleotide biosynthesis although there are major variations in these mechanisms from organism to organism [135]. A generalized and consolidated summary of these controls is shown in Fig. 7.5 from which

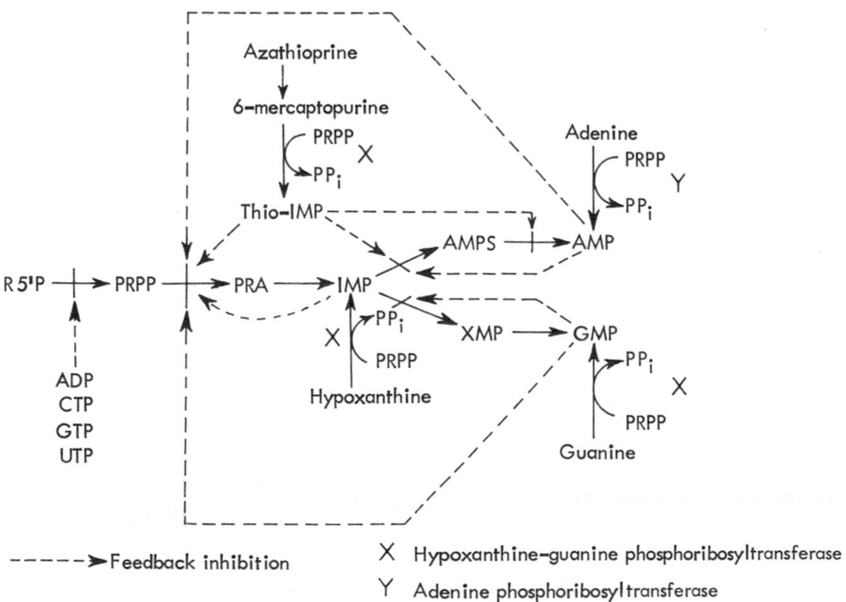

Fig. 7.5 Phosphoribosyl transferases and control of purine nucleotide biosynthesis.

it can be seen that both AMP and GMP exert feedback control on the formation of PRA from PRPP [60, 61]. In some systems, AMP also controls the formation of AMP from IMP whereas in others GMP controls the formation of XMP from IMP. GTP moreover is required as a co-factor in the synthesis of AMP from IMP. In addition to feedback inhibition most of the enzymes of the pathway are repressed by purine bases, nucleosides and nucleotides.

The enzyme HGPRT also catalyses the formation of the corresponding thio-IMP from the drugs 6-mercaptopurine and azathioprine (imuran). This analogue of IMP inhibits the formation of PRA and AMP and so prevents biosynthesis of the normal purine nucleotides [16, 17]. Because such drugs inhibit purine (and therefore nucleic acid) biosynthesis they are sometimes used as immunosuppressive or cancerostatic agents [18]. Azathioprine is also of value in the treatment of gout by inhibiting purine formation.

In the rare condition in children known as the *Lesch–Nyhan syndrome* there is a deficiency of the enzyme hypoxathine–guanine phosphoribosyltransferase [19–22]. Because of this there is an increase in the concentration of PRPP which in turn leads to increased synthesis of PRA and consequently to overproduction of purine nucleotides and ultimately of uric acid. The condition is associated with excessive uric acid synthesis and is resistant to the action of azathioprine presumably because this compound is only active after conversion into thio-IMP (Fig. 7.5), a reaction that requires the missing enzyme HGPRT.

7.4 THE BIOSYNTHESIS OF THE PYRIMIDINES

Whereas purine nucleotide biosynthesis proceeds by growth of the purine ring on PRPP, the formation of the pyrimidine nucleotides involves assembly of a pyrimidine derivative, orotic acid, and the subsequent combination of this moiety with PRPP. The complete series of enzymic reactions giving rise to the parent pyrimidine mononucleotide (uridine 5′-monophosphate, UMP) is shown in Fig. 7.6 [4–6, 24, 25].

The starting compounds are aspartic acid and carbamoylphosphate which combine under the influence of *aspartate carbamoyltransferase* to form carbamoylaspartate. Formation of the pyrimidine ring is then effected by the action of *dihydro-orotase* giving dihydro-orotic acid, dehydrogenation of which produces the important pyrimidine intermediate orotic acid. A phosphoribosyltransferase reaction then follows in which orotic acid accepts a ribose 5-phosphate group from PRPP. The

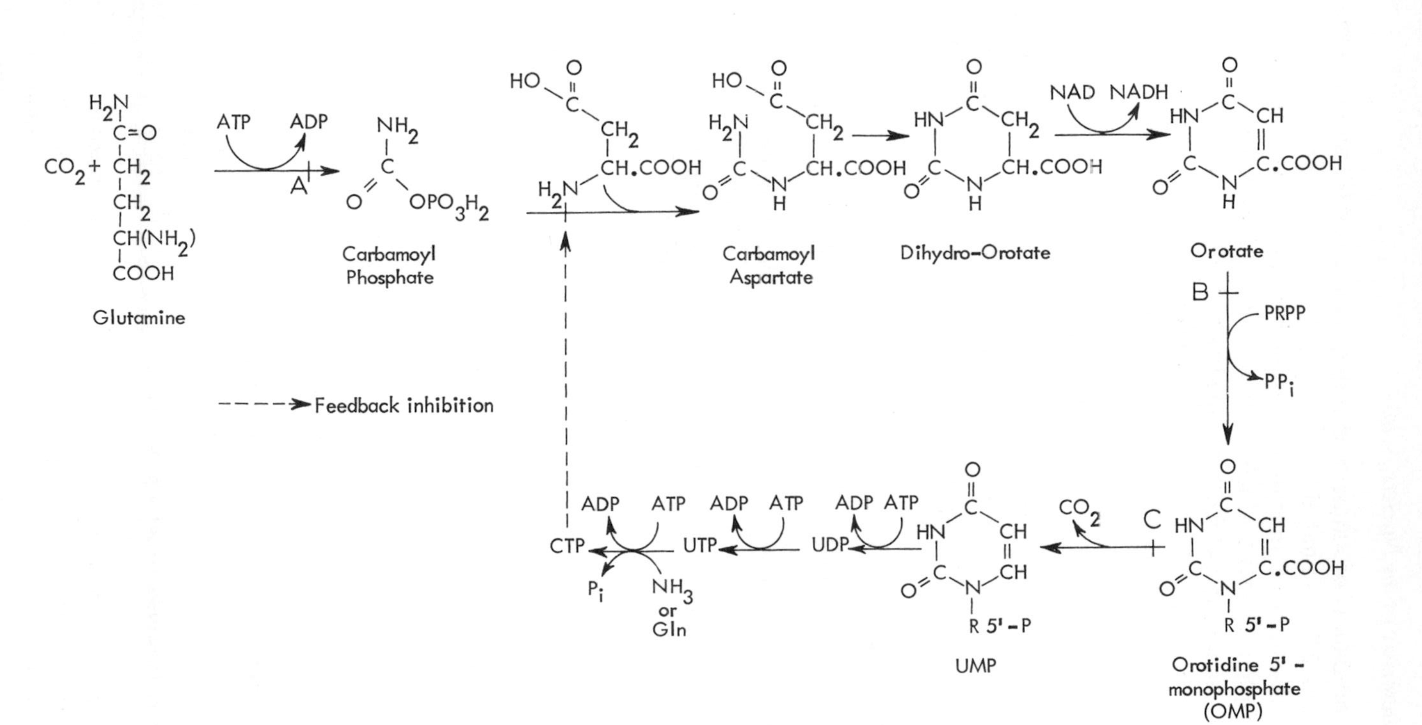

Fig. 7.6 Biosynthesis of pyrimidine nucleotides.

resulting product is orotidine 5′-monophosphate (OMP) and inorganic pyrophosphate is eliminated.

Decarboxylation of orotidine 5′-monophosphate gives uridine 5′-monophosphate (UMP) which is then converted by kinases through uridine 5′-diphosphate (UDP) into uridine 5′-triphosphate (UTP) and it is at this level of phosphorylation that conversion of uracil into cytosine takes place [26–29] under the influence of the enzyme *CTP synthetase* (EC 6.3.4.2):

$$UTP + NH_3 + ATP \rightarrow CTP + ADP + P_i$$

In eukaryotes, the first three enzymes in the pyrimidine biosynthetic pathway (carbamoylphosphate synthetase, aspartate carbamoyltransferase and dihydro-orotase) are all part of a single multifunctional protein of molecular weight 200000 [33, 35, 37, 41]. The carbamoyltransferase is allosterically controlled being activated by ATP and inhibited by CTP [54, 55, 99]. UMP exerts feedback inhibition on the synthesis of carbamoylphosphate [135, 99] and inhibits the decarboxylaton of OMP to UMP [63].

Inhibitors of pyrimidine biosynthesis include azaserine and DON (Fig. 7.6A) (see p. 183) which inhibit carbamoylphosphate synthesis in those systems utilizing glutamine as the amino group donor, 5-azaorotate, which inhibits the formation of OMP from orotic acid (B) and 6-azauridine which blocks the decarboxylation of OMP to UMP (C) [132].

Exposure of cultured cells to *N*-phosphono-acetyl-L-aspartate (a transition state analogue of aspartate carbamoyltransferase) leads to the selection of mutant cells which have amplified the gene coding for the multifunctional enzyme [33, 35, 37] (see p. 449).

7.5 THE BIOSYNTHESIS OF DEOXYRIBONUCLEOTIDES AND ITS CONTROL

The conversion of ribose into deoxyribose takes place at nucleotide level without breakage of the glycosidic linkage [29–31] since compounds such as uniformally labelled [^{14}C]cytidine are incorporated into the dCMP residues of DNA without change in the relative specific radioactivities of sugar and base.

Two related but readily distinguishable systems have been purified for the reduction of the ribosyl moiety of ribonucleotides to the corresponding deoxyribosyl derivative [31, 32]. In *Lactobacillus leichmannii* the reductase uses ribonucleoside triphosphates and requires a cobamide coenzyme. In *E. coli,* mammalian cells and almost all other systems investigated, the reductase uses ribonucleoside diphosphates and requires no coenzyme. The *L. leichmannii* enzyme is monomeric and has a molecular weight of 76000. The *E. coli* enzyme has two atoms of iron and two pairs of non-identical subunits, i.e. it has the constitution $\alpha\alpha'\beta_2$. α (the product of the nrd A gene) has a molecular weight of 80000 and has sulphydryl groups at the active site, α' differs from α in the *N*-terminal amino acid. β (the product of the nrd B gene) has a molecular weight of 39000 [42]. The mammalian enzyme has a similar constitution [130].

The mechanism for the reduction of CDP to dCDP in *E. coli* is known as the *thioredoxin system.* Thioredoxin is a sulphur-containing protein with some 108 amino acid residues [34]. It is reduced by NADPH under the influence of *thioredoxin reductase,* a flavoprotein containing FAD. The reduced thioredoxin in turn reduces CDP to the corresponding deoxy- derivative, becoming itself reoxidized to thioredoxin. Recent evidence has suggested that other reducing agents, e.g. glutathione, may be able to substitute for thioredoxin [42].

Ribonucleotide reductase reduces the four nucleotides, ADP, GDP, CDP and UDP to the corresponding deoxyribonucleotides. These are then phosphorylated by kinases to the triphosphates. dATP, dGTP and dCTP are used directly for DNA synthesis but dUTP is rapidly hydrolysed to dUMP by an active dUTPase. This prevents incorporation of dUTP into DNA (see p. 217). dUMP is converted by a series of steps into dTTP (see below).

There is a single catalytic site on ribonucleotide reductase for all four ribonucleoside diphosphates. The diphosphate to be reduced is determined by the conformation of the enzyme. (Evidence for two enzymes in mammalian systems can be explained on consideration of the allosteric controls [42–44, 129]). Enzyme conformation is altered by nucleotides bound at the allosteric sites. Although most studies have been performed using the enzyme from *E. coli*, the mammalian enzyme responds similarly. One of the allosteric sites (the l or low affinity site) binds ATP or dATP. When ATP is bound the enzyme is active but dATP inhibits all activity, possibly by causing aggregation [42, 130]. Deoxyadenosine is a potent inhibitor of DNA synthesis as it is rapidly converted into dATP which inhibits ribonucleotide reductase [63–65]. Binding to the other allosteric site (the h or high affinity site) is more complex. When ATP (also dATP in *E. coli*) is bound to the h site, reduction of CDP and UDP is enhanced. dGTP in the h site inhibits reduction of GDP, UDP and CDP but stimulates reduction of ADP. dTTP bound to the h site stimulates reduction of GDP but inhibits (possibly via dGTP) reduction of CDP and UDP (See Fig. 7.7).

The fine control exerted over the activity of ribonucleotide reductase is in line with its being a key enzyme in the programmed production of a balanced supply of deoxyribonucleoside triphosphates for synthesis of DNA [42, 125]. In support of this is the finding that the tissue level of ribonucleotide reductase correlates with the growth rate of the tissue and shows variations in activity during the cell cycle [45, 46].

Direct measurement of the deoxyribonucleoside triphosphate pool sizes in cells [47–49, 52, 53] shows them to be very low in non-dividing cells but to rise during S phase in parallel with increased ribonucleotide reductase activity. Pool sizes reach a maximum in G2 phase and then fall after mitosis.

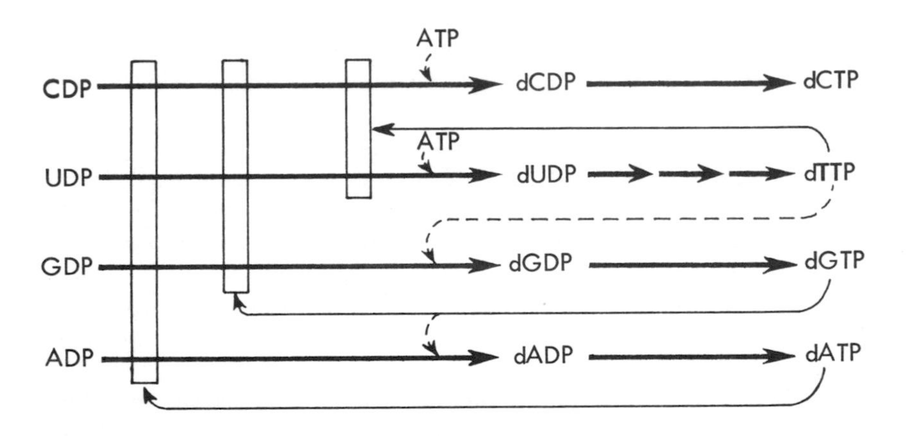

Fig. 7.7 Allosteric control of ribonucleotide reductase. (Open bars show inhibition: dashed line stimulation) (Based on Thelander and Reichard [42]).

Addition to cells of high concentrations of thymidine leads to the formation of a large intracellular pool of dTTP [48, 56–59]. This feeds back (either directly or through stimulating reduction of GDP) to inhibit further reduction of UDP and CDP bringing about a deficiency of dCTP and hence an inhibition of DNA synthesis. Addition of deoxycytidine, which is rapidly phosphorylated to dCTP, bypasses the block once again allowing DNA synthesis to occur [56, 58, 59, 131].

By treating cultured cells with normally lethal concentrations of deoxyribonucleosides an amplification of the gene coding for ribonucleotide reductase can be induced [63] (see p. 449).

Ribonucleotide reductase is inhibited by hydroxyurea [36, 42] which acts as a free radical scavenger [127, 128]. The action of hydroxyurea is irreversible in *E. coli* but is readily reversed in mammalian systems allowing its use as a chemotherapeutic agent and in cell synchronization [36].

7.6 THE BIOSYNTHESIS OF THYMINE DERIVATIVES

The essential step in the formation of thymine nucleotides is the methylation of deoxyuridine monophosphate (dUMP) to produce thymidine monophosphate (dTMP) (dUMP→dTMP) under the influence of thymidylate synthetase. The process is elaborate and takes place in several stages. The source of the additional carbon atom at C-5 is N^5, N^{10}-methylene tetrahydrofolic acid [38, 39]. The reaction is as follows:

$$N^5, N^{10}\text{-methylene tetrahydrofolate} + \text{dUMP} \rightarrow \text{dihydrofolate} + \text{dTMP}$$

The dihydrofolate is reduced again to tetrahydrofolate under the influence of dihydrofolate reductase:

$$\text{dihydrofolate} + \text{NADPH} + \text{H}^+ \rightarrow \text{tetrahydrofolate} + \text{NADP}^+$$

This reaction is powerfully inhibited by the folic acid analogues aminopterin and amethopterin (methotrexate) which therefore inhibit the formation of thymine derivates.

Thymidylate synthetase is low in non-growing cells but increases in activity in growing cells [126]. It is inhibited by 5-fluoro-dUMP. For this reason 5-fluorodeoxyuridine is an inhibitor of DNA synthesis.

Extracts of *E. coli* infected with a T-even bacteriophage contain the enzyme deoxycytidylate hydroxymethylase which brings about the formation of 5-hydroxymethyl deoxycytidylic acid from formaldehyde and deoxycytidylic acid in the presence of N^5, N^{10}-methylene tetrahydrofolic acid [40] (p. 264).

Another pathway leading to the synthesis of dTTP involves the deamination of dCMP to dUMP by the enzyme dCMP deaminase (dCMP aminohydrolase EC 3.5.4.12). This enzyme is present in only small amounts in non-dividing tissues but is active in rapidly growing tissues such as embryos, spleen and regenerating rat liver [68, 71–74, 77, 78, 64]. It thus shows a similar response to growth as thymidine kinase, thymidylate synthetase, ribonucleotide reductase and DNA polymerase (see p. 245). Although substrate for the enzyme may arise by phosphatase action on dCDP and dCTP the levels of circulating deoxycytidine probably are the major source of cellular dCMP.

dCMP deaminase is under allosteric control [65–67, 69, 72]. The enzyme is inhibited by dTTP and stimulated by dCTP and hence cellular dCMP is channelled towards the pyrimidine deoxynucleoside triphosphate present in limiting amounts.

7.7 AMINOPTERIN IN SELECTIVE MEDIA

Because of the part played by folic acid in purine and thymidylate biosynthesis, analogues of folic acid have become important in cell biology as ingredients in HAT medium (medium containing hypoxanthine, aminopterin and thymidine) [59, 70]. This growth medium is used for selection of animal cells able to bypass the aminopterin-imposed blocks in purine and thymidylate metabolism by incorporating exogenous hypoxanthine and thymidine. Mutant cells lacking the enzymes hypoxanthine phosphoribosyltransferase (HGPRT) or thymidine kinase (TK) cannot grow in HAT medium. However, fusion of a TK^- cell with an $HGPRT^-$ cell leads to a hybrid cell which can grow in HAT medium [75]. Similarly metabolic cooperation is shown by mixed cultures of TK^- and $HGPRT^-$ cells which exchange thymidylate and purine nucleotides across gap junctions and hence grow in HAT medium [76].

HAT medium is also used to select for transformants of TK^- cells treated with DNA containing the thymidine kinase gene from herpes simplex virus, linked by genetic-engineering techniques to any other DNA [89] (see Chapter 14).

7.8 FORMATION OF NUCLEOSIDE TRIPHOSPHATES

Kinases convert nucleosides into nucleoside monophosphates, diphosphates and triphosphates by transferring the α-phosphate from ATP to the substrate [6, 29]. The kinases which act on the pyrimidine nucleosides and deoxyribonucleosides enable the preformed materials to be utilized for nucleic acid synthesis by the so called 'salvage' pathway; e.g. thymidine→dTMP→dTDP→dTTP [50, 51]. The thymidine kinases are unique in that they are active in dividing tissues but low in non-growing tissues [79–83, 97] and activity shows regular changes during the cell cycle [97, 98, 124]. Some animal viruses code for an additional thymidine (or deoxypyrimidine nucleoside) kinase [84–88].

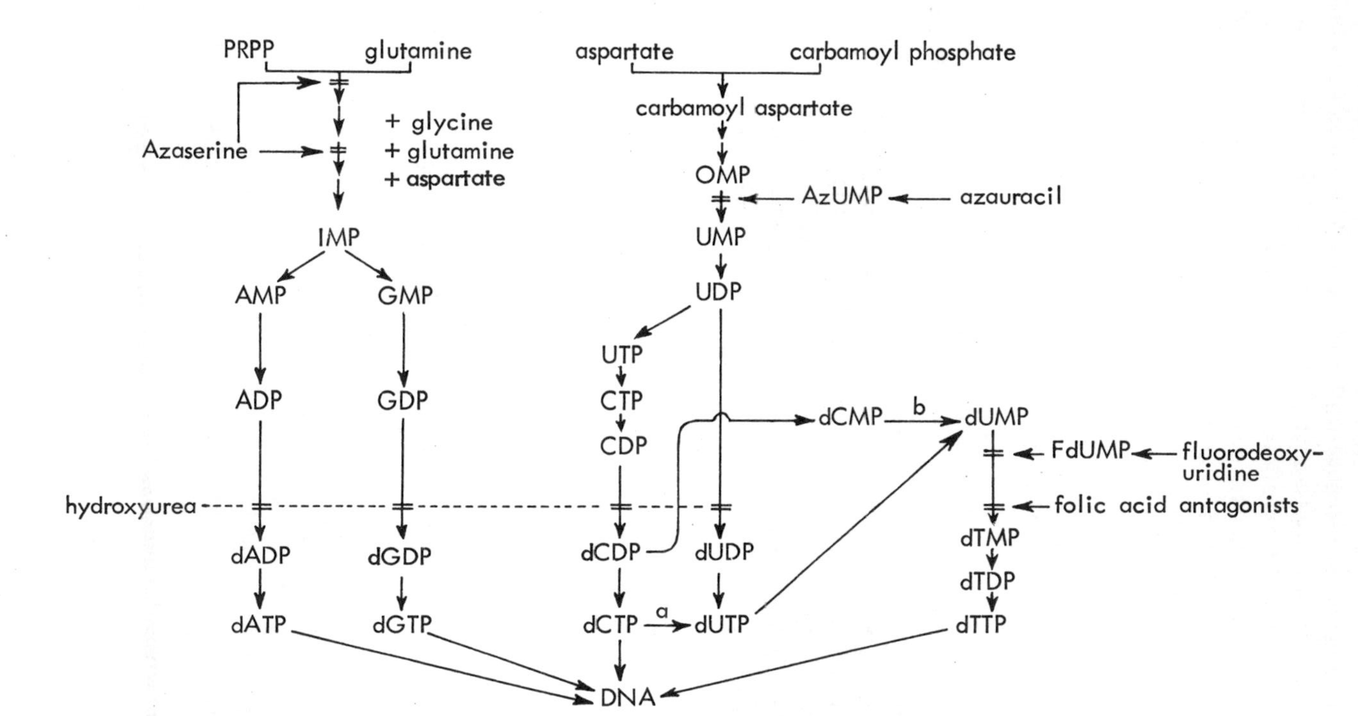

Fig. 7.8 Pathways involved in the biosynthesis of deoxyribonucleoside triphosphates: (a) bacterial systems only; (b) eukaryotes. The blocking action of some antimetabolites is indicated.

Thymidine kinase is also able to phosphorylate 5-fluorodeoxyuridine but the 5-FdUMP produced is an inhibitor of thymidylate synthetase and because of this fluorodeoxyuridine is an inhibitor of DNA synthesis.

Thymidine kinase is inhibited by dTTP [90–96] and by dCTP [94, 95] but there is also an element of forward promotion [92] which means that on incubating cells with thymidine the size of the dTTP pool produced is proportional to the external thymidine concentration [57, 59, 92].

Nucleotide interconversions are summarized in Fig. 7.8.

7.9 GENERAL ASPECTS OF CATABOLISM

RNA and DNA are hydrolysed by nucleases and diesterases first to oligonucleotides and eventually to mononucleotides and nucleosides, and the glycosidic linkages between the purine or pyrimidine bases and the sugar moieties are cleaved either hydrolytically or phosphorolytically to yield the free purine and pyrimidine bases [7, 100, 101]. The nature and function of nucleases and phosphodiesterases have been considered in Chapter 6, and the pyrophosphatases and phosphatases that attack nucleotides have been reviewed extensively [102–104].

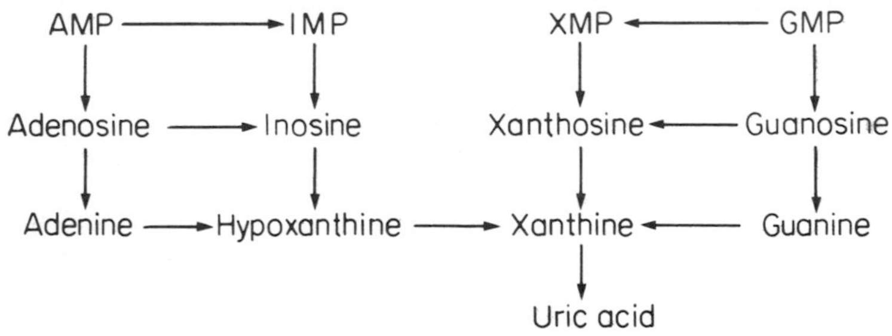

Fig. 7.9 The degradation of purines at the levels of nucleotides, nucleosides and bases.

It is recognized that cellular DNA tends to be strongly conserved and that the amount of degradation of DNA in normal circumstances is small. Some species of RNA have a relatively short life span and are therefore degraded quite rapidly. There is also evidence that nucleic acids may not undergo total degradation but that some of the intermediate products such as nucleotides and nucleosides may be reutilized so-called 'salvage' pathways.

7.10 PURINE CATABOLISM

The breakdown of purine nucleotides has been studied over many years. Adenine and its nucleosides and nucleotides can be deaminated hydrolytically under the influence of the enzymes adenine deaminase, adenosine deaminase, and adenylate deaminase to yield hypoxanthine, inosine, or inosine monophosphate respectively (Fig. 7.9) and guanine nucleotides are similarly attacked by guanine, guanosine or guanylate deaminases to yield xanthine or its ribose derivatives (Fig. 7.9). Hypoxanthine and xanthine are then oxidized under the influence of xanthine oxidase to yield uric acid (Fig. 7.10). Although the distribution of the enzymes involved is far from uniform in different species, this scheme of purine degradation appears to be of fairly general application, and experiments with ^{15}N have shown that, as might be expected, the administration of labelled purines to animals is followed by the appearance of the isotope in the excreted uric acid or in its further degradation products.

Uric acid itself is excreted by only a few mammals, since most non-uricotelic animals are provided with the enzyme uricase, which oxidizes uric acid to the much more soluble allantoin, and under certain conditions to other end-products as well [105]. The conversion of uric acid into allantoin appears to involve a number of intermediate compounds including the symmetrical compound hydroxyacetylene-diureine-carboxylic acid [7, 108–110].

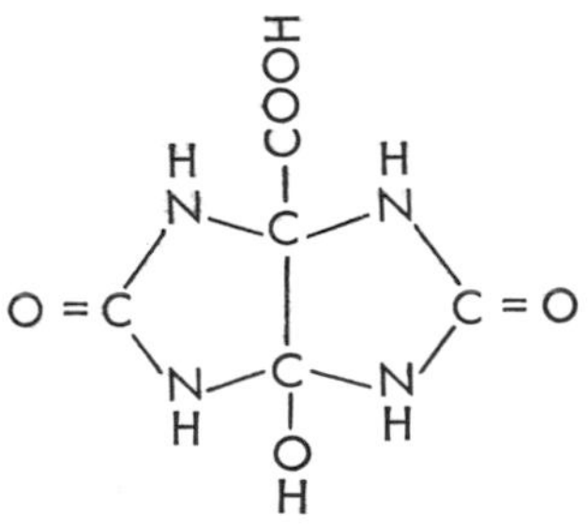

Hydroxyacetylene-diureine - carboxylic acid

Man and certain higher apes, however, are unable to bring about this step owing to absence of uricase from their tissus, and in them the end-product of purine metabolism is uric acid which is excreted in the urine along with very much smaller amounts of xanthine and hypoxanthine [106]. The Dalmatian coach-hound is peculiar in that it excretes uric acid in preference to allantoin, owing to lack of tubular reabsorption of uric acid in the kidney [7].

The substance allopurinol, which has a structure very similar to that of hypoxanthne, acts as a competitive inhibitor of xanthine oxidase, and so prevents uric acid formation. It is therefore sometimes used in the treatment of gout, a disease in which uric acid accumulates in the body.

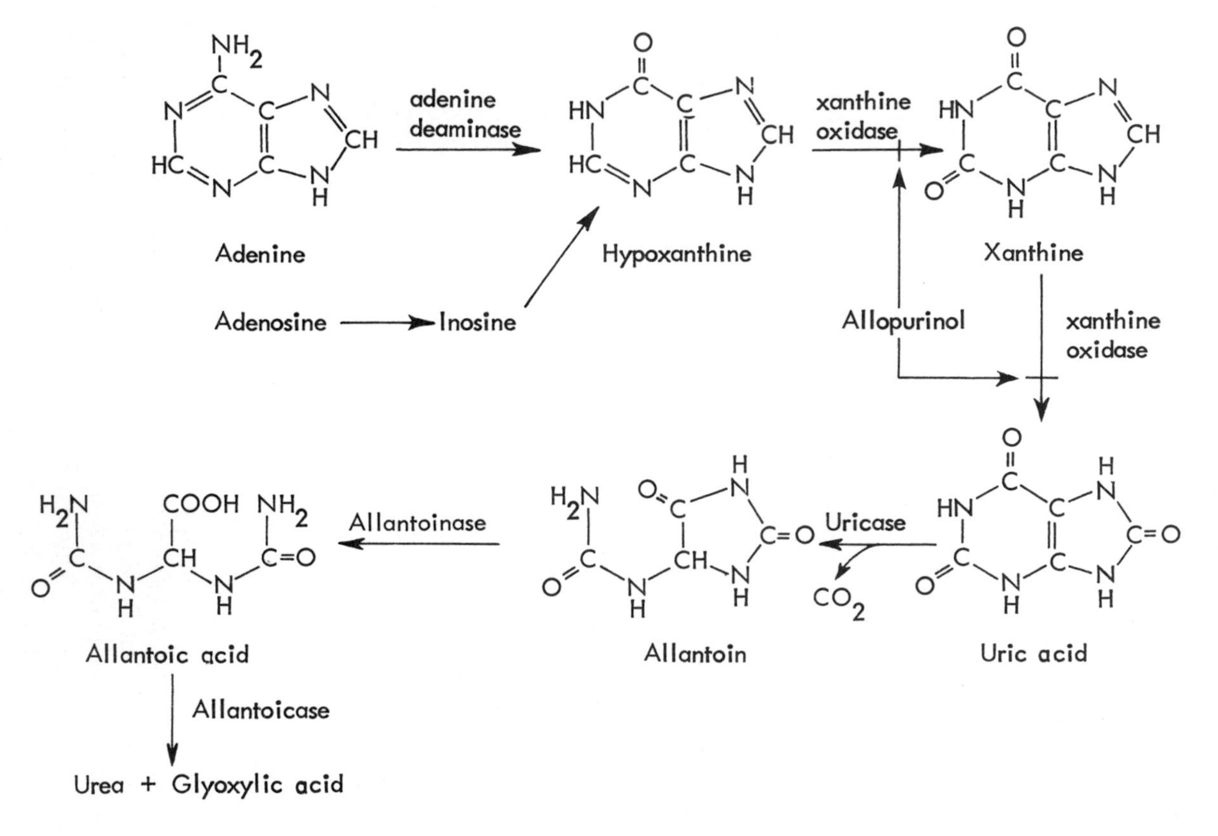

Fig. 7.10 The catabolism of purines.

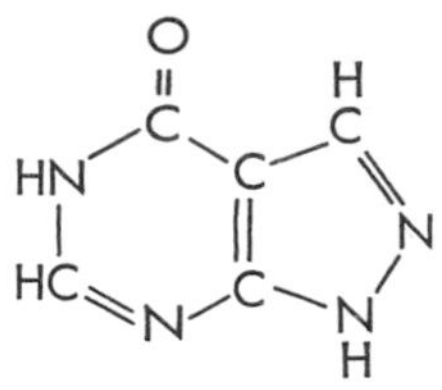

Allopurinol

Patients treated with allopurinol excrete xanthine and hypoxanthine in place of uric acid [106]. Mention has already been made of the use of azathioprine in the treatment of gout by inhibiting the synthesis of PRA from PRPP, IMP from hypoxanthine and AMP from IMP [132].

In fishes, in amphibia, and in more primitive organisms allantoin is broken down by allantoinase to allantoic acid, and this in turn may be degraded by allantoicase to urea and glyoxylic acid. The main nitrogenous excretory product in the spider is not uric acid but guanine. These aspects of comparative biochemistry are discussed in detail in the books by Baldwin [111], Florkin [112] and by Henderson & Paterson [7].

It is in the birds and the uricotelic reptiles that uric acid formation is most pronounced since in them uric acid rather than urea is the main nitrogenous excretory product. In most birds, uric acid production can be shown to take place in the liver, since hepatectomy is followed by cessation of uric acid synthesis and a rise in the blood ammonia level. The obvious inference that in birds and reptiles uric acid is derived ultimately from ammonia is supported by isotopic experiments. Urea does not act as a precursor of uric acid except in so far as it may give rise to ammonia. While the liver of the fowl or goose contains all the enzymes required for uric acid formation, that of the pigeon is lacking in xanthine oxidase. In the pigeon, therefore, hypoxanthine is produced in the liver, and is oxidized to uric acid in the kidney where xanthine oxidase is present.

Intravenous administration to normal human subjects of uric acid labelled with ^{15}N and examination of the excretion of isotope in the urine have shown that the injected uric acid is promptly diluted by a miscible pool of uric acid amounting to about 1 gram [113–115]. Since the rate of formation of uric acid calculated from the rate of fall in isotope concentration exceeds the rate of excretion of uric acid by 20 per cent or more, it would appear that some uric acid undergoes catabolic breakdown in man.

In the pathological condition known as gout [116], uric acid is deposited in the joints, particularly in the great toe, and under the skin

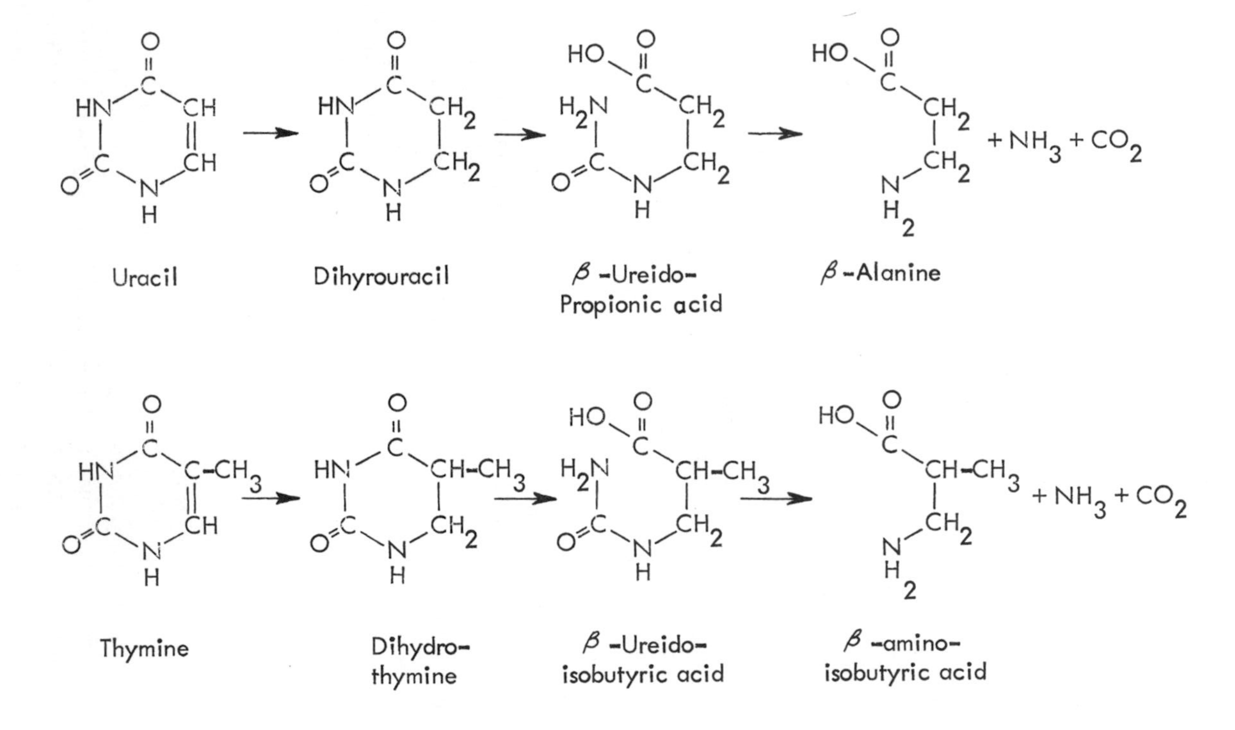

Fig. 7.11 The catabolism of uracil and thymine.

as nodules called tophi. In this disease the miscible pool of uric acid in the human body is increased to as much as 15 times the normal value [113, 117]. Administration of [^{15}N]-glycine to a gouty human subject has revealed a more rapid incorporation of isotope into the excreted uric acid as compared with the normal, although the excretory patterns for total nitrogen, urea, and ammonia are unchanged [118]. It has therefore been suggested that in gout the mechanism of transformation of dietary glycine to uric acid is more rapid than normal, so that overproduction occurs with consequent increase in the size of the miscible pool of uric acid.

7.11 PYRIMIDINE CATABOLISM

The catabolism of pyrimidine nucleotides, like that of purine nucleotides, involves dephosphorylation, deamination, and cleavage of glycosidic bonds, and many of the phosphatases that act upon purine nucleotides act also on the corresponding pyrimidine derivatives. As with purine nucleosides, so pyrimidine nucleosides may be hydrolysed to form pyrimidine bases and sugar, or they may be involved in phosphorolytic cleavage [7].

Cytosine can be deaminated by cytosine deaminase; this has been demonstrated in yeasts and other micro-organisms [119], and cytosine nucleosides are broken down to uridine nucleosides by cytidine deaminase which is widespread in animal tissues [120] as well as in bacteria.

The catabolic pathways of uracil [121] and for thymine [122, 123] in mammalian tissues involve reduction of the pyrimidines to the dihydro-derivatives, ring opening to give the appropriate ureido-acid, and the removal of ammonia and CO_2 to give β-alanine or its methylated derivative (Fig. 7.11). In some bacteria uracil and thymine undergo oxidative degradation via barbituric acid and methylbarbituric acid to urea and malonic or methylmalonic acids [7].

REFERENCES

1 Hartman, S.C. and Buchanan, J.M. (1959), *Annu. Rev. Biochem.*, **28,** 365.
2 Buchanan, J.M. and Hartman, S.C. (1959), *Adv. Enzymology,* **21,** 199.
3 Flaks, J.G. and Lukens, L.N. (1963), *Methods Enzymol.,* **6,** 52.
4 Warren, L. (1961), *Metabolic Pathways,* (ed. D.M. Greenberg), Academic Press, New York, Vol. 2, p. 459.
5 Schulman, M.P. (1961), *Metabolic Pathways,* (ed. D.M. Greenberg), Academic Press, New York, Vol. 2, p. 389.

6 Grav, H.J. (1967), *Methods in Cancer Research,* **3,** 243, Academic Press, New York.
7 Henderson, J.F. and Paterson, A.R.P. (1973), *Nucleotide Metabolism*, Academic Press, New York.
8 Colowick, S.P. and Kaplan, N.O. (1978), *Methods Enzymol.,* **51**.
9 Skipper, H.E., Mitchell, J.H. and Bennett, L.L. (1950), *Cancer Res.,* **10,** 510.
10 Skipper, H.E., Bennett, L.L. and Law, L.W. (1952), *Cancer Res.,* **12,** 677.
11 Rhoads, C.P. (ed.) (1955), *Anti-metabolites and Cancer* (American Association for the Advancement of Science).
12 Levenberg, B., Melnick, I. and Buchanan, J.M. (1957), *J. Biol. Chem.,* **225,** 163.
13 Kalckar, H. (1947), *Symp. Soc. Exp. Biol.,* **1,** 38.
14 Kornberg, A., Lieberman, I. and Simms, E.S. (1954), *J. Amer. Chem. Soc.,* **76,** 2027.
15 Murray, A.W. (1971), *Annu. Rev. Biochem.,* **40,** 811.
16 Kelley, W.N., Rosenbloom, F.M., Henderson, J.F. and Seegmiller, J.E. (1967), *Proc. Natl. Acad. Sci. USA,* **57**(6), 1735.
17 Kelley, W.N., Rosenbloom, F.M. and Seegmiller, J.E. (1967), *J. Clin. Invest.,* **46**(9), 1518.
18 Hitchings, G. (1967), *Fed. Proc. Fed. Am. Soc. Exp. Biol.,* **26,** 958.
19 Fujimoto, W.Y., Subak-Sharpe, J.H. and Seegmiller, J.E. (1971), *Proc. Natl. Acad. Sci. USA,* **68,** 1516.
20 Rubin, C.S., Dancis, J., Yip, L.C., Bowinski, R.C. and Balis, M.E. (1971), *Proc. Natl. Acad. Sci. USA,* **68,** 1461.
21 Boyle, J.A. (1970), *Science,* **169,** 688.
22 Kelley, W.N. (1968), *Fed. Proc. Fed. Am. Soc. Exp. Biol.,* **27,** 1047.
23 Seegmiller, J.E., Rosenbloom, F.M. and Kelley, W.N. (1967), *Science,* **155,** 1682.
24 Reichard, P. (1959), Adv. Enzymology, **21,** 263.
25 Crosbie, G.W. (1960), *The Nucleic Acids,* (ed. E. Chargaff and J.N. Davidson), Academic Press, New York, Vol. 3, p. 323.
26 Lieberman, I. (1955), *J. Amer. Chem. Soc.,* **77,** 2661.
27 Lieberman, I. (1956), *J. Biol. Chem.,* **222,** 765.
28 Kammen, H.O. and Hurlbert, R.B. (1959), *Cancer Res.,* **19,** 654.
29 Rose, I.A. and Schweigert, B.S. (1953), *J. Biol. Chem.,* **202,** 635.
30 Brown, G.B. (1954), *Ann. N.Y. Acad. Sci.,* **60,** 185.
31 Larsson, A. and Reichard, P. (1967), *Prog. Nucleic Acid Res. Mol. Biol.,* **7,** 303 (ed. J.N. Davidson and W.E. Cohn), Academic Press, New York.
32 Reichard, P. (1967), *The Biosynthesis of Deoxyribose,* Wiley, New York.
33 Stark, G.R. (1977), *TIBS* March 64.
34 Stryer, L., Holmgren, A. and Reichard, P. (1967), *Biochemistry,* **6,** 1016.
35 Kempe, T.D., Swyryd, E.A., Bruist, M. and Stark, G.R. (1976), *Cell,* **9,** 541.
36 Adams, R.L.P. and Lindsay, J.G. (1966), *J. Biol. Chem.,* **242,** 1314.
37 Coleman, P.F. Suttle, D.P. and Stark, G.R. (1977), *J. Biol. Chem.,* **252,** 6379.
38 Kornberg, A. (1957), *The Chemical Basis of Heredity* (ed. W.D. McElroy and B. Glass), John Hopkins, Baltimore.
39 Friedkin, M. (1963), *Annu, Rev. Biochem.,* **32,** 185.

40 Flaks, J.G. and Cohen, S.S. (1957), *Biochim. Biophys. Acta,* **25,** 667.
41 Makoff, A.J., Buxton, F.P. and Radford, A. (1978), *Mol. gen. Genet.,* **161,** 297.
42 Thelander, L. and Reichard, P. (1979), *Annu. Rev. Biochem.,* **48,** 133.
43 Peterson, D.M. and Moore, E.C. (1976), *Biochim. Biophys. Acta,* **432,** 80.
44 Cory, J.G., Mansell, M.M. and Whitford, T.W. (1976), *Adv. Enz. Regul.,* **14,** 45.
45 Turner, M.K., Abrams, R. and Lieberman, I. (1968), *J. Biol. Chem.,* **243,** 3725.
46 Murphree, S., Stubblefield, E. and Moore, C.E. (1969), *Exp. Cell Res.,* **58,** 118.
47 Walters, R.A. and Ratliff, R.L. (1975), *Biochim. Biophys. Acta,* **414,** 221.
48 Adams, R.L.P., Berryman, S. and Thomson, A. (1971), *Biochim. Biophys. Acta,* **240,** 455.
49 Walters, R.A., Tobey, R.A. and Ratcliff, R.L. (1973), *Biochim. Biophys. Acta,* **319,** 336.
50 Grav, H.J. and Smellie, R.M.S. (1963), *Biochem. J.,* **89,** 486.
51 Grav, H.J. and Smellie, R.M.S. (1965), *Biochem. J.,* **94,** 518.
52 Skoog, K.L., Nordenskjold, B.A. and Bjursell, K.G. (1973), *Eur. J. Biochem.,* **33,** 428.
53 Skoog, K.L. and Bjursell, G. (1974), *J. Biol. Chem.,* **249,** 6434.
54 Changeux, J.P. (1965), *Sci. Amer.,* **212**(4), 36.
55 Gerhart, J.C. and Pardee, A.B. (1964), *Fed. Proc. Fed. Am. Soc. Exp. Biol.,* **23,** 727.
56 Bjursell, G. and Reichard, P. (1973), *J. Biol. Chem.,* **248,** 3904.
57 Mitchison, J.M. (1971), *The Biology of the Cell Cycle,* Cambridge University Press.
58 Morris, N.R. and Fischer, G.A. (1960), *Biochim. Biophys. Acta,* **42,** 183.
59 Adams, R.L.P. (1980), *Cell Culture for Biochemists* (ed. Work, T.E. and Burdon, R.H.), Elsevier.
60 Wyngaarden, J.B. and Ashton, D.M. (1959), *Nature,* **183,** 747.
61 Bojarski, T.B. and Hiatt, H.H. (1960), *Nature,* **188,** 1112.
62 Creasey, W.A. and Handschumacher, R.E. (1961), *J. Biol. Chem.,* **236,** 2058.
63 Meuth, M. and Green H. (1974), *Cell,* **3,** 367.
64 Gelbard, R.S., Kim, J.H. and Perez, A.G. (1969), *Biochim. Biophys. Acta,* **182,** 564.
65 Scarano, E. (1960), *J. Biol. Chem.,* **235,** 706.
66 Scarano, E., Geraci, G. and Rossi, M. (1967), *Biochemistry,* **6,** 192.
67 Rossi, M., Geraci, G. and Scarano, E. (1967), *Biochemistry,* **6,** 3640.
68 Potter, V.R., Pitot, H.C., McElya, A.B. and Morse, P.A. (1960), *Fed. Proc. Fed. Am. Soc. Exp. Biol.,* **19,** 312.
69 Rossi, M., Dosseva, I., Pierro, M., Cacace, M.G. and Scarano, E. (1971), *Biochemistry,* **10,** 3060.
70 Littlefield, J.W. (1964), *Science,* **145,** 709.
71 Maley, G.F. and Maley, F. (1959), *J. Biol. Chem.,* **234,** 2975.
72 Maley, F. and Maley, G.F. (1960), *J. Biol. Chem.,* **235,** 2968.

73 Scarano, E., Talarico, M., Bonaduce, L. and de Petrocellis, B. (1960), *Nature*, **196,** 237.
74 Maley, G.F. and Maley, F. (1964), *J. Biol. Chem.*, **239,** 1168.
75 Littlefield, J.W. and Goldstein, S. (1970), *in vitro*, **6,** 21.
76 Pitts, J.D. (1971), *Growth Control in Cell Cultures*, Ciba Foundation Symp. (eds Wolstenholme, G.E.W. and Knight, J.), p. 89.
77 Potter, V.R. (1964), *Cancer Res.*, **24,** 1085.
78 Maley, G.F. and Maley, F. (1961), *Biochim. Biophys. Acta*, **47,** 181.
79 Weissman, S.M., Smellie, R.M.S. and Paul, J. (1960), *Biochim. Biophys. Acta*, **45,** 101.
80 Bollum, F.J. and Potter, V.R. (1959), *Cancer Res.*, **19,** 561.
81 Canellakis, E.S., Jaffe, J.J., Mantsavinos, R. and Krakow, J.S. (1959), *J. Biol. Chem.*, **234,** 2096.
82 Weissman, S.M., Paul, J., Thomson, R.Y., Smellie, R.M.S. and Davidson, J.N. (1960), *Biochem. J.*, **76,** 1P.
83 Gebert, R.A. and Potter, V.R. (1964), *Fed. Proc. Fed. Am. Soc. Exp. Biol.*, **23,** 268.
84 Kit, S. (1963), *Viruses, Nucleic Acids and Cancer*, p. 296. Williams and Wilkins, Baltimore.
85 Kit, S., Dubbs, D.R. and Piekarski, L.J. (1962), *Biochem. Biophys. Res. Commun.*, **8,** 72.
86 McAuslan, B.R. and Joklik, W.K. (1962), *Biochem. Biophys. Res. Commun.*, **8,** 486.
87 Keir, H.M. (1968), *Soc. Gen. Microbiol. Symp.*, **18,** 67.
88 Klemperer, H.G., Haynes, G.R., Shedden, W.I.H. and Watson, D.H. (1967), *Virology*, **21,** 120.
89 Mantei, N., Boll, W. and Weissmann, C. (1979), *Nature*, **281,** 35.
90 Potter, V.R. (1964), *Metabolic Control Mechanisms in Animal Cells*, National Cancer Institute, Monograph No. 13, p. 111.
91 Maley, F. and Maley, G.F. (1962), *Biochemistry*, **1,** 847.
92 Ives, D.H., Morse, P.A., Jr. and Potter, V.R. (1963), *J. Biol. Chem.*, **238,** 1467.
93 Breitman, T.R. (1963), *Biochim. Biophys. Acta*, **67,** 153.
94 Bresnick, E. and Karjala, R.J. (1964), *Cancer Res.*, **24,** 841.
95 Bresnick, E., Thompson, U.B., Morris, H.P. and Liebelt, A.G. (1964), *Biochem. Biophys. Res. Commun.*, **16,** 278.
96 Okazaki, R. and Kornberg, A. (1964), *J. Biol. Chem.*, **239,** 275.
97 Bello, L.J. (1974), *Exp. Cell Res.*, **89,** 263.
98 Subblefield, E. and Dennis, C.M. (1976), *J. Theor. Biol.*, **61,** 171.
99 Kantrowitz, E.R., Pastra-Landis, S.C. and Lipscomb, W.N. (1980), *TIBS* **5,** 124.
100 Smellie, R.M.S. (1955), *Nucleic Acids* (ed. E. Chargaff and J.N. Davidson), Vol. II, p. 393.
101 Potter, V.R. (1960), *Nucleic Acid Outlines*, Minneapolis, Burgess, p. 217.
102 Kielley, W.W. (1961), *The Enzymes*, 2nd Edn (ed. P.D. Boyer, H. Hardy and K. Myreback), Academic Press, New York. Vol. 5, p. 149.
103 Morton, R.K. (1965), *Comprehensive Biochemistry*, **16,** 55.

104 Bodansky, O. and Schwartz, M.K. (1968), *Adv. Clin. Chem.*, **11,** 277.
105 Canellakis, E.S. and Cohen, P.O. (1955), *J. Biol. Chem.*, **213,** 385,
106 Balis, E.W. (1968), *Fed. Proc. Fed. Am. Soc. Exp. Biol.*, **27,** 1067.
107 Sorensen, L.B. (1966), *Proc. Natl. Acad. Sci. USA,* **55,** 571.
108 Brown, G.B., Roll, P.M., Plentl, A.A. and Cavalieri, L.F. (1948), *J. Biol. Chem.*, **172,** 469.
109 Brown, G.B., Roll, P.M. and Cavalieri, L.F. (1947), *J. Biol. Chem.*, **171,** 835.
110 Dalgliesh, C.E. and Neuberger, A. (1954), *J. Chem. Soc.*, 3407.
111 Baldwin, E. (1949), *An Introduction to Comparative Biochemistry,* Cambridge University Press, London.
112 Florkin, M. (1949), *Biochemical Evolution,* Academic Press, New York.
113 Benedict, J.D., Forsham, P.H. and Stetten, D. (1949), *J. Biol. Chem.*, **181,** 183.
114 Buzard, J., Bishop, C. and Talbott, J.H. (1952), *J. Biol. Chem.*, **196,** 179.
115 Green, W., Bendich, A., Bodansky, O. and Brown, G.B. (1950), *J. Biol. Chem.*, **183,** 21.
116 Wyngaarden, J.B. (1966), *Adv. Metabolic Disorders,* **2,** 1.
117 Bishop, C., Garner, W. and Talbott, J.H. (1951, *J. Clin. Invest.*, **30,** 879.
118 Benedict, J.D., Roche, M., Yu, T.F., Bien, E.J., Gutman, A.B. and Stetten, D. (1952), *Metabolism,* **1,** 3.
119 O'Donovan, G.A. and Neuhard, J. (1970), *Bact. Rev.*, **34,** 278.
120 Wisdom, G.B. and Orsi, B.A. (1969), *Eur. J. Biochem.*, **7,** 223.
121 Schulman, M.P. (1954), *Chemical Pathways of Metabolism* (ed. D.M. Greenberg), Academic Press, New York. Vol. II, p. 223.
122 Canellakis, E.S. (1957), *J. Biol. Chem.*, **227,** 701.
123 Fink, K., Cline, R.E., Henderson, R.B. and Fink, R.M. (1956), *J. Biol. Chem.*, **221,** 425.
124 Eker, P. (1968), *J. Biol. Chem.*, **243,** 1979.
125 Elford, H.L., Freese, M. Passamani, E. and Morris, H.P. (1970), *J. Biol. Chem.*, **245,** 5228.
126 Conrad, R.H. and Ruddle, F.H. (1972), *J. Cell Sci.*, **10,** 471.
127 Thelander, L., Larsson, B., Hobbs, J. and Eckstein, F. (1976), *J. Biol. Chem.*, **251,** 1398.
128 Atkin, C.L., Thelander, L., Reichard, P. and Lang, G. (1973), *J. Biol. Chem.*, **248,** 7464.
129 Eriksson, S., Thelander, L. and Akerman, M. (1979), *Biochemistry,* **18,** 2948.
130 Engstrom Y., Eriksson, S., Thelander, L. and Akerman, M. (1979), *Biochemistry,* **18,** 2941.
131 Reynolds, E.C., Harris, A.W. and Finch, L.R. (1979), *Biochim. Biophys. Acta,* **561,** 110.
132 Roy-Burman, P. (1970), *Analogues of Nucleic Acids Components,* Springer Verlag, New York, p. 16.
133 Hartman, S.C. (1963), *J. Biol. Chem.*, **238,** 3036.
134 Abrams, R. and Bentley, M. (1959), *Arch. Biochem.*, **79,** 91.

135 Mandelstam, J. and McQuillen, K. (1973), *Biochemistry of Bacterial Growth*, 2nd Edn, Blackwell Scientific Publications, Oxford, p. 236.
136 Muller, M.M., Kaiser, E. and Seegmiller, J.E. (1977), *Advances in Experimental Medicine and Biology*, Vol. 76A and 76B, Purine Metabolism in Man—II, Plenum Press, New York.

Replication of DNA

8.1 INTRODUCTION

Each daughter cell produced at cell division contains an identical copy of the genetic material. Since DNA is now known to carry in its sequence of nucleotides the genetic information or 'code', the question of the way in which it is reproduced in the cell has attracted a great deal of attention. Like most biological phenomena, DNA replication can be studied at various levels of cellular disorganization. Section 8.2 deals with studies on intact prokaryotic and eukaryotic cells. Although results from such studies obviously reflect the true *in vivo* processes they are often difficult to analyse; this is partly because the immediate precursors of DNA are nucleotides to which the living cell is impermeable. In order to circumvent this problem and to have a system of complexity intermediate between the whole cell and the purified enzyme, permealized (e.g. toluenized) bacteria, cell lysates, and isolated nuclei have been investigated in an attempt to mimic the *in vivo* situation (see Section 8.3).

Studies with purified enzymes have indicated the detailed mechanism of the polymerization reaction and illustrated certain unexpected limitations on the ability of DNA polymerases to replicate DNA. These are considered in Section 8.4.

Our increased understanding of DNA replication has come about over the last few years as a result of the intensive use of bacterial mutants, especially temperature-sensitive mutants, i.e. bacteria which grow normally at low temperatures but cease to grow at higher temperatures.

The study of *E. coli* mutants defective in DNA synthesis has demonstrated that a number of proteins are essential for DNA replication. Some of these, together with the genes which specify them, are listed in Table 8.1. Many of these proteins have been purified and used with some success in attempts to reconstruct a DNA synthesizing system.

Table 8.1 *E. coli* genes whose products play a role in DNA synthesis [1].

Gene	Enzyme	Effect on DNA synthesis of changing to non-permissive temperature
dna A	Anti-repressor?	Current round of replication completed but no further initiation
dna B	DNA-dependent ATPase	Synthesis stops immediately; very short pieces produced
dna C/D	oRNA processor	Current round completed but no further initiation
dna E (pol C)	DNA polymerase III	Stops immediately
dna F (nrd)	Ribonucleotide reductase	Stops immediately
dna G	Primase	Stops immediately; cannot initiate Okazaki pieces
dna S (dut)	dUTPase	Short Okazaki Pieces formed
dna Z	DNA-dependent ATPase	
pol A	DNA polymerase I	Slow joining of Okazaki pieces; susceptible to u.v.
pol B	DNA polymerase II	Normal
lig	DNA ligase	Slow joining of Okazaki pieces
ssb	DNA binding protein	—

8.2 DNA REPLICATION – *IN VIVO* STUDIES

8.2.1. Semi-conservative replication

At the end of their paper [2] suggesting the double-stranded structure for DNA with its complementary base pairs, Watson and Crick wrote: 'It has not escaped our notice that the specific pairing we have postulated suggests a possible copying mechanism for the genetic material.'

Let us suppose that a short length of a DNA double helix has the nucleotide sequence shown in Fig. 8.1 (top), and further that the two strands can bc untwisted and separated from one another to form two single chains, as in Fig. 8.1. (middle), and that each base in the single strands can attach to itself the complementary deoxyribonucleotide by the same hydrogen bonding which exists in the intact DNA double helix. Finally, if these attached mononucleotides are polymerized to form a polynucleotide chain as in Fig. 8.1 (bottom), the end result will be the formation of two complete DNA double helices identical with each other and with the original molecule. One strand of each daughter molecule will be derived from the original DNA molecule; the other will be the product of the new synthesis.

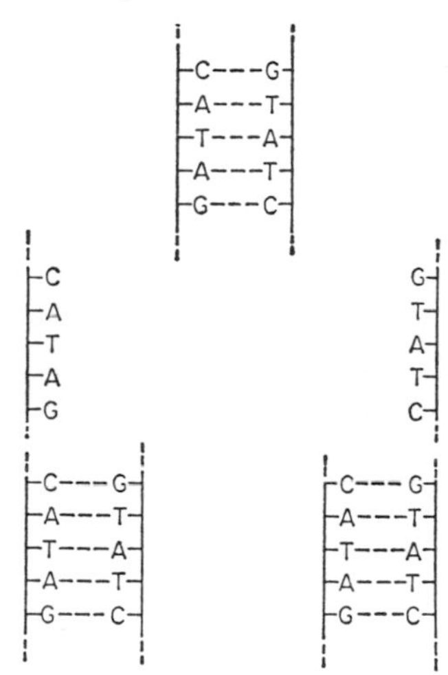

Fig. 8.1 Schematic illustration of the separation of the two strands of a portion of DNA with the formation of a new strand on each.

This mechanism is described as *semi-conservative* to distinguish it from the other possible mechanisms. In the *conservative* mechanism the two strands do not come apart but act together as a template to form a completely new double helical molecule; in this case one daughter molecule would be wholly new and the other totally derived from the parent. In the *dispersive* mechanism the parental molecule is partly degraded and the fragments are incorporated into two new daughter double helices. These possibilities are illustrated in Fig. 8.2.

Convincing evidence for semi-conservative replication of DNA was obtained by Meselson and Stahl [3] who grew *E. coli* in a medium containing $^{15}NH_4Cl$ of 96·5 per cent isotopic purity for fourteen generations so as to label the DNA very heavily with ^{15}N. The cells were then transferred to a medium containing $^{14}NH_4Cl$ and samples of bacteria were withdrawn at intervals for several generations. Each sample was lysed by means of sodium dodecyl sulphate and was centrifuged in a concentrated solution of caesium chloride at 140000*g* for 20 hours to enable the DNA to attain sedimentation equilibrium. The bands of

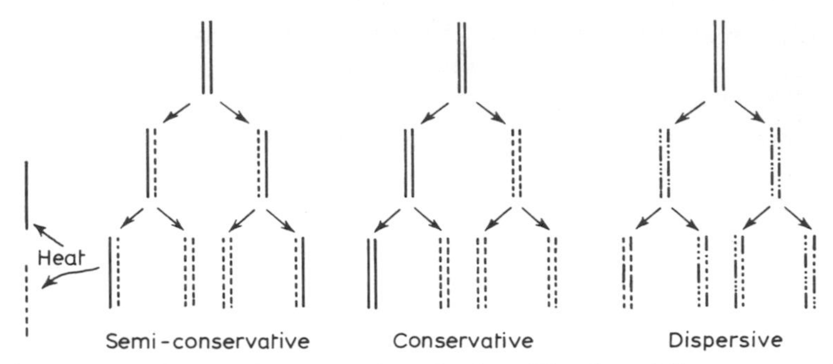

Fig. 8.2 Possible mechanisms of replication. The dotted lines represent filial strands. For details see text.

DNA were found in the CsCl gradient in the region of density 1·71 g/cm^3 and were well isolated from all other macromolecular components of the bacterial lysate. Ultraviolet absorption photographs taken during the course of the run revealed the position of the DNA bands.

At the start of the experiment the DNA appeared as one single band corresponding to the heavy ^{15}N-labelled nucleic acid (Fig. 8.3). Macromolecules containing half this level of ^{15}N then began to appear, and one generation time after the addition of ^{14}N these hybrid (light-heavy, ^{14}N-^{15}N) molecules alone were present. Subsequently a mixture of light-heavy (^{14}N-^{15}N) DNA and unlabelled (light-light, ^{14}N only) DNA was found. When two generation times had elapsed after the addition of ^{14}N, the half-labelled and unlabelled DNA molecules were present in equal amounts (Fig. 8.3). During subsequent generations the unlabelled DNA accumulated. Moreover, when the hybrid ^{14}N-^{15}N molecules were heated they separated to give a ^{14}N strand and a ^{15}N strand.

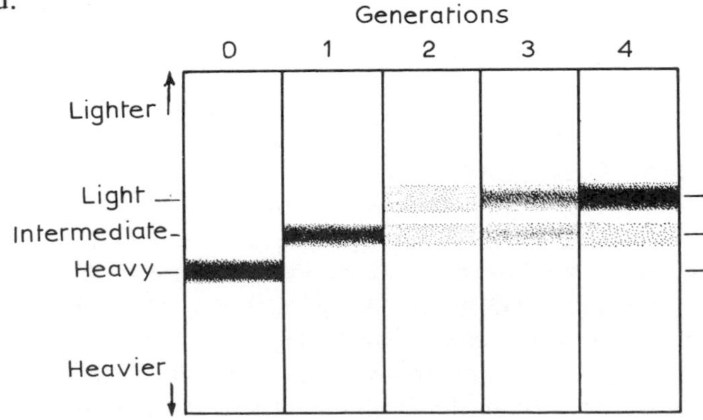

Fig. 8.3 The pattern of results from the Meselson and Stahl experiment. For explanation see text.

Such experiments indicate that in DNA synthesis each existing DNA molecule is split into two subunits. (Fig. 8.2.), each subunit going to a different daughter molecule. The other subunit of each daughter molecule is the product of new synthesis. These subunits do not undergo any fragmentation but remain intact for many generations.

Experiments at the chromosomal level also support the view that DNA replication is semi-conservative. When the replication of the *E. coli* chromosome is followed by autoradiography after labelling of the DNA with tritium, the amount of tritium per unit length of newly synthesized DNA is consistent with the presence of only one newly synthesized strand in a daughter chromosome [4].

In plants to which tritiated thymidine has been given as a specific precursor of DNA during the period of DNA synthesis, both the daughter chromatids are found to be labelled at the time of cell division. At the next round of duplication, however, after withdrawal of the ^{3}H-thymidine, these two chromosomes each produce one labelled and one unlabelled chromatid as would be expected [5].

8.2.2 Initiation and direction of replication

(a) *Site of initiation.* The results of the Meselson–Stahl experiment suggested that DNA replication would be found to be a sequential process and that successive rounds of replication would not begin at random positions on the chromosome. It is now clear that chromosomes, from most viruses and prokaryotes, whether containing linear or cyclic DNA, initiate their replication at one specific site. For small DNA molecules, or for fragments of larger molecules, these sites can be visualized using the electron microscope when they appear as double-stranded 'bubbles' (Plate IV and Fig. 8.8) [6–10]. On closer examination it is apparent that for a small region on one side of each fork the DNA is only single-stranded (see p. 217).

Cairns was able to visualize the whole of the replicating *E. coli* chromosome autoradiographically by growing the bacteria for several generations in the presence of tritiated thymidine (Fig. 8.4), and he showed that the chromosome exists as a continuous piece of double-stranded cyclic DNA [4].

That the initiation of the replication bubble occurs at a unique site has been shown by a number of different techniques for viral, plasmid, bacterial, and mitochondrial DNA. Thus initiation of replication of the linear T7 bacteriophage chromosome occurs 17 per cent from one end [8] and Schnös and Inman [11] were able to show that initiation of replication occurred at a particular site with reference to the partial denaturation map of bacteriophage λ. In the plasmid Col E1 [6] and in

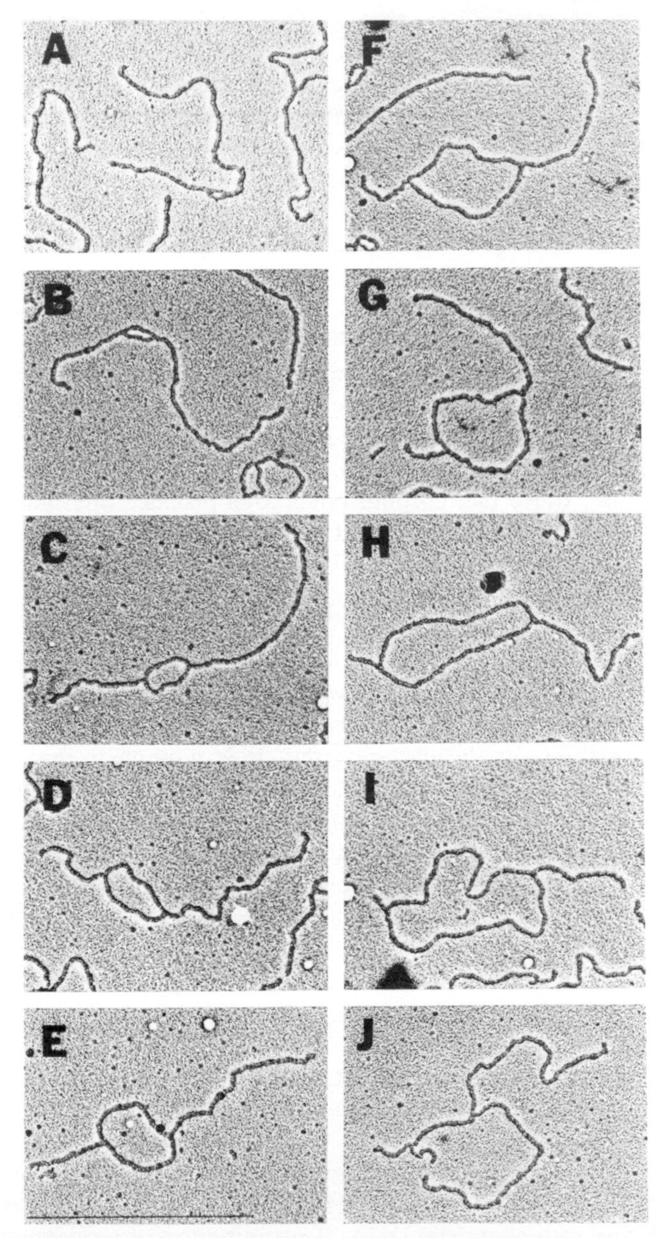

Plate IV Replicating SV40 DNA molecules. The molecules have been cut at a unique site with the restriction endonuclease EcoRI and have been arranged in increasing degree of replication (A through J) and oriented with the short branch at the left. (From Fareed, Garon and Salzman (see Chapter 8 ref. 10) with the authors' kind permission.)

the simian virus 40 [10] chromosome replication is always initiated at a fixed distance from the site of action of the Eco RI restriction endonuclease (Plate IV, see Chapter 6). In exponentially growing *E. coli* and *B. subtilis* genetic analysis has shown that, relative to non-growing cells, certain genes are present in greater than single copies [12, 13]. The interpretation of this finding is that genes replicated early will be present at twice the number of copies (or even four or eight times in rapidly growing cells) relative to late replicated genes (Fig. 4.2).

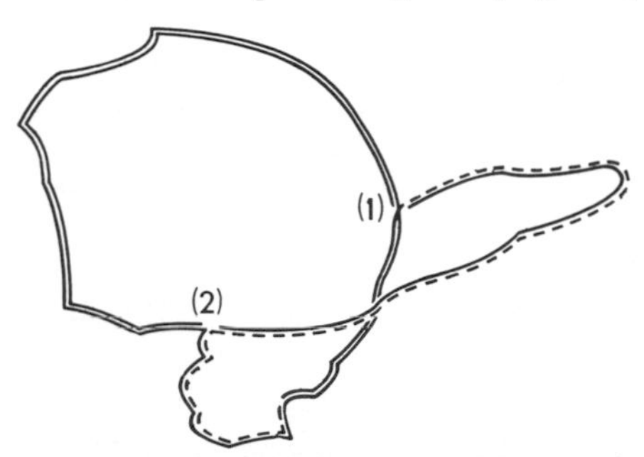

Fig. 8.4 Digrammatic representation of an autoradiograph of the chromosome of *E. coli* labelled with tritiated thymidine for two generations of replication [4]. The chromosome has been two-thirds duplicated and the two replication forks are indicated as (1) and (2). By courtesy of Dr John Cairns.

Isolation of complete adeno 2 DNA molecules after pulse labelling infected cells showed a predominance of radioactivity in those restriction fragments which map towards the ends of the genome. This and similar experiments lead to the conclusion that in this system replication was initiated at (or near) both ends of the linear genome [14].

(b) *Direction of replication.* Further studies of replicating bubbles (both by electron microscopy and by genetic analysis) show that in most cases the bubbles are expanding in both directions, i.e. the bubble is made up of two replicating forks moving away from a central point of initiation. There are a few exceptions to this, e.g. the plasmid Col E1 where replication is unidirectional [6] and the plasmid R6K in which one fork moves 20% to the left of the site of initiation and the other 80% to the right [15].

Replicating DNA from eukaryotes shows a large number of replication bubbles on electron micrographs [9, 16], and fibre autoradiographic studies suggest that DNA from mammalian cells is composed of many tandemly joined sections each about 15–60 μm long [17–19]. These sections or 'replicons' are separately replicated bidirectionally from their origin (Fig. 8.5).

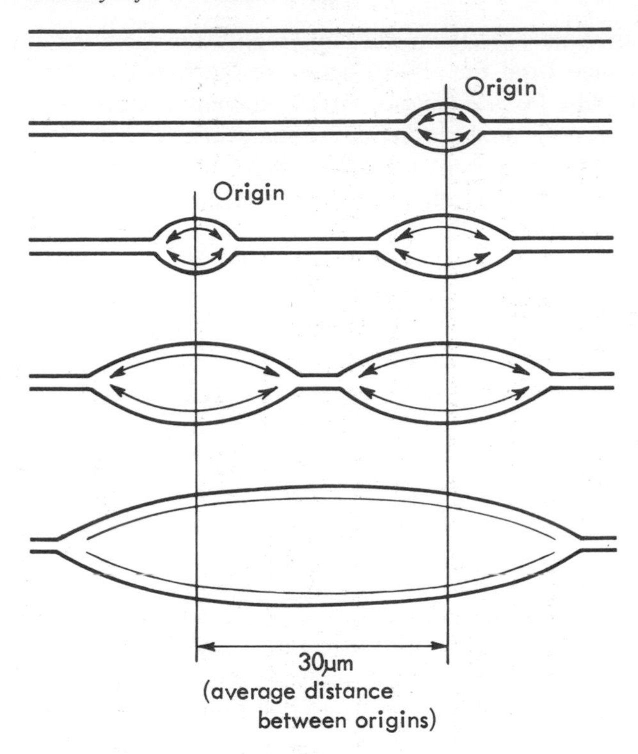

Fig. 8.5 Replication of mammalian DNA (2 replicons are shown) (After Huberman and Riggs [17]).

(c) *The structure of the origin.* Implicit in the fact that initiation of DNA synthesis occurs at a unique site on the chromosome is the suggestion that this site has a distinctive structure which allows the binding of the replicative machinery and the unwinding of the double helix.

It is considered probable that DNA synthesis is only initiated on a region of the DNA which is rendered single stranded [20] perhaps by reason of its being involved in transcription, and specific RNA transcripts known as origin RNA (oriRNA) have been implicated. Such RNA molecules may also serve as primers for subsequent DNA synthesis (see p. 218).

The replication origins of the small single-stranded bacteriophages fd, G4 and ØX174 [21–28], of several lambdoid bacteriophages [25, 29, 30], of the plasmid Col El [31], of *E. coli* [32], of the small animal viruses SV40, polyoma, BK [33–36] and several adeno viruses [37], of the single-stranded parvoviruses [38] and of mitochondrial DNA [39, 40] have been sequenced. Related viruses, e.g. the papovaviruses SV40, polyoma and BK [35] or the lambdoid bacteriophages [29] show considerable sequence homology and the replication origins of the prokaryotic DNAs contain sequences homologous to the interaction site of primase (the dna G protein, see p. 225) [22, 23, 28, 32]. All origins whether in regions coding for protein or not contain unusual sequences, e.g. direct repeats, palindromes and true palindromes which may enable these regions (or the RNA derived therefrom) to assume a hairpin configuration. The DNA may take up a hairpin or cruciform conformation in response to initiation proteins [37] which thereby allow an RNA polymerase to synthesize a primer on which DNA synthesis may be initiated. Alternatively, as with the single-stranded bacteriophages [21, 23, 26–28] a hairpin may be the only region of the DNA not covered by binding proteins and as such may be the only region accessible to a primase (see p. 257). Normally an RNA molecule, once synthesized is released from the DNA template which returns to its double-stranded form. If the non-transcribed strand of the DNA were able to stabilize itself by forming a hairpin this would have the effect of retaining the RNA strand at the replication origin to serve as a primer for DNA synthesis. It should also be borne in mind that rather than assuming a cruciform structure, four stranded regions of DNA may be formed which could act as structures recognized by specific initiation proteins [35, 41]. The region of phage G4 to which primase may bind prior to initiation of RNA synthesis shows considerable homology with the corresponding region on bacteriophage λ DNA [22, 23, 28, 37]. In bacteriophage λ however this region is not followed by a simple hairpin complementary to oriRNA but by what may exist as a complex cloverleaf arrangement which is the proposed binding site for the O protein (a phage-specific initiator protein). Following this is the region of DNA complementary to the primer for bacteriophage λ DNA synthesis (oopRNA).

(d) *Initiation in E. coli.* The origin of *E. coli* DNA replication (ori C) has been cloned and partially sequenced [42, 43].

It contains many termination codons; i.e. it does not code for proteins; with repeats and palindromes. There are structural analogies with the origins of Col El, lambda and bacteriophage G4 where primase is thought to bind.

E. coli temperature-sensitive mutants dna A and dna C fail to initiate new rounds of DNA replication when placed at restrictive temperature but complete those rounds already in progress. On cooling such mutants to the permissive temperature in the presence of tritiated thymidine the first region to become radioactive is a 1.3 kilobase fragment formed by treatment of the DNA with the restriction endonuclease Hind III [44]. The amount of incorporation is less in the case of dna A mutants than dna C mutants indicating that the dna A gene product is involved several minutes before the dna C gene product in the initiation process.

Initiation of replication in *E. coli* is sensitive to rifampicin and chloramphenicol and a protein located in the membrane which is normally made about 15 min before initiation is not made in dna A mutants [45]. OriRNA of *E. coli* has been isolated linked to high-molecular-weight DNA. It is not made in dna A mutants and is not linked to DNA in dna C mutants suggesting a role for these gene products in initiating oriRNA synthesis (dna A) and in the changeover to DNA synthesis (dna C) [46].

Chloroamphenicol-insensitive recessive mutants have been isolated which no longer require the dna A gene products for initiation suggesting its role may be to counteract an inhibitor of initiation absent in the insensitive mutant [47, 48].

8.2.3 Rate of replication

The rate of replication of DNA is dependent on the temperature and the supply of nutrients, particularly deoxyribonucleotides. The minimum time required to replicate the *E. coli* chromosome is about 40 minutes, implying a rate of synthesis of about 1700 base pairs per second [49]. As replication is bidirectional the rate at each fork is about 850 base pairs per second or 14 μm per minute. (Compare this with the rate of transcription which proceeds at 35–40 nucleotides per second in *E. coli*.) In mammalian cells the rate of fork movement is only 0·5–1·2 μm per minute or about 60 base pairs per second [50]. However, this slower rate is compensated for by the smaller size of the replication unit in eukaryotes (on average 50 μm as compared with 1100 μm in *E. coli*) and the fact that about 5000 replicating forks (out of a total of 35 000 per cell) are simultaneously active [51]. In certain early embryos such as that of the sea urchin the S-period lasts for only 20 minutes during which time the whole genome is replicated. This rapid synthesis of DNA is perhaps brought about both by increasing the rate of fork movement and by increasing the number of replicating sections simultaneously active (deoxyribonucleotide pools are much greater than in mammalian cells [18, 52]).

8.2.4 The replication fork

(a) *Discontinuous synthesis.* As described above, replication of DNA occurs by means of a 'fork' which grows along the DNA molecule resulting in semi-conservative replication. However, as the two DNA chains are antiparallel (i.e. one is running 5′→3′ and the other 3′→5′)

this poses the problem as to whether the two daughter strands are synthesized by two different mechanisms: one by adding nucleotides to a growing 3′-end and the other to a growing 5′-end. Addition to the 5′-end may involve deoxyribonucleoside 3′-triphosphates or, alternatively, may result in DNA molecules having terminal 5′-triphosphates (see Fig. 8.6). Such termini have not been found and, moreover, addition of nucleotides to a 5′-triphosphate terminated primer is incompatible with the proof-reading function of DNA polymerase (see p.239). This function requires the removal of a mismatched terminal residue. If this residue held the energy for the polymerization, i.e. if it were a triphosphate, then this energy would be dissipated on proof-reading and polymerization would halt.

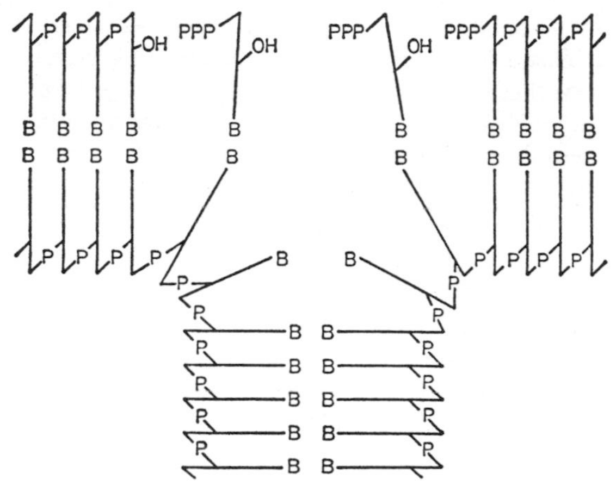

Fig. 8.6 A possible mechanism of DNA chain growth.

Other evidence argues against the 5′→3′ extension of DNA chains.

(1) Many DNA polymerases have been investigated and they all add deoxyribonucleoside 5′-phosphates on to a 3′-OH group on the growing DNA chain, i.e. the primer (see p. 229).

(2) No enzymes have been found which will generate or polymerize deoxyribonucleoside 3′-triphosphates, and neither have these putative substrates been found in cells.

(3) Both daughter strands have been shown to grow in the 5′→3′ direction (see p. 218).

The alternative explanation put forward by Okazaki [53] is that both chains are synthesized by the same mechanism but that one is made ‘backwards’ in short pieces which are subsequently joined together by DNA ligase (see Fig. 8.7). Although evidence for this mechanism is not apparent from the Cairns autoradiograph, careful investigation of electron micrographs (which give a resolution of about 100 nucleotides)

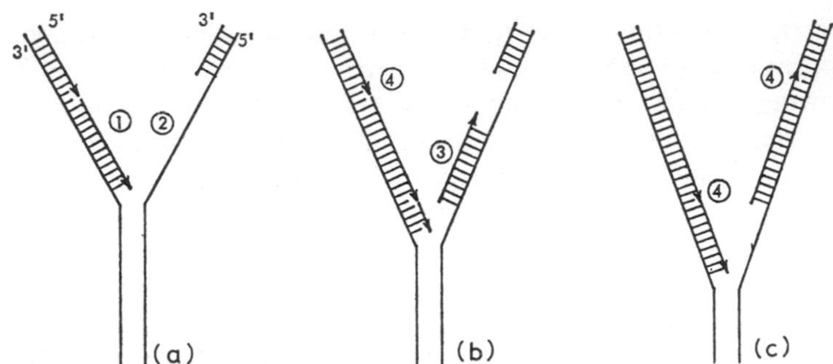

Fig. 8.7 A working model of DNA synthesis. Synthesis occurs in the 5′→3′ direction down the left-hand strand of the parental DNA molecule (1). This synthesis may be continuous or discontinuous and it exposes the right-hand strand of the parental DNA molecule. This single-stranded region (2) is probably stabilized by association with DNA-binding proteins. At some point DNA synthesis starts on the right-hand strand (3) which is copied backwards (i.e. 5′→3′) until the growing strand has filled the gap. DNA ligase then joins the newly synthesized pieces (4).

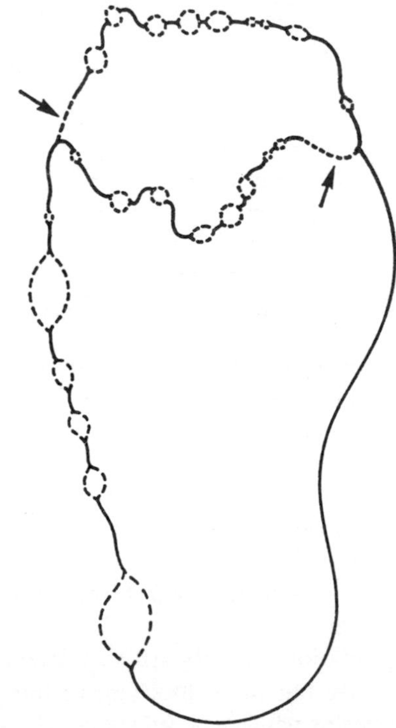

Fig. 8.8 A partial denaturation map of the replicating chromosome. Broken lines represent single-stranded regions of DNA, and the arrows indicate such regions occurring at the replication fork. Drawn from the electron micrographs of Schnös and Inman [11].

shows that on one side of a replication fork the DNA is single-stranded (Fig. 8.8) and this supports the 'discontinuous' mechanism. In this model the strand of DNA which is made continuously in the direction of fork movement is called the *leading strand* while that made discontinuously in the reverse mode is called the *lagging strand*.

(b) *Okazaki pieces*. The major support for discontinuous synthesis stems from the work of Okazaki with *E. coli* and T4 infected *E. coli*. He showed that the most recently synthesized DNA, i.e. that labelled with a brief pulse of tritiated thymidine, can be isolated *after denaturation* as short pieces now known as Okazaki pieces [53]. In this system the short pieces sediment at 8–10S on a gradient of alkaline sucrose, representing chain lengths of 1000–2000 nucleotides. In animal cells the fragments are much shorter, being only about 100–200 residues long, i.e. 4–5S [54, 55].

In order to detect Okazaki pieces it is essential to lower the temperature to slow down the rate of reaction and to give a *very brief* pulse of tritiated thymidine. Under these conditions considerably more than half the radioactivity is found in small pieces, showing that Okazaki pieces are made on both sides of the replication fork. (There is, however, some controversy over this; see p. 221.) A second possibility is that discontinuities are introduced into DNA during repair of regions of DNA containing uracil in place of thymine (see p. 294). It has been calculated that in *E. coli* during DNA replication one uracil will be accidentally incorporated into DNA for every 300 thymines which may lead to a cut in the DNA every 1200 bases; a size not too different from that of the Okazaki pieces [56]. However, the frequency of uracil incorporation in *B. subtilis* is very low and in those mutants which fail to excise it (ung^-) it is present less than once per 5000 bases [57].

Incorporation of uracil is accentuated in mutants lacking an active dUTPase (dut^-) [58]. Such cells have elevated levels of dUTP and this competes with and is incorporated into DNA in place of dTTP. A uracil *N*-glycosidase removes the uracil and the damaged DNA is repaired by an excision repair mechanism which involves cutting the DNA chain on the 5′ side of the damage (see p. 294). dut mutants produce short Okazaki fragments (sof) only 100–200 nucleotides long [59]. The discovery of such sof mutants was initially misinterpreted as indicating that Okazaki fragments were formed by joining of several much shorter pieces. In double mutants lacking the uracil *N*-glycosidase (ung^-) as well as the dUTPase the DNA contains significant amounts of uracil which is not excised [60]. Normal-sized Okazaki fragments are formed on the lagging strand of dut^-ung^- double mutants showing that the majority of these fragments are not normally produced as a result of repair but are apparently true intermediates in DNA replication [61, 62]. Moreover

Okazaki pieces accumulate in mutant *E. coli* cells deficient in DNA ligase or DNA polymerase I whether or not uracil excision occurs [63, 64] and this implies that they are true intermediates in DNA synthesis and that these two enzymes are involved in their subsequent incorporation into high-molecular-weight DNA.

(c) *Direction of chain growth.* When bacteriophage T4, growing at 8°C to reduce the rate of DNA synthesis, is incubated with [^{14}C]-thymidine for 150 seconds, and for the final 6 seconds with [^{3}H]-thymidine, Okazaki pieces can be isolated which apparently contain tritium at only the 3′-end [65, 66]. This was shown by degrading the isolated Okazaki pieces with exonuclease 1 of *E. coli* (which degrades single-stranded DNA from the 3′-end; see Chapter 6) when the ^{3}H label is released before the ^{14}C label. The complementary experiment using a nuclease from *B. subtilis* which acts from the 5′-end causes release of much of the ^{14}C before the ^{3}H is rendered acid-soluble.

These experiments demonstrate that Okazaki pieces are made in the 5′→3′ direction and support the discontinuous mechanism as outined in Fig. 8.8. However, Diaz and Werner [67] have recently thrown doubt on the interpretation of these results and suggested that the 10S Okazaki pieces arise through the joining of many smaller chains.

(d) *Initiation of Okazaki pieces.* In wild type *E. coli* rifampicin (an inhibitor of *E. coli* RNA polymerase) inhibits replication of phage M13 and certain plasmids. Replication is normal in mutants with a rifampicin-resistant RNA polymerase [68]. Moreover, some RNA synthesis has been shown to be essential for the *in vitro* conversion of M13 single strands into the duplex form [69]. Such evidence, considered with the fact that all DNA polymerases equire a 3′-hydroxyl priming end (see p. 229), suggested that this end may be provided by an oligoribonucleotide. That rifampicin does not inhibit continued replication of *E. coli* DNA (it does block the initiation of new rounds [70]) or the DNA of other phages such as ØX174 was interpreted to mean that in these cases a second RNA polymerase (resistant to the drug) was involved in the synthesis of the priming oligoribonucleotides. Indeed, the product of the dna G gene (see Table 8.1) is believed to be such an enzyme and has been shown *in vitro* under specific conditions to synthesize short lengths of RNA on single-stranded DNA of phage G4 [71] (see p. 257). Although not required for the initial stages of M13 replication, the dna G gene product is essential for duplication of the replicative form [72]. This and similar enzymes are called primases (see p. 255).

Okazaki fragments containing oligoribonucleotides at their 5′-end have now been isolated from both prokaryotic and eukryotic systems. However, the experimental proof of their existence has been subject to much controversy because non-covalent RNA–DNA interactions may have given spurious results in early experiments [73, 74]. However, methods (Fig. 8.9) to detect RNA covalently linked to the 5′-end of DNA molcules [73, 75] have shown that, although not all Okazaki

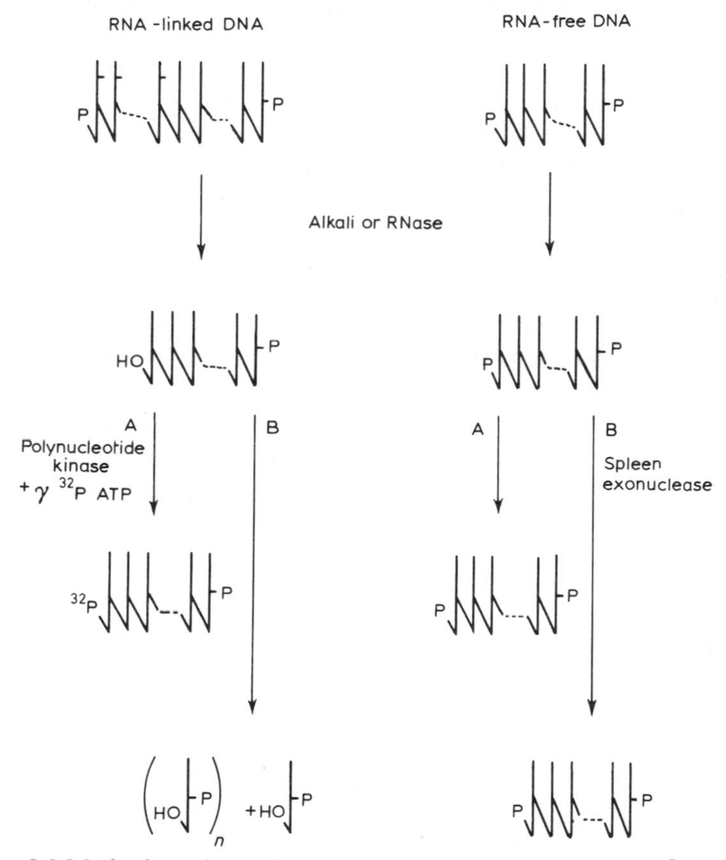

Fig. 8.9 Methods to detect RNA-linked Okazaki pieces produced [73, 75] *in vivo*. The isolated Okazaki pieces are first treated with polynucleotide kinase and non-radioactive ATP to ensure that all 5′-ends are phosphorylated. Subsequent treatment with alkali or ribonuclease removes any RNA from the 5′end and leaves a 5′-hydroxyl group. This group can be detected using polynucleotide kinase and [γ-^{32}P]-ATP (Method A). However, this method is not selective for nascent DNA. Method B overcomes this disadvantage by starting with tritiated Okazaki pieces resulting from a pulse labelling experiment. Spleen exonuclease is used under conditions where only 5′-OH terminated DNA (i.e. that initially linked to RNA) is degraded. These methods have been criticized on the grounds that treatment with alkali of DNA cleaved during repair (see p. 293) may also yield 5′-OH groups [76, 77].

pieces have ribonucleotides at their 5′-end, the shortest ones do. Moreover, RNA-linked Okazaki pieces accumulate in mutant *E. coli* cells deficient in either the polymerase or the 5′→3′ exonuclease of DNA polymerase I (see p. 237). Even so, using such a mutant, primers were detected on only about 30% of the Okazaki pieces [78] and in a wild-type *E. coli* only 3–6 of the total of 20–40 chains in the size range 3000–9000 bases chase into high-molecular-weight DNA [79]. When precautions are taken to eliminate all but covalent attachment of RNA to Okazaki pieces the size of the RNA segments obtained is only 1–3 nucleotides long and of no particular sequence [80, 81] which contrasts strongly with the 50–100 nucleotides erroneously attributed to the RNA primer [75, 82].

In animal cells about 40% of Okazaki pieces have RNA primers [83, 84] and the RNA primers of animal viruses are about 10 nucleotides long. They do not have a specific base sequence [55, 85, 86].

A review of the methods used to detect primers was published in 1976 [87].

The use of toluenized *E. coli* enables α-^{32}P-phosphate labelled *deoxyribo*nucleoside triphosphates to penetrate the cell membrane when transfer of radioactivity to a *ribo*nucleotide occurs [82]. This is definite evidence for a covalent attachment of RNA to DNA (see Fig.

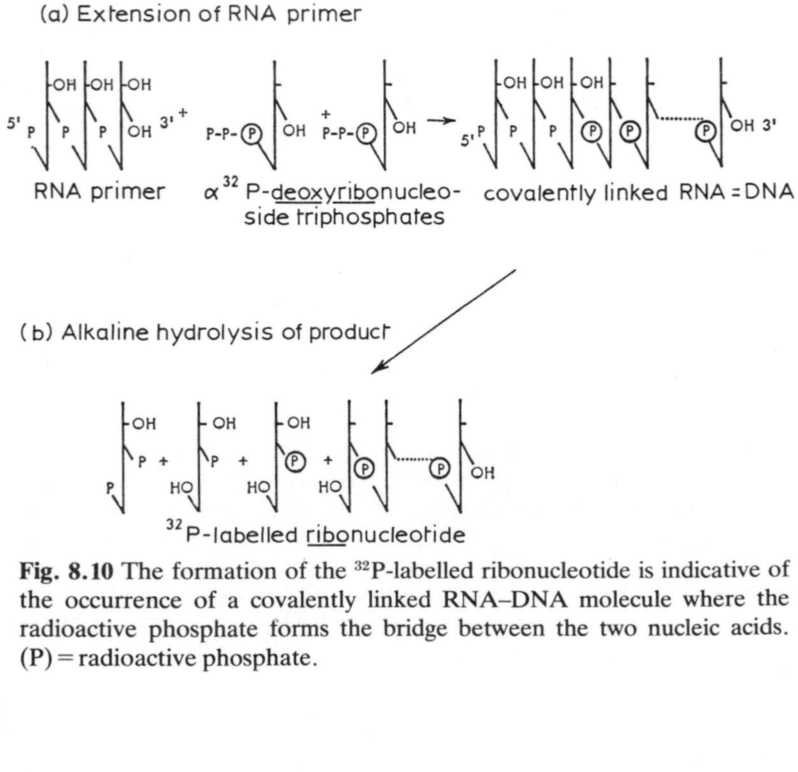

(a) Extension of RNA primer

RNA primer + α ^{32}P-deoxyribonucleoside triphosphates → covalently linked RNA = DNA

(b) Alkaline hydrolysis of product

^{32}P-labelled ribonucleotide

Fig. 8.10 The formation of the ^{32}P-labelled ribonucleotide is indicative of the occurrence of a covalently linked RNA–DNA molecule where the radioactive phosphate forms the bridge between the two nucleic acids. (P) = radioactive phosphate.

8.10). When DNA synthesis is studied in nuclei isolated from polyoma infected cells (see p. 80) the nascent DNA chains are found with a length of about ten ribonucleotides at their 5′-ends [55, 85]. When α-^{32}P-labelled deoxyribonucleoside triphosphates are injected into the slime mould *Physarum polycephalum* [88] there is transfer of the ^{32}P to ribonucleotides, which is indicative, as desribed above, of a covalent attachment of RNA and DNA.

Although this is strong evidence, in the systems studied, that each Okazaki piece is initiated by the synthesis of a short oligoribonucleotide which then serves as a primer for DNA polymerase, it should be borne in mind that many authors using other systems report failure to detect evidence of RNA primers. This may simply reflect the transient nature of the pripers. Indeed Okazaki's group have shown that many Okazaki pieces do not have RNA primers [73, 75] and RNA is not found in mature DNA. An exception to this is the DNA of the plasmid Col E1 grown in the presence of chloramphenicol In this case it is inferred that the drug prevents the normal excision of the RNA primers [89]. Mitochondrial DNA also frequently has a number of ribonucleotides left in the mature duplex [39].

Studies with the primase (see p. 255) indicate that it shows a very strong preference to initiate with adenosine followed by guanosine [90] and this fact taken together with the existence of primase binding sites near the origin of replication (see p. 213) suggests that initiation of Okazaki pieces may occur at particular sites on the lagging strand. This idea is supported by evidence from *E. coli* suggesting that the sites of initiation are at or very close to sites of adenine methylation [91]. However, the small bacteriophage P4 which only requires about 20 Okazaki pieces per round of replication shows no preferential initiation sites [92].

(e) *Continuous synthesis.* Discontinuous synthesis of one strand of DNA was postulated in order to overcome certain problems, but no problems appear to exist with the chain growing in the 5′→3′ direction, and so there is no *a priori* reason why this chain should not be made continuously. Indeed, this may well be the case when DNA synthesis is proceeding rapidly. Under these conditions nascent DNA is found in both large and small pieces. However, when deoxyribonucleotides are limiting and the rate of chain elongation slowed, both daughter strands appear to be made discontinuously. One possible explanation is that there is competition between the propagation of the growing chain and the initiation of new chains [93, 94].

Any initiation of an Okazaki piece on the leading strand is unlikely to be far in advance of the growing strand and so ligation will occur very rapidly, most probably before the Okazaki piece has grown to a detectable length. This is in contrast to the synthesis of an Okazaki piece on the lagging strand which (a) cannot be ligated until its synthesis is completed and (b) cannot be initiated until a region of single-stranded DNA of comparable length has been exposed on the lagging strand. Under conditions where primer removal and ligation is restricted (i.e. in the absence of a functional DNA polymerase I) Okazaki pieces are formed on both lagging and leading strands of the phage P2 [95]. However, under non-restrictive conditions discontinuities are not detected on the leading strand [96].

In vitro studies using the cellophane disc assay (see p. 225) have lead to the conclusion that discontinuities on the leading strand are most likely produced as a result of uracil excision [97] and that the normal continuous synthesis of the leading strand can be masked by uracil excision [98].

A similar system for elimination of uracil residues from DNA exists in eukaryotes [99].

8.2.5 Replication of chromatin in eukaryotes

Once the basic structure of chromatin in eukaryotic cells was known (see p. 89) two questions were asked. (1) what happens to the nucleosomal structures at the replication fork? and (2) to what extent is the structure of chromatin responsible for the mechanism of DNA synthesis which involves (a) Okazaki pieces (p. 217) and (b) replicons (p. 212)?

That nucleosomes quickly become associated with newly synthesized DNA was shown by the cleavage of nascent chromatin by micrococcal nuclease into approximately 200bp fragments [100–103]. However, the rate and extent of cleavage of nascent chromatin is greater than that of mature chromatin indicating a somewhat different structure which requires 15 minutes to mature, i.e. a time sufficient to synthesize 22–54000bp of DNA or approximately one replicon [17, 101, 104].

After a few experiments which were open to criticism and which yielded conflicting results it was finally established that new histones become associated with new DNA [105, 106] (rather than attaching at random to DNA and displacing histones made in a previous S-phase). Moreover assembly occurs in stages with histones H3 and H4 binding within 2 min, H2A and H2B between 2 and 10 min and histone H1 added more than 10 min after DNA synthesis is complete [106]. Nucleosomes are assembled from completely new histones and once assembled are stable for several cell generations [107].

Weintraub [108, 109] has pointed out that segregation of old nucleosomes at the replication fork may occur in one of three ways (Fig. 8.11). All the old nucleosomes may remain with one of the two parental strands when they separate at the replication fork-cooperative alignment (Fig. 8.11a). The old nucleosomes may be dispersed at random between the two strands of DNA (Fig. 8.11b). The old nucleosomes may dissociate into half nucleosomes; one half going to each side of the fork (Fig. 8.11c). The first possibility is favoured by the finding that not only are new nucleosomes composed of all new histones, but also cross-linking studies have shown that new nucleosomes are arranged adjacent to other new nucleosomes and are not dispersed amongst old nucleosomes [107].

When DNA synthesis is allowed to proceed in the presence of inhibitors of protein synthesis the chromatin that is made is particularly sensitive to micrococcal nuclease although that which remains is present

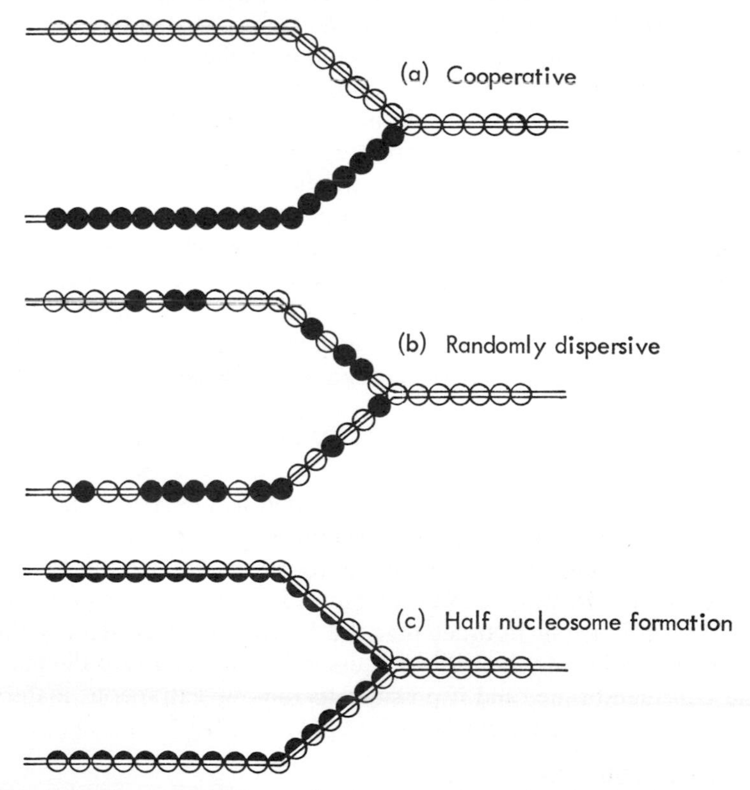

Fig. 8.11 Three possible mechanisms are shown for the distribution of nucleosomes at a replication fork (After Weintraub [108]) (O, Parental nucleosome, •, newly synthesized nucleosome).

as mononucleosomes [108]. This is consistent with a cooperative segregation of parental nucleosomes to one of the two forks with the other fork being devoid of nucleosomes [108]. Such structures are seen in the electron microscope [110] and Weintraub has shown that the nuclease-resistant daughter strands of SV40 minichromosomes replicating in the presence of cycloheximide hybridize to the template strand on the leading side of the fork, i.e. the old nucleosomes associate with the leading strand at replication [111].

Thus DNA fork movement may be considered to occur as a series of steps caused by relocation of nucleosomes on the leading side of the fork which is always made up of duplex DNA. At each relocation a region of about 200 bases of DNA on the lagging side of the fork is rendered single stranded providing a target for random initiation of RNA primers. For this reason the Okazaki pieces resulting have a mean size somewhat shorter than the 200 bases of nucleosomal DNA [112].

Replicons have a length approximately equal to that of chromatin loops [113] or the equivalent of 400 nucleosomes or about 70 turns of a solenoid [114]. As adjacent replicons tend to replicate at the same time in the cell cycle [17–19, 115–117] it is possible that a region of a chromosome becomes partially expanded and single-stranded DNA is exposed at the ends of the loops. Bidirectional replication would be initiated at such regions (see p. 212) and hence a replicon and a loop would be the same structure. Alternatively initiation of replicon synthesis may occur on the scaffold to which the loops are attached in which case each loop would be part of two adjacent replicons. As replicon size varies dramatically in developing amphibia and insects from 5 μm long in early cleavage to up to 350 μm long during spermatogenesis [18] chromatin structure would be expected to show parallel changes.

8.3 DNA REPLICATION—*IN VITRO* STUDIES

To help understand a process as complicated as DNA replication it is often an advantage to work with a system simpler than the intact cell; some reference has already been made to results obtained in this way. The geneticist indicates the involvement of a numbr of gene products in DNA replication, but in order that the biochemist may identify these proteins and delineate their role it is essential to circumvent the barrier of the cell membrane, and if possible to remove extraneous material. This has been done in a number of ways [118, 119].

8.3.1. Permeable cells

Treatment of *E. coli* with lipid solvents (ether [120] or toluene [121]) whilst inhibiting cell division appears to leave the cells and constituent enzymes intact yet renders the membrane permeable to small molecules including nucleotide precursors. Such cells when incubated with the four deoxyribonucleoside triphosphates, ATP, and Mg^{2+} continue semi-conservative replication. The products of such replication are the 10S Okazaki pieces which are only joined into high molecular weight DNA in the presence of NAD, the cofactor for *E. coli* DNA ligase. It was with toluenized cells and a α-^{32}P-labelled deoxyribonucleotides that Okazaki was able to show the covalent linkage of RNA and DNA (see p. 220).

8.3.2 Cell lysates

If the cell membrane is disintegrated a lysate results which, under certain conditions, is capable of maintaining DNA synthesis [122, 123]. It is very important not to dilute the macromolecular constituents of the cell but to maintain them near their *in vivo* concentration. One way in which this has been achieved is by lysis of cells on a cellophane disc held on an agar surface. Small molecules are then allowed to diffuse through the disc into the concentrated lysate. Semi-conservative DNA synthesis occurs in this and similar systems if the lysate is prepared from wild type *E. coli,* but when it is prepared from *E. coli* cells defective in DNA synthesis then the lysate reflects the *in vivo* capabilities of the cells. Thus lysates prepared from mutants temperature-sensitive in the genes dna B, dna D, or dna G (which fail to extend growing DNA chains at the non-permissive temperaturc; Tablc 8.1) only synthesize DNA at low temperatures. In lysates prepared from mutants temperature-sensitive in the genes dna A or dna C/D (which fail to initiate DNA synthesis at non-permissive temperatures) DNA synthesis fails to occur only if the cells are maintained at the non-permissive temperature for one generation time prior to lysis [124].

This system has proved very useful in the purification of the products of these dna genes [118, 119]. Thus fractions containing the dna G gene product from wild type cells, when added to a lysate from a mutant temperature-sensitive in dna G, were able to restore DNA synthesis at the non-permissive temperature. Although the products of the dna genes have all been at least partially purified by this complementation assay, functions for all these proteins have not yet been established. However, they have been used with some success in reconstruction experiments using fully defined components (see below).

Permealysed cells [125] and lysates from eukaryotic cells [126–128] also show a limited ability to continue DNA synthesis when provided with the four deoxyribonucleoside triphosphates, ATP, and Mg^{2+}. Eukaryotic cells can be rendered permeable to deoxyribonucleotides by treatment with hypotonic buffers. DNA synthesis continues in these cells for about 30 min before slowing down, probably due to completion of ongoing replicons coupled with an inability to initiate new replicons [129].

Cell lysates show similar characteristics [130] but isolated nuclei are much more limited in their capabilities [131–134] unless supplemented with cytosol [112]. In general, isolated nuclei appear capable of extending Okazaki pieces initiated *in vivo* but have only a limited capacity for ligating or initiating Okazaki pieces. DNA synthesis in isolated nuclei is sensitive to inhibitors of DNA polymerase α but not to inhibitors of DNA polymerase β or γ indicating a role for DNA polymerase α in this reaction [135, 136]. However, removal of all soluble DNA polymerase α activity from nuclei does not markedly reduce their DNA synthetic capacity, possibly indicating the presence of an enzyme complex [137, 138].

Cell lysates do initiate the syntheis of Okazaki pieces and these have been shown to have RNA primers 8–11 nucleotides in length [139]. The sequence of these oligoribonucleotides is diverse.

The disadvantage with lysates of eukaryotic cells is that very little is known about the nature of the DNA template or product. As with prokaryotes (see below), model systems have been used. Thus lysates of cells infected with SV40 or polyoma virus synthesize predominantly viral DNA *in vitro* [126, 127, 140–144] and moreover viral nucleoprotein complexes can be isolated which will continue to synthesize viral DNA in the presence of soluble cellular preparations [112, 145–147]. Soluble replication systems have also been prepared from adenovirus infected cells [148–150] and lysates made from cells infected with herpes simplex virus synthesize predominantly herpes simplex DNA *in vitro* [151, 152].

Soluble extracts from cells infected with an adenovirus coding for a temperature-sensitive DNA binding protein fail to support adenoviral DNA synthesis at restrictive temperatures, but can be complemented with wild-type DNA binding protein [153].

8.3.3 Soluble extracts

Although useful in the identification and purification of some of the factors necessary for DNA synthesis, the cellophane-disc assay is of limited value in defining the function of the various gene products and moreover it relies almost entirely on possession of temperature-sensitive mutants for every protein involved.

A simplication involves the use of soluble extracts from *E. coli* free of the bacterial DNA. When cells are lysed with lysozyme the bacterial chromosome can be pelleted by centrifugation at 200 000 **g**. The soluble preparation remaining is capable of performing specific stages of DNA synthesis when supplemented with simple DNA substrates, e.g. single-stranded cyclic DNA from the small phages ØX174 or M13. Once again possession of mutants is an invaluable aid but the advantage of this system over the crude lysate is that specific steps in DNA replication may be studied and hence information on the function of particular proteins may be obtained.

8.3.4 Reconstruction experiments

The ultimate aim of these experiments is to achieve the ordered replication of a complex DNA molecule such as the *E. coli* chromosome using totally defined components. This is far from being achieved. However, the requirements for the conversion of single-stranded bacteriophage DNA into the double-stranded replicative form (RF) and the replication of RF have been delineated and the reactions carried out using purified components. Progress was not all plain sailing, for in addition to the components identified with the aid of mutants and purified by the complementation assay (see above), addition of soluble extract was required. From this extract several further components have been purified for which, as yet, no mutants are known.

The detailed results of these reconstruction experiments are considered on p. 256.

8.4 ENZYMES OF DNA SYNTHESIS

8.4.1 Introduction

As we have said DNA synthesis occurs at a replication fork which progresses along the DNA molecule. A variety of proteins is required to bring about this process in an efficient manner. The major proteins are as follows:

(a) A DNA polymerase is required to add dNTPs to the growing leading strand and to growing Okazaki pieces on the lagging strand. In general DNA polymerases act alongside other helper proteins.

(b) A primase is required to help initiate Okazaki pieces.

(c) An exonuclease is required to remove the primers.

(d) A ligase is required to join the Okazaki pieces together.

(e) An unwinding protein is required to help unwind the duplex DNA at the replication fork.

(f) A nicking closing enzyme is required to relax the tension engendered by unwinding duplex DNA.

(g) A binding protein is required to stabilize single-stranded DNA exposed by progression of the leading strand prior to initiation of Okazaki pieces.

(h) A gyrase may be required to help unwind the double helix ahead of the replication fork or for initiation of replication.

The action of these proteins is summarized in Fig. 8.12 and the remainder of this section is devoted to the biochemistry of these proteins.

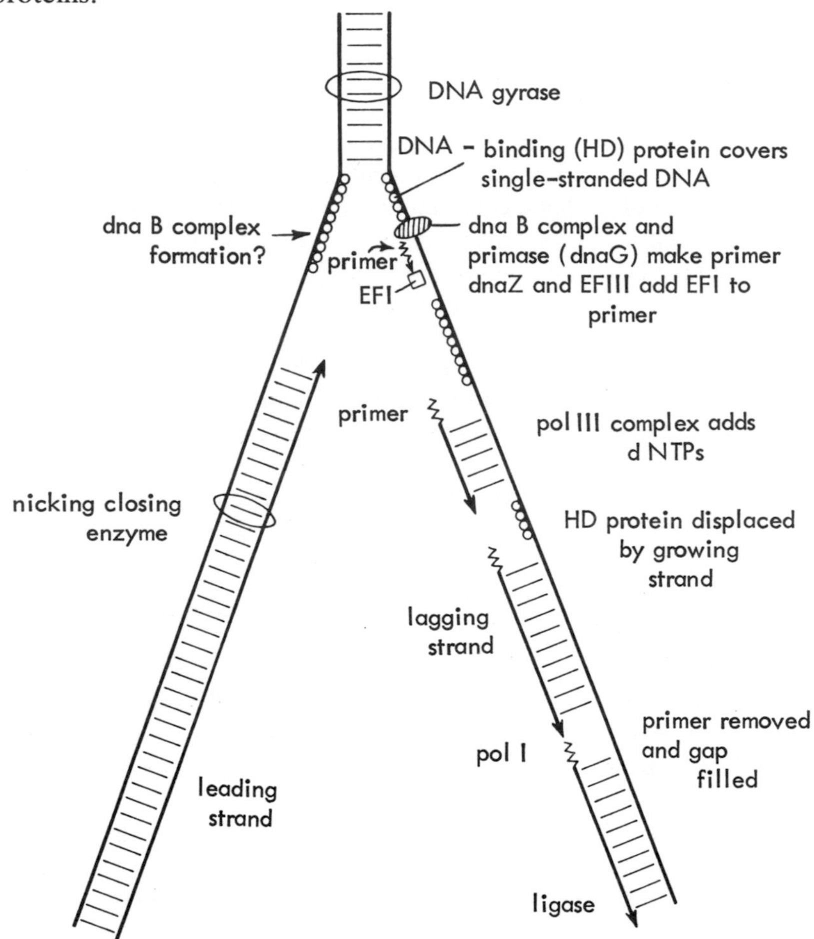

Fig. 8.12 Site of action of enzymes involved in DNA replication at the growing fork in *E. coli*.

8.4.2 DNA Polymerases

(a) Mechanism of action

DNA polymerase is the name given to an enzyme which catalyses the synthesis of DNA from its deoxyribonucleotide precursors. Such enzymes have been purified from various sources (bacterial, plant, and animal) but, apart from some details discussed below, the mechanism of the polymerization reaction is common to all preparations.

DNA polymerase catalyses the formation of a phosphodiester bond between the 3′-OH group at the growing end of a DNA chain (the primer) and the 5′-phosphate group of the incoming deoxyribonucleoside triphosphate. Growth is in the 5′→3′ direction, and the order in which the deoxyribonucleotides are added is dictated by base pairing to a template DNA chain. Thus, as well as the four triphosphates and Mg^{2+} ions, the enzyme requires both primer and template DNA. No DNA polymerase has been found which is able to initiate DNA chains.

The simplest case to consider is that in which a single-stranded template has bound to it a growing strand of primer terminating at the growing point in a 3′-hydroxyl group (Fig. 8.13(1)). Such a situation is also illustrated in Fig. 6.7 for a piece of double-stranded DNA partially degraded by exonuclease III.

The polymerase binds to the single-stranded template in the region of the 3′-hydroxyl end of the primer (Figs. 8.13 and 8.14). An incoming deoxyribonucleoside triphosphate containing a base which can pair with the corresponding base on the template becomes attached to the triphosphate binding site. The polymerase then catalyses a nucleophilic attack by the 3′-hydroxyl group of the primer on the α-phosphate of the deoxyribonucleoside 5′-triphosphate (Fig. 8.15). Inorganic

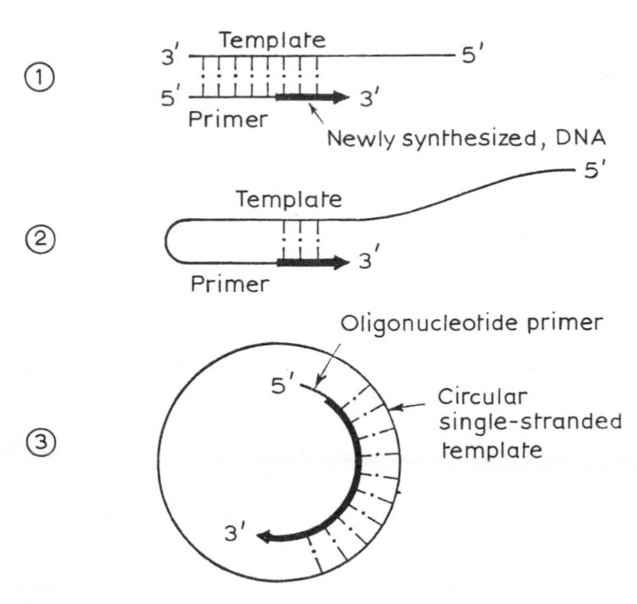

Fig. 8.13 Mechanism for the replication of single-stranded DNA.

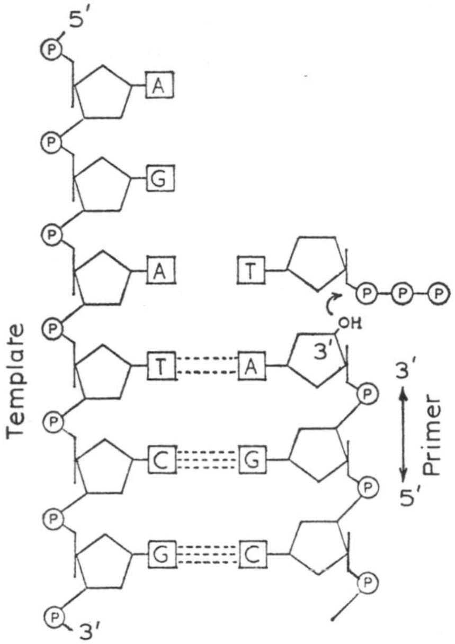

Fig. 8.14 Greater detail of the mechanism shown in Fig. 8.13(1).

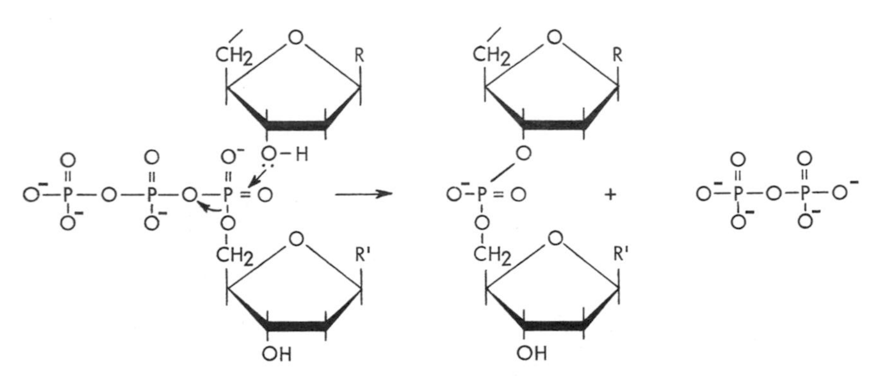

Fig. 8.15 The mechanism of the action of DNA polymerase.

pyrophosphate is released, a phosphodiester bond is formed, and the chain is lengthened by one unit. The enzyme moves along the template by the distance of one unit and the newly added nucleotide with its 3′-hydroxyl group now occupies the primer terminus site. The process is then repeated until the enzyme reaches the end of the template strand. *In vivo* the pyrophosphate is hydrolysed by a pyrophosphatase which drives the equilibrium in favour of DNA synthesis.

In the replication of single-stranded DNA the 3′-hydroxyl terminus may loop back on itself and serve as a priming strand as shown in Fig.

8.13(2), or short lengths of oligonucleotide may act as primers by becoming hydrogen bonded to the template.

When a cyclic single-stranded DNA is used as template, replication can occur only in the presence of short oligonucleotides which become attached by base-pairing to the template and serve as primers (Fig. 8.13(3)). DNA polymerase cannot use double-stranded DNA as a primer/template. However, if a nick or gap is introduced into one strand the exposed 3′-OH group acts as a primer and the new growing strand will fill the gap and then may continue displacing the complementary strand from the 5′-end. This can lead to complications since at some point the enzyme may leave the original template strand and begin to copy the complementary strand so that *in vitro* a branched structure is formed.

When native DNA is treated with exonuclease III, a partially single-stranded molecule is produced (Fig. 6.7). It can be repaired and restored to the native double-stranded form by the polymerase.

(b) E. coli *DNA polymerase I* (EC2.7.7.7)

Since the greatest amount of effort has been put into the study of DNA polymerase I from *E. coli* (the Kornberg enzyme) this will be considered in some detail and then comparisons will be made with other enzymes. Kornberg himself has recently written two books on the subject [154, 155].

Historical. In 1956 Kornberg and his collaborators used cell-free extracts from exponentially growing cultures of *E. coli* to demonstrate that, in the presence of ATP and Mg^{2+}, [^{14}C]-thymidine, a specific precursor for DNA, was incorporated into an acid-precipitable material which was judged to be DNA from the fact that it was no longer acid-precipitable after treatment with deoxyribonuclease. Using this incorporation method for assay purposes, the bacterial extract was fractionated and it soon became evident that crude extracts contained a kinase system which converted thymidine into thymidine triphosphate (dTTP). In subsequent work labelled dTTP was used as substrate and the enzyme responsible for its incorporation into DNA has been purified to homogeneity as judged by sedimentation, chromatography, and electrophoresis.

With a partially purified enzyme preparation it was shown that, for good incorporation of dTTP (labelled either with ^{14}C in C-2 of the thymine or with ^{32}P in the innermost phosphate group), the 5′-triphosphates of all four deoxyribonucleosides which normally occur in DNA had to be present together with Mg^{2+} ions and a denatured DNA template (calf thymus DNA was used in early experiments). Under these conditions excellent incorporation was observed as is illustrated in Table 8.2. Omission of any one triphosphate or Mg^{2+} or of DNA template, or pretreatment of the template with deoxyribonuclease, reduced the incorporation to a very low level. Deoxyribonucleoside diphosphates could not replace the triphosphates. With the purified enzyme it was also possible by chemical analysis to demonstrate net synthesis of DNA, and increases of DNA by a factor of 10–20 could be obtained indicating that 90–95 per cent of the isolated DNA was derived from the added deoxyribonucleoside triphosphates.

As well as DNA polymerase activity, the purified enzyme shows several other activities each of which is associated with the same enzyme molecule. Thus exonuclease (p. 239) activity (both 3′→5′ and 5′→3′) is purified along with the polymerase activity in a ratio which is not altered by fractionation.

Evidence for the copying of the template. The DNA polymerase is an enzyme with the unusual property of taking directions from a template and faithfully reproducing the sequence of nucleotides in the product. The evidence for this rests on several experimental observations.

Table 8.2 Incorporation of ^{32}P-dCTP into DNA (Bessman *et al.* [480]).

Additions	pmoles ^{32}P-DNA
dCTP (DNA omitted)	0.0
dCTP	2.5
dCTP + dGTP	5.1
dCTP + dGTP + dTTP	15.7
dCTP + dGTP + dTTP + dATP	3300

The incubation mixture (0.3 ml) contained 5 nmoles of dCTP (7.2×10^7 c.p.m. per mole) and 5 nmoles of each of the other deoxynucleoside triphosphates as indicated, together with 1.9 μmole of $MgCl_2$, 20 μmoles of glycine buffer (pH 9.2), 10μg of DNA, and 3μg of purified 'polymerase' from *E. coli.* The incubation was carried out at 37°C for 30 minutes.

(1) The most significant fact is that the enzyme can faithfully copy the nucleotide sequence of the single-stranded cyclic DNA bacteriophage ØX174. The product can be converted into biologically active material by subsequent cyclization by the enzyme polynucleotide ligase. This experiment was done in 1967 by Kornberg and his colleagues [156] who produced, *in vitro,* infective DNA identical with the natural material. These findings show that DNA polymerase I is an enzyme which has all the requirements necessary for the replication of DNA.

(2) When *E. coli* polymerase is used to prepare DNA under conditions such that only 5% of the sample produced comes from the template, the product has many of the same physical and chemical properties as DNA isolated from natural sources. The product appears to have a hydrogen-bonded structure similar to that of natural DNA and undergoes molecular melting (p. 38) in the same way.

It shows the same equivalence of adenine to thymine and guanine to cytosine that characterizes natural DNA. Moreover the characteristic ratio of A · T pairs to G · C pairs of a given DNA primer is imposed on the product whether the net DNA increase is 1% or 1000%. The base ratios in the product are not distorted when widely differing molar proportions of substrate are used. This is best illustrated by the use of the synthetic template poly(dA-dT) (see p. 238) when, from a mixture of all four deoxyribonucleoside triphosphates, only dAMP and dTMP are incorporated into the product.

(3) The nucleotide sequence in the template DNA is reproduced in the product. This has been established by the Kornberg group by using the technique of *nearest-neighbour sequence* analysis [157, 158] (see also p. 220). The partially purified *E. coli* enzyme is incubated with a particular template DNA and all four deoxyribonucleoside triphosphates, one of which, say dATP, is labelled with ^{32}P in the innermost phosphate. During the synthetic reaction this ^{32}P becomes the bridge between the nucleoside of that labelled triphosphate (A) and the nearest-neighbour nucleotide containing the base Z at the growing end of the polynucleotide chain (Fig. 8.16a). After the synthetic reaction is

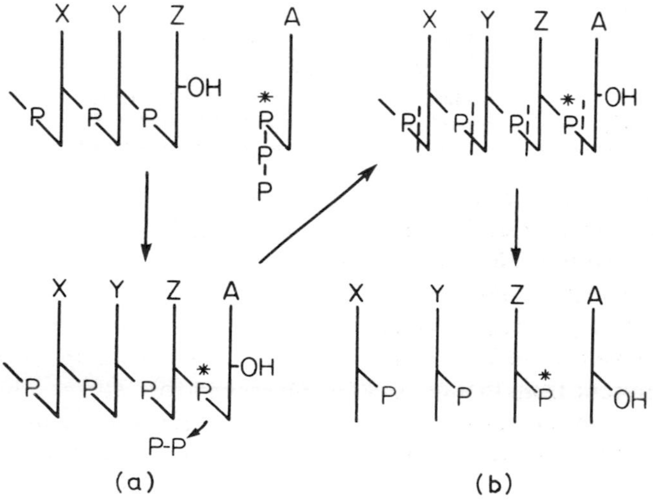

Fig. 8.16 Illustration of the method of nearest-neighbour sequence analysis.

complete the DNA is isolated and degraded with micrococcal DNase (p. 151) and spleen phosphodiesterase (p. 153) to yield the deoxyribonucleoside 3′-monophosphates. The ^{32}P is thus transferred to the 3′-carbon of the neighbouring nucleotide in the chain (Z in Fig. 8.16), i.e. the one with which the labelled triphosphate has reacted (Z might be any one of the four bases). The four deoxyribonucleoside 3′-monophosphates are isolated by paper electrophoresis, and their radioactivities measured to give the relative frequency with which the nucleotide originally labelled in position 5′ locates itself next to another nucleotide in the new chain. This procedure is carried out four times with a different labelled triphosphate each time so as to determine the relative frequencies of all sixteen possible nearest-neighbour (or dinucleotide) sequences [158]. The results of such an experiment with DNA from *M. phlei* as primer are shown in Table 8.3.

Table 8.3 Nearest-neighbour frequencies of *Mycobacterium phlei* DNA [159].

Labelled triphosphate	Deoxyribonucleoside 3′-phosphate isolated			
	Tp	Ap	Cp	Gp
dATP	TpA 0.012	ApA 0.024	CpA 0.063	GpA 0.065
dTTP	TpT 0.026	ApT 0.031	CpT 0.045	GpT 0.060
dGTP	TpG 0.063	ApG 0.045	CpG 0.139	GpG 0.090
dCTP	TpC 0.061	ApC 0.064	CpC 0.090	GpC 0.122
Sums	0.162	0.164	0.337	0.337

They illustrate several points:

(i) All sixteen possible nearest-neighbour sequences are present and they occur with widely varying frequencies.

(ii) The results show a very striking deviation from the nearest-neighbour frequencies predicted if the arrangement of mononucleotides were completely random. Thus the frequency of TpA in the first row is quite different from that of ApT in the second row whereas these two frequencies would have to be identical in a random assembly. The nucleotides have therefore been assembled in accordance with a definite pattern.

(iii) The sums of the four columns show the equivalence of A to T and of G to C in the product and indicate both the validity of the analytical method and the replication of the overall composition of the primer DNA.

(iv) The results indicate that base pairing occurs in the newly synthesized DNA and that its two strands are of opposite polarity. According to the Watson–Crick model the two strands of the double helix are of opposite polarity (p. 22), and it is presumed that each can act as a template for the formation of a new chain so as to give precise replication with the formation of two daughter helices identical with each other and with the parent helix. The results of nearest-neighbour sequence analysis support this mechanism. For example, the frequencies of ApA and TpT sequences are equivalent, and so are the frequencies of GpG and CpC. The matching of the other sequences depends upon whether the strands of the double helix are of similar or opposite polarity (Fig. 8.17).

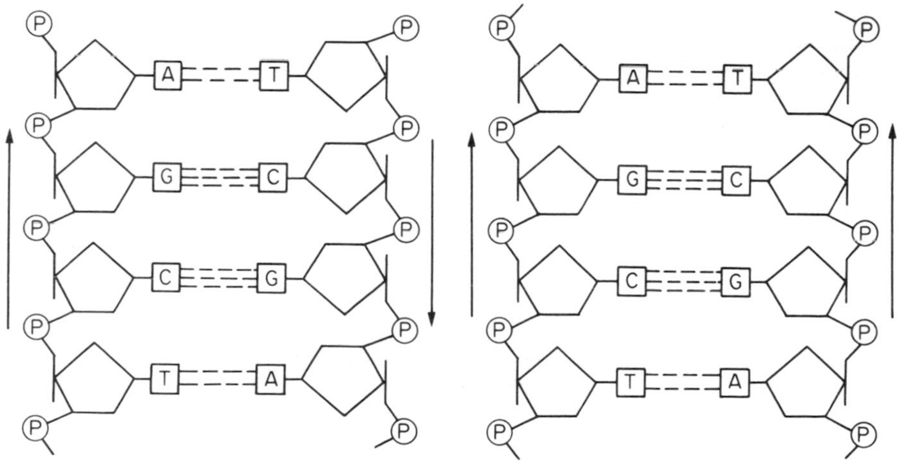

Fig. 8.17 Possible structures of the DNA molecule showing opposite polarity of strands (left) and the same polarity of strands (right).

If the strands are of opposite polarity the following matching sequences can be predicted:

CpA and TpG
GpA and TpC
CpT and ApG
GpT and ApC

whereas if the strands are of the same polarity the matching sequences would be:

TpA and ApT
CpA and GpT
GpA and CpT
TpG and ApC
ApG and TpC
CpG and GpC

The results in Table 8.3 favour the helix with strands of opposite polarity. The ApA and TpT, and the CpC and GpG sequences match similarly in both models.

(v) The nearest-neighbour frequencies measured by the method described above are those of the newly synthesized DNA. To verify that they are an accurate reflection of those in the original DNA template, an enzymically synthesized sample of calf thymus DNA in which only 5% of the total DNA consisted of the original template was itself used as template in a sequence analysis. The results showed good agreement between the sequence frequencies of the products primed by native DNA and by enzymically produced DNA (Table 8.4) whereas DNAs from other sources gave quite different results.

It can be concluded therefore that the polymerase yields a DNA product with strands of opposite polarity and that the sequence of bases is faithfully reproduced.

The chemical nature of DNA polymerase I. The purified Kornberg

Table 8.4 Nearest-neighbour frequencies of native and enzymically synthesized calf thymus DNA [158].

Nearest-neighbour sequence		Native calf thymus DNA as template		Enzymically synthesized calf thymus DNA as template	
ApA	TpA	0.089	0.053	0.088	0.059
ApG	TpG	0.072	0.076	0.074	0.076
ApC	TpC	0.052	0.067	0.051	0.064
ApT	TpT	0.073	6.087	0.075	0.083
GpA	CpA	0.064	0.064	0.063	0.078
GpG	CpG	0.050	0.016	0.057	0.011
GpG	CpC	0.044	0.054	0.042	0.055
GpT	CpT	0.056	0.067	0.056	0.068

enzyme is a protein of mol. wt 109 000 in the form of a single polypeptide chain [160]. This chain can be unfolded in guanidine-HCl-mercaptoethanol so as to denature the protein. When the reagent is diluted out renaturation occurs with restoration of activity.

The protein migrates as a single band on SDS-acrylamide gel electrophoresis; it contains only one sulphydryl group and one disulphide group; the residue at the amino terminal end is methionine. One *E. coli* cell contains about 400 molecules of enzyme, each spherical and of diameter 6.5 nm. It can form dimeric forms which can be visualized in the electron microscope [161].

Treatment of the enzyme with the protease *subtilisin* from *B. subtilis* breaks it into two fragments, a larger fragment of mol. wt 76 000 which retains polymerase activity and 3′→5′ nuclease activity (see below) but not the 5′→3′ nuclease activity, and a smaller fragment of mol. wt 34 000 which retains nuclease 5′→3′ activity in the presence of DNA [106–108].

The active centre of the enzyme. On the basis of binding experiments Kornberg has concluded that the multiple functions of the enzyme include the following operations [155, 162].

Extension of a DNA chain in the 5′→3′ direction by the addition to the 3′-hydroxyl terminus of mononucleotides from deoxyribonucleoside triphosphates at the rate of 1000 nucleotides per minute.

Hydrolysis of a DNA chain from the 3′-hydroxyl end in the 3′→5′ direction to yield 5′-monophosphates (the exonuclease II action referred to on p. 160).

Hydrolysis of a DNA chain from the 5′-phosphate (or 5′-hydroxyl) terminus in the 5′→3′ direction to yield mainly 5′-monophosphates (see p.160).

Pyrophosphorolysis of a DNA chain from the 3′-end; this is essentially the reversal of the polymerization reaction.

Exchange of inorganic pyrophosphate with the terminal pyrophosphate group of a deoxyribonucleoside triphosphate.

These last two reactions are simply the result of reversal of the polymerization reaction and are of doubtful *in vivo* significance since they require a high concentration of inorganic pyrophosphate.

Kornberg envisages the active centre of the enzyme as a specially adapted polypeptide surface comprising at least five major sites.

(1) A site for the binding of the template chain in the region where base pairs are formed and for a few nucleotides on each side of this.

(2) A site for the growing primer chain which is, of course, base-paired to the template.

(3) A site for the special recognition of the terminal 3′-hydroxyl group of the primer. This point is the start of the 3′→5′ hydrolytic cleavage.

(4) A triphosphate binding site for which all four triphosphates compete.

(5) A site which allows for the 5′→3′ cleavage of a 5-phosphoryl terminated chain. It is presumably this area that is broken off by subtilisin.

These sites determine the nature of the DNA which can bind to the enzyme. For example, linear single-stranded DNA binds readily on site (1) whereas an intact linear duplex such as the DNA of bacteriophage T7 does not bind if it has been prepared with great care so as to avoid internal breaks. An intact circular duplex such as plasmid DNA or ØX174 replicative form DNA (p. 256) does not bind to the enzyme until a 'nick' has been introduced in one of the strands by an appropriate nuclease yielding 3′-hydroxyl and 5′-phosphate termini. Such nicks are active points for replication whereas nicks introduced by micrococcal nuclease with 5′-hydroxyl and 3′-phosphoryl termini are not replication points although they bind the enzyme. One molecule of enzyme is bound at each nick in either case.

Recently evidence has been obtained that there are two distinct sites which recognize the 3′-OH group of the primer [163]. When the terminal nucleotide is correctly matched to the template the 3′-OH group resides in the polymerase site. However, if the terminal nucleotide is not correctly matched to the template the 3′-OH group is displaced into a site where it is susceptible to the 3′→5′ exonuclease (see p. 239).

Poly d(A-T) and Poly dG · dC. When DNA polymerase I is incubated without primer in the presence of dATP, dTTP, and Mg^{2+} an interesting polymer is formed containing adenine and thymine nucleotides [155, 164]. Polymer formation occurs only after a lag period of several hours and then takes place rapidly until 60–80% of the triphosphates have been utilized.

The product of the reaction contains equal amounts of adenine and thymine. Nearest-neighbour frequency analysis shows that the frequencies of ApT and TpA are each 0.500 whereas the sequences ApA and TpT are undetectable. The polymer therefore contains alternating residues of A and T.

The molecular weight calculated from the sedimentation value and reduced viscosity is between 2 and 8×10^6. The polymer melts sharply at 71°C with an increase of 37% in the absorbance at 260 nm. The process is completely reversible on cooling. Such physical data, including the X-ray diffraction pattern, suggest that the molecule is a long fibrous double-stranded structure with the strands joined by hydrogen bonds between adenine residues in one chain and thymine residues in the other.

A somewhat similar polymer [155, 157] is formed when *E. coli* DNA polymerase I is incubated with high concentrations of dGTP and dCTP in the presence of Mg^{2+}. Again, a lag period of several hours is found. This product contains guanine and cytosine, not necessarily in equal amounts, and nearest-neighbour sequence analysis shows that the frequencies of GpG and CpC are each 0.500. Mild acid hydrolysis releases all the dGMP but none of the dCMP. Sedimentation and viscosity measurements yield values similar to those found with the poly d(A-T) but the T_m value is much higher (83°C). These observations are consistent with the view that the molecule consists of two homopolymers, one containing only guanine and the other only cytosine, hydrogen-bonded throughout their lengths.

The 3′→5′ exonuclease and proof-reading. Before it was realized that this was an integral function of the polymerase the 3′→5 exonuclease activity was ascribed to an exonuclease II (see p. 258). It appears that this exonuclease functions to recognize and eliminate a non base-paired terminus on the primer DNA [165]. When *E. coli* DNA polymerase I is provided with 4 triphosphates and a primer/template with a mismatched end, the non-matching terminus is removed by the 3′→5′ exonuclease before polymerization begins. (In the absence of the triphosphates the exonuclease continues to remove nucleotides from the frayed ends of the DNA molecules [166].)

Such a nuclease by correcting errors occurring on polymerization, may be expected to increase dramatically the fidelity of the base pairing mechanism [154].

This proof-reading mechanism provides a justification for involving 5′-deoxyribonucleoside triphosphates in the 5′→3′ extension of a 3′OH on the growing DNA chain. Addition of nucleotides to a growing chain terminating in a triphosphate (see Fig. 8.7) would result, during proof reading in the removal of the energy required for further extension.

However, there are arguments for and against the significance of this proposed proof-reading mechanism. Certain *E. coli* mutants show an increased or decreased rate of mutation which can be correlated with a change in the relative activities of the polymerase and 3′→5′ activities of DNA polymerase I. However, as we shall see below, DNA polymerase I is believed to have a role not as the major polymerase involved in replication but as an enzyme involved in the repair of DNA. Furthermore, the rate of generation of dNMP by exonuclease action could account for only a minor increase in fidelity [167].

The error rates in eukaryote polymerases (see below) which do not have associated 3′→5′ exonucleases are comparable with that for *E. coli* DNA polymerases I [167–169].

On the other hand in *E. coli* infected with 'phage T4 (which codes for its own DNA polymerase) there are mutator and antimutator polymerases which show (a) decreased or increased rates of the pyrophosphate exchange reaction (see p. 237) and (b) increased or decreased rates of incorporation of non-complementary nucleotides, both of which have been correlated with changes in 3′→5′ exonuclease activities [170–172].

Rather than the 3′→5′ exonuclease of DNA polymerase playing a central role in proof-reading a different or additional proof-reading mechanism is thought to exist. This is kinetic proof reading first suggested by Hopfield [173] and involves the polymerizing enzyme binding the incoming triphosphate and then undergoing a change in conformation dependent on the particular nucleotide bound. Alternatively the template-bound enzyme may change conformation so as to accept only the correct nucleotide. This enhances the binding specificity for base pairing and would be an energy-dependent process [167, 169, 174–176]. It may account for some of the ATPases required for DNA replication (see below).

The 3′→5′ exonuclease of DNA polymerase I may then play a role in mismatch repair or excision repair in a manner similar to that of the 5′→3′ exonuclease (see p. 293) [163, 167].

The 5′→3′ exonuclease. This activity, which is present in the smaller fragment released from DNA polymerase I by *subtilisin* treatment (see above), cleaves base-paired regions of DNA, releasing oligonucleotides from 5′-ends. Because of its ability to jump several bases at a time, this nuclease can act on DNA molecules containing mismatched bases or distortions which render them unsuitable as substrates for polymerase [177]. It may thus serve a function, for instance, in the elimination of thymine dimers from DNA exposed to ultraviolet radiation (see p. 292).

When *E. coli* polymerase I binds to a nick on double-stranded DNA two reactions occur simultaneously. Polymerization extends the 3′-OH end and 5′→3′ exonuclease degrades the 5′-phosphate terminus. This results in *nick translation,* a process which may only end when the enzyme reaches the end of the DNA molecule [178]. Alternatively nick translation may end if a long 5′-terminus is released from the nick thereby rendering it no longer susceptible to the exonuclease. The branched product typical of reaction with native DNA primer/templates would then be formed (see p. 231). Normally DNA polymerase I is highly processive. For each DNA enzyme association event hundreds of nucleotides are polymerized before dissociation occurs [179]. However, a mutant of DNA polymerase I (*polA5*) shows all the normal properties of the wild-type enzyme but has a reduced rate of polymerization and nick translation caused by a lowered affinity of the enzyme for the DNA primer/template [180].

The role of the Kornberg enzyme in vivo. Although the DNA polymerase I is exceedingly effective in the copying of a single-stranded DNA template when provided with a primer, it is much less effective with double-stranded DNA. This observation and other considerations led to doubts as to the role of the Kornberg enzyme (DNA polymerase I) in the replication of DNA *in vivo* and to the suggestion that it is concerned merely in maintenance and repair of DNA (p. 293). These considerations may be summarized as follows:

(1) The purified enzyme cannot replicate double-stranded DNA semiconservatively to yield a biologically active product.

(2) The purified enzyme catalyses the incorporation of 1000 nucleotides per minute per molecule of enzyme whereas the estimated rate of incorporation *in vivo* is 100 times faster.

(3) Mutants of *E. coli* have been isolated which contain apparently normal Kornberg enzyme but are defective in DNA duplication. This demonstrates that other enzymes (which *may* include other polymerases) are required for *in vivo* DNA synthesis, and this may help to explain the deficiencies enumerated under (1) and (2) above where only the purified Kornberg polymerase was present.

The evidence against the Kornberg enzyme being *essential* for replication *in vivo* is based on the properties of a mutant of *E. coli* (*pol A1* or *pol* A^-) isolated by Cairns and de Lucia [181, 182]. This mutant and several others discovered later multiply normally but contain 1% or less of the Kornberg enzyme activity present in wild-type cells. Such mutants, however, show a reduced ability to join Okazaki fragments [183] (p. 217) and an increased sensitivity to ultraviolet [181] and ionizing radiation [184] and to alkylating reagents [181, 185]. This implies a role for the enzyme in 'gap' filling, and perhaps also in excision of RNA primers and mismatched base pairs. However, these functions are not completely lacking in mutants lacking DNA polymerase I, and it is suggested that they may also be performed to a limited extent by other enzymes, e.g. polymerases II and III and the rec B and C nuclease (see p. 295). Double mutants of pol A rec B are non-viable [186], suggesting an essential function which can be carried out by more than one enzyme.

(c) E. coli *DNA polymerase II*

The discovery of the *pol* A^- mutant of *E. coli* which grows well and replicates its DNA in the usual manner in spite of the absence of the Kornberg enzyme gave an impetus to the search for another enzyme apparatus which can synthesize DNA. Two further polymerases, designated polymerase II and III, were found in extracts of pol A^- cells.

These enzymes had not been detected previously because, in extracts of wild-type cells, they show little activity relative to DNA polymerase I when single-stranded or nicked DNA is used as template. DNA polymerase II and III show significant activity only with a 'gapped' DNA template (Table 8.5). The enzymes can be separated from one another by chromatography on phosphocellulose, DEAE-cellulose, or DNA-agarose [155].

Purified DNA polymerase II has a mol. wt of 90000–120000 and is homogeneous as judged by SDS–polyacrylamide gel electrophoresis. It synthesizes DNA in the 5'→3' direction and for maximal activity it requires all four triphosphates, Mg^{2+}, NH_4^+, and a native DNA template containing single-stranded gaps 50–200 bases long. The rate of reaction falls off with longer gaps, but may be restored by addition of *E. coli* DNA binding protein (see p. 251). The enzyme also requires a 3'-OH primer. It is sensitive to sulphydryl reagents and is not affected by antiserum to DNA polymerase I. The purified enzyme, however, like polymerase I, only synthesizes DNA at rates a fraction of those found *in vivo*. It is inhibited by the powerful antileukaemic agent Ara-CTP, the triphosphate of Ara-C (1-β-D-arabinofuranosylcytosine). The enzyme also possesses 3'→5' exonuclease activity, but no 5'→3' exonuclease activity [187].

The function of *E. coli* DNA polymerase II *in vivo* is unknown and mutants lacking the enzyme appear normal in all respects. However, double mutants lacking both polymerase I and polymerase II join Okazaki fragments even more slowly than do mutants lacking polymerase I alone [188].

(d) E. coli *DNA polymerase III*

In contrast to mutants with defective DNA polymerase I and II, *E. coli* with a temperature sensitive mutation in the gene for DNA polymerase III (*dnaE* or *pol C)* are not viable at the restrictive temperature [189] and lysates prepared from them are defective in DNA synthesis [187, 190]. Complementation of such lysates with DNA polymerase III purified from normal cells restores their DNA synthetic ability. This is strong evidence that, unlike DNA polymerases I and II, polymerase III is *essential* for DNA synthesis.

DNA polymerase III has been purified about 20000 fold and shown to be highly sensitive to salt in dilute solutions (i.e. normal assay conditions) but not when assayed by complementation (i.e. at high protein concentration) [154, 187, 191]. Although there appear to be only about 10 molecules per bacterial cell, its high rate of polymerization of nucleotides (Table 8.5) shows it to be capable of its proposed role in DNA synthesis.

Table 8.5 DNA polymerases of *E. coli*.

Polymerase (Gene)	Molecular weight	Molecules per cell	Nucleotides polymerized/s (a) per enzyme molecule (b) per bacterial cell	Direction of (a) Polymerization (b) exonuclease action	Template (all require 3′-OH Primer)
'I (*pol A*)	109 000	400	(a) 16–20 (b) 8000	(a) 5′→3′ (b) 3′→5′ and 5′→3′	Denatured Nicked Gapped
II (*pol B*)	120 000	17–100	(a) 2-5 (b) 500	(a) 5′→3′ (b) 3′→5′	Gapped
III (*pol C*)	180 000	10	(a) 250–1000 (b) 10 000	(a) 5′→3′ (b) 3′→5′ and 5′→3′	Gapped

The best template for DNA polymerase III is double-stranded DNA with many small gaps containing 3′-OH primary ends.

DNA polymerase III exists in a complex form in the cell and it has proved difficult to determine the molecular weight and subunit composition of the enzyme. The α subunit of the holoenzyme (that coded for by the *dnaE* or *polC* gene) has a molecular weight of 140000 [155, 192–196]. The polymerase III enzyme is also associated with two small subunits: ϵ (molecular weight 25000) and θ (molecular weight 10000) [155, 197] (see Table 8.6). It has both 3′→5′ exonuclease (which could be involved in proof-reading) and 5′→3′ exonuclease activities though the latter is only manifest *in vitro* on duplex DNA with a single-stranded 5′ tail [192, 194]. For this reason the enzyme cannot use a nicked primer/template but requires a gap.

Several mutator mutants of *E. coli* have been shown to have a temperature-sensitive DNA polymerase III [198] and fidelity of replication by this enzyme has been shown to be dependent on the

Table 8.6 Subunit composition of *E. coli* DNA polymerase III [155,197] and rat liver DNA polymerase α [220].

E. coli DNA polymerase III

Subunit	Mol. wt	Gene	Name		
α	140000	polC	polymerase III core enzyme (α, ε, θ)	polymerase III* (α–δ)	polymerase III holoenzyme (α–β)
ϵ	25000				
θ	10000				
τ	83000				
γ	52000	*dnaZ*	EFII		
δ	32000	*dnaX*	EFIII		
β	440000		EFI, copolymerase III*		

Rat liver DNA polymerase α

Mol. wt	
156000	core enzyme
64000	associated proteins forming an activating hetero-oligomer (64000–54000)
61000	
58000	
54000	

composition of the deoxyribonucleoside triphosphate pool available [199].

DNA polymerase III cannot use as template long single-stranded DNA molecules even when provided with a primer. However, a more complex form of the enzyme known by Kornberg as polymerase III* in the presence of other proteins (copolymerase III* or factor I) and lipids and ATP is able to extend an RNA primer hydrogen bonded to circular, single-stranded M13, G4, or ØX174 DNA [154, 155, 192, 196, 200, 201]. The reaction ceases when the growing DNA chain has passed around the circle and reached the 5′-end of the RNA primer. At this stage addition of DNA polymerase I will cause removal of the RNA and its replacement by DNA, and the final nick can be sealed by DNA ligase to yield a covalently closed, double-stranded, replicative form of the molecule (see Chapter 6). This reaction is considered in more detail on p. 257.

B. subtilis, a bacterium widely divergent from *E. coli,* also has three DNA polymerases which closely resemble those of *E. coli.* They differ in that polymerases I and II and also possibly III are devoid of nuclease activity and, in addition, DNA polymerase III is sensitive to 6-(*p*-hydroxyphenylazo)uracil (HPUra), an antibiotic active against gram-positive bacteria. DNA synthesis in lysates of *B. subtilis* is also inhibited by HPUra [202].

(e) DNA polymerases in eukaryotes

While the DNA polymerases of micro-organisms have been studied more intensively, similar enzymes are present in eukaryotic cells. In general they resemble the bacterial enzymes in their requirement for a template, a 3′-OH primer, and four deoxyribonucleoside triphosphates [203–205]. On fractionation of animal cells DNA polymerase activity is found in several fractions, but in early studies it was the enzyme recovered in the supernatant fraction which alone was investigated [206–208].

This soluble enzyme, DNA polymerase α, is most active in extracts of rapidly proliferating cells and has been extensively investigated in many tissues.

It is this enzyme (DNA polymerase α) which varies in activity with the growth rate of the cells (Fig. 8.18) [209–216].

Because of its lability and apparent heterogeneity, DNA polymerase α has not been purified to the same extent as *E. coli* DNA polymerase I [217] despite the fact that a 200-fold purification of the calf thymus enzyme was achieved in 1965 when it was separated from a terminal transferase (see below).

DNA polymerase α sediments in sucrose gradients as a broad peak in the 6–8S region and on gel filtration activity is found in a number of regions [218]. It has been suggested that the molecular weight of the A

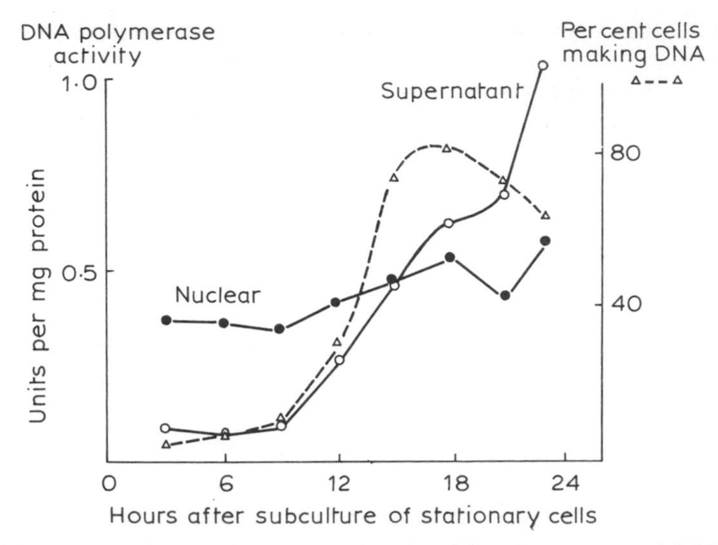

Fig. 8.18 Variation in the activities of nuclear (β) and supernatant (α) DNA polymerases with growth phase of cultured mouse L cells.

form is about 220000 and from this urea treatment releases a basic component to give the C form of molecular weight 155000. This is susceptible to terminal proteolysis to give the D form (140000) and the B form (103000) all of which are active [219]. The 155000 molecular weight enzyme is associated with four, non-identical subunits [220] (see Table 8.6).

DNA polymerase α has a pH optimum in the range 6.5–8.0 and is highly susceptible to inhibition by thiol-active reagents (i.e. *N*-ethylmaleimide and *p*-chloromercuribenzoate). It is also inhibited strongly by araCTP and by aphidicolin (a tetracycline diterpene tetraol) [221]. In both instances inhibition is competitive with dCTP. Strangely it is resistant to inhibition by 2′,3′-dideoxythymidine triphosphate (ddTTP) [136, 222]. Use of these inhibitors together with consideration of the high enzymic activity found in growing cells has led to the conclusion that this enzyme plays the major role in DNA replication in eukaryotes.

DNA polymerase α shows optimal activity with a gapped DNA template but shows a remarkable ability to use single-stranded DNA by forming transient hairpins [220]. It will use synthetic single-stranded templates provided with either oligoribo- or deoxyribo-nucleotide primers, e.g. poly(dT) · oligo (dA) or poly(dT) · oligo (A) [208, 223]. It will not bind to duplex DNA. It is processive, i.e. it adds about 11 nucleotides before it dissociates from the DNA [224, 225]. In contrast to the enzymes from *E. coli* no nuclease activity is associated with the eukaryotic enzymes with the possible exceptions of an enzyme isolated from mouse myeloma cells [226] or from herpes virus-infected cells [227].

Although polymerase α is normally recovered in supernatant fractions following cell homogenization, there is evidence to suggest that it may be located within the nucleus *in vivo* [228].

Unlike DNA polymerase α, the DNA polymerase recovered in the nuclear fraction (DNA polymerase β) shows little correlation between activity and the growth rate of the cells (Fig. 8.18). This enzyme, which has been purified to homogeneity, has a pH optimum of around 8.6–9.0 [229] and a molecular weight of 43000 [230]. However, in the final purification step used for the enzyme from Novikoff hepatoma or guinea-pig liver a polypeptide is removed from the enzyme whose molecular weight falls to 32000. Re-addition of the polypeptide greatly stimulates enzymic activity [231]. DNA polymerase β shows optimal activity with native DNA activated by limited treatment with DNase I (to produce single-stranded nicks and short gaps bearing 3′-OH priming termini) [232, 233] and shows negligible activity with denatured DNA. DNA polymerase β is relatively insensitive to thiol-active reagents, araCTP and aphidicolin but is strongly inhibited by ddTTP. The fact that DNA polymerase β never processes, i.e. it leaves the DNA primer after addition of a single nucleotide [179], may account for its insensitivity to araCTP. It resembles DNA polymerase α in being devoid of nuclease activities. Thus neither enzyme has the 'proof-reading' capacity of the bacterial enzymes, and in fact both will add on to a mismatched primer terminus [230, 234].

DNA polymerase β cannot use an oligoribonucleotide primer but, unlike polymerase α, given an oligodeoxyribonucleotide primer it is able to copy a ribonucleotide template [e.g. poly(A) · oligo(dT)] [209, 223]. DNA polymerase β is believed to play a role in repair of DNA and studies *in vitro* using nuclei from irradiated neuronal cells [214, 235] or from non-dividing mouse L929 cells treated with nuclease [236] show the repair synthesis to be sensitive to ddTTP.

Animal cells also have a small amount of DNA polymerase γ (mol. wt 119000) [223, 237, 238]. This enzyme, probably identical to the previously identified mitochondrial DNA polymerase [214, 237–244], shows greatest activity using an oligo deoxyribonucleotide primer and a ribonucleotide template (e.g. poly(A) oligo(dT). Other than a presumed role in replication of mitochondrial DNA polymerase γ has been implicated in the replication of adenovirus DNA [245–247].

Chloroplasts also possess the enzymic mechanisms for synthesizing their own DNA [248, 249].

In addition to replicative DNA polymerase activities, which require the presence of all four deoxyribonucleoside triphosphates, animal cell extracts contain a separable enzyme responsible for the addition of nucleotidyl units to the *ends* of polynucleotide chains. This second activity, which was originally described by Krakow in 1962 [250], does not require a template strand but catalyses the incorporation of nucleotide units from single triphosphates into terminal positions in the DNA primer molecule. It is not further stimulated by the addition of the other three triphosphates but it is stimulated by cysteine. It has been called 'terminal transferase enzyme' and may be used in the biosynthesis of homopolymers of deoxyribonucleotides [251].

In measuring the activities of these enzymes the assay mixture for the 'replicative' enzyme contains all four deoxyribonucleoside triphosphates, one of them, dTTP, being radioactively labelled. The assay mixture for the 'terminal' enzyme contains only one triphosphate; radioactive dTTP has been used routinely, but dATP is more effective with the purified enzyme.

It is possible to distinguish between the two types of enzyme by the use of actinomycin D which is well known as an inhibitor of both the replicative DNA polymers and the DNA-dependent RNA polymerase (p. 290). Its action is to block the surface of the priming strand of DNA by binding to guanine residues, and it is for this reason a powerful inhibitor of the replicative enzymes. On the other hand, it would be reasonable to suppose that actinomycin D would exercise a much less pronounced effect on 'terminal' incorporation since direct interference would arise only in primer molecules bearing deoxyguanylyl residues at or near the 3′-hydroxy-terminal residues. The results clearly indicate the sharp distinction between the 'terminal' and 'replicative' enzymes which can be revealed by actinomycin D [159].

Unlike the replicative enzymes, which are stimulated by low levels of EDTA, the terminal enzyme is completely inhibited by micromolar concentrations of EDTA which is believed to exert its effect by binding Zn^{2+}.

The terminal transferase has been purified to homogeneity from calf thymus. It can be dissociated by sodium dodecyl sulphate into two subunits of molecular weight 8000 and 26500 daltons [252].

The activity of the terminal transferase is very low in all tissues except thymus [253] and acute leukaemic lymphoblasts, i.e. lymphoid progenitor cells [254, 255]. To what extent this reflects the peculiar physiological functions of such cells and a role for the enzyme in the generation of immunological diversity is unclear [256].

8.4.3 DNA ligases

The ligase or joining enzymes catalyse the repair of a single-stranded phosphodiester bond cleavage of the type introduced by endonuclease.

They were first described in *E. coli* [257] and have since been described in both animal and plant cells.

DNA ligases catalyse the formation of a phosphodiester bond between the free 5′-phosphate end of an oligo- or poly-nucleotide and the 3′-OH group of a second oligo- or poly-nucleotide positioned next to it (Fig. 8.19). A ligase-AMP complex seems to be an obligatory intermediate and is formed by reaction with NAD in the case of *E. coli* and *B. subtilis* [258–260] and with ATP in mammalian and phage-infected cells [261–263] (Fig. 8.19). The adenyl group is then

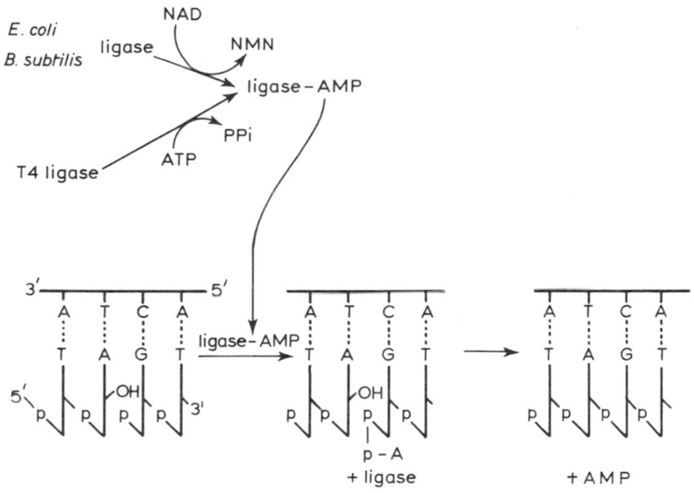

Fig. 8.19 Action of DNA ligases.

transferred from the enzyme to the 5′-phosphoryl terminus of the DNA. The activated phosphoryl group is then attacked by the 3′-hydroxyl terminus of the DNA to form a phosphodiester bond.

DNA ligase will close single-strand breaks in double-stranded DNA or in either strand of a polyribonucleotide–polydeoxyribonucleotide hybrid polymer; double-stranded ribopolymers are not substrates. Breakage of a single phosphodiester bond without removal of a nucleotide to give 5′-phosphoryl and 3′-OH termini is essential for repair by ligase activity (see Fig. 8.20). Reaction is independent of the base composition around the cleavage point.

The enzymes from *E. coli* or bacteriophage T4 will join short oligodeoxynucleotides in the presence of a long complementary strand; e.g. $d(T\text{-}G)_3$, $d(T\text{-}G)_4$, or $d(T\text{-}G)_5$ can be joined in the presence of poly(dC-dA).

(a) *Assay of DNA ligase.* DNA ligase has been assayed in a variety of systems, including the formation of covalently closed circles of double-stranded DNA, restoration of transforming activity of nicked DNA

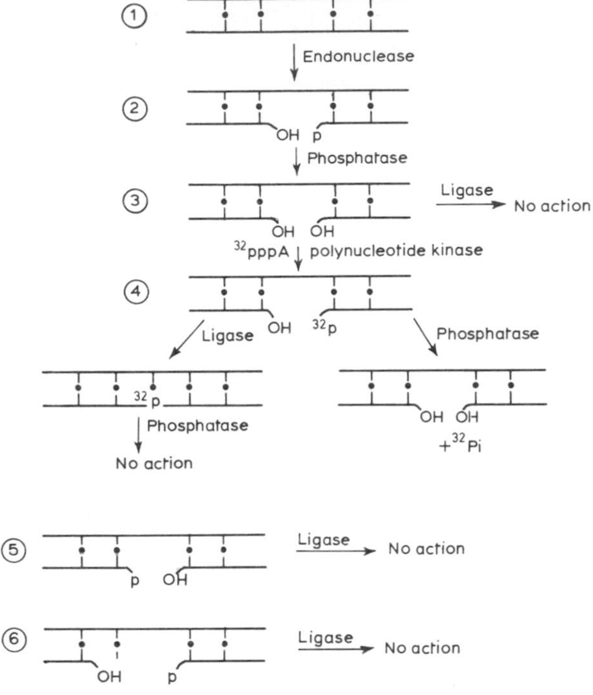

Fig. 8.20 An assay for DNA ligase. The substrate (4) is formed by treating double-stranded DNA (1) with DNase I to give single-stranded breaks (2), removing the 5′-phosphate residues with phosphatase (3) and replacing them with [^{32}P]-phosphate groups with the aid of polynucleotide kinase and labelled [^{32}P]-ATP. Forms (3), (5) and (6) are inactive as substrates for the ligase.

(e.g. [260]), and the formation of phosphatase-resistant radioactive phosphate [264] (see Fig. 8.20).

(b) *Role of DNA ligase.* The nature of the reaction catalysed by DNA ligase has made it an important feature of several models for DNA synthesis, for DNA repair, and for genetic recombination [265]. In all of these cases, its postulated role is in re-establishing continuity by joining a stretch of newly synthesized DNA to pre-existing DNA by the formation of a phosphodiester bond (see p. 228).

It has, however, proved difficult to define a situation in which DNA ligase could be shown to be absolutely necessary for any one of these functions. Nevertheless, since *E. coli* ligase-deficient mutants selected in different ways exhibit abnormal u.v. sensitivity [266], it appears that ligase has some function in the repair process. These mutants also accumulate Okazaki pieces at the restrictive temperature. That these pieces are eventually joined is probably a reflection of the presence of a few molecules of ligase still remaining in the mutant cells [267, 268]. There are normally about 300 molecules of ligase per cell, and it has been calculated that these are capable of sealing 7500 breaks per minute at 30°C [269]. Since only about 200 breaks per cell are formed on replication each minute there is a vast excess of enzyme.

Infection of *E. coli* by T4 leads to the synthesis of a T4 specified ligase which, in addition to joining two DNA chains, will also join RNA to DNA in an RNA–DNA hybrid [269–271]. Some ligase is essential for T4 development [266], and bacteriophages with a temperature-sensitive ligase show increased susceptibility to u.v. radiation [272] (see p. 294).

8.4.4 Helix-destabilizing (HD) proteins or DNA-binding proteins

These are a group of proteins brought together under this heading because of a common property. They have been isolated from many sources and they do not necessarily all perform the same function *in vivo*.

The first of these to be characterized was that coded for by gene 32 of bacteriophage T4 [273]. This and similar proteins isolated from T7 and uninfected *E. coli* [274] have the ability to convert double-stranded DNA into single-stranded form at a temperature 40°C below the normal melting temperature (Tm). The gene 32 protein has a molecular weight of 35 000 and it binds co-operatively to single-stranded regions of DNA, each protein covering some 10 nucleotide units. Thus if a limiting amount of gene 32 protein is mixed with excess single-stranded DNA some DNA molecules will become covered with protein and others remain protein free. There are from 300 to 800 molecules of binding protein per uninfected *E. coli* cell, which is enough to cover about 1600 nucleotides at each replication fork [274, 275]. A similar length of T4 DNA can be covered by the 10 000 molecules of gene 32 protein present in an infected cell. (There are about 60 T4 replication forks per infected cell [274].)

Purified polymerases are stimulated in a very specific manner by the presence of their complementary DNA binding proteins. Thus the T4 DNA binding protein will stimulate only the T4 polymerase and the T7 DNA binding protein will stimulate the T7 polymerase [275–277]. the *E. coli* DNA binding protein stimulates *E. coli* polymerases II and III* but does not stimulate polymerases I and III [278] (see p. 242). It is suggested that the polymerase and DNA binding protein form a specific complex that interacts with DNA and other proteins during replication [279].

The HD proteins from *E. coli* and T4-infected cells are believed to play a role in replication, recombination and repair [442, 250, 251] and an *E. coli* mutant (ssb–1) has been isolated showing a defective DNA binding protein and extracts fail to convert 'phage G4 single-strands into duplex form [282] (see p. 226). In contrast an HD protein coded for by gene 5 of bacteriophage M13 plays an essential role in preventing further replication by stabilizing the single-stranded progeny viral DNA [283].

DNA binding proteins have also been isolated from eukaryotic cells.

Thus helix-destabilizing proteins serve a passive role. Although *in vitro* they can bring about denaturation of DNA their role *in vivo* may simply be to stabilize single-stranded DNA.

8.4.5 DNA dependent ATPases (DNA unwinding proteins or DNA helicases (DH))

A number of proteins fall into this group. Like the HD proteins they have a property of being able to unwind DNA duplex, but these proteins do so at the same time as they hydrolyse ATP. There appear to be at least four DNA-unwinding proteins in *E. coli*. Two have been studied by Hoffman Berling's group [284–287]. The first (DHI) is a fibrous protein, mol. wt 180 000, similar to myosin. About 80 molecules bind to a single-stranded region of 200 nucleotides of DNA near a fork and travel down the DNA unwinding it [288]. There are about 500 molecules per cell. There are about 3000 molecules of the second enzyme (DHII mol. wt 75 000) and this protein acts like gene 32 HD protein in that it remains bound to the single strands of DNA produced by its action.

Rep is a protein essential for the replication of the small 'phages ØX174 and G4 and, presumably, it also plays a role in replication of the host cell's DNA. It may be identical with the second unwinding protein described above [289, 290]. It works in conjunction with *E. coli* DNA binding protein [274] to unwind partially single-stranded duplex DNA [291]. It binds a single-stranded region and translocates in the $3' \rightarrow 5'$ direction until it reaches the $5'$ end of the duplex region when unwinding occurs. Two molecules of ATP are hydrolysed per nucleotide unwound and DNA binding protein binds to the single-stranded regions produced [292]. This is a high price to pay for unwinding DNA as it involves the release of 14 kcal energy to break H-bonds whose energy varies from 1.2 kcal (A = T) to 5.0 kcal (G ≡ C) [341].

Another distinct unwinding protein (DHIII – mol. wt 2×20000) is similar to rep but moves along the bound strand in a $5' \rightarrow 3'$ direction. DHIII and rep may act in concert at the replication fork [293, 294].

The dna B gene product is another DNA dependent ATPase present in *E. coli* [295]. It is a pentamer or hexamer of subunit molecular weight about 50000 [296, 297]. There are only about 20 molecules per *E. coli* cell and they are involved in preparing DNA for the action of the primase (see p. 257). Once bound to the DNA the dna B protein appears capable of propelling itself along DNA chains at the expense of energy obtained from hydrolysis of ATP [298, 299].

The initial stage of transfer of DNA polymerase III to the RNA primer involved in replication of single-stranded 'phage DNA and host cell DNA also involves ATP hydrolysis. The proteins involved are the dna Z gene product and initiation factor III (IF III) [300], and copolymerase III star [301] but it is unlikely in this case that the expenditure of energy is concerned with unwinding the DNA double helix. These are proteins of molecular weight 52000, 32000 and 40000 respectively which form part of polymerase III holoenzyme [155, 197].

The gene 4 product of bacteriophage T7 is similar to rep except that it forms a complex with the T7 coded DNA polymerase (the gene 5 product). Together they are able to extend a 3′-OH at a gap in the duplex, unwinding the DNA ahead of the complex and hydrolysing 2 ATP molecules for every dNMP incorporated [302–305]. The gene 4 protein, in addition to acting as an unwindase, is also able to synthesize a specific primer for DNA synthesis (see below).

The products of genes 44 and 62 of phage T4 are believed to help in the movement of the T4 coded DNA polymerase along its template. The hydrolysis of ATP which they catalyse appears to increase the 'sticking' distance, i.e. by unwinding the DNA duplex the complex acts to increase the affinity of the polymerase for its template/primer [306].

A DNA-dependent ATPase has also been isolated from mammalian cells. Along with a DNA-binding protein it acts to stimulate activity of DNA polymerase α [307].

8.4.6 Nicking closing enzyme (Topoisomerase I)

When two DNA strands unwind as the replication fork moves forwards either the whole DNA molecule has to rotate at 10000 rev/min (an impossibility for closed cyclic DNA molecules) or some sort of swivel must be present. Cairns first proposed the idea of a swivel [4, 49] which might act by alternating action of an endonuclease and a ligase. However, the two functions are present in a single enzyme isolated from both *E. coli* and mouse embryo cells [307, 308]. The difference between a nicking closing enzyme and the earlier suggested mechanism is that with the nicking closing enzyme the cut ends are not released from the enzyme although rotation of the two strands about one another is allowed. The enzyme cuts the DNA to produce a 5′-OH group (which can be phosphorylated by polynucleotide kinase) and remains attached to the

3′-phosphate group [423]. The driving force for the reaction is the tension engendered in duplex DNA by unwinding or that present in isolated supercoiled DNA. In the latter instance the nicking closing enzyme will act until all the supercoils are removed and the DNA is released as a fully relaxed closed cyclic molecule [310].

That nicking closing does occur has been demonstrated by replicating DNA in the presence of $H_2{}^{18}O$ and looking for incorporation of ^{18}O into parental strands. The results are consistent with a nick every 100–500 bases and some nicks may lead to multiple swivelling [311].

8.4.7 DNA gyrase (Topoisomerase II)

This is a complex activity which *in vitro* will do the opposite of a nicking closing enzyme, i.e. it will introduce -ve supercoils into closed circular DNA but it requires the energy from ATP hydrolysis to achieve this [309, 312–315, 423].

It is in fact a 400000 mol. wt tetramer, the subunits of which are the products of two genes *nal* and *cou* (also known as *gyr A* and *gyr B*).

Cou product is a DNA dependent ATPase (mol. wt 90000) which is sensitive to the antibiotics coumermycin and novobiocin.

nal A product is a 100–110000 mol. wt nicking-closing enzyme sensitive to nalidixic acid and oxolinic acid.

Gyrase does not act at random but makes staggered cuts in DNA at specific sites.

T | G N Y N

A C N R N |

and, in the presence of oxolinic acid, it remains bound to the 5′ end of the cut [316].

DNA gyrase may play a role either by removing positive superhelical turns induced into DNA during replication or alternatively or additionally by putting negative super twists into the DNA ahead of the replication fork thus making unwinding easier. As treatment of *E. coli* with either coumermycin or nalidixic acid immediately stops DNA synthesis the contribution of DNA gyrase is essential [317, 318]. DNA gyrase is also required for replication of bacteriophage T7 DNA [319] and bacteriophage T4 appears to code for a corresponding activity [320]. The products of T4 genes 39, 52 and 60 form a complex with topoisomerase activity which is essential for initiation of replication of T4 DNA [321].

A model for the action of DNA gyrase can be constructed [322].

The DNA first wraps itself around the protein to form a structure similar to a nucleosome (see p. 91) [323]. ATP then binds and causes a conformational change which drives the translocation of the DNA relative to the enzyme leading to the formation of a positive superhelical loop on the DNA [324]. A nicking closing event (or double-stranded breaking, rejoining event) removes the positive superhelical turn [325] and on hydrolysis of ATP the enzyme returns to its original conformation leaving the DNA with a negative superhelical twist, i.e. partly unwound. Alternatively, overwound and underwound regions of duplex DNA could be produced and the overwound regions selectively relaxed. As the topological winding number changes by two units each time and as knotted and catenated molecules can be formed by gyrase action it is believed that the enzyme catalyses the formation of a transient double-stranded break with passage of another region of the DNA molecule (or another DNA molecule in catenane formation) through the gap [478, 479]. Decatenation (also catalysed by gyrase) may be an essential step in the physical separation of cyclic DNA molecules after replication (see p. 260). Single-stranded regions of DNA are required not only for initiation of DNA synthesis but also for recombination and transcription and DNA gyrase is believed to be required in these reactions [326–329].

There is no evidence for DNA gyrase in eukaryotic systems (except possibly in mitochondria [330]), where the histones may make it superfluous. However, a related enzyme activity has been found [472].

8.4.8 Primase

E. coli RNA polymerase is involved in synthesizing the primer to initiate replication of bacteriophage M13 and *E. coli* DNA. In all other instances investigated in *E. coli* systems other enzymes known as primases have taken over this function. The product of the dna G gene is involved in initiation of replication of the other small single-stranded DNA bacteriophages G4 and ØX174 and in the synthesis of the primers of the Okazaki pieces. The product of gene 4 of bacteriophage T7 serves a similar role synthesizing a specific tetranucleotide primer of sequence ppp ACCA and thereafter, in combination with the T7 coded DNA polymerase, acts to unwind the DNA in an ATP-dependent reaction [330, 331] (see above). The dna G protein has been purified and shown to be a rifampicin-resistant RNA polymerase of molecular weight 60–64 000 [20, 71]. As well as polymerizing ribonucleotides it can also use deoxyribonucleotides [90, 333] and make a mixed primer though the first nucleotide is always rA. In the presence of *E. coli* DNA-binding protein the primase will make a specific 29 residue RNA primer on single-stranded bacteriophage G4 DNA (Fig. 8.21). This is unusual,

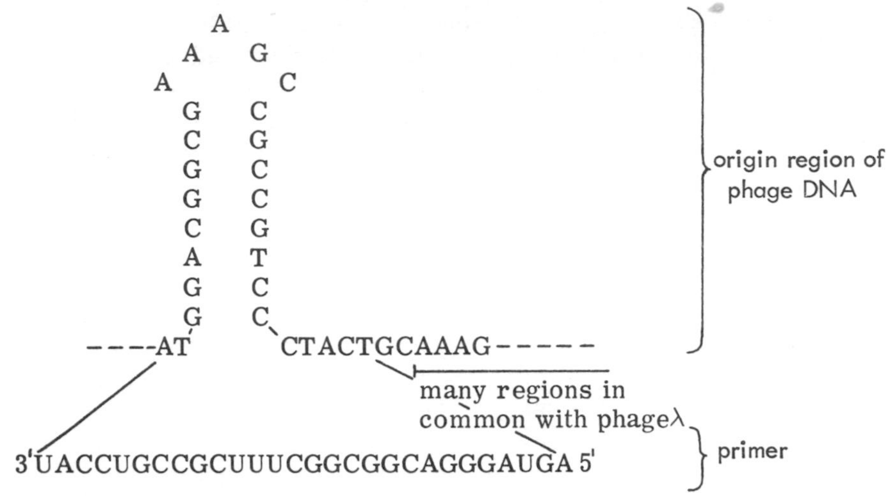

Fig. 8.21 Location of primer in phage G4 (this untranslated region is absent from ØX174), [22, 25, 334].

however, and interaction with other DNAs requires several other proteins (see below) [298, 299, 334]. Indirect evidence exists for a similar primase in eukaryotic cells [335, 336].

8.5 MODEL SYSTEMS FOR THE STUDY OF DNA SYNTHESIS

8.5.1 Single-stranded DNA viruses

The small bacteriophages fd (M13), G4 and ØX174 replicate in cells of *E. coli* and replication takes place in three stages [154, 155, 337, 338]:

(a) The original infecting cyclic DNA molecules (known as the viral or plus strand) is used as a template on which a complementary minus strand is synthesized (SS→RF).

(b) The resulting replicative form (RF) undergoes several cycles of replication by the so-called rolling circle mechanism (RF→RF).

(c) The RF molecules are used as templates from which progeny +strands are synthesized. These are subsequently packaged into the viral coat before the mature virus leaves the cell (RF→SS),

Although of similar size the replication of the three small bacteriophages mentioned above differs in a number of significant steps which can be explained as a result of differences in DNA sequence.

(a) SS→RF

On entering their host the DNA molecules become covered with the *E. coli* single-stranded DNA binding protein. In fd there remains exposed a hairpin loop which is able to bind *E. coli* RNA polymerase [337, 339–340, 376]. This enzyme synthesizes a short RNA which then acts as a primer for −strand synthesis [195, 342–345].

In G4 a similar hairpin loop is exposed but this has a binding site for the dna G protein – the primase [28, 333]. This enzyme is responsible for synthesizing the RNA primer in most prokaryotes including bacteriophage G4. The sequence of DNA around the origin of bacteriophage G4 and the sequence of the RNA primer synthesized by the primer is shown in Fig. 8.21 [22, 28, 71, 195, 346]. Studies with ØK suggest the primase binding site may be more complex [475].

In ØX174 no binding site for either RNA polymerase or primase is exposed. Comparison of the nucleotide sequences of the DNA of the closely related bacteriophages G4 and ØX174 reveals a deletion in the latter covering the primase binding site [23, 28]. In ØX174 a complex is formed involving a series of proteins identified as the products of the genes *dna B*, *dna C* and proteins i, n and n′ (called X, Y and Z by Hurwitz) [155, 340, 347, 348, 376].

This dna B complex may initially form at a specific site on ØX174 DNA (specified by protein n′) but the complex is able to move around the DNA molecule in a 5′→3′ direction using the energy released by ATP hydrolysis [340, 347, 376, 476] (Fig. 8.22). At various positions the complex interacts with dna G protein (the primase) which initiates an RNA primer which grows in the opposite direction to which the *dna B* complex is moving. Thus, once priming is initiated the *dna B* complex is free to move off again and it travels 200 nucleotides or more before again interacting with primase to initiate another primer [347, 349]. In all 6–8 primers are made per round of DNA synthesis [339, 347].

In all three bacteriophages the RNA primer formed now acts as the primer for DNA polymerase III action. Before DNA polymerase can act, however, elongation factor I (EFI, or copolymerase III) must be bound in a reaction involving the *dna Z* protein (also known as EFII), elongation factor III (EFIII), and ATP. DNA polymerase III then binds to EFI and deoxyribonucleotides are added and the −strand is synthesized [155, 300, 301, 348] (Fig. 8.22).

When the 3′-OH end of the growing −strand reaches the RNA primer it appears that DNA polymerase III is replaced by DNA polymerase I. This enzyme removes the RNA primer while completing the DNA chain in a nick-translation type of reaction (see p. 240). Finally the chain is completed by DNA ligase (Fig. 8.22).

(b) RF→RF

To initiate replication of RF requires in all three cases the intervention of a specific bacteriophage-coded protein. This protein makes an endo-

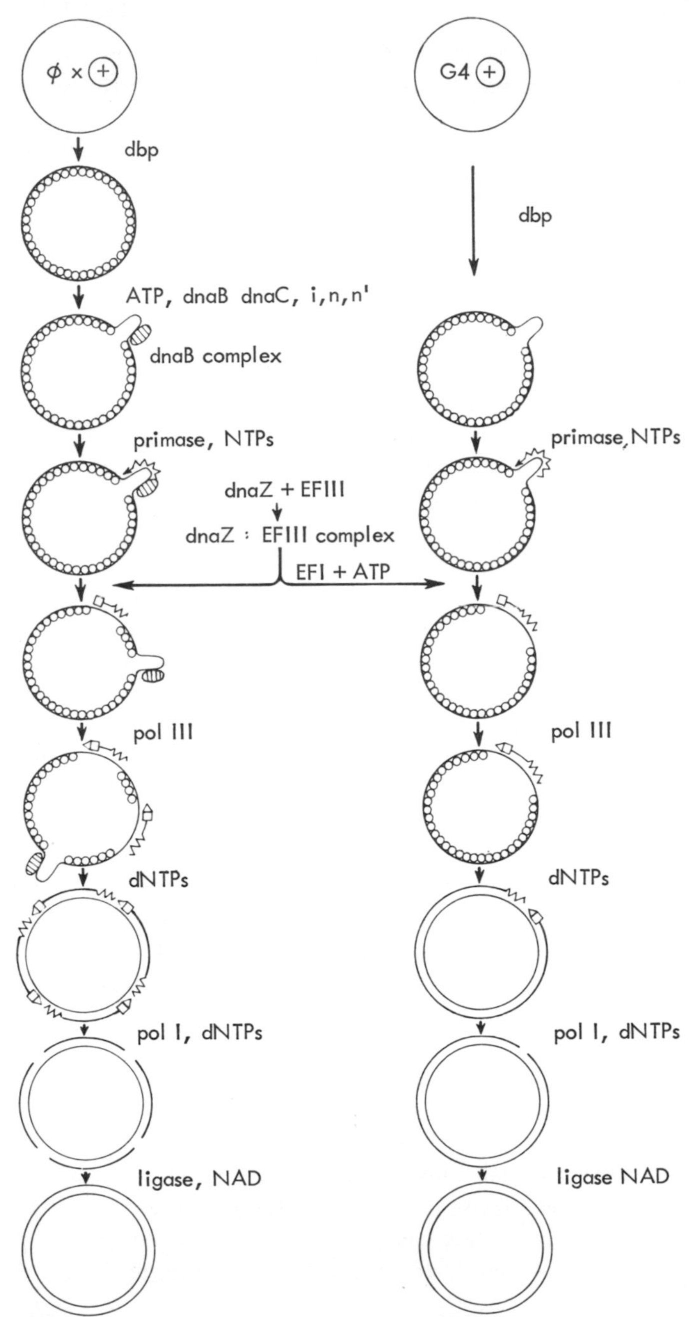

Fig. 8.22 Conversion of ØX174 and G4 single-stranded DNA to duplex form (M13 resembles G4 except *E. coli* RNA polymerase is used in place of primase). dbp, DNA binding protein; □, EFI; △, DNA polymerase III; ʌʌ, primer. After Kornberg [154, 155, 337].

nucleolytic cut in the + strand of the RF DNA (Fig. 8.23). The site of the cut is specific and occurs within the region of DNA coding for the endonuclease and there is evidence that the endonuclease will only act on the bacteriophage DNA molecule by which it was coded. The enzyme from bacteriophage ØX174 is called the cis A protein and it recognizes a site of 2-fold symmetry which may be present as a hairpin structure [350]. The binding site also contains a methylcytosine residue and blocking of DNA methylation in *E. coli* has been reported to interfere with replication of ØX174 [351, 352]. However, as ØX174 grown in mec$^-$ cells is not methylated this cannot be obligatory [353].

The unique site of action of the gene II nuclease from bacteriophage fd is between bases 5781 and 5782 which is adjacent to (24 bases 3′ on the + strand) the unique binding site of RNA polymerase (f in Fig. 8.23). In bacteriophage G4 the cis A protein acts on the opposite side of the DNA molecule (between nucleotides 506 and 507) to the primase binding site (nucleotides 3972–3992) (g in Fig. 8.23). The sequences recognized by the cis A protein of G4 and ØX174 are identical.

Once the cis A endonuclease has cut the + strand the enzyme remains

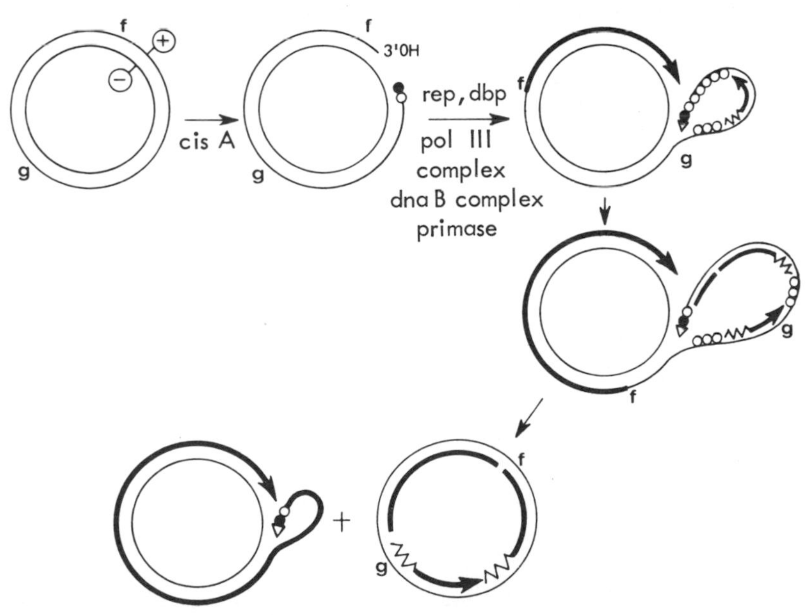

Fig. 8.23 RF replication of ØX174 DNA (for explanation see text). g and f correspond to the primase and RNA polymerase binding sites on single-stranded DNA from bacteriophage G4 and fd, respectively, and hence indicate the position of initiation of − strand synthesis with these bacteriophage DNAs. dbp, DNA binding protein (based on the work of Kornberg and others [154, 155, 337, 346, 347, 350, 354, 361, 363–365]).

attached to the 5′ end while the 3′ end can act as a primer for +strand synthesis by the DNA polymerase III complex (Fig. 8.23b). Unwinding of the DNA is accomplished by combined action of 'rep' – an *E. coli* coded ATP-dependent DNA unwinding protein, the cis A protein, and the single-stranded DNA binding protein [291, 354–358, 365]. DNA gyrase may also help to unwind the DNA duplex ahead of the replicating fork [359] (Fig. 8.23).

With fd the gene II protein does not remain bound to the 5′ end but rather there is an additional requirement for the host *dna A* protein [360–362].

As +strand synthesis proceeds in the 5′→3′ direction the 5′ end of the +strand is thus peeled off the rolling −strand when it associates with single-stranded DNA-binding protein (Fig. 8.23). The single +strand with its associated binding protein thus resembles the initial bacteriophage DNA. In fd it thus remains unaltered until +strand synthesis has gone almost a complete cycle when the RNA polymerase binding site is exposed and −strand synthesis is initiated [350] (Fig. 8.23). In bacteriophage G4 the primase binding site is exposed when +strand synthesis has gone half way round the molecule [363] (Fig. 8.23). In ØX174 dna B protein binds to the single-stranded +strand at several non random sites leading to synthesis of several short pieces of −strand DNA (Okazaki pieces) [354, 364] (Fig. 8.23).

When one round of +strand synthesis is complete the cis A protein at the 5′ end of the +strand is again brought adjacent to its site of action (Fig. 8.23) and it acts now as a nicking closing enzyme thereby releasing a complete cyclic +strand with one or more growing −strands and reinitiating +strand synthesis on a new rolling circle [354, 365].

(c) RF→SS

Later on in the infectious cycle bacteriophage coded proteins are synthesized which prevent reinitiation of −strand synthesis. The replicative form molecules thus continue to synthesize +strands which interact either with bacteriophage coat proteins (i.e. the products of ØX174 genes B, C, D, F, G and H) or in the case of M13 with the gene 5 protein which is exchanged for coat proteins when the viral DNA leaves the host cell [154, 155].

8.5.2 Plasmid (Col El) replication

This small double-stranded cyclic DNA molecule replicates from a unique origin but only in one direction [339]. Theta (θ) structures are formed and initiation does not involve permanent nicking of either strand [336, 367]. Up to three proteins bind near the origin and bring about supercoiling of the cyclic molecule [367] thereby exposing single-stranded regions to which *E. coli* RNA polymerase binds [368]. An RNA transcript is initiated 555 base pairs upstream from the origin (−555) but as transcription passes the origin the transcript forms a stable hybrid with the DNA [370]. This is processed by RNase H or RNase III (see p. 149) and can then act as a primer upon which DNA polymerase I synthesizes several hundred nucleotides complementary to the H strand. An early intermediate accumulates with a 6S piece of newly synthesized DNA. DNA polymerase III complex now takes over to complete the circle [339, 367, 370].

The lagging strand appears to be synthesized in the same manner as in ØX174 RF replication and it requires primase and dna B and dna C proteins and DNA gyrase [339].

This positive control over replication by the transcript of a region adjacent to the origin is found in other plasmids, but the transcript is usually translated, e.g. plasmid R6K synthesizes a protein (π) which interacts with the origin to initiate DNA replication [371]. In addition to the transcript referred to above, a small 110 nucleotide transcript (RNA I or R-100) is made in the other direction, starting at 445 base pairs upstream from the origin [369]. RNA I is able to form a complex secondary structure similar to tRNA and exerts a negative effect on initiation of DNA synthesis [481, 482]. Insertions into the region of the RNA I transcript lead to overproduction of the plasmid [369]. Mixed infections of an overproducer with a 2KB fragment from the replication origin of wild-type plasmid results in a suppression of the overproducer [468]. Similarly composite plasmids are known which exist at the copy number typical of the lower frequency component [372]. It is quite possible that elements of −ve and +ve control are present.

8.5.3 Double-stranded DNA viruses

Three bacteriophages (λ, T7 and T4) have been extensively studied together with the animal viruses SV40, polyoma and adeno. No common rule can be applied and each will be briefly considered.

(1) *Lambda.* The λ DNA molecule is linear but has sticky ends (see p. 66) and cyclizes immediately on entry into the cell [373–374]. Replication then is initiated at a unique origin and proceeds bidirectionally to give theta (θ) forms (cf. *E. coli* replication). However, at later stages of infection rolling circles are found.

In addition to host functions the products of λ genes O and P are required for initiation of DNA replication [375, 376]. The P gene product appears to take over the function of *E. coli dna C* gene product. It interacts with *E. coli dna B* protein and the complex interacts with lambda gene O product at a specific site (ori) rendered single-stranded by transcription by host RNA polymerase. The coding region for O protein includes ori and has been sequenced [466, 467]. The dna G primase then synthesizes a 4S RNA molecule, termed oop, which primes DNA polymerase III–catalysed DNA synthesis.

Alternatively oop RNA may be extended by RNA polymerase leading to synthesis of λ repressor and the state of lysogeny results where the λ genome becomes integrated into the host genome and the two replicate under host cell controls [375, 377, 378].

(2) Initiation of T7 DNA synthesis occurs 17% from one end of the molecule. A replication bubble forms which expands bidirectionally. When the short side is completely synthesized a Y-shaped replicating chromosome is formed (Fig. 8.24).

Concatamers, i.e. double-length molecules, are found later in infection [303, 379, 380].

Replication requires bacteriophage genes 1, 2, 3, 4, 5 and 6 [381, 382]. Gene 1 product is an RNA polymerase essential for the transcription of

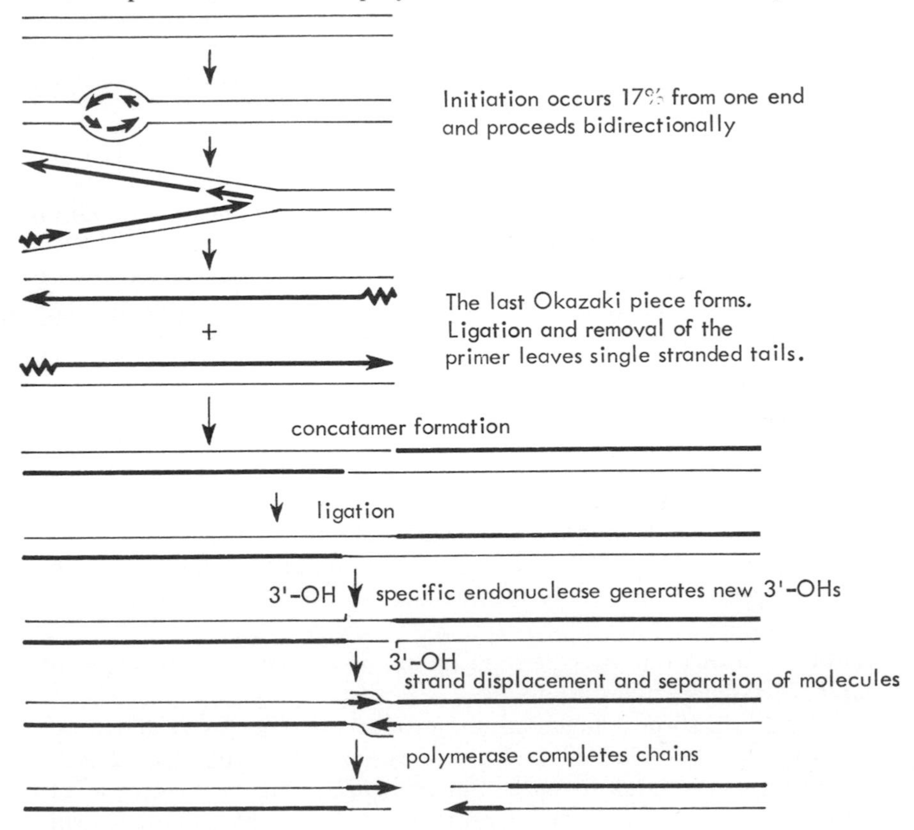

Fig. 8.24 Replcation of T7 DNA. After Watson [379].

genes 2–6 and gene 2 protein inactivates the host RNA polymerase [383]. Gene 3 protein is an endonuclease and gene 6 protein an exonuclease involved among other things, in the breakdown of host DNA.

In the presence of T7 gene 5 protein and the host protein thioredoxin (which together form a DNA polymerase [384, 385]) the T7 gene 4 protein catalyses an ATP-dependent unwinding of duplex DNA thereby promoting the DNA polymerase action [331, 386]. This complex is thus able to catalyse leading strand synthesis from a nick in duplex DNA and DNA binding protein and gyrase further promote this action. Gyrase either relieves the positive supertwist generated by replication or it may actively help unwinding by imposing a negative supercoiling ahead of the fork. There is some evidence for an initiation protein which allows initiation on intact duplex DNA but its mode of action is unclear [382]. In addition to its action as an unwinding protein the gene 4 protein is also a primase, and a very special primase. It synthesizes a tetranucleotide of almost unique sequence ppp $\mathrm{ACC}^{\mathrm{c}}_{\mathrm{A}}$ and like the *E. coli* primase it can use deoxyribonucleotides as well as ribonucleotides [81, 331, 332, 386, 387]. Although it can use any DNA [Except poly (dT) or poly (dG)] the product is template directed and the frequency of primer formation depends on the frequency of occurrence of the complementary tetranucleotides GGGT and TGGT. The deoxyribonucleotide at the primer/DNA junction can be any one of the four [388]. [388].

Thus as well as catalysing leading strand synthesis the gene 4 gene 5 protein complex can also catalyse initiation and elongation of Okazaki pieces on the lagging strand.

Gene 6 exonuclease or host DNA polymerase I 5′→3′ exonuclease can catalyse removal of RNA primers and the Okazaki pieces can be joined using either the host ligase or a ligase coded for by T7 gene 1.3.

Replication of linear duplex DNA molecules poses a major problem. The last Okazaki piece to be made on the lagging strand might start at the extreme 3′ end of the template, but when the primer is removed this will leave a short single-stranded tail at each end of the linear molecule; a tail which cannot be replicated by any mechanism considered so far.

James Watson [379] postulated a mechanism whereby the problem might be solved. T7 DNA has an identical sequence of about 260 nucleotides at each end of the molecule, i.e. the molecule is terminally redundant. Thus the unreplicated single-stranded tail of one molecule could H-bond to the similar region of another molecule giving rise to a concatamer of almost two unit lengths (see Fig. 8.24). The two pieces can be joined together by ligase action and two new nicks introduced several nucleotides 5′ of the original nick. The gene 3 endonuclease is a candidate for the enzyme capable of making these specific cuts. Nick translation (see p. 240) from the newly introduced 3′-OHs will allow the concatamer to separate into two halves and the replication to be completed.

(3) T4 replication is far more complex and initiation of DNA synthesis appears to occur at several points along the long linear molecule [306, 389]. Mutations in T4 genes 39, 52 and 60 lead to a delay in the onset of T4 replication and the DNA synthesis that does occur is dependent on the host cell gyrase [390]. These three genes code for an ATP-dependent topoisomerase [390, 391, 472].

Bruce Alberts group in particular has been trying to reconstruct the T4 replication system *in vitro*[392–394]. At least seven proteins, coded for by T4, have been purified. The gene 32 protein was the first DNA binding protein to be purified and studied and has served as the prototype for such proteins. The gene 43 protein is a DNA polymerase with both 3′→5′ and 5′→3′ exonuclease activities [393] and, like *E. coli* DNA polymerase III it will work with primed single-stranded DNA only in the presence of accessory proteins; the products of genes 44 and 62 (equivalent to *E. coli* dna Z and EFIII) and gene 45 (equivalent to *E. coli* EFI) [306, 395, 396]. Both the gene 44/62 protein complex and gene 41 protein are DNA dependent ATPases which hydrolyse one molecule of ATP for every 10 nucleotides incorporated.

The general mechanism envisaged is similar to that already considered for duplex DNA synthesis. *In vitro* the products of T4 genes 43, 32, 45 and 44/62 catalyse leading strand synthesis starting at nicks and displacing the other strand. The products of genes 41 and 58/61 synthesize short oligoribonucleotides on the lagging strand and hence prime lagging strand synthesis [306, 394–398, 473]. This multi-enzyme complex shows a high degree of fidelity [477].

When cells of *E. coli* are infected with the T-even bacteriophages, the economy of the cells is completely altered so as to lead to the production of new phage DNA which differs from the host DNA in containing hydroxymethylcytosine in place of cytosine (p. 70). These changes result in the production in the infected cell of a series of new and interesting enzymes [157, 232, 399–403].

(*a*) Within a few minutes of infection a hydroxymethylase (the product of phage T4 gene 42) appears which brings about the conversion of dCMP into hydroxymethyldeoxycytidine monophosphate (CH_2OH-dCMP or dHMCMP).

(*b*) At about the same time a *kinase* is produced which phosphorylates dHMCMP to the corresponding triphosphate dHMCTP.

Neither of these new enzymes is found in cells infected with bacteriophage T5 which does not contain HMC.

The kinases for dTMP and dGMP are also greatly increased, but not that for dAMP. This increase is due to the production of new enzymes which can be distinguished from the kinases of the host cell prior to infection.

(*c*) The formation of host DNA is prevented by the appearance of a pyrophosphatase (the product of phage T4 and 56) which converts dCTP into pyrophosphate and dCMP which then acts as a substrate for the hydroxymethylase.

(*d*) Five distinct glucosyltransferases are known to be induced after infection with T-even phages for the purpose of transferring glucose residues from uridine diphosphate glucose to the HMC of phage DNA in the proportions shown in Table 4.3. The glucosylase found in T2 phage-infected cells transfers a glucose residue to HMC in the α-configuration. Two glucosylating enzymes are produced after T4 infection; one adds a glycosyl group in α-linkage to HMC while the second also adds a glucose group but in β-configuration. After T6 infection two glucosyltransferases are also produced. One adds a monoglucosyl residue to HMC in α-linkage while the other reacts with the monoglucosylated groups on HMC to add a second glucose residue, the linkage between the residues being of the β-configuration.

(*e*) The products of T4 genes 46 and 47 are thought to be a nuclease which degrades DNA containing cytosine. T4 DNA which contains hydroxymethylcytosine is resistant.

In vivo the enzymes synthesizing the precursors for DNA synthesis are closely associated with the DNA polymerase [465, 470, 471, 474] and *in vitro* the complex of enzymes preferentially uses deoxynucleoside monophosphates to make T4 DNA [469].

(4) SV40 and polyoma. These similar small, animal tumour viruses contain a cyclic DNA molecule which replicates as a θ form [404] the parental strands remaining intact. Initiation occurs at a unique site and proceeds bidirectionally on a molecule covered in histones (the minichromosome – see p. 78). As the DNA is very short it cannot code for many proteins. Only the so called T-antigen may play a role in DNA synthesis and this binds near the origin of replication [405, 406]. Thus these viruses should prove ideal models for the study of replication in eukaryotic cells. Preparations have been obtained which continue to synthesize SV40 and polyoma DNA *in vitro* (see p. 226) and DNA polymerase α has been implicated in this process [222, 407, 408]. Okazaki pieces of about 4S are made on the lagging strand from primers of ten ribonucleotides in length [335, 409–411].

(5) Adenoviruses. These are animal viruses containing a linear duplex DNA molecule. To the 5′ end of each strand is covalently attached a protein of molecular weight 55000 [412, 413]. It appears that a

deoxycytidine residue attached to this protein can act as the primer for DNA synthesis which starts at both ends of the linear molecule [14, 413–416]. A similar protein may be involved in parvovirus replication [417]. This novel mechanism which requires both DNA polymerase α and DNA polymerase γ [245, 418] removes the necessity to synthesize concatamers as was the case with T7 replication.

(6) Mitochondrial DNA. In animal cells mitochondria contain a closed circular duplex DNA molecule which replicates unidirectionally from a unique origin [419, 420]. Prior to initiation the duplex undergoes partial unwinding by removal of about 48 Watson–Crick turns in the presence of a nicking closing enzyme [421, 422]. Initiation of DNA synthesis then occurs with synthesis of a 450 base length of DNA (the leading or heavy strand) to produce a 'D-looped' structure [240, 420, 422, 424]. This small piece of newly synthesized 7S DNA is unstable and may turn over several times [421, 425, 426] before it is extended. Extension occurs asymmetrically and unidirectionally until 60 per cent of the heavy strand (99 per cent in *Drosophila* mitochondria [427]) has been synthesized [422, 424], when light (lagging) strand synthesis is initiated. Thus although θ forms are seen in the electron microscope, one loop is a single-stranded parental H-strand displaced by synthesis of daughter H-strand.

In sea urchin oocytes mitochondrial DNA synthesis is similar to that found in mouse mitochondria but duplex synthesis occurs early with multiple initiations [428].

Mitochondrial DNA in *Tetrahymena* and *Paramecium* is not cyclic but is a linear duplex [429–432]. In the former replication is initiated near the centre of the molecule and 'eye'-shaped replication intermediates are seen in the electron microscope. Replication is bidirectional. In *Paramecium* replication proceeds unidirectionally from one end [429].

8.5.4 Retroviruses

The enzymes so far discussed copy DNA strands in the synthesis of DNA, but as long ago as 1963 Cavalieri [433] showed that an RNA template, poly(A-U), could serve as a template for the synthesis of a strand of DNA, poly(dT-dA), under the influence of DNA polymerase I from *E. coli.* Nevertheless excitement amounting almost to hysteria was created in 1970 by the announcement of the existence in certain RNA viruses of RNA-dependent DNA polymerases which use RNA as a template for the synthesis of DNA. These enzymes were discovered simultaneously by Temin [434] in the virus particles (virions) of Rous sarcoma virus (RSV) and by Baltimore [435] in Rauscher mouse leukaemia (R-MLV) virus. The observation was confirmed for more than half a dozen RNA viruses by Spiegelman [436], and the phenomenon of Teminism as it came to be called was hailed both as an example of the reversal of the Central Dogma, a suggestion which was vigorously contested by Crick [437], and as an important breakthrough in cancer research since the RNA viruses involved were oncogenic, i.e. capable of bringing about malignant change (see Chapter 4).

On treating the viral 70S RNA genome with agents which disrupt hydrogen bonds it dissociates into two genetically identical RNA molecules [438], and to several molecules of tRNA [439]. The unique RNA directed DNA polymerase (molecular weight 70000) is also present in the virion. The enzyme appears to use one of the tRNA molecules ($tRNA^{Try}$) [440] as a primer to synthesize DNA on the template 35S single-stranded RNA, and this distinguishes it from any cellular enzyme. With synthetic primer/templates the enzyme shows a marked preference for poly(rA) · oligo(dT) or poly(rC) · oligo(dG) [441]. Ribonucleoside triphosphates are without effect as substrates and the process is not sensitive to actinomycin D.

The immediate product of the reaction is a double-stranded RNA–DNA hybrid which is the result of the synthesis of a complementary strand of DNA on the single-stranded viral RNA as template.

The enzyme also shows ribonuclease H and DNA-dependent DNA polymerase activities which convert the hybrid into a duplex DNA. A particularly good template for this enzyme is the synthetic hybrid poly(dC) · poly(rG) [436]. The polymerase then proceeds to replicate the duplex DNA so as to provide more copies which may be integrated into the genome of the host cell. The enzyme replicates gapped double-stranded DNA templates better than single-stranded DNA templates but cannot use nicked double-stranded DNA.

A detailed model (Fig. 8.25) of reverse transcriptase action has recently been proposed [442, 443] to explain certain unexpected observations, e.g. the duplex DNA produced is longer at both ends than the viral RNA [444, 445] and during the reaction short molecules of both + and − strand DNA ('strong stop DNAs') are formed [446].

Avian reverse transcriptases are initiated with $tRNA^{trp}$ [440, 447–449] but the reverse transcription from Moloney murine leukaemia virus starts by using a $tRNA^{pro}$ which attaches to the viral RNA at about 150 nucleotides from the 5′ end. This $tRNA^{pro}$ serves as a primer for synthesis of −strand DNA. Synthesis stops when the DNA strand reaches the 5′ end of the template, i.e. after 150 nucleotides have been added [450]. This is minus strand strong stop DNA and it now 'jumps' to the 3′ end of the viral RNA molecule where it hybridizes to a

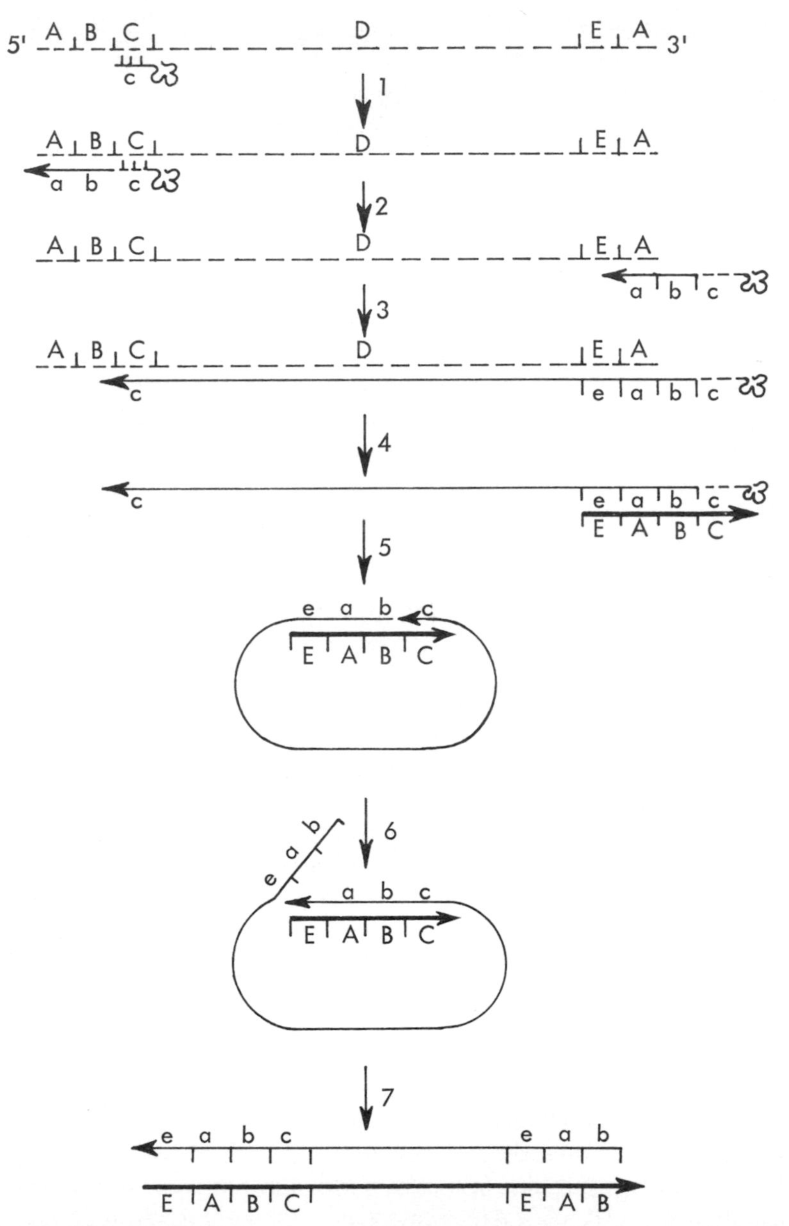

Fig. 8.25 Model for reverse transcription. 1. Synthesis of − strand strong stop DNA on tRNA primer. 2. 'jump' of strong stop DNA to 3′ end of RNA. 3. Elongation of − strand DNA to 8.2 kb. 4. Loss of RNA template and synthesis of + strand strong stop DNA. 5. Loss of tRNA primer and cyclization using 'sticky ends'. 6. Elongation of − strand DNA to 8.8 kb thereby displacing 5′-end of − strand. 7. Completion of + strand DNA (after Gilboa *et al.* [442]).

redundant sequence of 50–60 nucleotides, i.e. strong stop DNA extends by about 100 nucleotides beyond the viral RNA molecule (see Fig. 8.25). Minus strand strong stop DNA could now serve as a primer for DNA synthesis to copy the whole of the viral RNA strand. However, synthesis appears to stop at the site where strong stop DNA was first initiated, i.e. when the region complementary to $tRNA^{pro}$ has been replicated. The viral RNA appears to be degraded at this stage. By some mechanism not yet clear plus strand DNA synthesis is now initiated towards the 5′ end of the newly-synthesized minus DNA strand. Synthesis continues past the end of the minus DNA strand and terminates only when part of the $tRNA^{pro}$ molecule has been replicated. This is plus strand strong stop DNA. This produces a partial duplex with sticky ends which can cyclize thereby enabling the minus strand to be extended by nick translation and the plus strand strong stop DNA to be extended to the end of the template.

The presence of RNA-dependent DNA polymerase has now been reported in many viruses, oncogenic and otherwise [451–453], in spontaneous mammary carcinoma [454], and in virus-like particles isolated from human milk [454]. Enzyme activity can also be readily detected in normal mouse and human cells, adult and foetal, as well as in human tumour cells not associated with any RNA-containing virus [455] (see p. 8). The pattern which is emerging is of a group of enzymes all of which can produce DNA from a variety of RNA templates [456]. However, cellular RNA directed enzymes differ from the viral enzymes in preferring Mn^{2+} to Mg^{2+} and in their failure to use natural RNAs as templates.

The avian reverse transcriptase has been shown to be made up of two subunits α and β though β may be a precursor of α [457–459]. The most active form is $\alpha\beta$.

The enzyme shows only low fidelity and lacks associated exonuclease activity though it is capable of specifically hydrolysing the ribo strand of an RNA–DNA duplex [460–462].

A number of reviews on the subject have appeared [438, 450, 463, 464].

REFERENCES

1 Clark, A.J. and Ganesan, A. (1975), *Molecular Mechanisms for Repair of DNA—Basic Life Sciences,* Vol. 5A and 5B, p. 431 (ed. P.C. Hanawalt and R. B. Setlow), Plenum Press, New York.
2 Watson, J.D. and Crick, F.H.C. (1953), *Nature,* **171,** 737.
3 Meselson, M and Stahl, F.W. (1958), *Proc. Natl. Acad. Sci. USA,* **66,** 671.
4 Cairns, J. (1966), *Sci. Amer.,* **214**(1), 36.
5 Taylor, J.H. (1963), *Molecular Genetics* (ed. J. H. Taylor), Academic Press, New York, Part I, p. 65.

6 Lovett, M.A., Katz, L. and Helinski, D.R. (1974), *Nature,* **251,** 337.
7 Hirt, B. (1969), *J. Mol. Biol.,* **40,** 141.
8 Wolfson, J. and Dressler, D. (1972), *Proc. Natl. Acad. Sci. USA,* **69,** 2682.
9 Kriegstein, H.J. and Hogness, D.S. (1974), *Proc. Natl. Acad. Sci. USA,* **71,** 135.
10 Fareed, G.C., Garon, C.F. and Salzman, N.P. (1972), *J. Virol.,* **10,** 484.
11 Schnös, M. and Inman, R.B. (1970), *J. Mol. Biol.,* **51,** 61.
12 Masters, M. and Broda, P. (1971), *Nature New Biol.,* **232,** 137.
13 Hara, H. and Yoshikawa, H. (1973), *Nature New Biol.,* **244,** 200.
14 Weingartner, B., Winnacker, E.L., Tolun, A. and Pettersson, U. (1976), *Cell,* **9,** 259.
15 Lovett, M.A., Sparks, R.B. and Helinsky, D.R. (1975), *Proc. Natl. Acad. Sci. USA,* **72,** 2905.
16 Zakian, V.A. (1976), *J. Mol. Biol.,* **108,** 305.
17 Huberman, J.A. and Riggs, A.D. (1968), *J. Mol. Biol.,* **32,** 327.
18 Callan, H.G. (1972), *Proc. R. Soc. Lond. B,* **181,** 19.
19 Hand, R. (1978), *Cell,* **15,** 317.
20 Rowen, L. and Kornberg, A. (1978), *J. Biol. Chem.,* **253,** 758.
21 Gray, C.P., Sommer, R., Polka, C., Beck, E. and Schaller, H. (1978), *Proc. Natl. Acad. Sci. USA,* **75,** 50.
22 Fiddes, J.C., Barrell, B.G. and Godson, G.N. (1978), *Proc. Natl. Acad. Sci. USA,* **75,** 1081.
23 Godson, G.N., Barrell, B.G., Staden, R. and Fiddes, J.C. (1978), *Nature,* **276,** 236.
24 Suggs, S.V. and Ray, D.S. (1978), *Cold Spring Harbor Symp. Quant. Biol.,* **43,** 379.
25 Godson, G.N. (1978), *Cold Spring Harbor Symp. Quant. Biol.,* **43,** 367.
26 Langeveld, S.A., von Mansfeld, A.D.M., Baas, P.D., Jansz, S.H., von Arkel, G.A. and Weisbeck, P.J. (1978), *Nature,* **271,** 417.
27 Sanger, F., Air, G.M., Barrell, B.G., Brown, N.L., Coulson, A.R., Fiddes, J.C., Hutchison, C.A., Slocombe, P.M. and Smith, M. (1977), *Nature,* **265,** 687.
28 Sims, J., Capon, D. and Dressler, D. (1979), *J. Biol. Chem.,* **254,** 12615.
29 Grosschedl, R. and Hobom, G. (1979), *Nature,* **277,** 621.
30 Denniston-Thompson, K., Moore, D.D., Kruger, K.E., Furth, M.E. and Blattner, F.R. (1977), *Science,* **198,** 1051.
31 Tomizawa, J., Ohmari, H. and Bird, R.E. (1977), *Proc. Natl. Acad. Sci. USA,* **74,** 1865.
32 Sugimoto, K., Oka, A., Sugisaki, H. Takanami, M., Nishimura, A., Yasuda, Y. and Hirota, Y. (1979), *Proc. Natl. Acad. Sci. USA,* **76,** 575.
33 Subramanian, K.N. and Shenk, T. (1978), *Nucl. Acids Res.,* **5,** 3635.
34 Friedman, T., Estry, A., La Parte, P. and Deininger, P. (1979), *Cell,* **17,** 715.
35 Soeda, E., Arrand, J.R., Smoler, N. and Griffin, B.E. (1979), *Cell,* **17,** 357.
36 Soeda, E., Kimura, G. and Miura, K.I. (1978), *Proc. Natl. Acad. Sci. USA,* **75,** 162.
37 Tolun, A., Alestrom, P. and Pettersson, U. (1979), *Cell,* **17,** 705.
38 Astell, C.R., Smith, M., Chow, M.B. and Ward, D.C. (1979), *Cell,* **17,** 691.

39 Martens, P.A. and Clayton, D.A. (1979), *J. Mol. Biol.*, **135**, 327.
40 Crews, S., Ojala, D., Posakony, J., Nishiguchi, J. and Attardi, G. (1979), *Nature*, **277**, 192.
41 Lim, V.I. and Mazanov, A.L. (1978), *FEBS Lett.*, **88**, 118.
42 Hirota, Y., Yasuda, S., Yamada, M., Nishimura, A., Sugimoto, K., Sugisaki, H., Oka, A. and Takanami, M. (1978), *Cold Spring Harbor Symp. Quant. Biol.*, **43**, 129.
43 Messer, W., Meijer, M., Bergmans, H.E.N., Hansen, F.G., von Megenburg, K., Beck, E. and Shaller, H. (1978), *Cold Spring Harbor Symp. Quant. Biol.*, **43**, 139.
44 Marsh, R.C. and Worcel, A. (1977), *Proc. Natl. Acad. Sci. USA*, **74**, 270.
45 Gudas, L.J., James, R. and Pardee, A.B. (1976), *J. Biol. Chem.*, **251**, 3470.
46 Messer, W., Dankworth, L., Tippe-Schindler, R., Womack, J.E. and Zahn, G. (1975), *DNA Synthesis and its Regulation* (eds M. Goulian and P. Hanawalt), VI. III (Series ed. F. Fox). *ICN–UCLA Symposium on Molecular and Cellular Biology*, Benjamin, California.
47 Kogoma, T. (1978), *J. Mol. Biol.*, **121**, 58.
48 Gefter, M.L. (1975), *Annu. Rev. Biochem.*, **44**, 45.
49 Cairns, J. (1963), *J. Mol. Biol.*, **6**, 208.
50 Housman, D. and Huberman, J.A. (1975), *J. Mol. Biol.*, **94**, 173.
51 Lewin, B. (1974), *Gene Expression*, Vol. 2, John Wiley and Son.
52 Woodland, H.R. and Pestell, R.Q.W. (1972), *Biochem. J.*, **127**, 597.
53 Okazaki, R., Okazaki, T., Sakabe, K., Sugimoto, K., Kainuma, R., Sugino, A. and Iwatsuki, N. (1968), *Cold Spring Harbor Symp. Quant. Biol.*, **33**, 129.
54 Gautschi, J.R. and Clarkson, J.M. (1975), *Eur. J. Biochem.*, **50**, 403.
55 Magnussen, G., Pigiet, V., Winnacker, E.L., Abrams, R. and Reichard, P. (1973), *Proc. Natl. Acad. Sci. USA*, **70**, 412.
56 Schlomai, J. and Kornberg, A. (1978), *J. Biol. Chem.*, **253**, 3305.
57 Tamanoi, F., Machida, Y. and Okazaki, T. (1978), *Cold Spring Harbor Symp. Quant. Biol.*, **43**, 239.
58 Tye, B.-K., Nyman, P.O., Lehman, I.R., Hochhauser, S. and Weiss, B. (1977), *Proc. Natl. Acad. Sci. USA*, **74**, 154.
59 Tye, B.-K. and Lehman, I.R. (1977), *J. Mol. Biol.*, **117**, 293.
60 Warner, H.R. and Duncan, B.K. (1978), *Nature*, **272**, 32.
61 Olivera, B.M., Manlapaz-Ramos, P., Warner, H.R. and Duncan, B.K. (1979), *J. Mol. Biol.*, **128**, 265.
62 Tye, B.-K., Chien, J., Lehman, I.R., Duncan, B.K. and Warner, H.R. (1978), *Proc. Natl. Acad. Sci. USA*, **75**, 233.
63 Okazaki, R., Sugimoto, K., Okazaki, T., Imac, Y. and Sugino, A. (1970), *Nature*, **228**, 223.
64 Lehman, I.R., Tye, B.-K., and Nyman, P.O. (1978), *Cold Spring Harbor Symp. Quant. Biol.*, **43**, 221.
65 Okazaki, T. and Okazaki, R. (1969), *Proc. Natl. Acad. Sci. USA*, **64**, 1242.
66 Sugino, A. and Okazaki, R. (1972), *J. Mol. Biol.*, **64**, 61.
67 Diaz, A.T. and Werner, R. (1975), *J. Mol. Biol.*, **95**, 63.
68 Brutlag, D., Schekman, R. and Kornberg, A. (1971), *Proc. Natl. Acad. Sci. USA*, **68**, 2826.

69 Wickner, W., Brutlag, D., Schekman, R. and Kornberg, A. (1972), *Proc. Natl. Acad. Sci. USA,* **69,** 965.
70 Bagdasarian, M.M., Izakowska, M. and Bagdasarian, M. (1977), *J. Bact.,* **130,** 577.
71 Bouché, J.P., Zeckel, K. and Kornberg, A. (1975), *J. Biol. Chem.,* **250,** 5995.
72 Ray, D.S., Duebner, J. and Suggs, S. (1975), *J. Virol.,* **16,** 348.
73 Kurosawa, Y., Ogawa, T., Hirose, S., Okazaki, T. and Okazaki, R. (1975), *J. Mol. Biol.,* **96,** 653.
74 Sugino, A., Hirose, S. and Okazaki, R. (1972), *Proc. Natl. Acad. Sci. USA,* **69,** 1863.
75 Okazaki, R., Okazaki, T., Hirose, S., Sugino, A., Ogawa, T., Kurosawa, Y., Shinozaki, K., Tamanoi, F., Seki, T., Machida, Y., Fujiyama, A. and Kohara, Y. (1975), *DNA Synthesis and its Regulation* (eds M. Goulian and P. Hanawalt), Vol. III (Series ed. F. Fox), *ICN–UCLA Symposium on Molecular and Cellular Biology,* Benjamin, California.
76 Seidel, H. (1967), *Biochim. Biophys. Acta,* **138,** 98.
77 Thomas, K.R., Manlapaz-Ramos, P., Lundquist, R. and Olivera, B.M. (1978), *Cold Spring Harbor Symp. Quant. Biol.,* **43,** 231.
78 Miyamoto, C. and Denhardt, D.T. (1977), *J. Mol. Biol.,* **116,** 681.
79 Anderson, M.L.M. (1978), *J. Mol. Biol.,* **118,** 277.
80 Ogawa, T., Hirose, S., Okazaki, T. and Okazaki, R. (1977), *J. Mol. Biol.,* **112,** 121.
81 Okazaki, T., Kurasawa, Y., Ogawa, T., Seki, T., Shinozaki, K., Hirose, S., Fujiyama, A.A., Kohara, Y., Machida, Y., Tamanoi, F. and Hozumi, T. (1978), *Cold Spring Harbor Symp. Quant. Biol.,* **43,** 203.
82 Sugino, A. and Okazaki, R. (1973), *Proc. Natl. Acad. Sci. USA,* **70,** 88.
83 Kowalski, J. and Denhardt, D.T. (1979), *Nature,* **281,** 704.
84 Tseng, B.Y., Erickson, J.M. and Goulian, M. (1979), *J. Mol. Biol.,* **129,** 531.
85 Reichard, P., Eliasson, R. and Söderman, G. (1974), *Proc. Natl. Acad. Sci. USA,* **71,** 4901.
86 Pigiet, V., Eliasson, R. and Reichard, P. (1974), *J. Mol. Biol.,* **84,** 197.
87 Geider, K. (1976), *Curr. Top. Microbiol. Immunol.,* **74,** 58.
88 Huberman, J.A. and Horwitz, H. (1973), *Cold Spring Harbor Symp. Quant., Biol.,* **38,** 233.
89 Blair, D.G., Sherratt, D.J., Clewel, D.B. and Helinski, D.R. (1972), *Proc. Natl. Acad. Sci. USA,* **69,** 2518.
90 Rowen, L. and Kornberg, A. (1978), *J. Biol. Chem.,* **253,** 770.
91 Gomez-Eichelmann, M.C. and Lark, K.G. (1977), *J. Mol. Biol.,* **117,** 621.
92 Kahn, M. and Hanawalt, P. (1979), *J. Mol. Biol.,* **128,** 501.
93 Olivera, B.M. and Bonhoeffer, F. (1972), *Nature New Biol.,* **240,** 233.
94 Olivera, B.M., Lark, K.G., Herrmann, R. and Bonhoeffer, F. (1973), *DNA Synthesis in vitro* (eds R. D. Wells and R. B. Inman), University Park Press, p. 214.
95 Kurosawa, Y. and Okazaki, R. (1975), *J. Mol. Biol.,* **94,** 229.
96 Kuroda, R.K. and Okazaki, R. (1975), *J. Mol. Biol.,* **94,** 213.
97 Olivera, B.M. (1978), *Proc. Natl. Acad. Sci. USA,* **75,** 238.

98 Thomas, K.R., Manlapaz-Ramos, P., Lundquist, R. and Olivera, B.M. (1978), *Cold Spring Harbor Symp. Quant. Biol.*, **43**, 231.
99 Brynolf, K., Eliasson, R. and Reichard, P. (1974), *Cell*, **13**, 573.
100 Seale, R.L. (1975), *Nature*, **9**, 247.
101 Seale, R.L. (1976), *Cell*, **9**, 423.
102 Schlaeger, E.J. (1978), *Biochem. Biophys. Res. Commun.*, **81**, 8.
103 Adams, R.L.P. unpublished observations.
104 Burgoyne, L.A., Mobbs, J.D. and Marshall, A.J. (1976), *Nucl. Acids. Res.*, **3**, 3293.
105 Cremisi, C., Chestier, A. and Yaniv, M. (1977), *Cold Spring Harbor Symp. Quant. Biol.*, **42**, 409.
106 Worcel, A., Han, S. and Wong, M.L. (1978), *Cell*, **15**, 969.
107 Leffak, I.M., Grainger, K.R. and Weintraub, H. (1977), *Cell*, **12**, 837.
108 Weintraub, H. (1976), *Cell*, **9**, 419.
109 Weintraub, H., Worcel, A. and Alberts, B.A. (1976), *Cell*, **9**, 409.
110 Riley, D. and Weintraub, H. (1979), *Proc. Natl. Acad. Sci. USA*, **76**, 328.
111 Seidman, M.M., Levine, A.J. and Weintraub, H. (1979), *Cell*, **18**, 439.
112 De Pamphilis, M.L., Anderson, S., Bar-Shavit, R., Collins, E., Edenberg, H., Herman, T., Karas, B., Kaufmann, G., Krokan, H., Shelton, E., Su, R. Tapper, D. and Wassarman, P.M. (1978), *Cold Spring Harbor Symp. Quant Biol.*, **43**, 679.
113 Marsden, M.P.F. and Laemmli, U.K. (1979), *Cell*, **17**, 849.
114 Finch, J.T. and Klug, A. (1976), *Proc. Natl. Acad. Sci. USA*, **73**, 1897.
115 Kowalski, J. and Cheevers, W.P. (1976), *J. Mol. Biol.*, **104**, 603.
116 Edenberg, H.J. and Huberman, J.A. (1975), *Annu. Rev. Biochem.*, **44**, 245.
117 Sheinin, R. and Humbert, J. (1978), *Annu. Rev. Biochem.*, **47**, 277.
118 Wells, R.D. and Inman, R.B. (1973), *DNA Synthesis in vitro*, University Park Press.
119 Wickner, R.B. (1974), *Methods in Molecular Biology*, Vol. 7, *DNA Replication*, Dekker, New York.
120 Vosberg, H.P. and Hoffmann-Berling, H. (1971), *J. Mol. Biol.*, **58**, 739.
121 Moses, R.E. and Richardson, C.C. (1970), *Proc. Natl. Acad. Sci. USA*, **67**, 674.
122 Schaller, H., Otto, B., Nusslein, V., Huf, J., Herrmann, R. and Bonhoeffer, F. (1972), *J. Mol. Biol.*, **63**, 183.
123 Schekman, R., Wickner, W., Westergaard, O., Brutlag, D., Geider, K., Bertsch, L.L. and Kornberg, A. (1972), *Proc. Natl. Acad. Sci. USA*, **69**, 2691.
124 Nusslein, V. and Klein, A. (1974), in *Methods in Molecular Biology*, Vol. 7 (ed. R. B. Wickner), Dekker, New York.
125 Burgoyne, L.A. (1972), *Biochem. J.*, **130**, 959.
126 Hunter, T. and Francke, B. (1974), *J. Virol.*, **13**, 125.
127 Winnacker, E.L., Magnussen, G. and Reichard, P. (1972), *J. Mol. Biol.*, **72**, 523.
128 De Pamphilis, M.L., Beard, P. and Berg, P. (1975), *J. Biol. Chem.*, **250**, 4340.

129 Berger, N.A., Petzold, S.J. and Johnson, E.S. (1977), *Biochim. Biophys. Acta,* **478,** 44.
130 Fraser, J.M.K. and Huberman, J.A. (1977), *J. Mol. Biol.,* **117,** 249.
131 Wist, E., Krokan, H. and Prydz, H. (1976), *Biochemistry,* **15,** 3647.
132 Tseng, B.Y. and Goulian, M. (1975), *J. Mol. Biol.,* **99,** 317.
133 Krokan, H., Wist, E. and Prydz, H. (1977), *Biochim. Biophys. Acta,* **475,** 553.
134 Hershey, H.V. and Taylor, J.H. (1974), *Exp. Cell Res.,* **85,** 78.
135 Wist, E. and Prydz, H. (1979), *Nucl. Acids Res.,* **6,** 1583.
136 Waqar, M.A., Evans, M.J. and Huberman, J.A. (1978), *Nucl. Acids Res.,* **5,** 1933.
137 Butt, T.R., Wood, W.M., McKay, E.L. and Adams, R.L.P. (1978), *Biochem. J.,* **173,** 309.
138 Elford, H.L. (1974), *Arch. Biochem. Biophys.,* **163,** 537.
139 Tseng, B.Y. and Goulian, M. (1977), *Cell,* **12,** 483.
140 Francke, B. and Hunter, T. (1975), *J. Virol.,* **15,** 97.
141 Pigiet, V., Eliasson, R. and Reichard, P. (1974), *J. Mol. Biol.,* **84,** 197.
142 De Pamphilis, M.L., Beard, P. and Berg, P. (1975), *J. Biol. Chem.,* **250,** 4340.
143 De Pamphilis, M.L. and Berg, P. (1975), *J. Biol. Chem.,* **250,** 4348.
144 Kaufmann, G., Anderson, S. and De Pamphilis, M.L. (1977), *J. Mol. Biol.,* **116,** 549.
145 Su, R.T. and De Pamphilis, M.L. (1976), *Proc. Natl. Acad. Sci. USA,* **73,** 3466.
146 Edenberg, H.J., Waqar, M.A., Huberman, J.A. (1976), *Proc. Natl. Acad. Sci. USA,* **73,** 4392.
147 Edenberg, H.J., Waqar, M.A., Huberman, J.A. (1977), *Nucl. Acids Res.,* **4,** 3083.
148 Yamashita, T., Arens, M. and Green, M. (1977), *J. Biol. Chem.,* **252,** 7940.
149 Arens, M., Yamashita, T., Padmanabhan, R., Tsurno, T. and Green, M. (1977), *J. Biol. Chem.,* **252,** 7947.
150 Challberg, M.C. and Kelly, T.J. (1979), *Proc. Natl. Acad. Sci. USA,* **76,** 655.
151 Becker, Y. and Asher, Y. (1975), *Virology,* **63,** 209.
152 Roizman, B., Jacob, R.J., Knipe, D.M., Morese, L.S. and Ruvechan, W.T. (1978), *Cold Spring Harbor Symp. Quant. Biol.,* **43,** 809.
153 Horwitz, I.S., Kaplan, L.M., Abboud, M., Maritato, J., Chow, L.T. and Broker, T.R. (1978), *Cold Spring Harbor Symp. Quant., Biol.,* **43,** 769.
154 Kornberg, A. (1974), *DNA Synthesis,* Freeman, San Francisco.
155 Kornberg, A. (1980), *DNA Replication,* Freeman, San Francisco.
156 Kornberg, A. (1968), *Sci. Amer.,* **219**(4), 64.
157 Kornberg, A. (1961), *The Enzymatic Synthesis of DNA,* Wiley, London.
158 Josse, J., Kaiser, A.D. and Kornberg, A. (1961), *J. Biol. Chem.,* **236,** 864.
159 Becker, A. and Hurwitz, J. (1971), *Prog. Nucleic Acid Res. Mol. Biol.* (ed. J. N. Davidson and W. E. Cohn), **11,** 423.
160 Jovin, T.M., Englund, P.T. and Bertsch, L.L. (1969), *J. Biol. Chem.,* **244,** 2996.

161 Griffith, J., Huberman, J.A. and Kornberg, A. (1971), *Proc. Natl. Acad. Sci. USA,* **55,** 209.
162 Kornberg, A. (1969), *Science,* **163,** 1410.
163 Que, B.G., Downey, K.M. and So, A.G. (1979), *Biochemistry,* **18,** 2064.
164 Schachman, H.K., Adler, J., Radding, C.M., Lehman, I.R. and Kornberg, A. (1960), *J. Biol. Chem.,* **235,** 3242.
165 Brutlag, D. and Kornberg, A. (1972), *J. Biol. Chem.,* **247,** 241.
166 Galas, D.J. and Branscomb, E.W. (1978), *J. Mol. Biol.,* **124,** 653.
167 Loeb, L.A., Weymouth, L.A., Kunkel, T.A., Gopinathan, K.P., Beckman, R.A. and Dube, D.K. (1978), *Cold Spring Harbor Symp. Quant. Biol.,* **43,** 921.
168 Seal, G., Shearman, C.W. and Loeb, L.A. (1979), *J. Biol. Chem.,* **254,** 5229.
169 Agarwal, S.S., Dube, D.K. and Loeb, L.A. (1979), *J. Biol. Chem.,* **254,** 101.
170 Reha-Kranz, L.J. and Bessman, M.J. (1977), *J. Mol. Biol.,* **116,** 99.
171 Lo, K.Y. and Bessman, M.J. (1976), *J. Biol. Chem.,* **251,** 2475.
172 Lo, K.Y. and Bessman, M.J. (1976), *J. Biol. Chem.,* **251,** 2480.
173 Hopfield, J.J. (1974), *Proc. Natl. Acad. Sci. USA,* **71,** 4135.
174 Engler, M.J. and Bessman, M.J. (1978), *Cold Spring Harbor Symp. Quant. Biol.,* **43,** 929.
175 Travaglini, A.C., Mildran, A.S. and Loeb, L.A. (1975), *J. Biol. Chem.,* **250,** 8647.
176 Gillin, F.D. and Nossal, N.G. (1975), *Biochem. Biophys. Res. Comm.,* **64,** 457.
177 Cozzarelli, N.R., Kelly, R.B. and Kornberg, A. (1969), *J. Mol. Biol.,* **45.** 513.
178 Kelly, R.B., Cozzarelli, N.R., Deutscher, M.P., Lehman, I.R. and Kornberg, A. (1970), *J. Biol. Chem.,* **245,** 39.
179 Bambara, R.A., Vyemura, D. and Choi, T. (1978), *J. Biol. Chem.,* **253,** 413.
180 Matson, S.W., Capaldo-Kimball, F.N. and Bambara, R.A. (1978), *J. Biol. Chem.,* **253,** 7851.
181 de Lucia, P. and Cairns, J. (1969), *Nature,* **224,** 1164.
182 Gross, J. and Gross, M. (1969), *Nature,* **224,** 1166.
183 Okazaki, R., Arisawa, M. and Sugino, A. (1971), *Proc. Natl. Acad. Sci. USA,* **68,** 2954.
184 Kato, T. and Konda, S. (1970), *J. Bact.,* **104,** 871.
185 Smirnov, G.B., Favorskaya, Y.N. and Skavronskaya, A.G. (1971), *Mol. Gen. Genet.,* **111,** 357.
186 Emmerson, P.T. and Strike, P. (1973), *DNA Replication and the Cell Membrane* (ed. A. Kolber and M. Kohiyama).
187 Gefter, M.L. (1974), *Prog. Nucleic Acid Res. Mol. Biol.,* **14,** 101.
188 Tait, R.C. and Smith, D.W. (1974), *Nature,* **249,** 116.
189 Gefter, M.L., Hirota, Y., Kornberg, T., Wechsler, S.A. and Barnoux, C. (1971), *Proc. Natl. Acad. Sci. USA,* **68,** 3150.
190 Klein, A., Nusslein, V., Otto, B. and Powling, A. (1973), in *DNA Synthesis in vitro* (ed. R. D. Wells and R. B. Inman), University Park Press, p. 185.

191 Otto, B. (1973), *Biochem. Soc. Trans.*, **1,** 629.
192 Livingstone, D.M., Hinckle, D.C. and Richardson, C.C. (1975), *J. Biol. Chem.*, **250,** 461.
193 Kornberg, T. and Kornberg, A. (1974), in *The Enzymes* (ed. P. D. Boyer), Academic Press, New York, Vol. X, p. 119.
194 Lehman, I.R. (1974), in *The Enzymes* (ed. P.D. Boyer), Academic Press, New York, Vol. X, p. 237.
195 Kornberg, A. (1977), *Trans. Biochem. Soc.*, **5,** 359.
196 Wickner, W., Schekman, R., Geider, K. and Kornberg, A. (1973), *Proc. Natl. Acad. Sci. USA,* **70,** 1764.
197 McHenry, C.S. and Crow, W. (1979), *J. Biol. Chem.*, **254,** 1748.
198 Sevastopoulos, C.G. and Glaser, D.R. (1977), *Proc. Natl. Acad. Sci. USA,* **74,** 3947.
199 Fersht, A.R. (1979, *Proc. Natl. Acad. Sci. USA,* **176,** 4946.
200 Wickner, W. and Kornberg, A. (1974), *J. Biol. Chem.*, **249,** 6244.
201 Hurwitz, J. and Wickner, S. (1974), *Proc. Natl. Acad. Sci. USA,* **71,** 6.
202 Gass, K.B. and Cozzarelli, N.R. (1974), *Methods Enzymol.*, **29,** 17.
203 Bollum, F.J. (1974), in *The Enzymes* (ed. P. D. Boyer), Academic Press, New York, Vol. X, p. 145.
204 Loeb, L.A. (1974), in *The Enzymes* (ed. P.D. Boyer), Academic Press, New York, Vol. X, p. 174.
205 Weissbach, A. (1975), *Cell,* **5,** 101.
206 Fansler, B.S. (1974), *Int. Rev. Cytology,* Suppl. 4, 363.
207 Craig, R.K. and Keir, H.M. (1974), in *The Cell Nucleus* (ed. H. Busch), Academic Press, p. 35.
208 Bollum, F.J. (1975), *Prog. Nucleic Acid Res. Mol. Biol.*, **15,** 109.
209 Chang, L.M.S. and Bollum, F.J. (1972), *J. Biol. Chem.*, **247,** 7948.
210 Ove, P., Jenkins, M.D. and Lazlo, J. (1970), *Cancer Res.*, **30,** 535.
211 Chiu, J.F. and Sung, S.C. (1972), *Biochim. Biophys. Acta,* **262,** 397.
212 Chang, L.M.S., Brown, M. and Bollum, F.J. (1973), *J. Mol. Biol.*, **74,** 1.
213 Lindsay, J.G., Berryman, S. and Adams, R.L.P. (1970), *Biochem. J.*, **119,** 839.
214 Hubscher, U., Kuenzle, C.C., Limacher, W.K., Sherrer, P. and Spadari, S. (1978), *Cold Spring Harbor Symp. Quant. Biol.*, **43,** 625.
215 Wang, H.F. and Popenoe, E.A. (1977), *Biochim. Biophys. Acta,* **474,** 98.
216 Loeb, L.A. and Agarwal, S.S. (1971), *Exp. Cell Res.*, **66,** 299.
217 Fichot, O. Pascal, M., Mechali, M. and de Recondo, A.M. (1979), *Biochim. Biophys. Acta,* **561,** 29.
218 Holms, A.M., Hesselwood, I.P. and Johnston, I.R. (1974), *Eur. J. Biochem.*, **43,** 487.
219 Hesslewood, I.P., Holmes, A.M., Wakeling, W.F. and Johnston, I.R. (1978), *Eur. J. Biochem.*, **84,** 123.
220 Mechali, M., Abadiedebat J. and de Recondo, A.M. (1980), *J. Biol. Chem.*, **255,** 2114.
221 Ogura, M., Suzuki-Hari, C., Nagano, H., Mano, Y. and Ikegami, S. (1979), *Eur. J. Biochem.*, **97,** 603.
222 Edenberg, H.T., Anderson, S. and de Pamphilis, M.L. (1978), *J. Biol. Chem.*, **253,** 3273.

223 Spadari, S. and Weissbach, A. (1975), *Proc. Natl. Acad. Sci. USA,* **72,** 503.
224 Korn, D., Fisher, P.A., Battey, J. and Wang, T.S.-F. (1978), *Cold Spring Harbor Symp. Quant Biol.,* **43,** 613.
225 Fisher, P.A., Wang, T.S-F. and Korn, D. (1979), *J. Biol. Chem.,* **254,** 6128.
226 Chen, Y.-C., Bohn, E.W., Plank, S.R. and Wilson, S.H. (1979), *J. Biol. Chem.,* **254,** 11678.
227 Knopf, K.W. (1979), *Eur. J. Biochem.,* **98,** 231.
228 Foster, D.N. and Gurney, T. (1973), *J. Cell Biol.,* **59,** 103a.
229 Chang, L.M.S. (1975), *Methods Enzymol.,* **29,** 81.
230 Wang, T.S.-F., Sedwick, W.D. and Korn, D. (1974), *J. Biol. Chem.,* **249,** 841.
231 Kunkel, T.A., Tcheng, J.E. and Meyer, R.R. (1978), *Biochim. Biophys. Acta,* **520,** 302.
232 Aposhian, H.V. and Kornberg, A. (1962), *J. Biol. Chem.,* **237,** 519.
233 Loeb, L.A. (1969), *J. Biol. Chem.,* **244,** 1672.
234 Chang, L.M.S. (1973), *J. Biol. Chem.,* **248,** 6983.
235 Waser, J., Hubscher, V., Kuenzle, C.C. and Spadari, S. (1979), *Eur. J. Biochem.,* **97,** 361.
236 Adams, R.L.P. and Kirk, D. unpublished results.
237 Spadari, S. and Weissbach, A. (1974), *J. Mol. Biol.,* **86,** 11.
238 Spadari, S. and Weissbach, A. (1974), *J. Biol. Chem.,* **249,** 5809.
239 Ter Schegget, J., Flavell, R.A. and Borst, P. (1971), *Biochim. Biophys. Acta,* **254,** 1.
240 Kasamatsu, H., Robberson, D.L. and Vinograd, J. (1971), *Proc. Natl. Acad. Sci. USA,* **68,** 2252.
241 Meyer, R.R. and Simpson, M.V. (1970), *J. Biol. Chem.,* **245,** 3426.
242 Fry, M. and Weissbach, A. (1973), *Biochemistry,* **12,** 3602.
243 Hubscher, U., Kuenzle, C.C. and Spadari, S. (1979), *Proc. Natl. Acad. Sci. USA,* **76,** 2316.
244 Bertazzoni, U., Scovassi, A.I. and Brun, G.M. (1977), *Eur. J. Biochem.,* **81,** 237.
245 Abboud, M.M. and Horwitz, M.S. (1979), *Nucl. Acids Res.,* **6,** 1025.
246 Ito, K., Arens, M. and Green, M. (1976), *Biochim. Biophys. Acta,* **447,** 340.
247 Krokan, H., Schaffer, P. and de Pamphilis, M.L. (1979), *Biochemistry,* **18,** 4431.
248 Tewari, K.K. and Wildman, S.G. (1967), *Proc. Natl. Acad. Sci. USA,* **58,** 689.
249 Spencer, D. and Whitfield, P.R. (1967), *Biochem. Biophys. Res. Comm.,* **28,** 538.
250 Krakow, J.S., Coutsogeorgopoulos, C. and Canellakis, E.S. (1962), *Biochim. Biophys. Acta,* **55,** 639.
251 Bollum, F. (1965), *J. Biol. Chem.,* **240,** 2599.
252 Chang, L.M.S. and Bollum, F.J. (1971), *J. Biol. Chem.,* **246,** 909.
253 Chang, L.M.S. (1971), *Biochem. Biophys. Res. Commun.,* **44,** 124.
254 Srivastava, B.I.S. (1974), *Cancer Res.,* **34,** 1015.
255 Srivastava, B.I.S. (1975), *Res. Commun. Chem. Path. Pharm.,* **10,** 715.

256 Coleman, M.S., Hutton, J.J., de Simone, P. and Bollum, F.J. (1974), *Proc. Natl. Acad. Sci. USA*, **71,** 4404.
257 Gellert, M. (1967), *Proc. Natl. Acad. Sci. USA*, **57,** 48.
258 Olivera, B.M. and Lehman, I.R. (1967), *Proc. Natl. Acad. Sci. USA*, **57,** 1426.
259 Zimmerman, S.B., Little, J.W., Oshinsky, C.K. and Gellert, M. (1967), *Proc. Natl. Acad. Sci. USA*, **57,** 1841.
260 Laipis, P.J., Oliver, B.M. and Ganesan, A.T. (1969), *Proc. Natl. Acad. Sci. USA*, **62,** 289.
261 Weiss, B. and Richardson, C.C. (1967), *Proc. Natl. Acad. Sci. USA*, **57,** 1021.
262 Lindah, T. and Edelman, G.M. (1968), *Proc. Natl. Acad. Sci. USA*, **61,** 680.
263 Tsukada, K. and Ichimura, M. (1971), *Biochem. Biophys. Res. Commun.*, **42,** 1156.
264 Gefter, M.L., Becker, A. and Hurwitz, J. (1967), *Proc. Natl. Acad. Sci. USA*, **58,** 240.
265 Goulian, M. (1971), *Annu. Rev. Biochem.*, **40,** 855.
266 Gellert, M. and Bullock, M.L. (1970), *Proc. Natl. Acad. Sci. USA*, **67,** 1580.
267 Konrad, E.B., Modrich, P. and Lehman, I.R. (1973), *J. Mol. Biol.*, **77,** 519.
268 Gottesman, M.M., Hicks, M.L. and Gellert, M. (1973), *J. Mol. Biol.*, **77,** 531.
269 Lehman, I.R. (1974), in *The Enzymes* (ed. P. D. Boyer), Vol. X, p. 237.
270 Mate, K. and Hurwitz, J. (1974), *J. Biol. Chem.*, **249,** 3680.
271 Sano, H. and Feix, G. (1974), *Biochemistry*, **13,** 5110.
272 Baldy, M.W. (1969), *Cold Spring Harbor Symp. Quant. Biol.*, **33,** 333.
273 Alberts, B.M. and Frey, L. (1970), *Nature*, **227,** 1313.
274 Weiner, J.H., Bertsch, L.L. and Kornberg, A. (1975), *J. Biol. Chem.*, **250,** 1972.
275 Sigal, N., Delius, H., Kornberg, T., Gefter, M.L. and Alberts, B. (1972), *Proc. Natl. Acad. Sci. USA*, **69,** 3537.
276 Huberman, J.A., Kornberg, A. and Alberts, B.M. *J. Mol. Biol.*, **62,** 39.
277 Reuben, R.C. and Gefter, M.L. (1973), *Proc. Natl. Acad. Sci. USA*, **70,** 1846.
278 Geider, K. and Kornberg, A. (1974), *J. Biol. Chem.*, **249,** 3999.
279 Barry, J. and Alberts, B.M. (1972), *Proc. Natl. Acad. Sci. USA*, 69, 2717.
280 Breschkin, A.M. and Mosig, G. (1977), *J. Mol. Biol.*, **112,** 279.
281 Christiansen, C. and Baldwin, R.L. (1977), *J. Mol. Biol.*, **115,** 441.
282 Meyer, R.R., Glassberg, J. and Kornberg, A. (1979), *Proc. Natl. Acad. Sci. USA*, **76,** 1702.
283 Van Darp, B., Schneck, P.K. and Staudenbauer, W.L. (1979), *Eur. J. Biochem.*, **94,** 445.
284 Abdel-Monem, M., Chanel, M.C. and Hoffmann-Berling, H. (1977), *Eur. J. Biochem.*, **79,** 33.

285 Abdel-Monem, M., Chanal, M.C. and Hoffmann-Berling, H. (1977), *Eur. J. Biochem.*, **79**, 39.
286 Kuhn, B., Abdel-Monem, M. and Hoffmann-Berling, H. (1978), *Cold Spring Harbor Symp. Quant. Biol.*, **43**, 63.
287 Kuhn, B., Abdel-Monem, M. Krell, H. and Hoffmann-Berling, H. (1978), *J. Biol. Chem.*, **254**, 11343.
288 Abdel-Monem, M., Lauppe, H.F., Kartenbeck, J., Durwald, H. and Hoffmann-Berling, H. (1977), *J. Mol. Biol.*, **110**, 667.
289 Scott, J.F., Eisenberg, S., Bertsch, L.L. and Kornberg, A. (1977), *Proc. Natl. Acad. Sci. USA*, **74**, 193.
290 Scott, J.F. and Kornberg, A. (1978), *J. Biol. Chem.*, **253**, 3292.
291 Yarranton, G.T. and Gefter, M.L. (1979), *Proc. Natl. Acad. Sci. USA*, **76**, 1658.
292 Kornberg, A., Scott, J.F. and Bertsch, L.L. (1978), *J. Biol. Chem.*, **253**, 3298.
293 Yarranton, G.T., Das, R.H. and Gefter, M.L. (1979), *J. Biol. Chem.*, **254**, 11997.
294 Yarranton, G.T., Das, R.H. and Gefter, M.L. (1979), *J. Biol. Chem.*, **254**, 12002.
295 Wickner, S., Wright, M. and Hurwitz, J. (1974), *Proc. Natl. Acad. Sci. USA*, **71**, 783.
296 Ueda, K., McMacken, R. and Kornberg, A. (1978), *J. Biol. Chem.*, **253**, 261.
297 Reha-Krantz, L.J. and Hurwitz, J. (1978), *J. Biol. Chem.*, **253**, 4043.
298 Arai, K. and Kornberg, A. (1979), *Proc. Natl. Acad. Sci. USA*, **76**, 4308.
299 McMacken, R., Ueda, K. and Kornberg, A. (1977), *Proc. Natl. Acad. Sci. USA*, **74**, 4190.
300 Wickner, S. (1976), *Proc. Natl. Acad. Sci. USA*, **73**, 3511.
301 Wickner, W. and Kornberg, A. (1973), *Proc. Natl. Acad. Sci. USA*, **70**, 3679.
302 Hinckle, D.C. and Richardson, C.C. (1975), *J. Biol. Chem.*, **250**, 5523.
303 Kolodner, R. and Richardson, C.C. (1977), *Proc. Natl. Acad. Sci. USA*, **74**, 1525.
304 Kolodner, R. and Richardson, C.C. (1978), *J. Biol. Chem.*, **253**, 574.
305 Kolodner, R., Masamune, Y., Le Clerc, J.E. and Richardson, C.C. (1978), *J. Biol. Chem.*, **253**, 566.
306 Alberts, B., Morris, C.F., Mace, D., Sinha, N., Bittner, M. and Moran, L. (1975), in *DNA Synthesis and its Regulation* (ed. M. Goulian, P. Hanawalt and C. F. Fox), W.A. Benjamin, Menlo Park, CA, p. 241.
307 Cobianchi, F., Riva, S., Mastomei, GK., Spadari, S., Pedrali-Noy, G. and Falaschi, A. (1978), *Cold Spring Harbor Symp. Quant. Biol.*, **43**, 639.
308 Wang, J.C. (1971), *J. Mol. Biol.*, **55**, 523.
309 Champoux, J.J. and Dulbecco, R. (1972), *Proc. Natl. Acad. Sci. USA*, **69**, 143.
310 Shure, M. and Vinograd, J. (1976), *Cell*, **8**, 215.
311 Strayer, D.R. and Boyer, P.D. (1978), *J. Mol. Biol.*, **120**, 281.
312 Sugino, A., Peebles, C.L., Kreuzer, K.N. and Cozzarelli, N.R. (1977), *Proc. Natl. Acad. Sci. USA*, **74**, 4767.

313 Mizuuchi, K., O'Dea, M.H. and Gellert, M. (1978), *Proc. Natl. Acad. Sci. USA,* **75,** 5960.
314 Gellert, M. Mizuuchi, K., O'Dea, M.H., Itoh, T. and Tomizawa, J.I. (1977), *Proc. Natl. Acad. Sci. USA,* **74,** 4772.
315 Peebles, C.L., Higgins, N.P., Kreuzer, K.N., Morrison, A., Brown, P.O., Sugino, A. and Cozzarelli, N.R. (1978), *Cold Spring Harbor Symp. Quant. Biol.,* **43,** 41.
316 Morrison, A. and Cozzarelli, N.R. (1979), *Cell,* **17,** 175.
317 Gellert, M., O'Dea, M.H., Itoh, T. and Tomizawa, J. (1976), *Proc. Natl. Acad. Sci. USA,* **73,** 4474.
318 Sumida-Yasumoto, C., Yudelevich, A. and Hurwitz, J. (1976), *Proc. Natl. Acad. Sci. USA,* **73,** 1887.
319 De Wyngaert, M.A. and Hinckle, D.C. (1979), *J. Virol.,* **29,** 529.
320 McCarthy, D. (1979), *J. Mol. Biol.,* **127,** 265.
321 Liu, L.F., Liu, C.-C. and Alberts, B.M. (1979), *Nature,* **281,** 456.
322 Denhardt, D.T. (1979), *Nature,* **280,** 196.
323 Liu, L.F. and Wang, J.C. (1978), *Cell,* **15,** 979.
324 Sugino, A., Higgins, N.P., Brown, P.O., Peebles, C.L. and Cozzarelli, N.R. (1978), *Proc. Natl. Acad. Sci. USA,* **75,** 4838.
325 Gellert, M. Mizuuchi, K., O'Dea, M.H., Ohmari, H. and Tomizawa, J. (1978), *Cold Spring Harbor Symp. Quant. Biol.,* **43,** 35.
326 Beattie, K.L., Wiegand, R.C. and Radding, C.M. (1977), *J. Mol. Biol.,* **116,** 783.
327 Smith, C.L., Kubo, M. and Imamoto, F. (1978), *Nature,* **275,** 420.
328 Mizuuchi, K., Gellert, M. and Nash, H.A. (1978), *J. Mol. Biol.,* **121,** 375.
329 Marians, K.J., Ikeda, J.-E., Schlagman, S. and Hurwitz, J. (1977), *Proc. Natl. Acad. Sci. USA,* **74,** 1965.
330 Castara, F.J. and Simpson, M.V. (1979), *J. Biol. Chem.,* **254,** 11 193.
331 Scherzinger, E., Lanka, E., Morelli, G., Seiffert, D. and Yuki, A. (1977), *Eur. J. Biochem.,* **72,** 543.
332 Scherzinger, E., Lanka, E. and Hillenbrand, G. (1977), *Nucl. Acids Res.,* **4,** 4151.
333 Wickner, S. (1977), *Proc. Natl. Acad. Sci. USA,* **74,** 2815.
334 Bouché, J.P., Rowen, L. and Kornberg, A. (1978), *J. Biol. Chem.,* **253,** 756.
335 Eliasson, R. and Reichard, P. (1979), *J. Mol. Biol.,* **129,** 393.
336 Reichard, P., Rowen, L., Eliasson, R., Hobbs, J. and Eckstein, F. (1978), *J. Biol. Chem.,* **253,** 7011.
337 Kornberg, A. (1978), *Cold Spring Harbor Symp. Quant. Biol.,* **43,** 1.
338 Sims, J., Koths, K. and Dresler, D. (1978), *Cold Spring Harbor Symp. Quant. Biol.,* **43,** 349.
339 Wickner, S. (1978), *Annu. Rev. Biochem.,* **47,** 1163.
340 Sumida-Yasumoto, C., Ikeda, J.-E., Benz, E., Marians, K.J., Vicuna, R., Sugrue, S., Zipurzky, S.L. and Hurwitz, J. (1978), *Cold Spring Harbor Symp. Quant. Biol.,* **43,** 311.
341 Alberts, B. and Sternglanz, R. (1977), *Nature,* **269,** 655.

342 Gray, C.P., Sommer, R., Polke, C., Beck, E. and Schaller, H. (1978), *Proc. Natl. Acad. Sci. USA,* **75,** 50.
343 Geider, K., Beck, E. and Schaller, H. (1978), *Proc. Natl. Acad. Sci. USA,* **75,** 645.
344 Schaller, H. (1978), *Cold Spring Harbor Symp. Quant. Biol.,* **43,** 401.
345 Beck, E., Sommer, R., Averswald, E.A., Kurz, C., Zinc, B., Osterburg, G., Schaller, H., Sugimoto, K., Sugisaki, H., Okamoto, T. and Takanami, M. (1978), *Nucl. Acids Res.,* **5,** 4495.
346 Martin, D.M. and Godson, G.N. (1977), *J. Mol. Biol.,* **117,** 321.
347 Meyer, R.R., Shlomai, J., Kobori, J., Baters, D.L., Rowen, L., McMacken, R., Ueda, K. and Kornberg, A. (1978), *Cold Spring Harbor Symp. Quant. Biol.,* **43,** 289.
348 Wickner, S. and Hurwitz, J. (1976), *Proc. Natl. Acad. Sci. USA,* **73,** 1053.
349 McMacken, R. and Kornberg, A. (1978), *J. Biol. Chem.,* **253,** 3313.
350 Meyer, T.F., Geider, K., Kurz, C. and Schaller, H. (1979), *Nature,* **278,** 365.
351 Friedman, J. and Razin, A. (1976), *Nucl. Acids Res.,* **3,** 2665.
352 Friedman, J., Friedman, A. and Razin, A. (1977), *Nucl. Acids Res.,* **4,** 3483.
353 Hattman, S., Gribbin, C. and Hutchison, C.A. (1979), *J. Virol.,* **32,** 845.
354 Eisenberg, S., Griffith, J. and Kornberg, A. (1977), *Proc. Natl. Acad. Sci. USA,* **74,** 3198.
355 Koths, K. and Dressler, D. (1978), *Proc. Natl. Acad. Sci. USA,* **75,** 605.
356 Duguet, M., Yarranton, G. and Gefter, M. (1978), *Cold Spring Harbor Symp. Quant. Biol.,* **43,** 335.
357 Ikeda, J.E., Yudelevich, A. and Hurwitz, J. (1976), *Proc. Natl. Acad. Sci. USA,* **73,** 2669.
358 Eisenberg, S. and Kornberg, A. (1979), *J. Biol. Chem.,* **254,** 5328.
359 Sumida-Yasumoto, C., Yudelevich, A. and Hurwitz, J. (1976), *Proc. Natl. Acad. Sci. USA,* **73,** 1887.
360 Meyer, T.F. and Geider, K. (1970, *J. Biol. Chem.,* **254,** 12642.
361 Geider, K. and Meyer, T.F. (1978), *Cold Spring Harbor Symp. Quant. Biol.,* **43,** 59.
362 Mitra, S. and Stallions, D.R. (1976), *Eur. J. Biochem.,* **67,** 37.
363 Godson, G.N. (1977), *J. Mol. Biol.,* **117,** 353.
364 Denhardt, D.T. (1975), *J. Mol. Biol.,* **99,** 107.
365 Eisenberg, S., Scott, J.F. and Kornberg, A. (1978), *Cold Spring Harbor Symp. Quant. Biol.,* **43,** 295.
366 Katz, L., Williams, P.H., Sato, S., Leavitt, R.W. and Helinsky, D.R. (1977), *Biochemistry,* **16,** 1677.
367 Blair, D.G. and Helinsky, D.R. (1975), *J. Biol. Chem.,* **250,** 8785.
368 Backman, K., Betlach, M., Boyer, H.W. and Yanofsky, S. (1978), *Cold Spring Harbor Symp. Quant. Biol.,* **43,** 69.
369 Conrad, S.E. and Campbell, J.L. (1979), *Cell,* **18,** 61.
370 Itoh, T. and Tomizawa, J. (1978), *Cold Spring Harbor Symp. Quant. Biol.,* **43,** 409; (1980), *Proc. Natl. Acad. Sci. USA,* **77,** 2450.
371 Kolter, R., Inuzuka, M. and Helinski, D.R. (1978), *Cell,* **15,** 1199.

372 Cabello, F., Timmis, K. and Cohen, S.N. (1976), *Nature*, **259,** 285.
373 Takahashi, S. (1975). *J. Mol. Biol.*, **94,** 385.
374 Carter, B.J., Shaw, B.D. and Smith, M.G. (1969), *Biochim. Biophys. Acta*, **195,** 494.
375 Hayes, S. and Szybakski, W. (1975), in *DNA Synthesis and its Regulation* (eds M. Goulian, P. Hanawalt and C. F. Fox), W.A. Benjamin Inc., p. 486.
376 Wickner, S.H. (1978), *Cold Spring Harbor Symp. Quant. Biol.*, **43,** 303.
377 Honigman, A., Hu, S.-L., Chase, R. and Szybalski, W. (1976), *Nature*, **262,** 112.
378 Walz, A., Pirrotta, V. and Ineichen, K. (1976), *Nature*, **262,** 665.
379 Watson, J.D. (1972), *Nature New Biol.*, **239,** 197.
380 Wolfson, J. and Dressler, D. (1979), *J. Biol. Chem.*, **254,** 10490.
381 Masker, W.E. and Richardson, C.C. (1976), *J. Mol. Biol.*, **100,** 543.
382 Richardson, C.C., Romano, L.J., Kolodner, R., Le Clerc, J.E., Tamanoi, F., Engler, M.J., Dean, F.B. and Richardson, D.S. (1978), *Cold Spring Harbor Symp. Quant. Biol.*, **43,** 427.
383 De Wyngaert, M.A. and Hinkle, D.C. (1979), *J. Biol. Chem.*, **254,** 11247.
384 Mark, D.F. and Richardson, C.C. (1976), *Proc. Natl. Acad. Sci. USA*, **73,** 780.
385 Hori, K., Mark, D.F. and Richardson, C.C. (1979), *J. Biol. Chem.*, **254,** 11591.
386 Hillenbrand, G., Morelli, G., Lanka, E. and Scherzinger, E. (1978), *Cold Spring Harbor Symp. Quant. Biol.*, **43,** 449.
387 Romano, L.J. and Richardson, C.C. (1979), *J. Biol. Chem.*, **254,** 10476.
388 Romano, L.J. and Richardson, C.C. (1979), *J. Biol. Chem.*, **254,** 10482.
389 Delius, H., House, C. and Lozinski, A.W. (1971), *Proc. Natl. Acad. Sci. USA*, **68,** 3049.
390 Liu, C.F., Liu, C.C. and Alberts, B.M. (1979), *Nature*, **281,** 456.
391 Stetler, G.L., King, G.J. and Huang, W.M. (1979), *Proc. Natl. Acad. Sci. USA*, **76,** 3737.
392 Morris, C.F., Sinha, N.K. and Alberts, B.M. (1975), *Proc. Natl. Acad. Sci. USA*, **72,** 4800.
393 Liu, C.C., Burke, R.L., Hibner, U., Barry, J. and Alberts, B. (1978), *Cold Spring Harbor Symp. Quant. Biol.*, **43,** 469.
394 Morris, C.F., Hama-Inaba, H., Mace, D., Sinha, N.K. and Alberts, B. (1979), *J. Biol. Chem.*, **254,** 6787.
395 Piperno, J.R. and Alberts, B.M. (1978), *J. Biol. Chem.*, **253,** 5174.
396 Piperno, J.R., Kallen, R.G. and Alberts, B.M. (1978), *J. Biol. Chem.*, **253,** 5180.
397 Nossal, N.G. (1979), *J. Biol. Chem.*, **254,** 6026.
398 Nossal, N.G. and Peterlin, B.M. (1979), *J. Biol. Chem.*, **254,** 6032.
399 Somerville, R., Ebisuyaki, K. and Greenberg, G.R. (1959), *Proc. Natl. Acad. Sci. USA*, **45,** 1240.
400 Mathews, C.K., Brown, F. and Cohen, S.S. (1964), *J. Biol. Chem.*, **239,** 2957.
401 Zimmerman, S.B., Kornberg, S.R. and Kornberg, A. (1962), *J. Biol. Chem.*, **237,** 512.

402 Goulian, M., Lucas, Z.J. and Kornberg, A. (1968), *J. Biol. Chem.*, **243,** 627.
403 Prashad, N. and Hosoda, J. (1972), *J. Mol. Biol.*, **70,** 617.
404 Jaenisch, R., Mayer, A. and Levine, A. (1971), *Nature New Biol.*, **233,** 72.
405 Varshavsky, A.J., Sundin, O. and Bohn, M. (1979), *Cell,* **16,** 453.
406 Tjian, R. (1978), *Cell,* **13,** 165.
407 Otto, B. and Fanning, E. (1978), *Nucl. Acids Res.*, **5,** 1715.
408 Digges, I., Eason, R. and Adams, R.L.P. (1977), unpublished work.
409 Eliasson, R. and Reichard, P. (1978), *J. Biol. Chem.*, **253,** 7469.
410 Reichard, P. and Eliasson, R. (1978), *Cold Spring Harbor Symp. Quant. Biol.*, **43,** 271.
411 Eliasson, R. and Reichard, P. (1978), *Nature,* **272,** 184.
412 Rekosh, D.M.K., Russell, W.C. and Bellet, A.J.D. (1977), *Cell,* **11,** 283.
413 Winnacker, E.L. (1978), *Cell,* **14,** 761.
414 Flint, S.J., Berget, S.M. and Sharp, P.A. (1976), *Cell,* **9,** 559.
415 Kelly, J.J. and Lechner, R.L. (1978), *Cold Spring Harbor Symp. Quant. Biol.*, **43,** 721.
416 Challberg, M.D. and Kelly, T.J. (1979), *J. Mol. Biol.*, **135,** 999.
417 Revie, D., Tseng, B.Y., Grafstrom, R.H. and Goulian, M. (1979), *Proc. Natl. Acad. Sci. USA,* **76,** 5539.
418 Brison, O., Kedinger, C. and Wilhelm, J. (1977), *J. Virol.*, **24,** 423.
419 Robberson, D.L., Clayton, D.A. and Morrow, J.F. (1974), *Proc. Natl. Acad. Sci. USA,* **71,** 4447.
420 Gillum, A.M. and Clayton, D.A. (1978), *Proc. Natl. Acad. Sci. USA,* **75,** 677.
421 Berk, A.J. and Clayton, D.A. (1974), *J. Mol. Biol.*, **86,** 801.
422 Berk, A.J. and Clayton, D.A. (1976), *J. Mol. Biol.*, **100,** 85.
423 Champoux, J.J. (1978), *J. Mol. Biol.*, **118,** 441.
424 Robberson, D.L., Kasamasu, H., and Vinograd, J. (1972), *Proc. Natl. Acad. Sci. USA,* **69,** 736.
425 Bogenhagen, D. and Clayton, D.A. (1978), *J. Mol. Biol.*, **119,** 49.
426 Bogenhagen, D. and Clayton, D.A. (1978), *J. Mol. Biol.*, **119,** 69.
427 Goddard, J.M. and Wolstenholme, D.R. (1978), *Proc. Natl. Acad. Sci. USA,* **75,** 3886.
428 Matsumoto, L., Kasamatsu, H., Piko, L. and Vinograd, J. (1977), *J. Cell Biol.*, **63,** 146.
429 Goddard, J.M. and Cummings, D.M. (1977), *J. Mol. Biol.*, **109,** 327.
430 Arnberg, A.C., Van Bruggen, E.F.J., Clegg, R.A., Upholt, W.B. and Borst, P. (1974), *Biochim. Biophys. Acta,* **361,** 266.
431 Clegg, R.A., Borst, P. and Weijers, P.J. (1974), *Biochim. Biophys. Acta,* **361,** 277.
432 Arnberg, A.C., Van Bruggen, E.F.J., Borst, P., Clegg, R.A., Schutgens, R.B.H., Weigers, P.J. and Goldbach, R.W. (1975), *Biochim. Biophys. Acta,* **383,** 359.
433 Cavalieri, L.F. and Carroll, E. (1970), *Biochem. Biophys. Res. Commun.*, **41,** 1055.
434 Temin, H.M. and Mizutani, S. (1970), *Nature,* **226,** 1211.

435 Baltimore, D. (1970), *Nature,* **226,** 1209.
436 Spiegelman, S., Burny, A., Das, M., Keydar, J., Schlom, J., Travnicek, M. and Watson, K. (1970), *Nature,* **228,** 430.
437 Crick, F. (1970), *Nature,* **227,** 561.
438 Wang, R.M. (1978), *Annu. Rev. Microbiol.,* **32,** 561.
439 Erikson, E. and Erikson, R.L. (1971), *J. Virol.,* **8,** 254.
440 Harada, F., Sawyer, R.C. and Dahlberg, J.E. (1975), *J. Biol. Chem.* **250,** 3487.
441 Baltimore, D. and Smoler, D. (1971), *Proc. Natl. Acad. Sci. USA,* **68,** 1507.
442 Gilboa, E., Mitra, S.W., Goff, S. and Baltimore, D. (1979), *Cell,* **18,** 93.
443 Baltimore, D., Gilboa, E., Rottenberg, E. and Yoshimura, F. (1978), *Cold Spring Harbor Symp. Quant. Biol.,* **43,** 869.
444 Hsu, T.W., Sabran, J.L., Mark, G.E., Guntaka, R.V. and Taylor, J.M. (1978), *J. Virol.,* **28,** 810.
445 Varmus, H.E., Shank, P.R., Hughes, S.E., Kung, H.-J., Heasley, S., Majors, J., Vogt, P.K. and Bishop, J.M. (1978), *Cold Spring Harbor Symp. Quant. Biol.,* **43,** 851.
446 Varmus, H.E., Heasley, S., Kung, H.-J., Oppermann, H., Smith, V.C., Bishop, J.M. and Shank, P.R. (1978), *J. Mol. Biol.,* **120,** 55.
447 Flint, J. (1976), *Cell,* **8,** 151.
448 Haseltine, W., Kleid, D.G., Panet, A., Rottenberg, E. and Baltimore, D. (1976), *J. Mol. Biol.,* **106,** 109.
449 Coffin, J.M. and Haseltine, W.A. (1977), *J. Mol. Biol.,* **117,** 805.
450 Bishop, J.M. (1978), *Annu. Rev. Biochem.,* **47,** 35.
451 Parks, W.P., Todaro, G.J., Scholnick, E.M. and Aaronson, S.A. (1971), *Nature,* **229,** 258.
452 Gallo, R.C. and Sarin, P.S. (1971), *Nature,* **232,** 140.
453 Schlom, J., Harter, D.H., Burny, A. and Spiegelman, S. (1971), *Proc. Natl. Acad. Sci. USA,* **68,** 182.
454 Schlom, J. and Spiegelman, S. (1971), *Proc. Natl. Acad. Sci. USA,* **68,** 1613.
455 Scolnick, E.M., Aaronson, S.A. and Todaro, G.J. and Parks, W.P. (1971), *Nature,* **229,** 318.
456 Spiegelman, S. (1971), *Proc. R. Soc., B,* **177,** 87.
457 Papas, T.S., Pry, T.W. and Marciani, D.J. (1977), *J. Biol. Chem.,* **252,** 1425.
458 Hizi, A. and Joklik, W.K. (1977), *J. Biol. Chem.,* **252,** 2281.
459 Hizi, A., Leis, J.P. and Joklik, W.K. (1977), *J. Biol. Chem.,* **252,** 2290.
460 Chang, L.M.S. and Bollum, F.J. (1971), *Biochemistry,* **10,** 536.
461 Seal, G. and Loeb, L.A. (1976), *J. Biol. Chem.,* **251,** 975.
462 Battula, N., Dube, D.K. and Loeb, L.A. (1975), *J. Biol. Chem.,* **250,** 8404.
463 Gillespie, D., Saxinger, W.C. and Gallo, R.C. (1975), *Prog. Nucleic Acid Res. Mol. Biol.,* **15,** 1 (ed. W. E. Cohn).
464 Wu, A.M. and Gallo, R.C. (1976), *Crit. Rev. Biochem.,* **3,** 289.
465 Chiu, C.-S., Tomack, P.K. and Greenberg, G.R. (1976), *Proc. Natl. Acad. Sci. USA,* **73,** 757.
466 Furth, M.E. and Yates, J.L. (1978), *J. Mol. Biol.,* **126,** 227.

467 Sherer, G. (1978), *Nucl. Acids Res.*, **5,** 3141.
468 Shepard, H.M., Gelfand, D.H. and Polisky, B. (1979), *Cell,* **18,** 267.
469 Reddy, G.P.V. and Mathews, C.K. (1978), *J. Biol. Chem.*, **253,** 3461.
470 Chao, J., Leach, M. and Karam, J. (1977), *J. Virol.*, **24,** 557.
471 Reddy, G.P.V., Singh, A., Stafford, M.E. and Mathews, C.K. (1977), *Proc. Natl. Acad. Sci. USA,* **74,** 3152.
472 Liu, L.F., Liu, C.-C. and Alberts, B.M. (1980), *Cell,* **19,** 697.
473 Nossal, N.G. (1980), *J. Biol. Chem.*, **255,** 2176.
474 Wirak, D.O. and Greenberg, G.R. (1980), *J. Biol. Chem.*, **255,** 1896.
475 Sims, J. and Benz, E.W. (1980), *Proc. Natl. Acad. Sci. USA,* **77,** 900.
476 Shlomai, J. and Kornberg, A. (1980), *Proc. Natl. Acad. Sci. USA,* **77,** 799.
477 Hibner, U. and Alberts, B.M. (1980), *Nature,* **285,** 300.
478 Mizuuchi, K., Fisher, L.M., O'Dea, M.H. and Gellert, M. (1980, *Proc. Natl. Acad. Sci. USA,* **77,** 1847.
479 Kreuzer, K.N. and Cozzarelli, N.R. (1980), *Cell,* **20,** 245.
480 Bessman, M.J., Lehmann, I.R., Simms, E.S. and Kornberg, A. (1958), *J. Biol. Chem.*, **233,** 171.
481 Hashimoto-Gotoh, T. and Timmis, K.N. (1981), *Cell,* **23,** 229.
482 Stuitje, A.R., Spelt, C.E., Veltkamp, E. and Nijkamp, H.J.J. (1981), *Nature,* **290,** 264.

DNA repair and recombination

9.1 INTRODUCTION

Damage to DNA resulting in the insertion of incorrect bases or distortion of the normal double-helical structure must be corrected if the cell is to survive. This chapter considers some of the ways in which damage may arise and the various strategies the cell may adopt to rectify the damage. If the damage is quickly rectified little harm is done, but if the DNA is replicated before repair is complete alternative repair mechanisms must be called into play and these usually require recombination events.

9.2 MUTATIONS AND MUTAGENS

Alterations in the base pattern of DNA may arise in various ways. For example, existing bases may be replaced by others, or they may be deleted, or new bases may be inserted in the DNA chain. Occasional mistakes in the normal duplication of DNA give rise to *spontaneous mutations* but such mistakes are surprisingly rare [1, 2]. The frequency of such mutations depends on conditions of temperature, pH, composition of growth medium, and the like, but it can be greatly increased by exposure of cells to ultraviolet and ionizing radiations (p. 291) or to certain types of chemical which are known collectively as *mutagens*. Such substances include base analogues, some dyes of the acridine series, alkylating agents, certain antibiotics, urethane, hydroxylamine, and nitrous acid. This last substance has been used very effectively in studying mutations in certain viruses such as TMV (p. 390).

Mutagenic substances are the subject of an extensive literature [1, 3–11] which reflects the considerable effort now being devoted to attempts to inhibit cell division, especially in neoplastic tissues, by the use of compounds which might be expected to inhibit nucleic acid biosynthesis. Research in this field has been stimulated by the hope of finding a basis for an improved therapy for cancer. Reference has already been made (Chapter 7) to the use of such compounds as azaserine and the folic acid antagonists in preventing the synthesis of the purine and pyrimidine nucleotides. Some of the other substances which have been used to prevent nucleic acid biosynthesis and to bring about mutations artificially are discussed below.

9.2.1 Base and nucleoside analogues

Some of the artificially produced base analogues are incorporated into RNA and DNA and may have powerful mutagenic effects [1, 9–11]. Among the most important analogues are the halogenated pyrimidines, and those bases where nitrogen has been substituted for a —CH= group (see Fig. 9.1.).

The action of these unnatural bases seems, at least in some cases, to be twofold:

(1) They generally block some stage in the biosynthesis of the normal purine and pyrimidine nucleotides. Thus 8-azaguanine inhibits the biosynthesis of GMP and 6-mercaptopurine blocks the conversion of

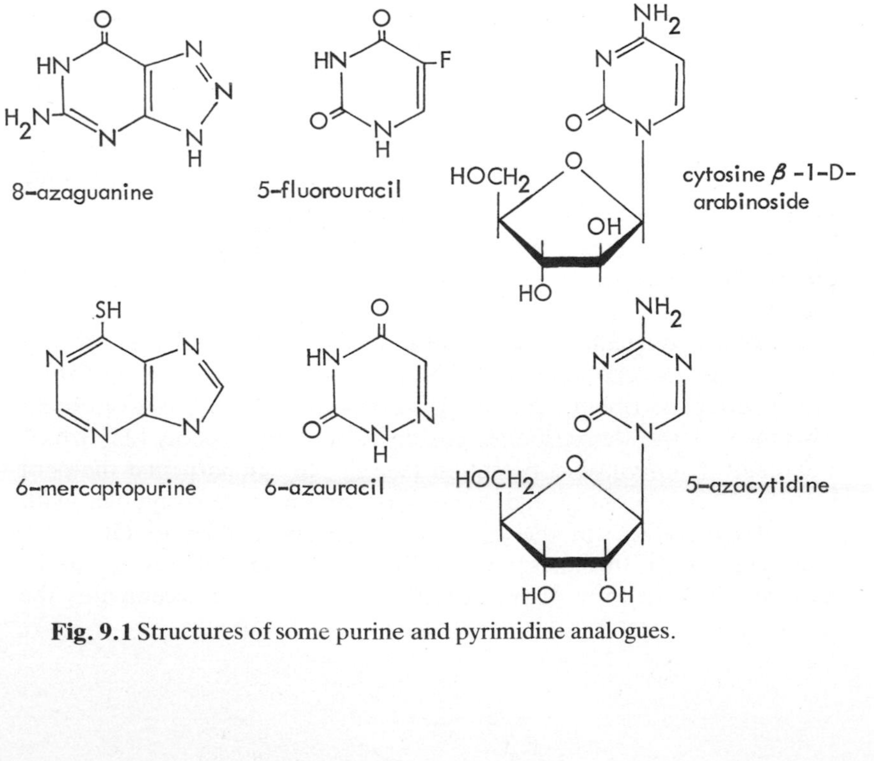

Fig. 9.1 Structures of some purine and pyrimidine analogues.

IMP into AMP [12]. In general, these inhibitions are brought about only after the inhibitor itself has been converted into its nucleotide. Thus 6-azauracil is converted first into its nucleoside (Aza-U) then into its nucleotide (Aza-UMP) which inhibits the action of orotidine 5′-phosphate decarboxylase (Fig. 7.6) and so prevents pyrimidine biosynthesis [9, 10, 13]. 5-Fluorouracil, which has proved to be a potent inhibitor of the growth of certain tumours, is converted first into its ribonucleotide (F-UMP) and then into its deoxyribonucleotide (F-dUMP) which exerts its main effect by inhibiting conversion of dUMP into dTMP [9, 10, 14] (p. 191) and hence inhibiting DNA synthesis (Fig. 7.8).

When 5-fluorodeoxyuridine is added to cells in culture it is converted into 5F-dUMP which blocks DNA synthesis. As no other processes are effected, cells progress around the cycle and accumulate at the beginning of the S-phase (p. 83). The inhibition can be overcome by the addition of thymidine with the result that a population of cells in synchronized growth result [9, 15].

(2) They are themselves, after conversion into nucleotides, incorporated to varying degrees into RNA and/or DNA although the incorporation may take an abnormal form. Thus 8-azaguanine can be incorporated at the expense of guanine into the RNA of TMV [16] and, to a much larger extent, into the RNA of *B. cereus* [17]. Only very small amounts are incorporated into the DNA.

5-Azacytidine is incorporated into RNA but this rapidly interferes with protein synthesis [11, 18]. 5-Azadeoxycytidine is incorporated into DNA but this renders the cells non-viable [19] or alters the growth characteristics [20]. 5-Bromouracil can replace thymine in DNA where it normally base-pairs with adenine. However, in its rare enol-state (which bromodeoxyuridine (in the enol form) occurs leading to GC→AT instead of adenine so bringing about the base-pair transition A-T into G-C. DNA containing 5-bromouracil instead of thymine is very susceptible to breakage at light-induced bromouracil dimers [21] (see below).

The main mutagenic effects of bromodeoxyuridine, however, do not arise from its incorporation into DNA in place of thymidine but rather from its effect on ribonucleotide reductase (see p. 188). This leads to an imbalance in the deoxyribonucleoside triphosphate pools [22] which may result in base misincorporation (see p. 244). In particular the pool size of dCTP is reduced and some substitution of deoxycytidine with bromodeoxyuridine (in the enol form) occurs leading to GC→AT transitions as the misincorporated bromodeoxyuridine reverts to its more stable keto form [116]. Addition of thymidine accentuates the mutagenic effect of bromodeoxyuridine by further decreasing the dCTP pool size whereas deoxycytidine antagonizes the mutagenic effects.

2-Aminopurine can also base pair with either cytosine or thymine and, because it inhibits adenosine deaminase it also increases the pool size of dATP which in turn increases the size of the dCTP pool relative to the dTTP pool leading to misincorporation of cytosine opposite adenine or 2-aminopurine [116].

The D-arabinosyl nucleosides are effectively analogues of deoxyribonucleosides (e.g. cytosine β-D-arabinoside is incorporated in place of deoxycytidine into DNA where it causes chain termination or a marked reduction in the rate of further chain extension [9–11, 21, 23].

9.2.2 Alkylating agents

The alkylating agents exert a variety of biological effects including mutagenesis, carcinogenesis, and tumour growth inhibition [1, 4, 24, 26, 53]. They all carry one, two, or more alkyl groups in reactive form and include the well-known compounds sulphur mustard or di-(2-chloroethyl) sulphide and nitrogen mustard or methyl-di-(2-chloroethyl) amine.

The action of the alkylating agents on DNA is complex. They are known to react with purine bases, particularly with guanine at the N-7 atom, and the bifunctional alkylating agents (i.e. those with two reactive alkyl groups) may thus bring about cross-linking between the opposing strands in the DNA molecule. Alkylation of purines in position 7 also gives rise to unstable quaternary nitrogens so that the alkylated purine may separate from the deoxyribose leaving a gap which might interfere with DNA replication or cause the incorporation of the wrong base [1, 24]. The phosphate groups may also be alkylated. The phosphate triester so formed is unstable and may hydrolyse between the sugar and the phosphate so that the DNA chain is broken.

The relationship between DNA alkylation and carcinogenesis (tumour induction) is tenuous though certain alkylations do cause miscoding (i.e. are mutagenic). This is not so for 7-methylguanine but

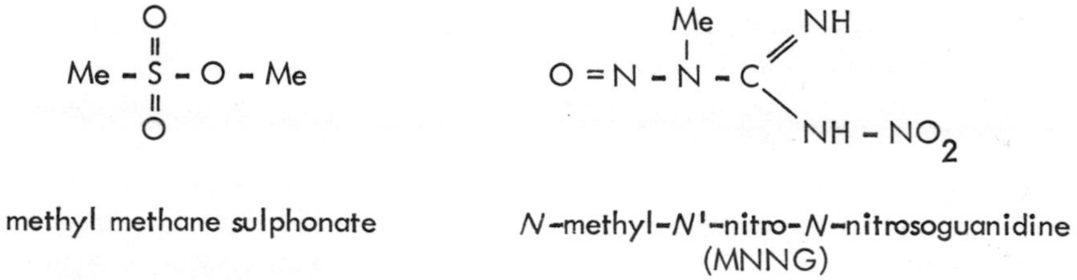

Fig. 9.2 Alkylation agents.

applies to the presence of 3-methylcytosine in the RNA of tobacco mosaic virus [4]. Good alkylating agents e.g. dimethyl sulphate and methylmethane sulphonate are poor carcinogens but the reverse is true for *N*-methyl-*N*′-nitro-*N*-nitrosoguanidine (MNNG) (Fig. 9.2). 2-*O*-methylated bases may cause miscoding as may also 6-*O*-methylguanine and 3-*N*-methyladenine [26].

9.2.3 Antibiotics and allied agents

Some of the carcinostatic antibiotics have been of great value in the study of the nucleic acid biosynthesis [25]. One of the most useful is actinomycin D (Fig. 9.3) which forms complexes with the deoxyguanosine residues in DNA and so blocks it as a template. Actinomycin D therefore inhibits both the DNA polymerase and the DNA-dependent RNA polymerase (p. 309), the former being much less

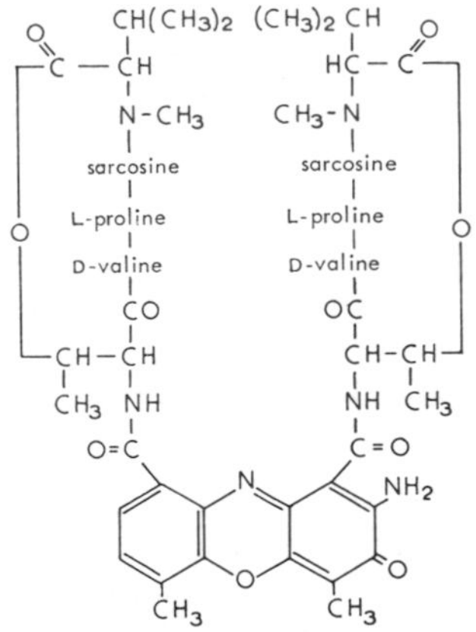

Fig. 9.3 Structure of actinomycin D.

sensitive to its action than the latter [27]. At a concentration of actinomycin of 1.0 μM, for example, the DNA-dependent RNA polymerase is almost completely inhibited whereas the DNA polymerase is only slightly affected [28]. At lower concentrations the synthesis of the different sorts of RNA is affected to different extents, ribosomal RNA being particularly sensitive [29]. In the complex formed between actinomycin and DNA the drug is pictured as being intercalated between alternating dG-dC nucleotide pairs with the cyclic peptide in the groove hydrogen-bonded to guanine [8].

Mitomycin C inhibits bacterial DNA synthesis by causing covalent cross-linking of the complementary DNA strands [10, 30]. The drug is reduced in the cell to produce an active bifunctional alkylating agent which cross-links guanine residues [31].

Phleomycin attaches covalently to thymine residues in DNA [10, 32] to inhibit DNA polymerase action [33]. As with bleomycin, which binds in a similar way, this results in single-strand breaks in the DNA [10, 34].

The antibiotic streptonigrin inhibits both DNA and RNA synthesis by inhibiting the respective polymerases. Edeine and nalidixic acid inhibit bacterial DNA synthesis, the latter by inhibiting DNA gyrase. Antibiotic action is discussed further on p. 421 and is reviewed in [10, 35].

9.2.4 Dyes

Proflavine (Fig. 9.4), one of the acridine seris of dyes, inhibits DNA-dependent RNA biosynthesis by its molecules becoming intercalated between adjacent nucleotide-pair pairs in the DNA molecule [36, 37]. Its mutagenic action on phages has been applied effectively in the study of coding triplets (p. 386). The acriflavines (also acridine dyes),

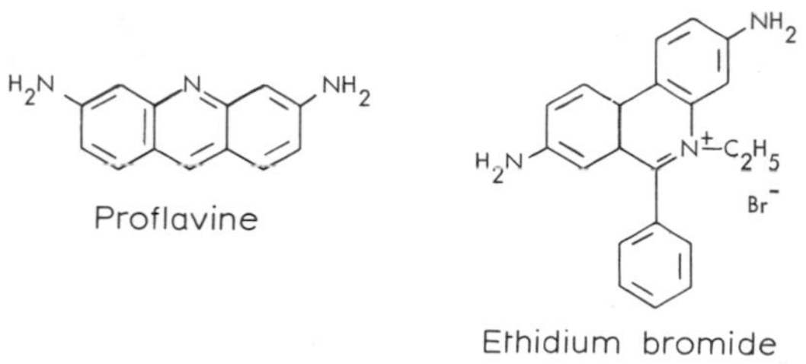

Fig. 9.4 Structures of proflavine and ethidium bromide.

ethidium bromide (Fig. 9.4) and propidium di-iodide also intercalate between DNA nucleotide pairs and in so doing bring about some unwinding of the double helix. This unwinding is resisted in closed circular supercoiled molecules (see p. 28) which therefore bind less dye and are therefore considerably denser than open circle (nicked) molecules from which they can be separated in gradients of caesium chloride containing the dye [38].

9.2.5 The effects of ionizing radiations

It has long been known that rapidly growing tissues are much more sensitive to the action of X-rays than are adult tissues, and it is generally recognized that irradiation exerts a pronounced inhibitory influence on the process of cell division. It might therefore be expected that irradiation would exert an appreciable effect on the metabolism and biosynthesis of the nucleic acids.

The effect of ionizing radiations on living cells is the subject of a vast literature which cannot be discussed here but those aspects of the problem which concern nucleic acid metabolism have been reviewed by several authors [39, 40].

Pelc and Howard [41] first showed that incorporation of ^{32}P into DNA of bean *(Vicia faba)* root cells was inhibited if the root tips were irradiated during the G_1 period but that much higher doses were required if the exposure was during the S period (see p. 83). In growing cells there is an association between the onset of DNA synthesis and resistance to irradiation [42]. Thereafter survival rates remain approximately constant throughout this period and then decline in G_2. Although DNA synthesis is an important factor in radiation sensitivity, it is not the only one as is shown by changes in sensitivity which occur following the addition in G_1 phase of inhibitors of DNA synthesis [43]. Similar results found with bone marrow cells by Lajtha and his colleagues led them to postulate the existence of a 'system connected with but not identical with DNA synthesis which is more radiosensitive than the process of DNA synthesis' [44], and studies with regenerating rat liver have given strong support of this opinion. The most sensitive period in the cell cycle is the period immediately prior to mitosis. Cells irradiated at this time with as little as 9 rads show a delay in entering mitosis [45].

What may be of importance is the distinction between rapidly growing cells which appear to possess throughout the cell cycle all the enzymes required for DNA synthesis, and non-growing (G_0) cells which must synthesize these enzymes when stimulated to grow. Regenerating liver and primary cell cultures fall into this second category and the initiation of DNA synthesis is very sensitive to X-radiation in such systems.

9.2.6 Ultraviolet radiation

Large doses of ultraviolet radiation can damage living cells by causing the formation of chemical bonds between adjacent pyrimidine nucleotides in the DNA. Two pyrimidine bases joined in this way in one strand form what is known as a *dimer*, and of the three possible types of pyrimidine

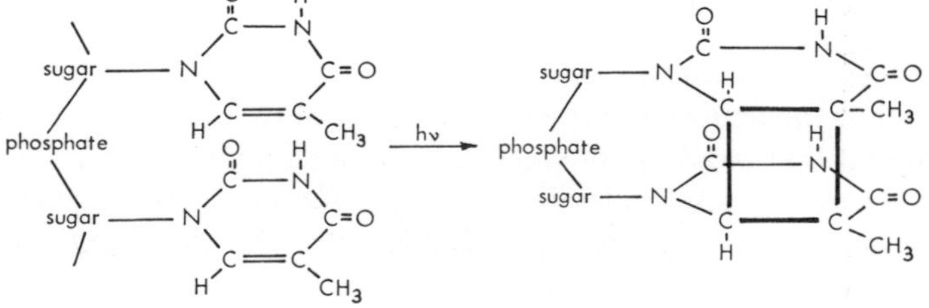

Fig. 9.5 The formation of a thymine dimer under the influence of ultraviolet light.

dimer, the thymine dimer, is formed most readily (Fig. 9.5). The presence of such dimers blocks the action of the DNA polymerase and so prevents replication [46–48].

9.3 RAPID REPAIR MECHANISMS

It is important when DNA is damaged, or when mistakes occur in its synthesis, that it is rapidly repaired. There are a number of mechanisms of repair depending partly on the type of damage incurred.

9.3.1 Photoreactivation

When bacteria damaged by ultraviolet light are exposed to an intense source of visible light (wavelengths between 320 and 370 nm) a large proportion of the damaged cells recover. This process is known as *photoreactivation* and is due to the activation by visible light of an enzyme which cleaves the pyrimidine dimers and restores the two bases to their original form. This photo-reactivating enzyme has been obtained in pure form [49], and in *E. coli* it is the product of the *phr* gene [50]. A similar enzyme occurs in eukaryotes [51, 52].

9.3.2 Excision repair

This more complex repair mechanism acts by excising the damaged section of DNA and then repairing the resultant gaps. The reaction takes place in a number of separate stages of which the first is unique to repair of DNA (Fig. 9.6). (i) An endonuclease recognizes the local distortion and breaks the adjoining phosphodiester bond so as to introduce a nick on the 5′-side of the dimer with a 3′-hydroxyl terminus at the nick. The nature of the endonuclease will be considered in more detail below. (ii) A second enzyme excises a short stretch of the DNA strand including the dimer. (iii) DNA polymerase uses the intact complementary strand as template to synthesize a piece of DNA to fill the gap.

In *E. coli* DNA polymerase I may be the enzyme which first excises the portion of the affected strand (including, on average, 30 nucleotides [54]) by virtue of its 5′→3′ nuclease action (p. 240) and then fills the gap [55–57]. *pol* A^- mutants show increased u.v. sensitivity [56] and, although they do manage to repair lesions, the patch size is considerably bigger than in the wild type [58–60]. This may be because DNA polymerases II and III take over the role of polymerase I or that repair is occurring by a recombination mechanism (see below) [61]. (iv) The repair is completed by ligase action [48, 62–64].

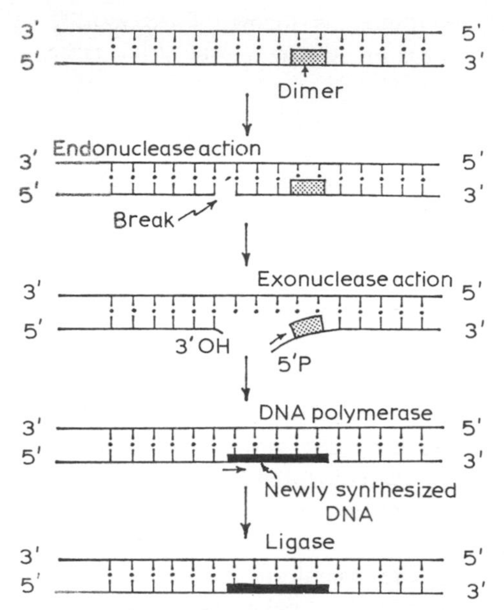

Fig. 9.6 The repair of DNA damaged by ultraviolet light. The action of u.v. has been to produce a dimer which is excised by the sequential action of an endonuclease and an exonuclease leaving a short gap which is filled in by the action of DNA polymerase. The final join is effected by polynucleotide ligase.

The endonucleolytic step may be performed by one of a number of different enzymes depending on the nature of the damage. Thymine dimers are recognized by a specific endonuclease which appears to recognize the particular distortion [65, 66]. In *E. coli* this is endonuclease III which involves he genes *uvr*A *uvr*B and *uvr*C in its synthesis. The product of *uvr*C appears to control the action of DNA ligase thus preventing premature closure of the nick [67].

Endonuclease V of *E. coli* recognizes DNA containing uracil (see p. 217) and cuts to the 5′ side of the unusual base [68]. However, a more common mechanism for repairing DNA containing uracil or hypoxanthine or alkylated bases is to first remove the base from the sugar phosphate backbone by means of an *N*-glycosidase [68–75]. The depurinated or depyrimidinated DNA is then recognized by an endonuclease III which involves the genes *uvr*A, *uvr*B and *uvr*C in its *coli* this apurinic nuclease is endonuclease II which is identical with exonuclease III, i.e. has 3′, 5′-exonuclease and phosphatase activities [77]. It has a molecular weight of 28–32 000 and is closely associated with *N*-glycosidase [76]. The role of *N*-glycosidases has been reviewed by Lindahl [78].

Recently a repair mechanism involving the insertion of a purine into depurinated DNA has been reported [79, 80].

The mechanism of repair synthesis is defective in the skin fibroblasts of patients suffering from the condition known as *xeroderma pigmentosum* [81–83]. The repair endonuclease is missing and such people are therefore abnormally sensitive to exposure to sunlight and tend readily to develop skin cancer [51, 84–87].

9.4 POST REPLICATION REPAIR

If repair is quickly and accurately effected before the damaged DNA is replicated the integrity of the genetic message is maintained. Such excision repair is very effective, of high fidelity and leads to the insertion of only small repair patches. If excision repair is slow or defective, as in uvr^- cells or in patients with *xeroderma pigmentosum* or *actinic keratosis*[88] then the cell tries to replicate the damaged region of DNA. When the advancing polymerase reaches a lesion such as a thymine dimer it stops and a gap is left in the daughter strand opposite the lesion. In *E. coli* such damage induces the so called sos functions [52, 89]. Repair now requires a recombination event between the two daughter DNA molecules (see p. 296). In this type of repair a preformed patch supplied by another copy of the DNA is used to replace the defective segment [90]. In *E. coli* recombination requires the products of the genes *rec* A, *rec* B, and *rec* C and leads to large repair patches. The *rec* B and *rec* C genes specify a nuclease (exonuclease V) which can act as an endo- or an exo-nuclease [91, 92] or an ATP-dependent DNA unwinding protein [93, 94]. The *rec* BC nuclease has two subunits, α and β, and both *rec* B and *rec* C are involved in the synthesis of β [95]. The *rec* A gene product has many functions among which is an ability to catalyse denaturation of DNA in the presence of single-stranded DNA [96–98].

An alternative mechanism of post-replicational repair is an error-prone system induced by damaged DNA and this may be the origin of many of the mutations occurring as a result of damage to DNA. In *E. coli* u.v.-induced mutations are dependent on *rec*A and also *lex*A and their origin is consistent with a decreased fidelity of DNA polymerase allowing nucleotide incorporation across gaps [51, 52, 89]. This decreased fidelity is not, however, restricted to the regions of DNA actually damaged by the u.v. radiation but occurs generally within the cell [99].

The non-viability of *pol* A *rec* A double mutants and some *pol* A *rec* B double mutants points to the existence of two alternative mechanisms of dark repair of damaged DNA. Only in the absence of both mechanisms (i.e. in the double mutant) does damaged DNA accumulate resulting in cell death.

There are many *E. coli* mutants known which show an increased frequency of mutations. Some of these have already been considered under proof reading. Others are involved in mismatch correction. If a section of DNA duplex contains a mismatch but both bases are normal (e.g. a T opposite a C) some signal must exist if the cell is to be able to correct the damage. It has been suggested that one of the functions of DNA methylation (see p. 164) is to define which is the parental strand. Methylation occurs shortly after synthesis of DNA but for a finite time the daughter strand is unmethylated and could be subject to mismatch repair. Dam mutants which have a much reduced level of DNA adenine methylation show increased rates of mutation [99, 100, 118].

Other mechanisms for DNA repair have also been postulated, e.g. repair by strand displacement and branch migration at the replication fork so that the intact daughter strand may act as a template to bridge the gap opposite the damage [101].

9.5 RECOMBINATION

Recombination is the production of new DNA molecule(s) from two parental DNA molecules, such that the new DNA molecules carry genetic information derived from both parental DNA molecules. Recombination involves a physical rearrangement of the parental DNA and may occur during (1) a mixed infection of two viruses or plasmids carrying different genetic markers, (2) crossing over at meiosis in eukaryotic cells, (3) integration of bacteriophage, viral or plasmid DNA into a host cell chromosome, (4) transformation or (5) post-replication repair.

Usually recombination occurs between regions of DNA which are wholly or largely complementary in sequence. Occasionally illegitimate recombination occurs as in the integration of transposons (see p. 64) where very little complementarity exists between the incoming DNA and the recipient genome. In site-specific recombination (see below) complementarity exists over only a very limited region.

9.5.1. General recombination [107]

This can take place anywhere along the length of two complementary DNA molecules. It has been studied in fungi where the products of meiosis are recovered in four or eight cells, and in *E. coli* infected with two bacteriophages carrying different genetic markers when multiple recombination events may occur. In *E. coli* general recombination is dependent on the host *rec* system as well as other enzymes and proteins involved in DNA replication. With bacteriophage T4 gapped hybrid intermediates are formed (with gaps of up to 300 nucleotides long [102]) but these are not found in recombination with bacteriophage λ where the bacteriophage *red* genes are also involved [103, 104] (see below).

The *rec* A protein, among other functions is able to catalyse the annealing of single-stranded DNA to a homologous duplex at the expense of ATP hydrolysis [96–98, 105, 117]. To do this it acts as an unwinding protein and leads to D loop formation (see Fig. 9.7). This function is probably vital for integration of single-stranded DNA arising from conjugal transfer (p. 61) or bacterial transformation (p. 59) and is probably followed by endonuclease or exonuclease action to eliminate single-stranded regions. The *rec* BC nuclease (exonuclease V) may have this function (see p. 160).

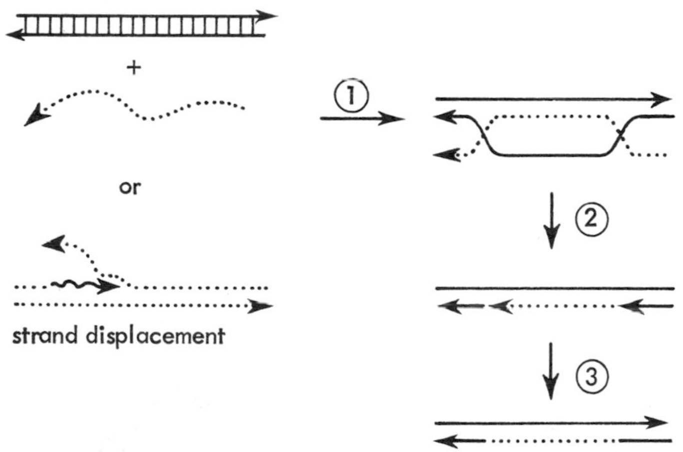

Fig. 9.7 Recombination involving D loops: (1) D loop formation catalysed by *rec* A protein, (2) endonuclease and/or exonuclease action, (3) ligase action.

Alternatively single strands may be produced by a strand-displacement mechanism acting from a nick in duplex DNA (see Fig. 9.7). High rates of recombination are found in mutants defective in ligase, DNA polymerase I or dUTPase [106], situations which lead to an increased frequency of nicks or gaps in DNA (see p. 217). In this scheme more complex heteroduplex recombinants can be formed by a crossing over of strands and branch migration [107, 108].

Holliday [109] was the first to postulate such models (Fig. 9.8) which envisage formation of regions of hybrid or heteroduplex DNA following endonuclease action and pairing of the broken strands with their complement in the other duplex. The sites of the second breaks determine whether the original parental duplexes are restored (with an intermediate region of hybrid DNA) or whether the hybrid DNA links duplex regions from the two parents. If the hybrid DNA is not a perfect match then (a) correction may occur or (b) segregation will occur at the next round of DNA synthesis.

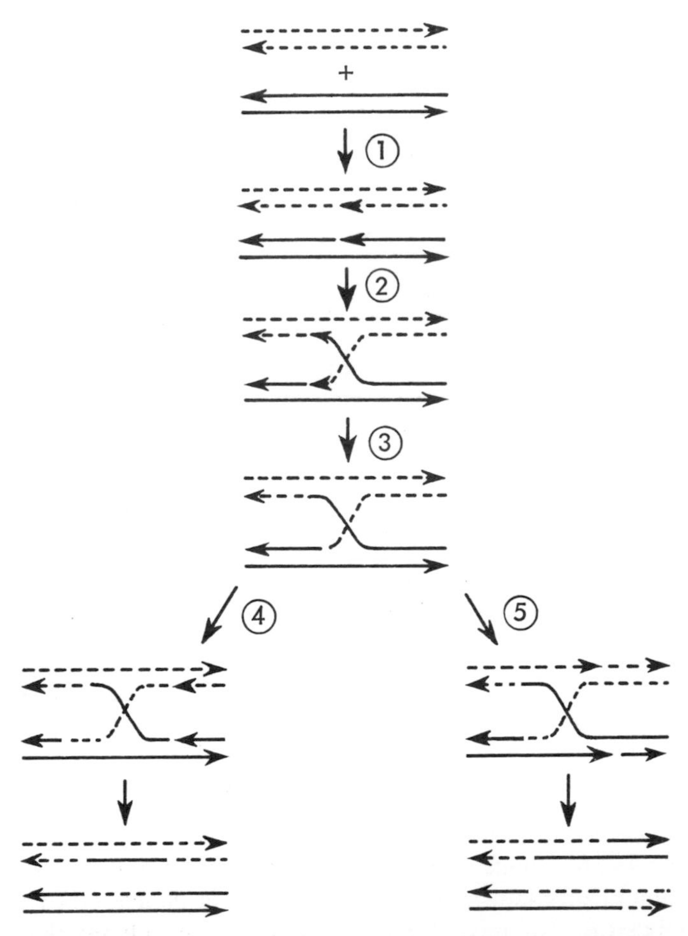

Fig. 9.8. Holliday model of recombination. Identical strands of two duplexes are nicked (1) and pair with the complementary strands from the other duplex (2). Ligase seals the new joint (3). A second nick in the same strands (4) leads to insertions of a short length of 'foreign' DNA into an otherwise homologous duplex. If the second nick occurs in the other strands (5) then the resulting duplexes are both derived from parts of each parental duplex which are joined by a region of hybrid DNA.

The *red* genes of bacteriophage λ code for a 5′→3′ exonuclease (*red* α) and the β-protein (*red* β) of unknown function [104]. The exonuclease can act on redundant single strands but acts preferentially on 5′-phosphate ends of duplex DNA to produce molecules with 3′ single-stranded tails. Annealing of homologous single-stranded regions followed by exonuclease action leads to formation of a hybrid molecule by a non-reciprocal method i.e. only one complete DNA molecule results (Fig. 9.9). If the tails are not removed by exonuclease [103] but endonuclease and polymerase intervene (as in bacteriophage T7 replication, see p. 263) a reciprocal recombination can occur [110].

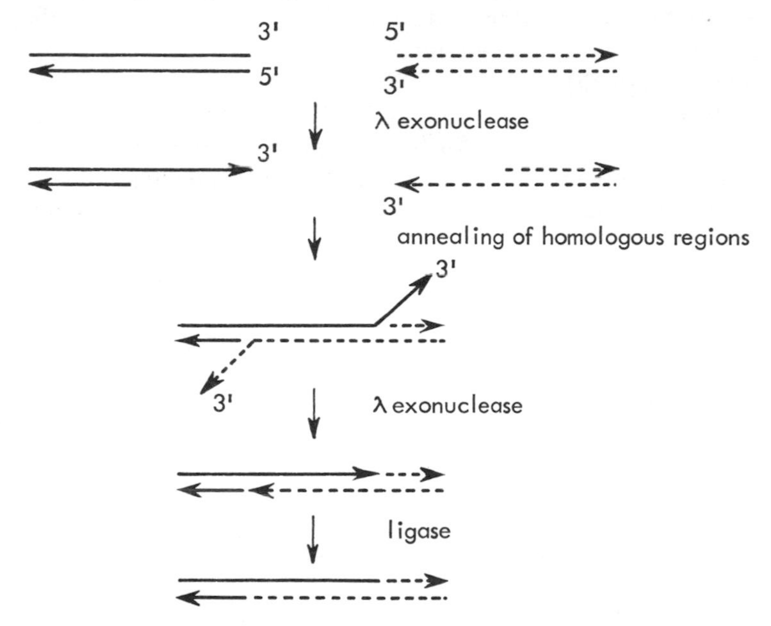

Fig. 9.9 Action of λ exonuclease to produce 'lap-jointed' molecules.

9.5.2 Site specific recombination

Although this has been mostly studied in the lambdoid bacteriophages it may also occur with the F-factor plasmids which integrate into the bacterial chromosome at a limited number of sites to produce the Hfr state [111] (see p. 63).

In the lysogenic state bacteriophage λ is integrated at a specific site in the bacterial chromosome where it is aligned in a specific orientation. Integration does not reqire the *rec* or *red* systems of general recombination but requires specific sequences on both the bacteriophage and bacterial DNA (the *att* sites) and a bacteriophage-coded protein. The bacteriophage *int* gene codes for an integrase; a 42000 molecular weight protein with nicking closing activity which binds specifically to DNA carrying the *att* site [112]. The bacteriophage *xis* gene codes for a protein which is required, along with the integrase, for excision, but whose function is unclear.

Bacteriophage Ø80 integrates into the *E. coli* genome at the site *att* 80 (near *lac*) and bacteriophage λ at the site *att* λ (between *gal* and *bio*). Both the site on the bacteriophage DNA and on the bacterial DNA are made up of three parts: POP′ and BOB′ respectively. The O region is a sequence of 15 bases which is homologous in the two genomes and when recombination occurs as first suggested by Campbell [113–115] the two *att* sites are split and recombined to give BOP′ and POB′, i.e. O is retained (Fig. 9.10).

Incorrect excision leads to defective bacteriophage carrying genetic markers adjacent to *att* i.e. for bacteriophage λ either *gal* or *bio*. Such defective transducing bacteriophage, e.g. λ *dgal* can integrate into rec^+ cells at a *gal* site and have proved very useful to viral geneticists (see p. 75).

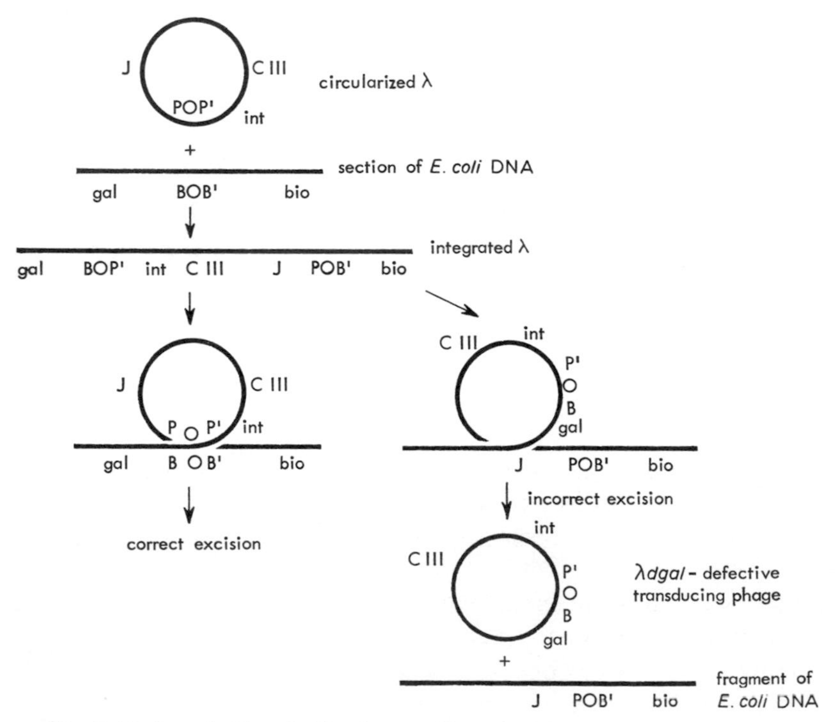

Fig. 9.10 Campbell model for site specific recombination on bacteriophage λ.

REFERENCES

1 Freeze, E. (1963), *Molecular Genetics* (ed. J.H. Taylor), Academic Press, New York. Part 1, p. 207.
2 Pauling, L. (1964), *Bull. N.Y. Acad. Med.*, **40,** 334.
3 Neuss, N., Gorman, M. and Johnson, I.S. (1967), *Mehods in Cancer Res.*, Vol. III, p. 633 (ed. H. Busch).
4 Singer, B. (1975), *Prog. Nucleic Acid. Res. Mol. Biol.*, **15,** 219 (ed. W.E. Cohn), Academic Press, New York.
5 Bennett, L.L. and Montgomery, J.A. (1967), *Methods in Cancer Res.*, Vol. III, p. 549 (Ed. H. Busch).
6 Stock, J.A. (1975), *Biology of Cancer,* 2nd Edn (eds. E.J. Ambrose and F.S.C. Roe), Ellis Horwod Ltd, p. 279.
7 Berenblum, I. (1974), *Carcinogenesis as a Biological Problem, Frontiers of Biology,* Vol. 34, North Holland Publ. Co.
8 Sobell, H.M. (1973), *Prog. Nucleic Acid Res. Mol. Biol.*, **13,** 153 (eds. J.N. Davidson and W.E. Cohn).
9 Walker, R.T., De Clercq, E. and Eckstein, F. (1979), *Nucleoside Analogues,* Plenum Press, New York.
10 Kersten, H. and Kersten, W. (1974), *Inhibitors of Nucleic Acid Synthesis,* Chapman and Hall, London.
11 Suhadolnik, R.J. (1979), *Prog. Nucleic Acid Mol. Biol.*, **22,** 193 (ed. W.E. Cohn), Academic Press, New York.
12 Lasnitski, J., Matthews, R.E.F. and Smith, J.D. (1954), *Nature,* **173,** 346.
13 Lipmann, F. (1963), *Prog. Nucleic Acid Res.*, **1,** 135 (eds. J.D. Davidson and W.E. Cohn), Academic Press, New York.
14 Loeb, L.A., Fransler, B., Williams, R. and Mazia, D. (1969), *Exp. Cell Res.*, **57,** 298.
15 Priest, J.H., Heady, J.E. and Priest, R.E. (1967), *J. Cell Biol.*, **35,** 483.
16 Matthews, R.E.F. (1953), *Nature,* **171,** 1065.
17 Hilmoe, R.J. and Hepplell, L.A. (1957), *J. Am. Chem. Soc.*, **79,** 4810.
18 Zain, B.S., Adams, R.L.P. and Imrie, R.C. (1973), *Cancer Res.*, **33,** 40.
19 Adams, R.L.P., unpublished results.
20 Taylor, S.M. and Jones, P.A. (1979), *Cell,* **17,** 771.
21 Roy-Bowman, P. (1970), *Analogues of Nucleic Acid Components,* Springer Verlag, New York.
22 Kaufman, E.R. and Davidson, R.L. (1978), *Proc. Natl. Acad. Sci. USA,* **75,** 4982.
23 Hunter, T. and Francke, B. (1975), *J. Virol.*, **15,** 759.
24 Lawley, P.D. and Brookes, P. (1961), *Nature,* **192,** 1081.
25 Goldberg, I.H. and Friedman, P.A. (1971), *Annu. Rev. Biochem.*, **40,** 775.
26 Lijinsky, W. (1976), *Prog. Nucleic Acid Res. Mol. Biol.*, **17,** 247 (ed. W.E. Cohn), Academic Press, New York.
27 Reich, E. and Goldberg, I.H. (1964), *Prog. Nucleic Acid Res.*, **3,** 184 (eds. J.N. Davidson and W.E. Cohn), Academic Press, New York.
28 Hurwitz, J. and August, J.T. (1963), *Prog. Nucleic Acid Res.*, **1,** 59 (eds. J.N. Davidson and W.E. Cohn), Academic Press, New York.

29 Penman, S., Vesco, C. and Penman, M. (1968), *J. Mol. Biol.*, **34,** 49.
30 Iyer, V.N. and Szybalski, W. (1963), *Proc. Natl. Acad. Sci. USA*, **50,** 355.
31 Goldberg, I.H. and Friedman, P.A. (1971), *Annu. Rev. Biochem.*, **40,** 775.
32 Pietsch, P. and Garrett, H. (1968), *Nature*, **219,** 488.
33 Falaschi, A. and Kornberg, A. (1964), *Fed. Proc. Fed. Am. Soc. Exp. Biol.*, **23,** 940.
34 Stern, R., Rose, J.A. and Friedman, R.M. (1974), *Biochemistry*, **13,** 307.
35 Franklin, T.J. and Snow, G.A. (1981), *Biochemistry of Antimicrobial Action*, 3rd Edn., Chapman and Hall, London.
36 Goldberg, I.H., Reich, E. and Rabinowitz, M. (1963), *Nature*, **199,** 44.
37 Lerman, L.S. (1964), *J. Cell Comp. Physiol.*, **64,** Suppl. 1, 1.
38 Radloff, R., Bawer, and Vinograd, J. (1967), *Proc. Natl. Acad. Sci. USA*, **57,** 1514.
39 Kanazir, D. (1969), *Prog. Nucleic Acid Res. Mol. Biol.*, **9,** 117 (eds. J.N. Davidson and W.E. Cohn), Academic Press, New York.
40 Ord, M.G. and Stocken, L.A. (1968), *Proc. R. Soc. Edin., B*, **70,** 117.
41 Pelc, S.R. and Howard, A. (1955), *Radiat. Res.*, **3,** 135.
42 Terasima, T. and Tolmach, L.J. (1963), *Science*, **140,** 490.
43 Sinclair, W.K. (1967), *Proc. Natl. Acad. Aci. USA*, **58,** 115.
44 Lajtha, L.G. (1960), *The Nucleic Acids* (eds. E. Chargaff and J.N. Davidson), Academic Press, New York, Vol. 3, p. 527.
45 Puck, T.T. and Steffen, J. (1963), *Biophys. J.*, **3,** 379.
46 Setlow, R.B. (1968), *Prog. Nucleic Acid Res. Mol. Biol.* **9,** 257 (eds. J.N. Davidson and W.E. Cohn), Academic Press, New York.
47 Sueoka, N. (1967), *Molecular Genetics*, Part II, p. 1 (ed. J.H. Taylor), Academic Press, New York.
48 Hanawalt, P.C. (1972), *Endeavour*, **31,** 83.
49 Sutherland, B.M., Chamberlin, M.J. and Sutherland, J.C. (1973), *J. Biol. Chem.*, **248,** 4200.
50 Setlow, J.K and Setlow, R.B. (1963), *Nature*, **197,** 560.
51 Lehman, A.R. and Bridges, B.A. (1977), *Essays in Biochemistry*, **13,** 71.
52 Hanawalt, P.C., Cooper, P.K., Ganesan, A.K. and Smith, C.A. (1979), *Annu. Rev. Biochem.*, **48,** 783.
53 Roberts, J.J. (1975), *Biology of Cancer* (eds. E.J. Ambrose and F.J.C. Roe), Ellis Horwood Ltd., p. 180.
54 Setlow, R.B. and Carrier, W.L. (1964), *Proc. Natl. Acad. Sci. USA*, **51,** 226.
55 Kelly, R.B., Atkinson, M.R., Huberman, J.A. and Kornberg, A. (1969), *Nature*, **224,** 495.
56 Monk, M., Peacey, M. and Gross, J.D. (1971), *J. Mol. Biol.*, **58,** 623.
57 Kato, T. and Kondo, S. (1970), *J. Bact.*, **104,** 871.
58 Cooper, P.K. and Hanawalt, P.C. (1972), *J. Mol. Biol.*, **67,** 1.
59 Glickman, B.W. (1974), *Biochim. Biophys. Acta*, **335,** 115.
60 Dorson, J.W., Deutsch, W.A. and Moses, R.E. (1978), *J. Biol. Chem.*, **253,** 660.
61 Grossman, L. (1974), *Adv. Radiat. Biol.*, **4,** 77.
62 Hornsey, S. and Howard, A. (1956), *Annu. N.Y. Acad. Sci.*, **63,** 915.

63 Kushner, S.R., Kaplan, J.C., Ono, H. and Grossman, L. (1971), *Biochemistry,* **10,** 3325.
64 Kaplan, J.C., Kushner, S.R. and Grossman, L. (1971), *Biochemistry,* **10,** 3315.
65 Waldstein, E.A., Peller, S. and Setlow, R.B. (1979), *Proc. Natl. Acad. Sci. USA,* **76,** 3746.
66 Radman, M. (1976), *J. Biol. Chem.,* **251,** 1438.
67 Seeberg, E. (1978), *Proc. Natl. Acad. Sci. USA,* **75,** 2569.
68 Gates, F.T. and Linn, S. (1977), *J. Biol. Chem.,* **252,** 1647.
69 Talpaert-Barle, M., Clerici, L. and Campagnari, F. (1979), *J. Biol. Chem.,* **254,** 6387.
70 Wist, E., Uhjem, O. and Krokan, H. (1978), *Biochim. Biophys. Acta,* **520,** 253.
71 Tamanoi, F. and Okazaki, T. (1978), *Proc. Natl. Acad. Sci. USA,* **75,** 2195.
72 Karran, P. and Lindahl, T. (1978), *J. Biol. Chem.,* **253,** 5877.
73 Laval, J. (1977), *Nature,* **269,** 829.
74 Kirtikar, D.M. and Goldthwait, D.A. (1974), *Proc. Natl. Acad. Sci. USA,* **71,** 2022.
75 Kirtikar, D.M., Dipple, A. and Goldthwait, D.A. (1975), *Biochemistry,* **14,** 5548.
76 Verly, W.G. and Rassart, E. (1975), *J. Biol. Chem.,* **250,** 8214.
77 Weiss, B. (1976), *J. Biol. Chem.,* **251,** 1896.
78 Lindahl, T. (1979), *Prog. Nucleic Acid Res. Mol. Biol.,* **22,** 135 (ed. W.E. Cohn), Academic Press, New York.
79 Livneh, Z., Elad, D. and Sperling, J. (1979), *Proc. Natl. Acad. Sci. USA,* **76,** 1089.
80 Deutsch, W.A. and Linn, S. (1979), *J. Biol. Chem.,* **254,** 12099.
81 Regan, J.D. (1971), *Science,* **174,** 147.
82 Cleaver, J.E. (1970), *J. Invest. Dermatol.,* **54,** 181.
83 Bootsma, D., Mulder, M.P., Pot, F. and Cohen, J.A. (1970), *Mutation Res.,* **9,** 507.
84 Tanaka, K., Sekiguchi, M. and Okada, Y. (1975), *Proc. Natl. Acad. Sci. USA,* **72,** 4071.
85 Smith, C.A. and Hanawalt, P.C. (1978), *Proc. Natl. Acad. Sci. USA,* **75,** 2598.
86 Marx, J.L. (1978), *Science,* **200,** 518.
87 Park, S.D. and Cleaver, J.E. (1979), *Nucl. Acids Res.,* **7,** 1151.
88 Abo-Darub, J.M., Mackie, R. and Pitts, J.D. (1978), *Bull. Cancer,* **63,** 357.
89 Witkin, E.M. and Wermundsen, I.E. (1978), *Cold Spring Harbor Symp. Quant. Biol.,* **43,** 881.
90 Cole, R.S. (1973), *Proc. Natl. Acad. Sci. USA,* **70,** 1064.
91 Tomizawa, S.I. and Ogawa, H. (1972), *Nature New Biol.,* **239,** 14.
92 Lieberman, R.P. and Oishi, M. (1974), *Proc. Natl. Acad. Sci. USA,* **71.** 4816.
93 Rosamond, J., Telander, J.M. and Linn, S. (1979), *J. Biol. Chem.,* **254,** 8646.
94 McKay, V. and Linn, S. (1976), *J. Biol. Chem.,* **251,** 3716.
95 Lieberman, R.P. and Oishi, M. (1974), *Proc. Natl. Acad. Sci. USA,* **71,** 4816.
96 Bridges, B.A. (1979), *Nature,* **277,** 514.

97 Shibata, T., Das Gupta, C., Cunningham, R.P. and Radding, C.M. (1979), *Proc. Natl. Acad. Sci. USA,* **76,** 1638.
98 Cunningham, R.P., Shibata, T., Das Gupta, C. and Radding, C.M. (1979), *Nature,* **281,** 191.
99 Radman, M., Villani, G., Boiteux, S., Kinsella, A.R., Glickman, B.W. and Spadari, S. (1978), *Cold Spring Harbor Symp. Quant. Biol.,* **43,** 937.
100 Marinus, M.G. and Morris, N.R. (1973), *J. Bact.,* **114,** 1143.
101 Higgins, N.P., Kato, K. and Strauss, B. (1976), *J. Mol. Biol.,* **101,** 417.
102 Tomizawa, J.I. (1967), *J. Cell Physiol.,* **70** (Suppl. 1), 201.
103 Cassuto, E. and Radding, C. (1971), *Nature New Biol.,* **229,** 13.
104 Manly, K.F., Signer, E.R. and Radding, C.M. (1969), *Virology,* **37,** 177.
105 Weinstock, G.M., McEntee, K. and Lehman, I.R. (1979), *Proc. Natl. Acad. Sci. USA,* **76,** 126.
106 Konrad, E.B. (1977), *J. Bact.,* **130,** 167.
107 Radding, C.M. (1978), *Annu. Rev. Biochem.,* **47,** 847.
108 Potter, H. and Dressler, D. (1978), *Proc. Natl. Acad. Sci. USA,* **75,** 3698.
109 Holliday, R. (1964), *Genet. Res.,* **5,** 282.
110 Boon, T. and Zinder, N.D. (1971), *J. Mol. Biol.,* **58,** 133.
111 Broda, P. (1967), *Genet. Res.,* **9,** 35.
112 Kotewicz, M., Chung, S., Takeda, Y. and Echols, H. (1977), *Proc. Natl. Acad. Sci. USA,* **74,** 1511.
113 Campbell, A. (1962), *Adv. Gen.,* **11,** 101.
114 Campbell, A. (1969), *Episomes,* Harper and Row, New York.
115 Hsu, P-L., Ross, W. and Landy, A. (1980), *Nature,* **285,** 85.
116 Hopkins, R.L. and Goodman, M.F. (1980), *Proc. Natl. Acad. Sci. USA,* **77,** 1801.
117 McEntee, K., Weinstock, G.M. and Lehman, I.R. (1980), *Proc. Natl. Acad. Sci. USA,* **77,** 857.
118 Glickman, B.W. and Radman, M. (1980), *Proc. Natl. Acad. Sci. USA,* **77,** 1063.

RNA biosynthesis: the transcription apparatus

In the process of DNA transcription the positioning of nucleotide units in the RNA molecules that are being made is under the control of the DNA which acts as template. The means by which this template dictates such a sequence involves both base-pairing interactions and specific interactions between proteins and nucleic acids. Additionally each RNA chain is initiated at a specific site on the DNA template and subject to termination at another unique type of site on the template. In other words there are defined units of transcription. It is a selective process. Specific signals in the DNA template are recognized by the transcription apparatus. Initiation is governed by promoter regions in the DNA, and a region governing termination is designated a terminator.

Transcription is mediated by DNA-dependent RNA polymerases, which have now been isolated from a wide variety of sources, eukaryotic and prokaryotic. The properties of the enzyme from *E. coli* have, however, been the most widely explored. This purified enzyme can carry out the selective transcription of certain DNAs *in vitro*.

The products of DNA-dependent RNA polymerases have ribonucleotide sequences complementary to one of the strands of the DNA which was used as 'template'. A guanine residue in the DNA template strand dictates the insertion of a cytosine nucleotide in the RNA strand under construction, whilst a cytosine in the DNA causes a guanine nucleotide to appear in the new RNA strand. Similarly, a thymine in the DNA results in an adenine nucleotide in the RNA, and an adenine in the DNA results in a uracil nucleotide in the RNA.

10.1 THE MODE OF ACTION AND STRUCTURE OF BACTERIAL RNA POLYMERASES (FOR REVIEWS SEE [1–9])

The first indication for the existence of a DNA-dependent RNA synthesizing enzyme in bacterial cells using the four ribonucleoside 5′-

triphosphates as substrates came to light in the early 1960s. This enzyme, RNA polymerase, was initially purified from *Escherichia coli, Micrococcus luteus* and *Azotobacter vinelandi.* The enzyme from *E. coli* has been the one most extensively studied although the RNA polymerases from quite a large range of other bacteria have now been investigated.

The holoenzyme from *E. coli* (3000 to 6000 per cell) is a complex zinc-containing protein which can be dissociated, for instance in 6M-urea, into a number of polypeptide chain subunits as follows,

Two α-chains of mol. wt. 39000
One β-chain of mol. wt. 155000
One β'-chain of mol. wt. 165000
One molecule of σ (sigma)-factor, mol. wt. 95000

and can be represented as $\alpha_2\beta\beta'\sigma$. Without σ-factor the holoenzyme is termed 'core' enzyme or $\alpha_2\beta\beta'$. Chromatography on phosphocellulose has been used by Burgess and his colleagues [5] to separate the holoenzyme into the 'core' enzyme and the σ-factor. Considerable genetical data [10] and molecular reconstruction studies [11] confirm that β', β, α and σ are all functional subunits of the enzyme. RNA polymerases from other bacteria also comprise similar subunits although there are some slight differences in molecular weights.

The core enzyme will catalyse the synthesis of RNA chains from random sites on a DNA template *in vitro* quite well when 'foreign' DNA (for instance that from calf thymus) is added, but ineffectively when the DNA added is that from *E. coli* or T4 bacteriophage. Addition of σ-factor, however, restores the synthetic activity of the core enzyme by activating selective initiation of RNA synthesis [7]. It complexes with the core enzyme to yield holoenzyme which interacts specifically with promoter regions on the DNA template. Using T4 bacteriophage DNA as *in vitro* template the σ-factor addition specifically stimulates the core enzyme into transcribing those bacteriophage genes which correspond precisely to those normally expressed *in vivo* during the early stages of T4 bacteriophage infection of *E. coli.*

10.1.1 Binding and initiation

The sequence of molecular events for accurate initiation of RNA synthesis can be envisaged as follows [7]. After a series of unproductive random interactions with the DNA template, the holoenzyme 'recognizes' a specific structure, or sequence, within a promoter region. This enables the holoenzyme to bind to the DNA in this region at least an order of magnitude more tightly than in the non-specific interactions mentioned above. The basic promoter contains three sets of sequence information organized spatially along the DNA [9, 12]. An idealized promoter sequence has been suggested by Pribnow [9] and is shown in Fig. 10.1. It is a composite of over 20 known promoter sequences (see [9]). Pribnow suggests [9] for instance that the RNA polymerase might recognize the promoter by making sequence specific contacts with the native DNA helix; the sigma factor at what he terms R_σ and the 'core' component at a region designated R_c. From the use of chemical and photochemical probes it seems that the polymerase contacts with different promoters are strikingly homologous in space [85] and are actually more similar than the sequence homologies. The RNA polymerase appears to twist the DNA in order to bring regions of different promoters into homologous positions [85]. All contacts lying upstream from the R_c region (or 'Pribnow box') could well be recognized in the initial binding step as the polymerase seems to touch down on only *one* face of the DNA helix. With the untwisting of the DNA (from the Pribnow box to a region surrounding the initiation point, I) in order to form the RS (rapid start) complex [9], contacts not recognized initially (since they were on the other side of the DNA helix) would now be accessible to the RNA polymerase.

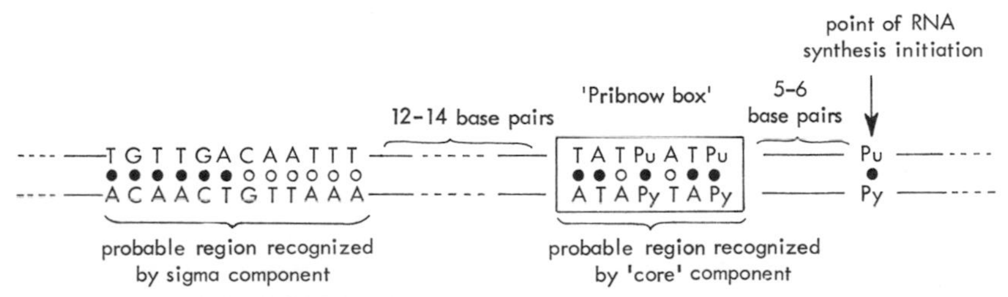

Fig. 10.1 An idealized promoter sequence for *E. coli* RNA polymerase holoenzyme [9]. Solid dots between bases represent the base pairs which are most invariant (see the text).

Whilst the RNA polymerase does not move laterally along the DNA during this transition it is nevertheless positioned to commence the synthesis of an RNA molecule (Fig. 10.2), using only *one* of the DNA strands. The process is initiated through the specific acceptance by the polymerase 'core' of the first and second substrate ribonucleoside 5'-triphosphates complementary to the nucleotide bases at the initiation point, I. These are coupled to give a dinucleoside tetraphosphate [13].

$$\begin{matrix} \text{pppA} & & & \text{pppApX} & \\ \text{or} & + \text{pppX} & \longrightarrow & \text{or} & + \text{PP}_i \\ \text{pppG} & & & \text{pppGpX} & \end{matrix}$$

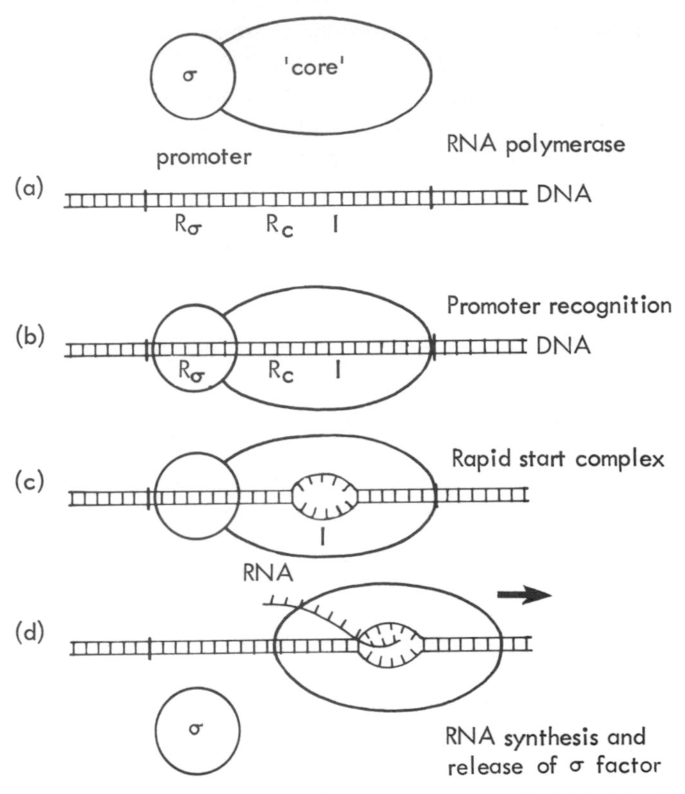

Fig. 10.2 A diagrammatic representation of the stages in initiation of RNA synthesis.

In practice the polymerase only initiates natural transcripts with either ATP or GTP, thus I can be defined precisely as the thymine or cytosine (pyrimidine) residue in the template strand that pairs with the initiating ribonucleoside 5′-triphosphate.

Different promoters are used more or less frequently by the DNA polymerase as start signals for transcription. Kinetic studies indicate that the rate of RS complex formation determines promoter strength. A noteable weak promoter is that for the *E. coli lac* operon repressor gene. This is probably only transcribed once or twice in every cell generation and DNA sequence data indicate that the sequence postulated to interact with the sigma factor has only 5 out of the 12 ideal base pairs [12] indicated in Fig. 10.1.

These initial reactions (as distinct from later chain elongation reactions) can be blocked by the antibiotic rifampicin, and genetic data indicate that rifampicin-resistant mutants carry the resistant phenotype in the β-subunit [14]. The means whereby rifampicin blocks initiation is not yet clear although it does appear to competitively inhibit the binding of the polymerase substrates GTP and ATP to the β-subunit [2]. Other antibiotics affecting the β-subunit are the streptolydigins and streptovaricins (see below). Once the initial internucleotide link between either GTP or ATP and the next nucleotide specified by the template has been made, the σ-factor is released, thus possibly reducing the affinity of the polymerase for the promoter site and allowing the remaining core to move along the DNA template, spinning off a strand of RNA as it proceeds. The process can be visualized in the electron microscope [15, 16].

10.1.2 Elongation of RNA chains and direction of transcription

Basically the new RNA chain grows by the successive addition of nucleoside monophosphate residues from substrate nucleoside 5′-triphosphates to the initial dinucleoside tetraphosphate at its 3′-OH terminus, e.g.

$$\text{pppGpX} \xrightarrow{n\text{YTP}} \text{pppGpX(pY)}_n + n\text{PP}_\text{i}$$

For instance, the RNA being made in Fig. 10.3 has a triphosphate group at position 5′ on the first nucleotide and a free hydroxyl group at position 3 on the other, or growing, end. Alkaline hydrolysis of this particular RNA will yield a molecule of the nucleoside cytosine (C) from the 3′-hydroxyl end of the molecule, uridine and adenosine monophosphates (Up and Ap) and a molecule of guanosine tetraphosphate (pppGp) from the 5′-end. Synthesis proceeds from the 5′-end to the 3′-end of the RNA molecule. This can also be shown to be the case *in vivo.* The sequence of nucleotides added to the growing chain appears to be determined by the sequence of nucleotides in the DNA strand used as template (see Fig. 10.3). However, the selection of the incoming nucleotide may be determined primarily by its ability to fit exactly along with the template base into a site on the enzyme molecule rather than by any ability to interact with the template base alone [8].

Whilst the antibiotics rifampicin and streptovaricin bind to the β-subunit and only block initiation, streptolydigin binds to the β-subunit and interferes with the elongation steps. Another agent which prevents elongation is actinomycin D. However, it does so not by binding to the enzyme but by complexing with deoxyguanosine residues in the DNA template and thus preventing movement of the core along the template [17].

As the RNA chain is formed it peels off the DNA template and, in *E.*

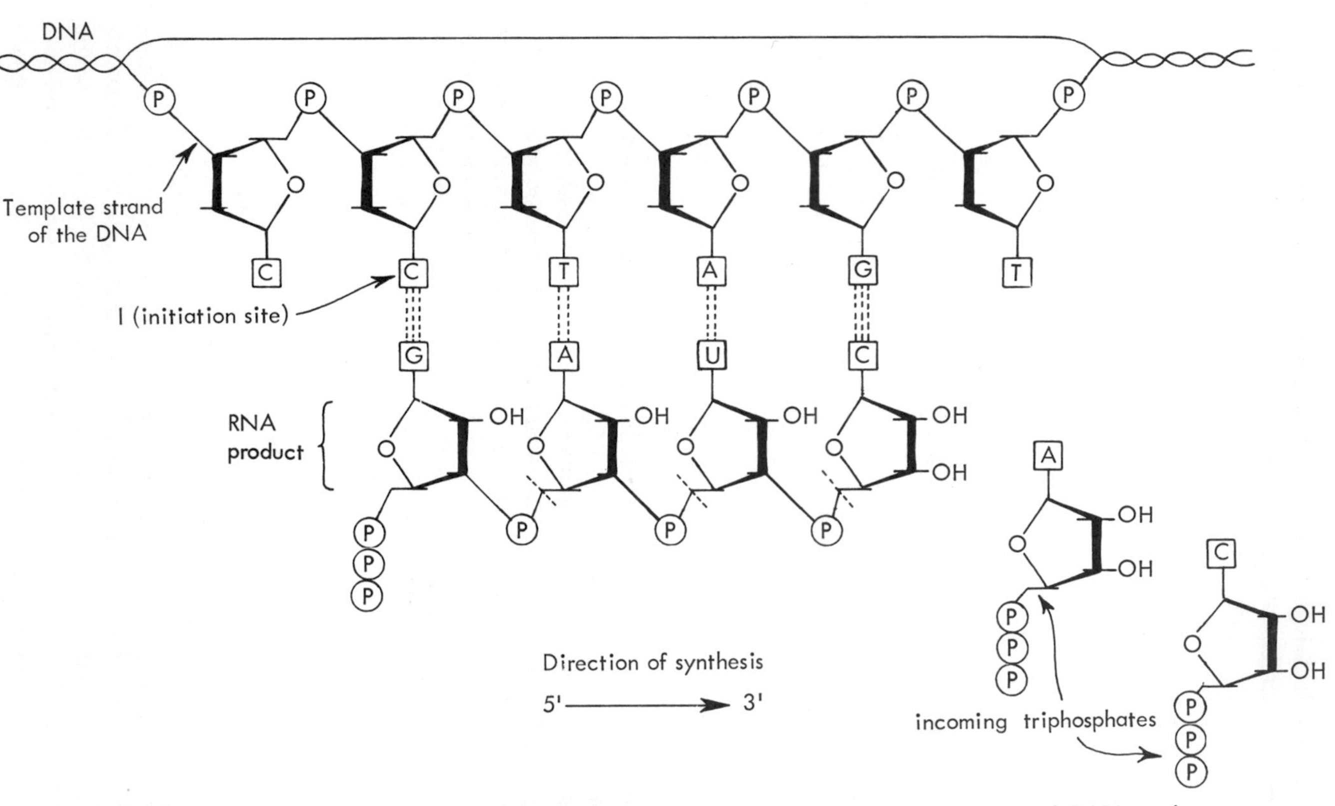

Fig. 10.3 A diagrammatic representation of the biosynthesis of RNA on one strand of DNA acting as template. The broken lines indicate sites of hydrolysis with alkali.

coli at least, immediately becomes associated with the ribosomes so that polysomes (see Chapter 12) may be formed before transcription is complete [18].

10.1.3 Chain termination

The selective termination of RNA chains *in vitro* is less well understood. It appears in the first place that transcription can be terminated directly by virtue of specific sequences on the DNA template acting directly on the core enzyme. In principle the process of termination involves (a) the cessation of RNA chain elongation, (b) the release of the newly formed RNA, and (c) the release of the RNA polymerase from the DNA.

All *strong* terminators for *E. coli* RNA polymerase have three common structural features [12, 19] namely a set of AT base pairs preceded by a set of GC base pairs overlapping with one half of a symmetric DNA sequence (inverted repeat sequence). The symmetric DNA sequence can vary as long as it includes some of the GC pairs and some of the AT pairs [20–24] (see Fig. 10.4). Termination usually occurs somewhere within the set of AT base pairs and the core polymerase dissociates from the DNA. This may result from a reduction in the number of contacts which can be established between the core polymerase and the particular array of nucleotide residues at a strong terminator.

Another essential element in the basic termination mechanism may be the nascent RNA chain. As shown in Fig. 10.4 the terminated RNA can assume a 'hairpin' structure (as a consequence of the transcription of the inverted repeat DNA sequence at the terminator). This RNA

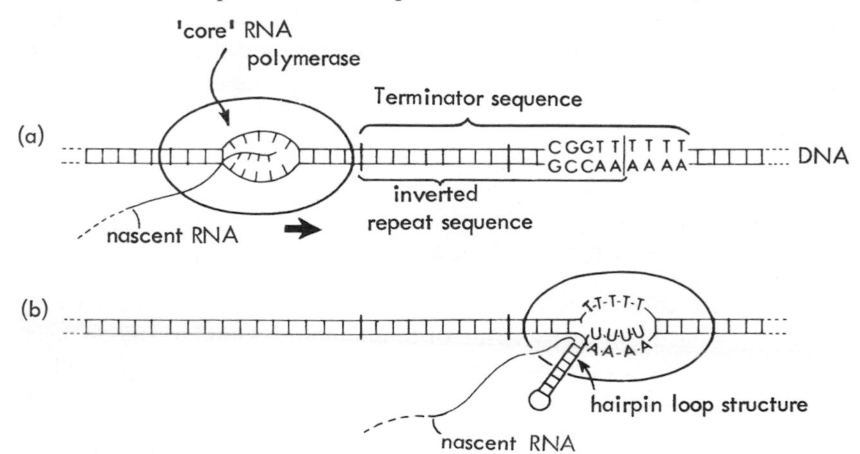

Fig. 10.4 Schematic illustration of a terminator sequence (a) and the formation of ternary terminator complex (b).

hairpin loop structure may act in some way to 'brake' the elongation reaction [9].

It might fit into a niche in the core enzyme-DNA complex, or bind to the core polymerase itself through some structure specific interaction. The question of the release of the short 3′-terminal oligo (U) region from the DNA template still remains. However, the remaining RNA · DNA hybrid here would only involve a small number of rU · dA base pairs which are known to be exceptionally unstable [86]. Possibly the dissociation of the core polymerase from the DNA permits strand closure and the elimination of this short weakly bonded RNA tail out of the double helix [9].

Whilst a strong terminator comprises a spatially organized set of base pairs that results in the discharge of the polymerase from the DNA, weak terminators do not contain this crucial set of base pairs and depend on an additional factor, rho, to achieve this end. This factor [25–27], from *E. coli* at least, is an oligomeric protein (monomer mol. wt 5000) which also exhibits an ATPase activity which is obligatory for its action [28, 29]. In rho-dependent termination there is a requirement for the interaction of the factor with both the nascent RNA chain and the core RNA polymerase as it is halted at the terminator. The essential braking at this point is envisaged as occurring through an RNA hairpin structure as described for strong terminators. Using the energy of ATP hydrolysis it is possible that the rho factor pushes the RNA polymerase off the weak terminator sequence, perhaps removing the nascent RNA from the complex at the same time.

The process of termination can be regulated. For example anti-terminator proteins such as the λ bacteriophage N-gene product can block rho-dependent termination. Specifically N acts to override the termination of transcripts originating at two λ vegetative promoters p_L and p_R which normally terminates at the rho-dependent terminators t_{L1} and t_{R1} [26, 27, 30].

10.2 THE EUKARYOTIC NUCLEAR DNA-DEPENDENT RNA POLYMERASES (FOR REVIEWS SEE [31–36])

In mammalian cells the RNA polymerase activity detected in nuclei was initially difficult to study by virtue of being tightly bound to the nuclear chromatin complex. Somewhat specialized techniques were required for its solubilization in reasonable yields before purification, similar to that achieved for the *E. coli* enzyme, could even commence.

Depending on the origin of the cells used, different approaches to solubilization were employed. Some procedures were mild and involved merely incubating, or homogenizing nuclei, in slightly alkaline buffer. A moderately drastic treatment involved brief sonication in low-ionic-strength medium. In a more drastic approach, sonication was carried out in a medium of high ionic strength [31].

After solubilization from chromatin, chromatography of the enzyme activity from a variety of animal tissues on DEAE-cellulose, or DEAE-Sephadex, revealed the presence of multiple forms of RNA polymerase. This multiplicity of RNA polymerase has also been observed in lower eukaryotes such as an aquatic fungus, yeast, and maize. The enzyme activities are usually eluted from the columns using a linear gradient of ammonium sulphate or potassium chloride. Depending on the tissue used, two to three discrete peaks of activity are resolvable by this approach [31].

The classes of RNA polymerase separated chromatographically between 0.10M and 0.37M salt concentrations are referred to usually as I, II and III in order of their elution. To add another level of complexity, classes I, II and III, which are the major species detected in most eukaryotic cells, have each been further resolved into at least two classes (e.g. I_A and I_B etc.).

Class I RNA polymerases have been established to be of nucleolar origin, whereas classes II and III are of nucleoplasmic origin. Additionally these enzymes operate optimally under somewhat different conditions. The nucleolar enzymes work best at low ionic strength and utilize Mg^{2+}. On the other hand, Mn^{2+} and higher ionic strengths are required for maximum activity of the nucleoplasmic enzymes. Moreover, the activity of nucleoplasmic class II enzymes is inhibited by α-amanitin (a toxin from the poisonous mushroom *Amanita phalloides*) at concentrations as low as 3×10^{-8}M, whereas the nucleolar activity is not affected even at much higher doses [31]. The nucleoplasmic class III enzymes are also affected by α-amanitin but only at much higher concentrations (10^{-4}M).

α-Amanitin is specific for the eukaryotic polymerases as rifampicin is specific for the prokaryotic enzyme (and for the mitochondrial polymerases, as will be seen later). The mushroom toxin appears to bind to the RNA polymerase rather than to the DNA template. Whereas rifampicin inhibits initiation by the bacterial polymerase, α-amanitin blocks RNA synthesis after initiation, presumably at the level of chain elongation. In fact its action resembles more the action of another bacterial RNA polymerase inhibitor streptolydigin which has been recently shown to block elongation in bacteria.

How then do these enzymes relate structurally to the prokaryotic enzyme described in Section 10.1? All the eukaryotic nuclear DNA-dependent RNA polymerases are macromolecular multi-subunit enzymes having molecular weights close to 500 000 [33]. In no instance has a specific function in the polymerizing process been ascribed to one particular subunit. In all cases the enzymes consist of two high molecular-weight subunits (in excess of 100 000 daltons) and a number of smaller subunits (less than 100 000 daltons). The larger subunits of the three main classes of enzymes are probably of different size and there has been little conservation of subunit sizes between different animal species. It appears at the present time that the three major forms are probably discrete entities and are primarily separate gene products although some of the observed microheterogeneity observed could arise from preparative artefacts [33–35].

Since factors such as sigma and rho play an important role in the regulation of transcription a search for such factors was made in eukaryotic systems. So far no polypeptide strictly analogous to these factors has been identified in any of the eukaryotic enzymes although exogenous factors [82] have been isolated which alter the elongation [37–40] or initiation rates [41–43].

Despite these advances, the role of these various polymerases in cellular RNA synthesis remains to be clarified. At least in *E. coli*, two types of data, (a) that certain temperature-sensitive mutants in RNA synthesis can be shown to have extremely heat-labile RNA polymerases, and (b) that rifampicin, which blocks all the initiation of cellular RNA synthesis, can be shown to bind to the β-subunit, and that resistant mutants carry the resistant phenotype in the same subunit, both support the view that the *E. coli* enzyme studied *in vitro* is also responsible for the *E. coli* RNA synthesis *in vivo*. In eukaryotes there is much less direct genetic evidence.

On the other hand considerable indirect evidence obtained with nuclei and subnuclear fractions along with the use of the inhibitor α-amanitin have implicated the nucleolar enzymes (class I) in the synthesis of the ribosomal precursor RNA whereas the nucleoplasmic enzymes (class II and class III) are responsible for the synthesis of mRNA precursors and low molecular weight RNAs (5S RNA and tRNA) respectively [33–36]. Attempts to provide direct evidence in support of these hypotheses from initiation studies using the appropriate purified soluble polymerases and DNA have pointed, not surprisingly to the involvement of other factors.

Purified Class III RNA polymerase will not specifically transcribe the genes for 5S RNA or tRNAs from naked DNA [44]. Specificity of transcription is observed, however, if the template DNA is retained in its natural association with histones and non-histone proteins as the chromatin component of nuclei (see Chapter 4). Such findings indicate that certain determinants required for specific initiation of transcription have an association with nuclear components. Recently *soluble* cell-free systems have been developed from mammalian cells in which the genes for low-molecular-weight RNAs are accurately transcribed by endogenous class III RNA polymerases [45, 81]. The fractionation of these systems has led to the partial purification of various components which are necessary for the accurate transcription by purified RNA polymerase III [45]. A similar approach involving *soluble* cell-free systems from cultured human cells has revealed that accurate initiation of genes encoding mRNA speies requires not only class II RNA polymerase but additional soluble factors (proteins?) [46, 87, 88, 89].

Whilst the use of cell-free systems has emphasized the complexity of eukaryotic transcription systems the problem has been approached in a different way involving the microinjection of DNA fragments containing specific genes into the nuclei of individual *Xenopus laevis* oocyte nuclei [47–50]. In this technique, developed by Gurdon and his colleagues [51] each oocyte is injected with up to 10^9 copies of a particular gene along with labelled ribonucleoside 5′-triphosphates to follow transcription. The remarkable feature of this system is that RNA is transcribed from these *injected* genes with an impressive specificity and accuracy. For instance, the endogenous class III RNA polymerase of *Xenopus laevis* can recognize the initiation and termination signals of injected 5S RNA genes from *Xenopus borealis* [48] and tRNA genes from *Saccharomyces cerevisiae* (bakers' yeast) [52], and the class II polymerases the genes for specific viral proteins in injected simian virus 40 (SV40) DNA [47]. Thus amongst other features which will be discussed later (Chapter 11) the system has potential in the evaluation of specific DNA regions involved in transcriptional initiation and termination in eukaryotes.

As an alternative to direct microinjection into oocyte nuclei as a 'living test-tube' to study transcription of cloned DNAs, other groups have used oocyte nuclear extracts instead [53–56]. These appear to recognize the appropriate initiation and termination signals in added *Xenopus* 5S RNA genes [53, 54] or added yeast or *Drosophila* tRNA genes [55, 56].

10.3 TRANSCRIPTION AND EUKARYOTIC CHROMATIN STRUCTURE

Both chain initiation and elongation RNA transcripts are inhibited but not abolished by the presence of nucleosomes in DNA [57]. The former effect may be due to the preferential association of histones with the more flexible AT rich regions of the duplex which are characteristic of promoter sites [58]. Nonetheless, transcription can occur *in vitro* on SV40 minichromosomes without extensive dissociation of the core histones [57]. The location of the nucleosomal DNA on the outside of the particle makes it possible to envisage the passage of the RNA polymerase molecule along the DNA with minimal structural perturbation. It has been suggested that the nucleosome may undergo transitory unfolding during RNA synthesis to form two half nucleosomes [59].

There is a considerable body of evidence which suggests that transcribing chromatin lacks histone H1 and as a consequence it adopts a more open structure since histone H1 is required for the formation of the compact solenoidal superstructure [60]. Access to the DNA may be further facilitated by modification of the remaining core histones since it can be shown that acetylated histones form nucleosomes which are more susceptible to nuclease digestion [61]. Alternatively, or in addition, it is possible to show that certain of the high-mobility-group proteins, particularly HMG 14 and HMG 17, are associated preferentially, though not exclusively with transcriptionally active chromatin [62]. It is thought that these proteins replace histone H1 on the internucleosomal DNA and thus contribute to the stabilization of the more open superstructure.

It is possible to differentiate experimentally between transcriptionally active and inactive chromatin by two techniques. The presence of histone H1 renders nucleosomes insoluble at physiological ionic strength so that a simple salt fractionation of DNase II [64] or micrococcal nuclease [62] digested material can separate the nucleosomes associated with transcribing DNA from nucleosomes derived from the bulk of the chromatin. This method has the advantage that the transcribing nucleosomes themselves can be isolated and characterized. The more open superstructure of transcribing genes also results in their showing preferential sensitivity to DNase I, so that if digestion with this enzyme is limited to a small percentage of the total DNA, the active genes are excised and the residual material can be demonstrated, by C_0t value analysis, to be depleted in sequences known to be transcriptionally competent in the tissue under analysis [63, 64]. In this second method the transcribing genes are totally digested and the nucleosomes associated with them are not recovered intact, but the proteins released during digestion can be purified and examined for their location on the puffing regions of polytene chromosomes [65] or for their effects in a reconstituted system [66]. In chromatin for example from oestradiol sensitive human breast tumour [83] and hen oviduct [84] cells, there is a noticeable enrichment of oestradiol–receptor complexes in these transcriptionally active fractions.

Weintraub [66] has made the important distinction between proteins associated with the recognition of the sequence to be transcribed and those involved in the propagation of transcription. All the proteins which have to date been suggested to have preferential association with transcriptionally active chromatin, are neither tissue nor cell specific and thus fall into the second category. Those proteins which are responsible for the recognition event would be expected to be present in much smaller quantities and consequently to be less readily detected by current methodology.

10.4 MITOCHONDRIAL AND CHLOROPLAST DNA-DEPENDENT RNA POLYMERASES

Compared with the nuclear DNA-dependent RNA polymerases, relatively little is known about the properties and functions of the mitochondrial polymerases. Soluble preparations have now been obtained from a number of sources, and the enzymes from *Neurospora crassa* and yeast have been extensively purified [67, 68].

That from rat liver appears closely associated with the mitochondrial membrane and detergents are required for its extraction [69]. The yeast enzyme resembles eukaryotic nuclear polymerases in size with major subunits of 135 000 and 195 000 plus smaller components. It is inhibited by α-amanitin but not by rifampicin [70]. In contrast and like the prokaryotic RNA polymerases, the enzymes from rat liver, or *N. crassa* mitochondria, are sensitive to rifampicin but not α-amanitin [70].

RNA polymerase has also been isolated from plant chloroplasts [71, 72]. That from maize chloroplasts [73] has a high molecular weight (about 500 000 daltons) and a multisubunit structure (large subunits 180 000 and 140 000 daltons) like the nuclear enzymes. However, peptide maps comparing the subunits of maize chloroplast and nuclear class II polymerase show that at least the two large subunits are not identical [74]. A general conclusion is that RNA polymerase from higher plant chloroplasts is insensitive to rifampicin. Maize chloroplast polymerase will preferentially transcribe chloroplast DNA provided an additional soluble factor is added and the template is in a supercoiled form [90].

10.5 POLYNUCLEOTIDE PHOSPHORYLASE (EC 2.7.7.8)

Historically the first clear indication of the mechanism by which RNA molecules might be synthesized enzymically was actually obtained in

1955 by Ochoa and his colleagues [75–77] who isolated from the micro-organism *Azotobacter vinelandii* an enzyme which catalyses the synthesis of high-molecular-weight polyribonucleotides from nucleoside 5′-diphosphates with the release of orthophosphate. The reaction is reversible, requires magnesium ions but *no template.* It reaches equilibrium when 60–80 per cent of the nucleoside diphosphate has disappeared, and may be represented:

$$n\text{NDP} \leftrightharpoons (\text{NMP})_n + n\text{P}_i$$

where N stands for adenine, hypoxanthine, uracil or cytosine. The enzyme involved has been named *polynucleotide phosphorylase* and has been extensively reviewed [76–77]. It is widely distributed in bacteria and has also been found in plant tissues. There is at present little convincing evidence for its occurrence in animal tissues. It can readily be purified from bacterial sources [79] and has proved to be of great value in the preparation of polynucleotides in the laboratory.

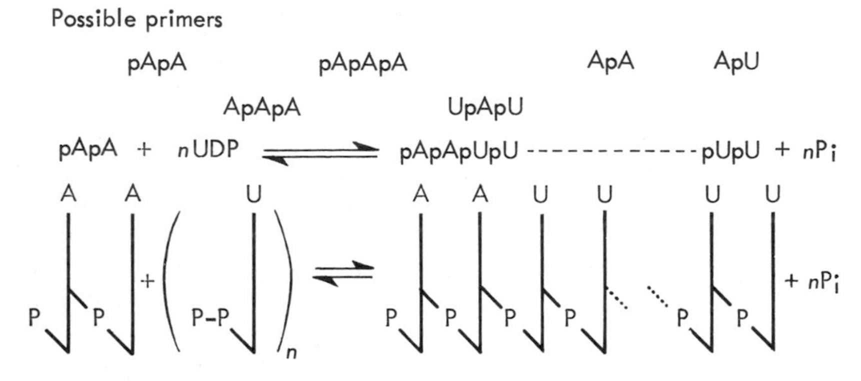

Fig. 10.5 Primers in the polynucleotide phosphorylase reaction.

Crude preparations of polynucleotide phosphorylase require no primer but, with highly purified preparations, polynucleotide formation occurs only after an initial lag period which can be eliminated by the addition of small amounts of polynucleotide or even of certain oligonucleotides such as triadenylic acid (pApApA) or diadenylic acid (pApA). Oligonucleotide primers are incorporated into the nearly made polynucleotide.

The essential features of the primer are that it should contain at least two nucleoside residues, one of which carries a free C-3′-hydroxyl group on which a new phosphodiester bond can be formed (Fig. 10.5).

The first bond formed is a phosphodiester bridge between the 5′-phosphate of a UMP residue and the free terminal C-3′-hydroxyl of pApA, and the chain is extended by similar condensations, the primer being incorporated into the product.

The purified enzyme contains about 3 per cent of nucleotide in the form of a complex oligonucleotide which probably acts as a built-in primer and is responsible for the definite but sluggish reaction which takes place in the absence of added primer.

The reversal of the polymerization reaction, phosphorolysis, in which the polynucleotide is incubated with the enzyme in the presence of an excess of inorganic phosphate to yield the nucleoside diphosphates by stepwise removal of mononucleotide units, has also been studied. The biosynthetic polymers are readily phosphorolysed, and so are oligonucleotides which act as primers, but, as might be expected, dinucleotides and dinucleoside monophosphates are not. Tobacco mosaic virus RNA and highly polymerized yeast RNA are phosphorolysed readily, but yeast RNA treated with alkali is phosphorolysed slowly. The formation of multi-stranded chains as between poly(A) and poly(U) results in a slow rate of phosphorolysis. The transfer RNA of the cell cytoplasm is also incompletely phosphorolysed, 70–80 per cent being left unchanged presumably because of the secondary structure of tRNA. The phosphorolysis appears to affect mainly the terminal groups.

The function of polynucleotide phosphorylase in the cell has been the subject of much discussion [80]. It is possible that it is primarily responsible for the degradation of RNA to yield nucleoside diphosphates, which are the immediate precursors of deoxyribonucleotides (see Chapter 7), and it may even control the level of inorganic phosphate in the cell. It may also be involved in the degradation of messenger RNA, though this is now considered unlikely.

REFERENCES

1 Hurwitz, J. (1970), *Studies on the DNA-Dependent Synthesis of RNA with RNA Polymerase, Harvey Lectures,* **64,** 157, Academic Press, New York, London.
2 Silverstri, L. (1970), *RNA-polymerase and Transcription,* North-Holland Publishing Co., Amsterdam.
3 Sethi, V.S. (1971), *Prog. Biophys. Mol. Biol.* (eds J. A. V. Butler and D. Noble), **23,** 67.
4 Bautz, E.K.F. (1972), *Prog. Nucleic Acid Res. Mol. Biol.,* **12,** 129.
5 Burgess, R.R. (1971), *Annu. Rev. Biochem.,* **40,** 711.
6 Chamberlin, M.J. (1974), *Annu. Rev. Biochem.,* **43,** 721.

7 Travers, A. (1974), *Cell,* **3,** 97.
8 Chamberlin, M.J. (1976), in *RNA polymerase,* p. 11 (eds R. Losick and M. Chamberlin), Cold Spring Harbor Laboratory.
9 Pribnow, D. (1979), in *Biological Regulation and Development,* Vol. 1 (ed. R. F. Goldberger), Plenum Press, New York and London.
10 Scaife, J. (1976), in *RNA polymerase,* p. 207 (eds R. Losick and M. Chamberlin), Cold Spring Harbor Laboratory.
11 Zillig, W., Palm, P. and Heil, A. (1976), in *RNA polymerase,* p. 101 (eds R. Losick and M. Chamberlin), Cold Spring Harbor Laboratory.
12 Gilbert, W. (1976), in *RNA polymerase,* p. 193 (eds R. Losick and M. Chamberlin), Cold Spring Harbor Laboratory.
13 Krakow, J.S., Rhodes, G. and Jovin, T.M. (1976), in *RNA polymerase,* p. 127 (eds R. Losick and M. Chamberlin), Cold Spring Harbor Laboratory.
14 Maitra, U. and Hurwitz, J. (1965), *Proc. Natl. Acad. Sci. USA,* **54,** 815.
15 Miller, O.L., Beatty, B.R., Hamkalo, B.A. and Thomas, C.A. (1970), *Cold Spring Harbor Symp. Quant Biol.,* **35,** 505.
16 Hamkalo, B.A. and Miller, O.L. (1973), *Annu. Rev. Biochem.,* **42,** 379.
17 Goldberg, I.H. and Friedman, P.A. (1971), *Annu. Rev. Biochem.,* **40,** 775.
18 Das, H.K., Goldstein, A. and Lowrey, L.I. (1967), *J. Mol. Biol.,* **24,** 23.
19 Roberts, J.W. (1976), in *RNA polymerase,* p. 247 (eds R. Losick and M. Chamberlin), Cold Spring Harbor Laboratory.
20 Squires, C., Lee, F., Bertrand, K., Squires, C.L., Bronson, M.J. and Yanofsky, C. (1974), *J. Mol. Biol.,* **103,** 351.
21 Lee, F., Squires, C. and Yanofsky, C. (1976), *J. Mol. Biol.,* **103,** 383.
22 Rosenberg, M., Weissman, S. and de Crombrugghe, B. (1975), *J. Biol. Chem.,* **250,** 4755.
23 Pieczenik, G., Barrell, B.G. and Gefter, M.L. (1972), *Arch. Biochem. Biophys.,* **152,** 152.
24 Sanger, F., Air, G.M., Barrell, B.G., Brown, N.L., Coulson, A.R., Fiddes, J.C., Hutchinson, C.A., Slocombe, P.M. and Smith, M. (1977), *Nature,* **265,** 687.
25 Roberts, J. (1969), *Nature,* **224,** 1168.
26 Roberts, J. (1970), *Cold Spring Harbor Symp. Quant. Biol.,* **35,** 121.
27 Roberts, J. (1975), *Proc. Natl. Acad. Sci. USA,* **72,** 3300.
28 Lowery-Goldhammer, C. and Richardson, J.P. (1974), *Proc. Natl. Acad. Sci. USA,* **71,** 2003.
29 Howard, B. and de Crombrugghe, B. (1975), *J. Biol. Chem.,* **251,** 2520.
30 Herskowitz, I. (1974), *Ann. Rev. Genetics,* **7,** 389.
31 Jacob, S.T. (1973), *Prog. Nucleic Acid Res. Mol. Biol.,* **13,** 93.
32 Chambon, P. (1974), in *The Enzymes,* Vol. 10, p. 261 (ed. P. D. Boyer), Academic Press, New York.
33 Chambon, P. (1975), *Annu. Rev. Biochem.,* **49,** 613.
34 Rutter, W.J., Goldberg, M.I. and Perriard, J.C. (1974), *Biochemistry of Cell Differentiation, MTP International Review of Science, Biochemistry Series 1,* p. 267 (ed. J. Paul), Butterworths, London.
35 Beebee, T. and Butterworth, P.H.W. (1977), *Biochem. Soc. Symp.,* **42,** 75.

36 Roeder, R.G. (1976), in *RNA polymerase,* p. 285 (eds R. Losick and M. Chamberlin), Cold Spring Harbor Laboratory.
37 Stein, H. and Hansen, P. (1971), *Cold Spring Harbor Symp. Quant. Biol.,* **35,** 709.
38 Siefert, K.H., Juhasz, P.P. and Benecke, B.J. (1973), *Eur. J. Biochem.,* **33,** 181.
39 Sekimizii, K., Kobayashi, N., Mizumo, D. and Natori, S. (1976), *Biochemistry,* **15,** 5064.
40 Revie, D. and Dahmus, M.E. (1979), *Biochemistry,* **18,** 1813.
41 Spindler, J.R. (1979), *Biochemistry,* **18,** 4042.
42 Chuang, R.Y. and Chuang, L.F. (1975), *Proc. Natl. Acad. Sci. USA,* **72,** 2935.
43 Shea, M. and Kleinsmith, L.J. (1973), *Biochem. Biophys. Res. Commun.,* **50,** 473.
44 Bitter, G.S. and Roeder, R.G. (1979), *Nucleic Acids Res.,* **7,** 433.
45 Weil, P.A., Segall, J., Harris, B., Ng, S-Yu and Roeder, R.G. (1979), *J. Biol. Chem.,* **254,** 6163.
46 Weil, P.A., Luse, D.S., Segall, J. and Roeder, R.G. (1979), *Cell,* **18,** 469.
47 Mertz, J.E. and Gurdon, J.B. (1977), *Proc. Natl. Acad. Sci. USA,* **74,** 1502.
48 Brown, D.D. and Gurdon, J.B. (1977), *Proc. Natl. Acad. Sci. USA,* **74,** 1176.
49 Kressman, A., Clarkson, S.G., Pirotta, V. and Birnstiel, M.L. (1978), *Proc. Natl. Acad. Sci. USA,* **75,** 1176.
50 Cortese, R., Melton, D., Tranquilla, T. and Smith, J.D. (1978), *Nucleic Acids Res.,* **5,** 4593.
51 Gurdon, J.B. (1977), *Proc. R. Soc., B,* **198,** 211.
52 Melton, D.A., De Robertis, E. and Cortese, R. (1980), *Nature,* **284,** 143.
53 Birkenmeier, E.H., Brown, D.D. and Jordan, E. (1978), *Cell,* **15,** 1077.
54 Sakonju, S., Bogenhagen, D.F. and Brown, D.D. (1980), *Cell,* **19,** 13.
55 Ogden, R.C., Beckman, J.J., Abelson, J., Kay, H.J., Soll, D. and Schmidt, O. (1979), *Cell,* **17,** 399.
56 Schmidt, O., Mao, J., Silverman, S., Hoverman, B. and Soll, D. (1978), *Proc. Natl. Acad. Sci. USA,* **75,** 4819.
57 Wasylyk, B., Thevenin, G., Oudet, P. and Chambon, P. (1979), *J. Mol. Biol.,* **128,** 411.
58 Wasylyk, B. Oudet, P. and Chambon, P. (1979), *Nucleic Acids Res.,* **7,** 703.
59 Pardon, J.F., Cotter, R.I., Lilley, D.M.J., Worcester, D.M., Campbell, A.M., Wooley, J.C. and Richards, B.M. (1978), *Cold Spring Harbor Symp. Quant. Biol.,* **42,** 11.
60 Gottesfeld, J.M. and Partington, G.A. (1977), *Cell,* **12,** 953.
61 Nelson, D., Perry, M.E. and Chalkley, R. (1979), *Nucleic Acids Res.,* **6,** 561.
62 Goodwin, G.H., Mathew, C.G.P., Wright, C.A., Venkov, C.D. and Johns, E.W. (1979), *Nucleic Acids Res.,* **7,** 1815.
63 Weintraub, H. and Groudine, M. (1976), *Science,* **93,** 848.
64 Garel, A. and Axel, R. (1976), *Proc. Natl. Acad. Sci. USA,* **73,** 3966.
65 Mayfield, J., Serunian, L., Silver, L. and Elgin, S.C.R. (1978), *Cell,* **14,** 539.

66 Weisbrod, S. and Weintraub, H. (1979), *Proc. Natl. Acad. Sci. USA,* **76,** 630.
67 Kunztzel, H. and Schafer, K.P. (1971), *Nature,* **231,** 265.
68 Scragg, A.H. (1971), *Biochem. Biophys. Res. Commun.,* **45,** 701.
69 Saccone, C., Gallerani, R., Gadieta, M.N. and Greco, M. (1971), *FEBS Lett.,* **18,** 339.
70 Eccleshall, R.R. and Criddle, R.S. (1974), in *The biogenesis of mitochondria. Transcriptional, translational and genetic aspects,* p. 31 (eds A. M. Kroon, and C. Saccone), Academic Press, New York.
71 Bottomley, W., Smith, H.J. and Bogorad, L. (1971), *Proc. Natl. Acad. Sci. USA,* **68,** 2412.
72 Polya, G.M. and Jagendorf, A.T. (1971), *Arch. Biochem., Biophys.,* **146,** 649.
73 Smith, H.J. and Bogorad, L. (1974), *Proc. Natl. Acad. Sci. USA,* **71,** 4839.
74 Kidd, E.H. and Bogorad, L. (1979), *Proc. Natl. Acad. Sci. USA,* **76,** 4890.
75 Grunberg-Manago, M., Oritz, P.J. and Ochoa, S. (1956), *Biochim. Biophys. Acta,* **20,** 269.
76 Ochoa, S. (1960), *Les Nucleoproteins,* p. 241, Onzieme Conseil de Chimie, Instut de Chimie Solvay, Stoeps, Bruxelles.
77 Enzymes in Polynucleotide Metabolism (1959), *Ann. N.Y. Acad. Sci.,* **81,** 511.
78 Hilmoe, R.J. (1959), *Ann. N.Y. Acad. Sci.,* **81,** 660.
79 Ochoa, S. Basilio, C. and Krakow, J.S. (1963), *Methods Enzymol.,* **6,** 3.
80 Grunberg-Manago, M. (1963), *Prog. Nucleic Acid. Res. Mol. Biol.,* **1,** 93.
81 Wu, G-J. (1980), *J. Biol. Chem.,* **255,** 251.
82 Sawadogo, M., Sentenac, A. and Fromageot, P. (1980), *J. Biol. Chem.,* **255,** 12.
83 Scott, R.W. and Frankel, F.R. (1980), *Proc. Natl. Acad. Sci. USA,* **77,** 1291.
84 Hemminki, K. and Vauhkonen, M. (1977), *Biochem. Biophys. Acta,* **474,** 109.
85 Siebenlist, U., Simpson, R.B. and Gilbert, W. (1980), *Cell,* **20,** 269.
86 Martin, F.H. and Tinoco, I. (1980), *Nucleic Acids Res.,* **8,** 2295.
87 Luse, D.S. and Roeder, R.G. (1980), *Cell,* **20,** 691.
88 Wu, G.U. (1978), *Proc. Natl. Acad. Sci. USA,* **75,** 2175.
89 Wasylyk, B., Kedinger, C., Corden, J., Brison, O. and Chambon, P. (1980), *Nature,* **285,** 367.
90 Jolly, S.O. and Bogorad, L. (1980), *Proc. Natl. Acad. Sci. USA,* **77,** 822.

Patterns of transcription and RNA processing

11

11.1 USE OF RNA–DNA HYBRIDIZATION

Hybrid formation between RNA molecules and sequences of complementary nucleotides in denatured DNA is a widely used technique for determining structural and metabolic relationships between particular DNA sequences in the genome and the cellular RNA molecules that result from their transcription (see also Chapter 3). The discovery that base-paired double-stranded structures could be formed between DNA and RNA molecules rapidly followed the discovery of DNA renaturation [1, 2]. Early work with nucleic acids from microorganisms suggested that the reaction is highly specific. Analysis of hybrid formation [3, 8] can be facilitated using either DNA and RNA radioactively labelled. As was the case for the kinetic analysis of DNA strand reassociation (see Chapter 3), the DNA to be examined is first sheared to small fragments of uniform size which are then denatured. The resulting single-stranded fragments are incubated in neutral buffer (0.12 M sodium phosphate) along with appropriate RNA molecules. Hydroxyapatite chromatography can then be used to separate hybrid duplex molecules from single-strands of unhybridized DNA or RNA. Alternatively any remaining single-stranded nucleic acids can be removed by digestion with single-strand specific nucleases such as S_1 nuclease from *Aspergillus* (see Chapter 6) (Fig. 11.1).

Another analytical technique is the use of nitrocellulose filters which retain denatured DNA molecules by virtue of the fact that they aggregate and collect at the filter surface [9]. Any RNA molecules hybridized to denatured DNA strands will be retained as well, whereas free RNA will pass through the filter. Treatment of the filter with pancreatic RNase (see Chapter 6) before analysis removes any portions of RNA molecules not in true hybrid with DNA. In a modification of

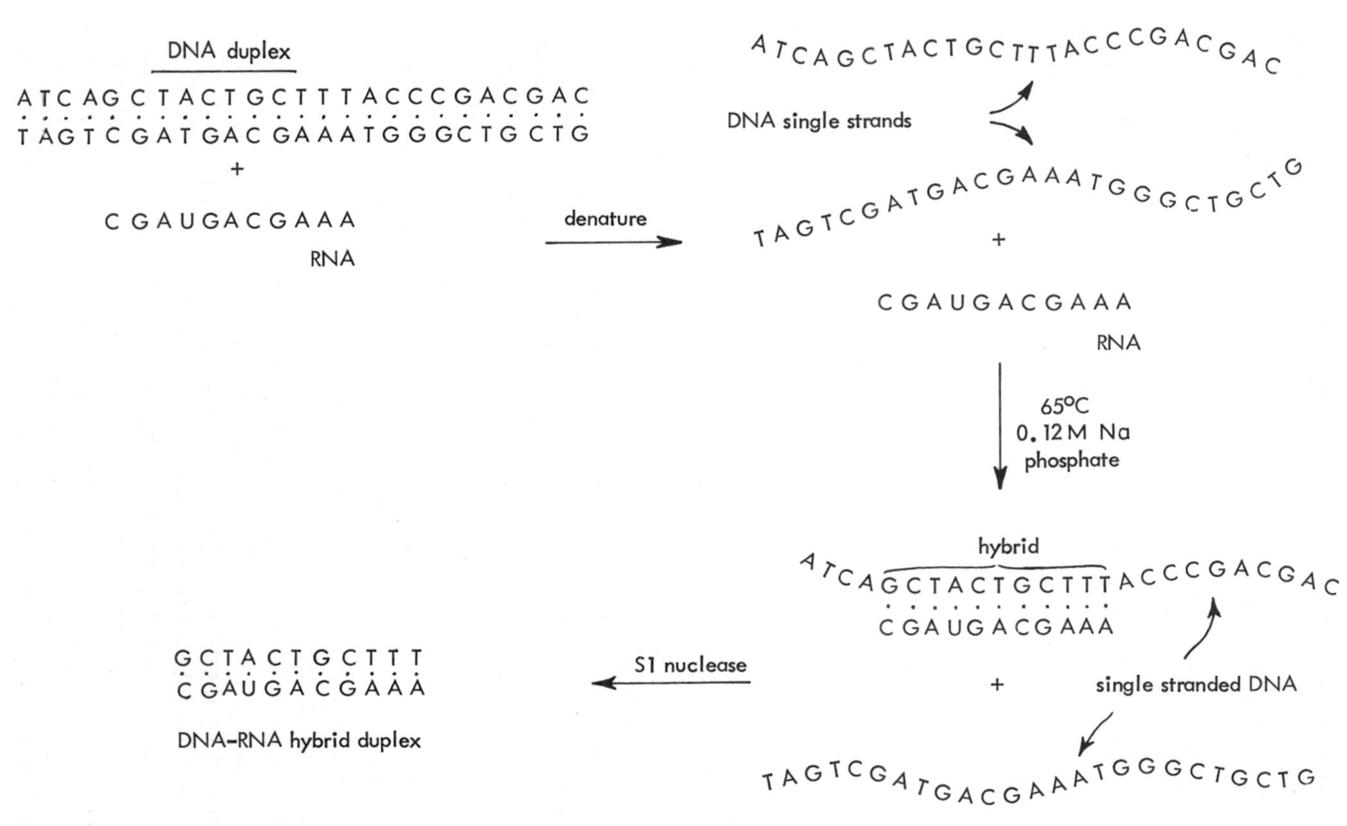

Fig. 11.1 Formation of DNA–RNA hybrids.

this nitrocellulose technique, denatured DNA can be immobilized on the nitrocellulose filters prior to incubation with the RNA solution [10]. A recent variant of this immobilization approach has been the actual *covalent* attachment of the RNA (or DNA) to diazobenzyloxymethylated cellulose paper [11] and the subsequent incubation of the paper plus RNA with various radioactively labelled DNAs [11, 12].

Hybridization between DNA and RNA of eukaryotic origin suffers from the same restrictions as the reassociation of cellular DNA itself. Any unique sequence is present in only a very low concentration and thus hybrid formation involving unique sequences (e.g. messenger RNAs) requires a high product of nucleic acid concentration and time at relatively high temperatures (usually 25°C or so below the melting temperature of the RNA–DNA hybrid, this being the temperature at which hybrid formation is at a maximum [13]). This presents practical problems due to the relative lability of RNA under such conditions. Great care has to be taken to remove contaminating nucleases and heavy metal ions from the hybridization reaction mixtures. However, it is found that addition of formamide [14] allows the hybridizations to bc carried out at much lower temperatures (around 40°C). One per cent formamide lowers the melting point of RNA–DNA duplexes by 0.7 degrees.

Initially hybridization reactions were carried out using small amounts of DNA and large excesses of radioactively labelled RNAs. The aim was to measure the amount of genomic DNA that was transcribed into various cellular RNAs. In principle at least as the concentration of RNA is increased the radioactivity bound to DNA should increase until a plateau of saturation is reached when all the available sites on DNA are hybridized with RNA. However, depending on the type of RNA under investigation very large amounts of RNA may be necessary to achieve true saturation (particularly where transcripts of unique sequences are involved). Moreover, the transcripts of one repetitive sequence in total cellular DNA can hybridize with related sequences leading to falsely high results. However, hybridization of the RNA with only non-repetitive DNA sequences [15] (previously isolated by hydroxyapatite chromatography) gives a better indication of how much non-repetitive DNA is represented in cellular transcripts. Since poly(A)-containing mRNA can be separated from ribosomal and transfer RNAs (by affinity-chromatography methods) higher concentrations of unique transcripts can be readily attained in the hybridization reactions.

Such RNA-driven reaction conditions can also be used to estimate sequence complexity, the kinetics of the hybridization depends on the sequence complexity of the RNA population involved [16]. The greater the number of different RNA sequences the longer the reaction takes.

$R_0t_{1/2}$ derived from the product of the RNA concentration and the time of incubation needed for half saturation provides a measure of the complexity of the RNA preparation. The method can be used to estimate the number of different RNA sequences present in a preparation of known molecular weight provided some suitable single RNA species of known molecular weight is used as a reference [17, 18]. Thus since the $R_0t_{1/2}$ of the transition observed when an excess of rabbit haemoglobin β-chain mRNA (about 2×10^5 daltons) is hybridized with its homologous complementary DNA (cDNA) is 3×10^{-4} mol s l^{-1}, the number of sequences of 2×10^{-5} daltons in an RNA preparation that had a single transition with a $R_0t_{1/2}$ of 0.06 is $0.06/3 \times 10^{-4}$ i.e. 200. As can be seen from Fig. 11.2 hybridization of an excess of HeLa cell mRNA (polyadenylated class) with its homologous cDNA actually yields quite

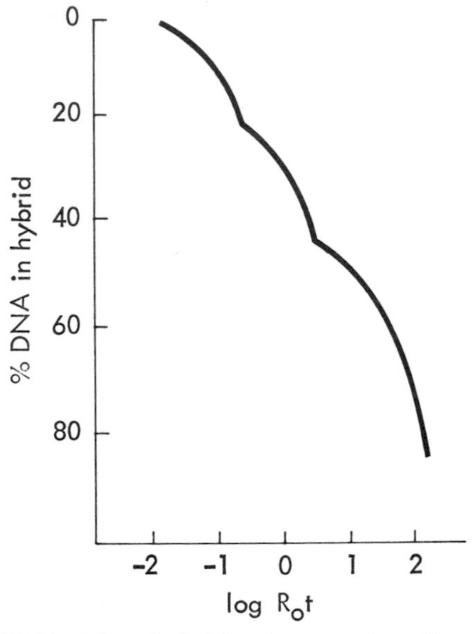

Fig. 11.2 An RNA-driven hybridization reaction. Complementary DNA made to HeLa cell mRNA is hybridized with an excess of HeLa cell mRNA (see [17]). Three transitions are observed with mid points ($R_0t_{1/2}$) of 0.05, 0.9 and 45 mol s l^{-1} respectively.

a complex pattern which can be resolved into three transitions. It has been interpreted as indicating that 22 per cent of HeLa cell mRNA comprises 17 different sequences of average molecular weight 6×10^5, 28 per cent has 370 different sequences and 50 per cent has 33000 different sequences [17]. Moreover such an approach also indicated that around 1.4 per cent of 'single-copy' HeLa cell DNA is complementary to polyadenylated mRNA [17]. The corresponding data from rat liver and mouse leukaemia (Friend) cells yielded values of 1.8 per cent and 1.9 per cent respectively [341, 342]. Hybridization reactions where the DNA molecules are in vast excess can be used to determine which frequency classes of DNA are represented in population of RNA molecules. The RNA molecules are made radioactive and are present in only tracer amounts. Hybrid formation between RNA and DNA, and the re-association of DNA strands takes place simultaneously but the amount of DNA withdrawn into RNA–DNA hybrid does not change the concentration of DNA sequences. Since the rates of association depend only on the initial DNA content such hybridization carried out in solution are called *DNA-driven reactions*. The $C_0t_{1/2}$ for the hybridization of a specific RNA indicates the frequency class of DNA from which the RNA was transcribed. As can be seen from Fig. 11.3, whilst the bulk of mouse cell mRNA is transcribed from unique DNA sequences, a small proportion arises from middle repetitive sequences.

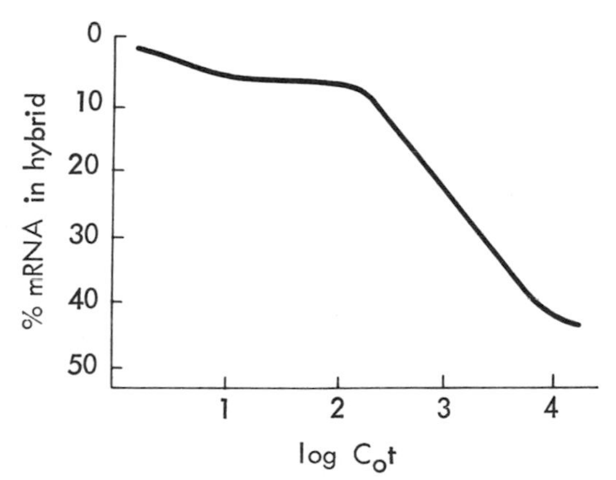

Fig. 11.3 A DNA-driven hybridization reaction. ^{3}H-labelled mouse cell mRNA is hybridized with a 2000-fold excess of mouse cell DNA (mean size 0.3 kb).

As will be evident from many other sections of this book the process of hybridization now has very many applications, not only in the study of nucleic acid structure and function but also in the rapidly expanding area of nucleic acid technology.

11.2 THE CONCEPT OF POST-TRANSCRIPTIONAL PROCESSING

Whilst most of the cellular RNA of eukaryotes is cytoplasmic, the nucleus does contain some RNA (about 5 per cent of the total). It turns out, however, that its existence is mainly ephemeral, and there is now substantial evidence indicating that it includes several species which are intermediates in biosynthetic pathways leading to the formation of tRNAs, ribosomal RNAs, and messenger RNAs of the cytoplasm. The primary structure of these intermediates is modified in certain instances by cellular enzyme systems subsequent to the completion of their transcription from the DNA template.

Such modifications include (a) 'trimming' or 'tailoring', i.e. alteration to the length of the primary transcription product by scission mechanisms, or (b) the reduction of the length by the removal of internal sequences in a reaction known as 'splicing' or (c) the alteration of primary nucleotide sequences as a result of base or sugar modification, or (d) the addition of specific nucleotide sequences. This chemical editorial work carried out by the cell is a prerequisite in the formation of certain nucleic acid species both in eukaryotic and in prokaryotic cells, and the molecular events involved can be collectively described as *post-transcription processing events.*

11.3 THE BIOSYNTHESIS OF TRANSFER RNA

The use of various RNA–DNA hybridization techniques has indicated that *E. coli* has around 60 specific sequences, or genes, complementary to tRNA molecules [19]. The number in eukaryotes is much greater (yeast 320-400, *Drosophila* 750, *Xenopus* 8000 [20]). However, these higher numbers actually represent a relatively small number of genes which exist in multiple copies. For instance, in *Xenopus laevis* 43 basic species are coded for, so that an average tRNA genes exists in 200 copies [21].

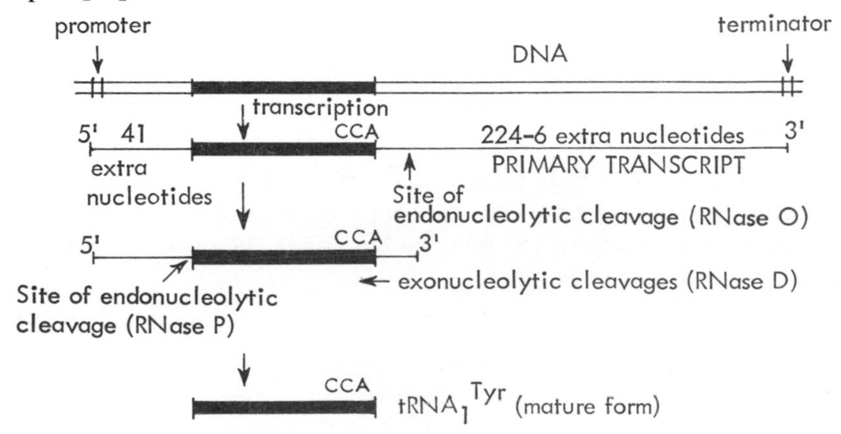

Fig. 11.4 *E. coli* tRNA$_1^{Tyr}$ transcription unit, its transcription, the precursors and their post-transcriptional processing.

Much of our knowledge of the basic steps in tRNA biosynthesis comes from the early studies of the *E. coli* $tRNA_1^{Tyr}$ system [19, 22–24] (Fig. 11.4). Transcription starts at a promoter site [47] some 41 nucleotides upstream from the sequence of DNA actually specifying the mature tRNA and continues some 224–226 nucleotides beyond the site specifying the –CCA 3′-end and terminates at a rho-dependent termination site. At least two types of nucleolytic reaction are required to convert the large primary transcript into a molecule of tRNA dimensions. The enzyme RNase P will endonucleolytically remove the extra 41 nucleotides from the 5′-end. This enzyme has a broad specificity and is probably involved in the synthesis of all *E. coli* tRNAs. It appears that this nuclease recognizes general features of tRNA structure rather than specific sequences. It is a ribonucleoprotein complex, 80 per cent RNA and 20 per cent protein (the major RNA component being about 300 nucleotides) [25].

The processing events at the 3′-end of the $tRNA_1^{Tyr}$ are less-well understood [26–28]. Both an endonuclease and exonuclease have been implicated. It has been suggested that an endonuclease (possibly RNase O) recognizes a hairpin-like site seven nucleotides from the 3′-end of the tRNA sequence and the remaining seven nucleotides are removed by an exonuclease, possibly RNase D [355]. As is the case for RNase P, this activity seems to recognize the tertiary structure of tRNA and the 3′–CCA residues to perform this trimming function.

Temperature-sensitive mutants of *E. coli* defective in RNase P accumulate a variety of tRNA precursors some of which are monomeric in form like the $tRNA_1^{Tyr}$ precursor, and other precursors containing two, three or even five tRNA sequences [28]. These multimeric precursors can be different tRNAs or tandem arrays of the same tRNA [29]. Bacteriophage T4 contains genes for 8 tRNAs which are linked in clusters. The DNA sequence of a cluster of seven tRNA genes has been determined and it appears that these tRNAs are transcribed as part of a long multimeric primary transcript [24]. A schematic version of the post-transcriptional processing of a multimeric precursor is shown (Fig. 11.5).

In eukaryotes the tRNA genes are arranged in very disparate ways. Some are scattered throughout the genome [30–34]. Some are clustered in an obviously regular manner at single chromosomal regions. Such clusters can contain several different kinds of tRNA genes, some present in multiple copies [35 36]. An additional complexity is that when certain tRNA genes from yeast were examined the actual DNA sequence in the gene was not colinear with the tRNA products [37–39]. Instead they contained interruptions in the coding sequences which have been termed *intervening sequences*. The $tRNA^{Tyr}$ genes from yeast

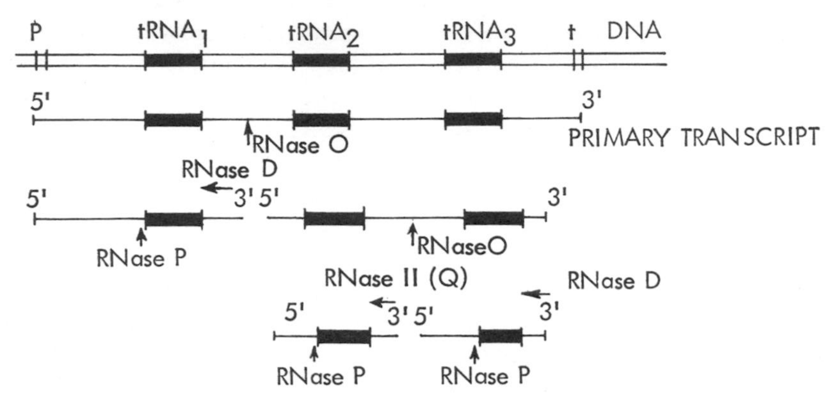

Fig. 11.5 Post-transcriptional processing of a hypothetical multimeric tRNA precursor.

contain an intervening sequence of 14 base pairs (bp) just following the anticodon whilst $tRNA^{Phe}$ genes contain an intervening sequence of 18–19 bp. A particular temperature-sensitive mutant of yeast (ts 136) accumulates precursors to these tRNAs at non-permissive temperatures. Although these do not turn out to be the primary transcripts they contain as part of their sequence a transcript of the intervening sequence present in the gene [40, 41]. Such molecules proved useful as substrates in the successful search for an enzymic 'splicing' activity which could clip out the intervening sequence and 'ligate' the ends of the surrounding sequences to form the mature tRNA [40, 41]. This novel enzymic activity requires magnesium and monovalent cations as well as ATP. In the absence of ATP, the intervening sequence is excised but the free ends are not ligated. From studies on the expression of plasmid-cloned yeast $tRNA^{Tyr}$ genes, microinjected directly into the nuclei of *Xenopus* oocytes (see previous Chapter), both precursors and mature $tRNA^{Tyr}$ can be detected and on the basis of such studies an outline of yeast $tRNA^{Tyr}$ biosynthesis has been proposed [42] (see Fig. 11.6).

The transcription of a cloned silkworm $tRNA_2^{Ala}$ gene in extracts of *Xenopus laevis* oocytes has also been studied [43]. This gene however has *no* intervening sequence but the primary product has a 5′-leader sequence of three extra nucleotides and a 3′-trailer sequence of 22 nucleotides which is endonucleolytically removed and replaced by a —CCA sequence. RNase P-like activity is believed to remove the three extra 5′-nucleotides [44]. The $tRNA^{Met}$ and $tRNA^{Leu}$ genes found in a 3.18 kb repeating unit in *Xenopus* [45] also have no intervening sequences and also lack the trinucleotide sequence coding for the 3′-CCA sequence which is added post-transcriptionally (see later). A $tRNA^{Phe}$

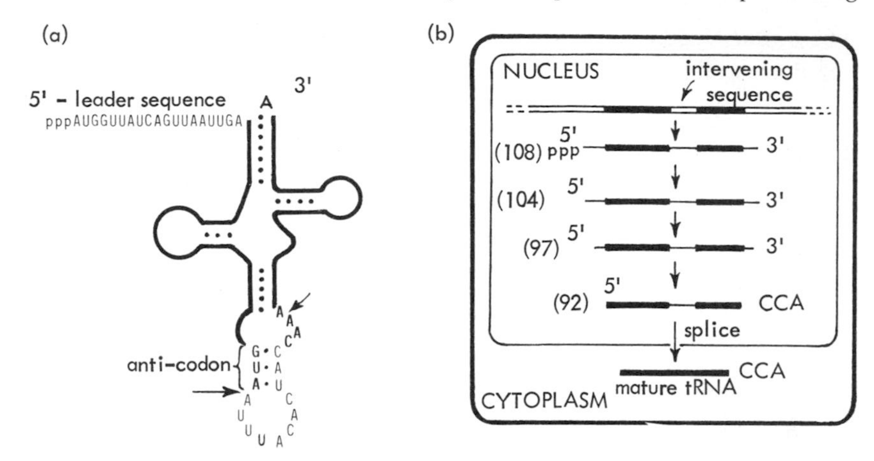

Fig. 11.6 Processing of yeast tRNA^{Tyr} precursors into mature tRNA^{Tyr}. (a) Schematic illustration of 108 base precursor with 19 nucleotide long 5′-leader sequence, 14 nucleotide intervening sequence (between arrows) and 5′-terminal triphosphate. (b) Schematic diagram of processing of steps. The numbers in brackets indicate the nucleotides in each of the intermediates. The 5′-leader is removed in three stages, the last of which is accompanied by the excision of the 3′-trailer end and addition of the 3′-CCA end group. Nucleotide modifications take place between 108 and 104, 97 and 92 and 92 and mature tRNA.

gene from *N. crassa* has been examined which is quite similar to that from yeast. It has an intervening sequence of 16 bp one nucleotide 3′ to the anticodon position. Whereas the coding sequence is 90 per cent identical to that from yeast the intervening sequences are only 51 per cent homologous [343].

The success of the amphibian oocyte nuclear transcription system has indicated a lack of eukaryotic species specificity regarding the splicing mechanisms and genomic regulatory regions for eukaryotic tRNA genes. Surprisingly in the case of *Xenopus* tRNA^{Met} genes the sequence information required to direct specific transcription initiation extends into the region coding for the tRNA itself. Sequences near both the 5′ and 3′ ends of the actual coding region are required [46]. It should be appreciated however that the Stokes radius of RNA polymerase III is of the same order of magnitude as the length of the tRNA structural gene. Just beyond the 3′ end of each tRNA coding region is an adenine rich tract [45]. This has been found at the 3′ end of other eukaryotic genes transcribed by RNA polymerase III and may act as a terminator signal in a fashion analogous to that described for *E. coli* terminators described in the previous Chapter. Studies on tRNA genes from

Drosophila suggest that the 5′-flanking sequences also contain regulatory elements [372].

11.3.1 The terminal addition of nucleotide units to tRNA

All transfer RNAs have the common trinucleotide sequence pCpCpA at their 3′ ends which may readily be removed to yield an acceptor molecule whose end groups may be represented as X and Y (Fig. 11.7). Two CMP moieties are first attached sequentially to the 3′-hydroxyl of the ribose in the terminal nucleotide Y. The near terminal CMP now accepts an AMP residue by a similar pyrophosphoryl cleavage of ATP. The final sequence at the 3′ end of the polynucleotide chain is therefore pXpYpCpCpA [48–52].

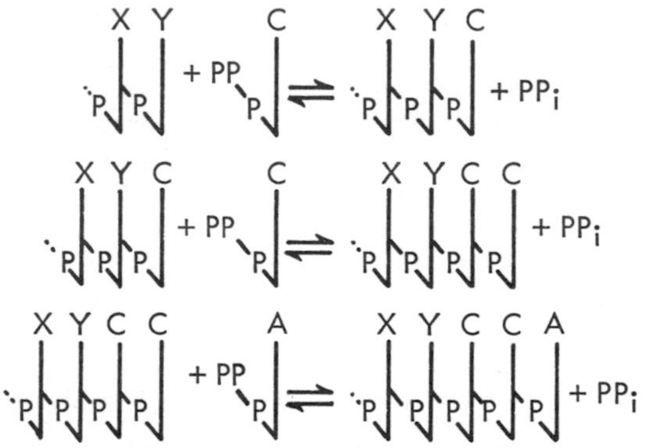

X Y + PPC ⇌ X Y C + PP_i

X Y C + PPC ⇌ X Y C C + PP_i

X Y C C + PPA ⇌ X Y C C A + PP_i

Fig. 11.7 Addition of terminal units to tRNA to give the final 3′-sequence—pXpYpCpCpA.

In the cell this sequence appears to have a metabolic turnover independent of the remainder of the transfer RNA molecule. The three nucleotides are apparently continuously removed and replaced [53, 54]. The cellular role of these nucleotide additions to tRNA is not understood, but it may serve in some way to regulate protein synthesis, perhaps by controlling the levels of functional tRNAs.

11.3.2 Nucleotide modification in tRNAs

During the post-transcriptional processing events involving the removal of the leader and trailer sequences several unusual modified nucleotides appear e.g. dihydrouracil, pseudouridine methyl-containing nucleotides. The presence or absence of the leader sequence can have an effect on the timing of these particular modifications.

Other modified nucleosides found in tRNA can be classed as follows: (i) those located in the first position of the anticodon, e.g. uridine-5-oxyacetic acid, 5-methylaminomethyl-2-thiouridine, inosine. The first of these can recognize A, G, or U in the wobble position of the codon (see Chapter 12) whereas inosine can recognize C, U, and A; (ii) those located next to the 3′-OH end of the anti-codon, e.g. 6-(isopent-2-enyl)adenosine, 2-methylthio-6-(isopent-2-enyl)adenosine, 6-methyladenosine, 2-methyladenosine, 1-methylguanosine. Such modified nucleosides adjacent to the anticodon may facilitate the formation of precise codon–anticodon base pairs by stabilizing the three-dimensional structure of the anticodon loop; (iii) those located elsewhere in the tRNA molecule, e.g. 7-methylguanosine, 4-thiouridine, 2-dimethylguanosine, 1-methyladenosine, and 5-methylcytosine. These are generally present in single-stranded parts of the clover-leaf structure, and although their role is unknown they may serve to stabilize the three-dimensional structure by preventing incorrect base-pairing [55].

The tRNA methylases catalyse the transfer of an intact methyl group from *S*-adenosyl-L-methionine to a C, N, or O atom of a purine or pyrimidine base or of ribose. The activity was first described by Borek [57] and is present predominantly in the soluble or cytoplasmic fraction of cells, although tRNA synthesis is nuclear [56–62].

Several tRNA methylases are present in any one cell system. For example, Hurwitz and his colleagues [58] separated six activities from *E. coli*. Eight activities have been studied in yeast [68] and a growing number have been purified from higher eukaryotic sources [63–68], including mitochondria [69]. tRNA methylases will methylate the homologous substrate only if it is methyl-deficient [57, 70] and can act on heterologous substrates to variable extents.

Several neoplastic tissues, both experimentally produced and spontaneous [59, 71–73], contain elevated levels of tRNA methylase activity with some changes in specificity [74, 75]. However, whether this differential activity is a reflection of new tRNA synthesis or of increased levels of methylation in tRNA remains to be established. Hormone treatment [78] and virus [79] and bacteriophage [76, 77] infection affect tRNA methylase activity. Bacteriophage T2 causes changes in base-specific methylation after infection [76] and bacteriophage T4 has a similar affect. A reduced activity is the result of induction in a lysogenic strain of *E. coli* K12 λ^+ [77].

Induction by ultraviolet irradiation of *E. coli* K12 λ^+ leads to the development of a dialysable inhibitor of methylation [77]. The activity seems to be directed against uracil methylase.

Animal tissues have been shown to contain a complex inhibitory activity to tRNA methylases, distinct from the above activity. It is largely absent from certain tumour and embryonic tissues and is influenced by hormones (for reviews see [80, 81]).

Another type of modification studied at the enzymic level is the formation of 4-thiouridine, another constituent of some tRNAs.

A tRNA sulphur transferase has been isolated from *E. coli* which catalyses the transfer of sulphur from L-cysteine into tRNA uracil *in vitro*. ATP and Mg^{2+} are required [56]. Isopent-2-enyladenosine, on the other hand, is formed in tRNA by the action of isopent-2-enyl pryophosphate tRNA-transferase. No cofactors are required. Pseudouridine formation in tRNA probably occurs by the enzymic modification of uridine residues in the RNA chain. Enzymes capable of carrying out this conversion have been detected in *E. coli* [56] and in the cytoplasm of mammalian cells [56]. The exact mechanism is not known. Whether it involves the cleavage of an *N*-glycoside bond in uridine, rotation of a uracil residue, and the formation of a *C*-glycosidic linkage remains to be seen.

11.4 BIOSYNTHESIS OF RIBOSOMAL RNA

11.4.1 The prokaryotic situation

In *E. coli* there are at least seven ribosomal RNA (rRNA) transcription units. These are dispersed in the genome and contain, in single transcription units, the sequences coding for 16S rRNA, 23S rRNA, 5S RNA and one or more tRNAs [82]. The tRNA sequences are located either in the spacer region between the 16S and 23S RNA genes or at the 3′-end beyond the 5S gene (Fig. 11.8). These units are transcribed in their entirety to yield a 30S primary transcript which accumulates in mutants of *E. coli* deficient in endonuclease RNase III [83, 84]. Treatment of the 30S transcript with RNase III gives rise to intermediate precursors for 16S(p16S_{III}) 23S(p23S_{III}) 5S(p5S_{III}) as well as fragments containing tRNAs [85]. Examination of the 5′- and 3′-ends of p16S_{III} indicates that together they can form 26 continuous base pairs and it is suggested that the substrate for the RNase III-catalysed production of p16S_{III} is a hairpin with a loop comprising the entire 1600 nucleotides of 16S rRNA [86]. (A similar structure can be constructed for the generation of p23S_{III} [87].) p16S_{III} is then acted upon by RNase M16 which generates the normal 5′-end of mature 16S rRNA with the help of ribosomal protein S4 (p.415) [88]. Another ribonuclease activity generates the normal 3′-end of 16S RNA. Similar types of enzyme appear to be involved in the production of 23S RNA from p23S_{III}. 5S RNA is generated from p5S_{III} as a result of RNase E removing the extra sequences at the 3′-end [89] and an *exonuclease* removing the three extra nucleotides at the 5′-end. In *B. subtilis* a 5S precursor has been isolated with 21 extra nucleotides at the 5′-end and 42 extra at the 3′-end [90]. An activity termed RNase M5 [91] removes *both* these extra sequences to generate the mature species. Sequence studies indicate that this ribonuclease must recognize some feature of 5S tertiary structure.

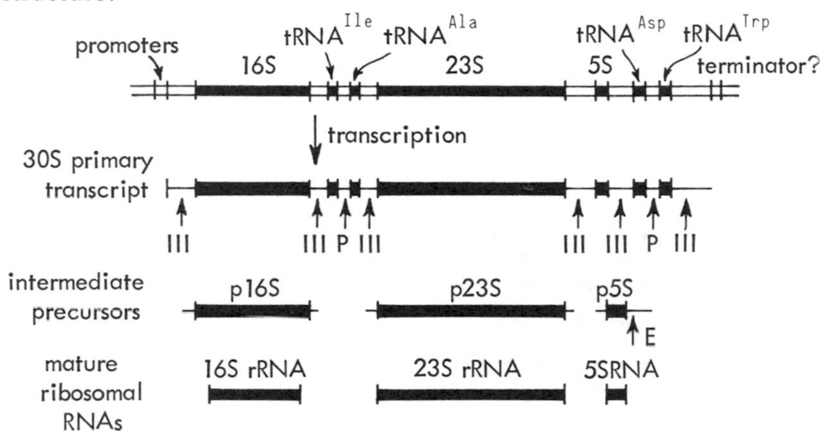

Fig. 11.8 Schematic illustration of the gene arrangements in one of the ribosomal transcription units (rrnX) in *E. coli*, their transcription and post-transcriptional processing. (The arrows indicate the approximate regions of cleavage by RNase III, RNase P and RNase E.)

The intermediate tRNA precursors generated from the 30S transcript are probably further processed by using RNase P and the exonuclease activity described in the previous section on tRNA biogenesis.

The promoter regions for some of the ribosomal transcription units in *E. coli* have been sequenced and in the case of units *rrn D* and *rrn X* two promoters are in fact found (P_1 and P_2) some 110 nucleotides apart [92]. RNA synthesis initiates with GTP at a point 284 bases from the ribosomal 16S sequence in *rrr D* and with ATP at 285 bases from the 16S sequence in *rrn X*. At P_2 in each unit RNA synthesis surprisingly starts with CTP 176 bases from the 16S unit. Two promoters, similarly disposed, are found in the unit *rrn E* [93]. Transcription termination at least in unit *rrn C* seems similar to the termination encountered in other *E. coli* operons [94] (see Chapter 10).

11.4.2 The eukaryotic situation

In most eukaryotes, the 5S RNA genes are separate in the genome from the genes for the rRNAs (18S and 28S) [95]. Yeast and *Dictyostelium* are known exceptions in which the 5S RNA and the larger rRNA genes are interspersed. The 5S and large rRNA genes occur in multiple copies in all eukaryotes [96]. In several eukaryotes (e.g. *Tetrahymena, Physarum*) the multiple rRNA gene copies are predominantly or entirely extrachromosomal [97]. In *X. laevis* there is a single autosomal nucleolus organizer region comprising 400–500 tandemly arranged copies of a DNA repeating unit whose structure is shown in Fig. 11.9. Only part of each unit is transcribed to yield a 40S primary transcript which includes the sequences for 18S, 5.8S and 28S rRNAs. This particular stretch of the repeating unit is constant in length and is almost exactly the same in nucleotide sequence in different repeating units. However, the transcribed region is flanked by a non-transcribed spacer sequence which can vary from about 2.7 kb to 9 kb in different repeating units [98].

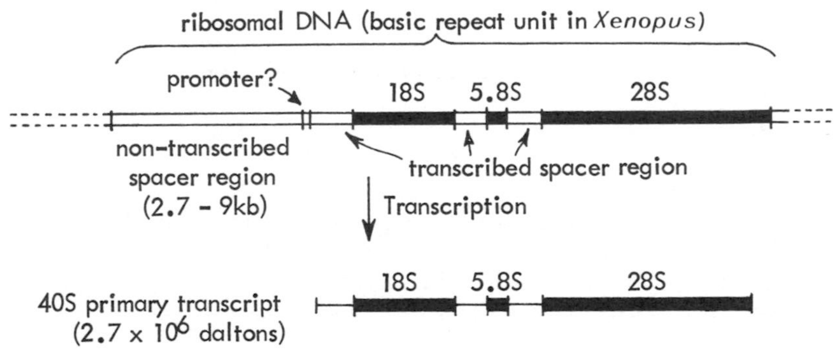

Fig. 11.9 Schematic illustration of a repeat unit of ribosomal DNA in *Xenopus* with its 40S primary transcription products.

Despite this difference in the length of the non-transcribed spacer, spacers differ little in sequence because they are internally repetitive. The basic internal repeats are similar throughout the non-transcribed spacer and may be as short as 15 nucleotides, although higher-order periodicities of around 750 nucleotides can be discerned. With regard to the structure of possible regulatory sites, a sequence AGGGGAAGAC has been detected which agrees with partial sequence data for the 5′-end of the 40S primary transcript. Whereas any putative promoter sequence must lie close to the transcriptional start point, a puzzling feature of the 130 nucleotide sequence around this point is that it is duplicated about 1000 base pairs upstream in the non-transcribed spacer region [99, 100]!

In *Drosophila melanogaster* there are two nucleolus organizers both near centromeric heterochromatin, one on the X-chromosomes the other on the Y [101]. Each organizer contains about 250 rRNA genes. There are two types of repeating unit. One is similar to that in *Xenopus* (see Fig. 11.10). The second type has the same basic organization, but the sequence encoding the 26S rRNA is internally interrupted by an intervening sequence, not present in the mature 26S rRNA. The basic transcription unit thus comprises sequences encoding the 18S, 5.8S and 26S rRNAs and an additional 30 nucleotide-long 2S RNA which is ultimately found associated with the mature 26S rRNA [102] (see Fig. 11.10). The intervening sequence in the 26S coding sequence occurs 2.5 kb from the 5′-end of the 26S region. There are at least two major classes of intervening sequence (type 1, 0.5 to 6 kb; type 2, 1.5 to 4 kb). Each class consists of discrete sub-classes differing in length and sequence [103]. Almost 60% of the repeating units comprising the nucleolus organizers on the X-chromosome contains type 1 sequence but such intervening sequences are either rare in the rDNA of the Y-chromosome or not present at all [103]. Sequences homologous to

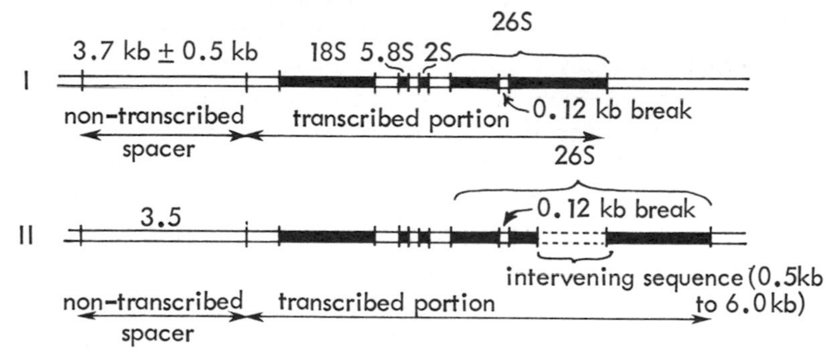

Fig. 11.10 Two types of rDNA repeat structures (I and II) found in *Drosophila melanogaster* (note that mature 26S rRNA actually contains a break in its structure and lacks a 0.12 kb sequence which is present in the DNA).

type 1 intervening sequences are widespread in the *Drosophila* genome [104] (i.e. without the nucleolus organizer). Type 2 intervening sequences occur in about 16% of the repeating unit of rDNA from both X and Y chromosomes. Repeating units with and without intervening sequences are interpersed in the X-chromosome rDNA but are clustered on the Y-chromosome. Whilst it would appear that the intervening sequences are in fact transcribed [105, 106], the units containing intervening sequences may only be transcribed rarely or their transcripts processed very rapidly. At least two alternative processing pathways for rRNA can be demonstrated in *Drosophila* [360]. The basic arrangement of the 17S and 26S rRNA genes in yeast is similar to the tandem array encountered for the 18S and 28S genes in *Xenopus* but the 5S RNA genes are actually physically linked to the other rRNA genes (see Fig. 11.11). However, the DNA repeat unit in a cluster of 100–140 [361] actually contains *two* transcriptional units, one codes for a common precursor (37S) for 17S, 5.8S and 26S rRNA and is transcribed by RNA polymerase I whereas a second is transcribed from the *opposite* DNA strand by RNA polymerase III and codes for *mature* 5S RNA [107].

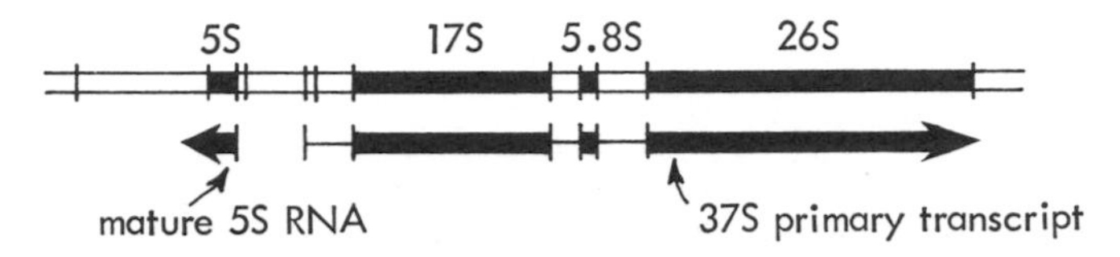

Fig. 11.11 rDNA repeat from *Saccharomyces carlsbergiensis* and its transcription products.

In *Tetrahymena* the macronuclear genes coding for ribosomal RNA also occur in multiple copies [108] but are located on linear extra-chromosomal molecules. Each rDNA molecule is an inverted repeat or palindromic structure containing two cistrons for a primary precursor RNA molecule arranged around a central axis of symmetry (Fig. 11.12) with the 17S rRNA coding regions closer to the centre. More surprisingly, in certain strains the 26S rRNA coding region contains an intervening sequence, not represented in the mature 26S rRNA molecules [109, 110]. A quite similar situation exists in *Physarum* where the ribosomal genes exist in a palindromic DNA molecule. However, two intervening sequences occur in the 26S rRNA coding regions [111], and spliced out in a random order even before transcription of the primary transcript is complete [362].

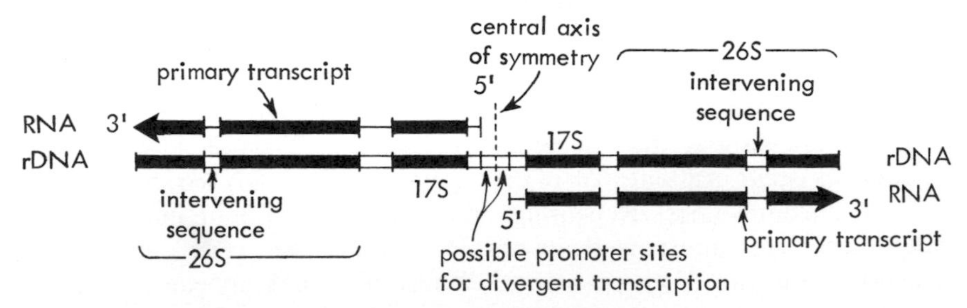

Fig. 11.12 Topographical map of the rDNA molecule of *Tetrahymena*.

When the primary ribosomal transcripts from *Tetrahymena* that contain the sequences of both 17S and 26S regions are analysed by R-loop mapping techniques (p.347) it appears that the intervening region whose sequence is now known [113] is part of the primary transcript [109, 112] and thus post-transcriptional removal of the intervening sequence is an early processing event in *Tetrahymena*, and can be demonstrated *in vitro* in isolated nuclei [114].

In mammals, repeating units of a type similar to that found in *Xenopus* are found. In the calf, for instance, the repeat unit is somewhat longer as a result of a large non-transcribed spacer component [115]. In human cells there is considerable polymorphism with regard to the length of the spacer but the variation is of a discrete nature rather than the continuous polymorphism in length found in *Xenopus* [116, 117].

With regard to the post-transcriptional processing of the various primary transcripts from the eukaryotic ribosomal transcription units, relatively little is known of the detailed enzymological steps involved. In human cells (HeLa) a 45S primary transcript of 4.1×10^6 daltons is found in the nucleolus where it undergoes a series of molecular 'tailoring' events which reduce its molecular dimensions to give rise eventually to the 28S and 18S ribosomal RNA species found in cytoplasmic ribosomes. Whilst the 45S species is found exclusively in the nucleolus, after longer labelling times radioactivity begins to appear in other species of RNA, first in 41S and 32S RNA (which are also confined to the nucleolus) and 20S RNA and then in 18S RNA which passes rapidly to the cytoplasm. Finally, radioactivity appears in 28S RNA initially in the nucleolus, then in the nucleoplasm, and shortly afterwards in the cytoplasm. A precursor–product relationship (see Fig. 11.13) between these RNAs has been proposed [118] based on a variety of data including recent electron-microscopic examination of the various RNA species [119].

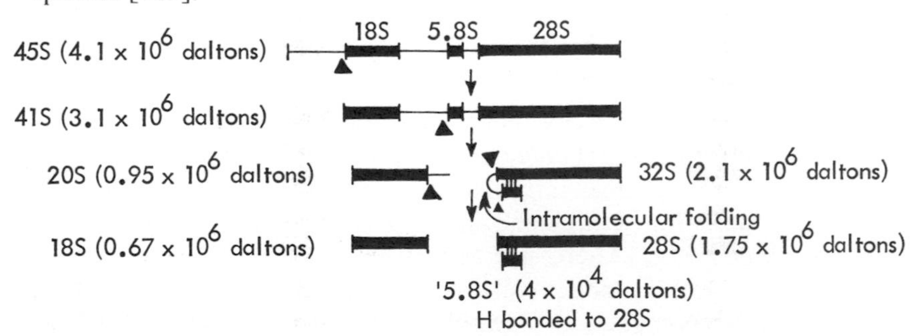

Fig. 11.13 The structure of HeLa cell rRNA precursors, and their sites of nucleolar cleavage (▲).

The '5.8S RNA' fragment appears to be generated simultaneously with the 32S→28S RNA transition. This has been interpreted by some as indicating the possibility that, in the conversion, a loop of 'spacer' RNA located on the precursor chain between the 28S and 5.8S species is degraded [120].

11.4.3 Modification of ribosomal RNA

In *E. coli,* the lighter species of rRNA has been shown to contain approximately 20 per cent more methyl groups than the heavier species [123, 124]. In bacterial rRNA, methylation of bases is about four times more frequent than methylation of sugar [123], whereas in plants [125] and mammals [126], many more sugar residues than base residues are methylated. In HeLa cells 70 and 46 nucleotides are methylated in 28S and 18S rRNA respectively [252]. In general, the formation of N^6-dimethyladenine seems to distinguish 18S from 28S rRNA in mammalian cells [127].

Methylation of 45S ribosomal precursor RNA (p. 339) takes place at the site of its synthesis in the nucleolus [128], and HeLa cells deprived of methionine produce undermethylated rRNA but do not form ribosomes [129]. Resistance to the antibiotic kasugamycin has been correlated with the lack of methylated base in ribosomal RNA [130] in bacteria.

The ribosomal RNA methylases, which act at the polynucleotide level using *S*-adenosylmethionine in the same way as the tRNA methylases, were first isolated from *E. coli* [124, 131, 132] and since then have been demonstrated in a variety of systems. Both sugar and base moieties can be modified *in vitro* [133]. Recently these methylases have been located as component parts of nascent ribosomal particles [134] in *E. coli* cells. In mammalian cells they appear to be associated with nucleoli [135], possibly in some association with nascent ribosomal particles by analogy with the *E. coli* situation.

11.4.4 Eukaryotic 5S ribosomal RNA synthesis

With the exceptions already mentioned, 5S RNA originates independently in eukaryotes [121] in the non-nucleolar part of the nucleus [122] and later becomes permanently associated with the large ribosomal subunit. In *X. laevis* for example the arrangement of 5S RNA genes is quite complex [136, 137]. There are two distinct 5S RNA gene families comprising separate tandem arrays. These families encode slightly different 5S RNA sequences, one being expressed only during oogenesis (oocyte 5S RNA), another throughout the developmental cycle except in early embryogenesis. The oocyte 5S DNA which contains some 24000 units has been extensively studied and each contains a variable region of AT-rich spacer DNA made up of repetitive elements, a single gene, and, at a distance of 73 nucleotides, a related sequence called a 'pseudogene' which appears not to be transcribed [138, 139] (Fig. 11.14a). Different repeating units, even adjacent ones, frequently differ in spacer length by multiples of 15 base pairs. The other, 'somatic', 5S

RNA genes are much less abundant [137] but the repeating unit of *Xenopus borealis* somatic 5S RNA is about 850 nucleotides long and consists of a single gene copy and a GC-rich non-repetitive spacer that is the same length in different repeating units [133] (Fig. 11.14b). In *Drosophila* there is a single cluster of 160 5S RNA gene copies at band 56F on chromosome 2 [140]. The repeat unit is shown in Fig. 11.14(c) and there is some noticeable length heterogeneity in the spacer region [346].

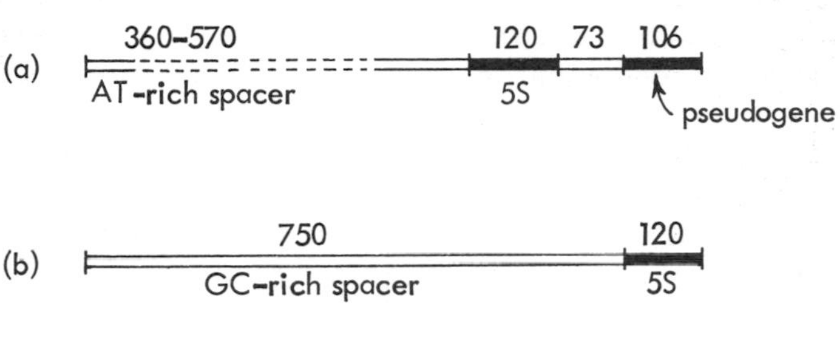

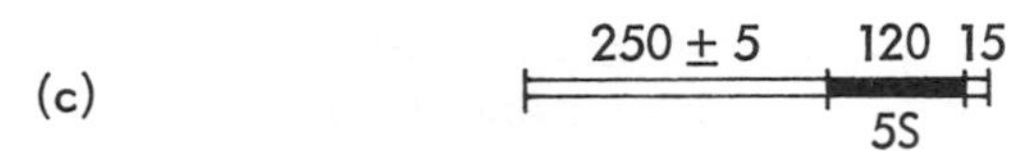

Fig. 11.14 5S DNA repeat units (a) *Xenopus laevis* oocyte 5S DNA, (b) *Xenopus borealis* somatic 5S DNA, (c) *Drosophila melanogaster* 5S DNA (numbers indicate base pairs in each region).

A search for primary transcripts of the 5S genes revealed a putative 5S precursor in *Drosophila* cells exposed to high temperature. This comprised the entire nucleotide sequence of 5S RNA supplemented by an extra 15 nucleotides which is very rapidly removed by processing at normal culture temperatures [141]. However, it appears that in the higher eukaryotes so far examined, 5S RNA is transcribed directly from the genomic sequences and thus no post-transcriptional processing appears to be necessary. A search for putative promoter regions utilizing the *in vitro* transcription systems derived from *Xenopus* oocyte nuclei (p. 315) have revealed that sequences of the 5S repeating unit can be removed right up to the structural gene region itself without imparing initiation. Indeed it appears that a control region beginning some forty nucleotides *within* the gene directs RNA polymerase III to initiate transcription approximately 50 nucleotides upstream from the border of this region [142]. The 3′-border of this control region resides between nucleotides 80 and 83 of the structural gene [143]. A factor necessary for the accurate transcription of *Xenopus* 5S genes *in vitro* has been isolated from soluble extracts of *Xenopus* ovaries. It binds to intragenic regions from positions 45 to 96 on somatic and oocyte-type 5S genes, in a manner independent of the presence of purified RNA polymerase III [344].

11.4.5 Spacer DNA

Some spacer DNA may have a regulatory role, for example, recognition sequences for cleavage of RNA precursors; specific signals for initiation and termination of transcription; sequences that mediate the correct timing of gene expression during development; sequences that serve as origins of DNA replication or differential replication that occurs in amplification. Such sequences may occur in each repeating unit or only periodically in the gene cluster. However, as spacer DNA seems freer to change in length and in sequence than the associated genes, only a small fraction of it is likely to be devoted to regulatory information requiring strict sequence conservation. On the other hand there is evidence to suggest that internal duplications and deletions possibly involving unequal crossing-over events [345] have contributed to the rapid variation in spacer structure [137]. It may be that spacer sequences simply exist as a result of the continuous production and elimination of tandem gene clusters. Alternatively spacers may have some function independent of their precise length and sequence.

11.4.6 Ribosomal RNA and ribosome production

The biosynthesis of ribosomes in bacteria is interconnected with the growth rate and appears to be regulated by altering the rate at which genes for the ribosomal components are transcribed and the products processed [144]. Our knowledge of the structure and transcription of the genes for the RNA components of the ribosomes is more advanced than that of the protein components. However, the genes for certain ribosomal proteins (L 11, L 1, L 10, L 7/12) have been located adjacent to the two genes specific for the β and β' subunits of RNA polymerase [145, 146]. Binding of ribosomal proteins to the 30S primary transcript occurs early in the transcription process [147] and the substrate for RNase III is, therefore, probably a ribonucleoprotein although the actual specificity of this nuclease appears to be independent of the associated ribosomal protein. Some of the other nucleolytic processing events are blocked by chloramphenicol treatment of the cells and thus may involve recognition sites provided by certain protein components. Ribosomal proteins are known to bind to specific sites on mature ribosomal RNA. Complete reconstitution of ribosomal subunits from constituent proteins and RNA is possible [148] and the process proceeds with the formation of intermediate particles whose physico-chemical properties are close to those of the precursor particles observed in the cell. Since the reconstitution process was found to be cooperative the presence of a few proteins may be critical in the production of mature ribosomes.

In normal growing *E. coli*, certain precursor ribosome particles have been observed, p30S (precursor to the small subunit) and $p_1$50S and $p_2$50S (precursors to the large subunit) [149]. p30S contains p16S_{III} RNA and some proteins of the small subunit. Both $p_1$50S and $p_2$50S contain p26S_{III} RNA and 5S RNA but their protein composition is different, $p_1$50S is a precursor of $p_2$50S, the transition involving eight new ribosomal proteins to the 16 or 17 already bound to $p_1$50S [150].

In eukaryotes the ribosomal proteins appear to be synthesized in the cytoplasm, but it is reasonably certain that they become associated with the ribosomal precursor RNAs in the nucleolus where the actual assembly of ribosomes takes place as indicated in Fig. 11.15. It is believed that the cleavage of the precursors takes place within nascent ribosomal particles [151].

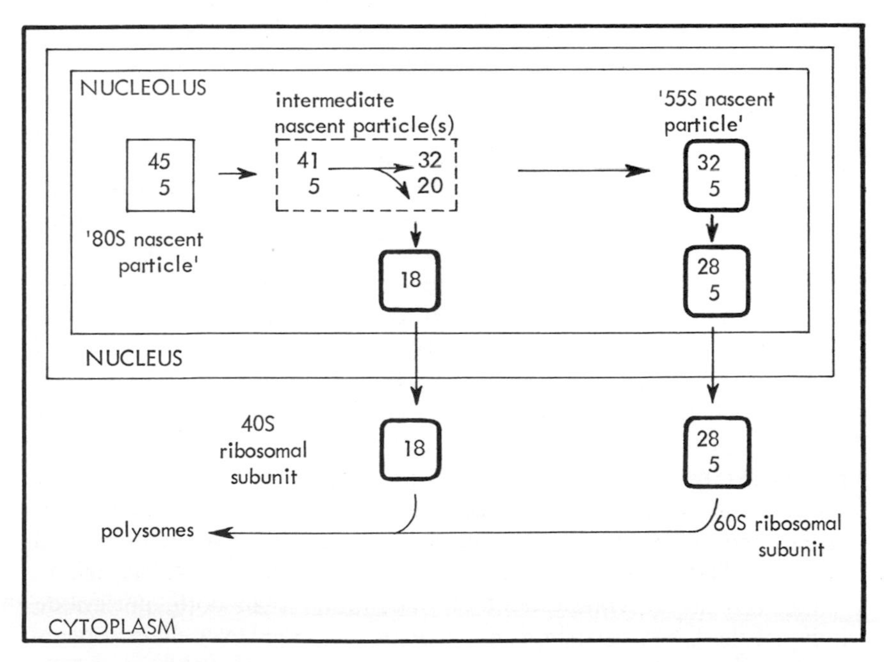

Fig. 11.15 An outline of eukaryotic ribosome formation illustrating the processing of RNA taking place within 'nascent ribosomal particles'. The numbers within the particles refer to the sedimentation coefficients of their constituent RNAs. Note the occurrence of 5S RNA even within '80S nascent particles' [120].

11.5 MESSENGER RNA BIOSYNTHESIS

The initial observation which led to the discovery of messenger RNA was made in 1953 by Hershey [152] who observed the rapid formation of new RNA molecules in cells of *E. coli* infected with bacteriophage T2. Volkin and Astrachan [153, 154] in 1956 labelled this RNA with ^{32}P and, from the distribution of label among the mononucleotides released on alkaline hydrolysis, concluded that it differed in composition from that of *E. coli* and resembled the DNA of the infecting bacteriophage. This suggested that it had been formed on a DNA template.

The physical characteristics of the RNA were investigated by Spiegelman and his colleagues [38, 155–157] who demonstrated specific hybrid formation between this RNA and bacteriophage T2-DNA and concluded that the rapidly labelled RNA formed after infection was in fact a T2-specific RNA with base sequences complementary to those in the T2-DNA. About the same time Jacob and Monod [158, 159] concluded that this new RNA synthesized after bacteriophage infection became attached to pre-existing ribosomes in the bacterial cell and could be detached in a caesium chloride gradient after lowering the magnesium concentration. That the ribosomes involved had been synthesized before infection was proved by labelling *E. coli* cells with ^{13}C and ^{15}N and infecting these 'heavy' cells with T2 bacteriophage in a 'light' medium containing ^{12}C and ^{14}N, when the T2-specific RNA was found attached to 'heavy' or 'old' ribosomes as was also the nascent T2-specific protein labelled by short-term exposure to radioactive amino acids. The T2-specific RNA had acted as a 'messenger' from the T2-DNA to the ribosomes where it directed the formation of bacteriophage protein. The existence of a similar unstable RNA in non-infected bacterial cells was soon demonstrated in several laboratories [160, 161]. Whilst the base composition of the new T2-mRNA was similar to that of T2-DNA, the total content of guanine and cytosine was approximately equal to the corresponding total for the bacteriophage DNA, but the individual bases occur unequally in the RNA (20 per cent for guanine and 17 per cent for cytosine). This observation, suggested Bautz and Hall [162], would be accounted for if the mRNA was always synthesized as the complement of one, rather than both, nucleotide chains of the DNA. Evidence that only one specific DNA chain appears to be made use of when a gene functions was also obtained from a study of the effect of 5-fluorouracil on the expression of certain rII mutants of bacteriophage T4 [163].

Additional support for the view that only one strand of the DNA helix is used as template came from studies with other bacteriophages. In bacteriophages α and SP8 the two strands of their DNA differ in density sufficiently for it to be possible to separate the individual 'heavy' and 'light' strands in caesium chloride density gradients. The RNA formed in the bacteriophage-infected bacterial cell, however, forms a molecular hybrid with only one of these strands, the 'heavy' one [164–166]. Similar studies using the component strands of ØX174 replicative form led to the same conclusions [167].

As mentioned in Chapter 5, various approaches combining molecular hybridization and gel electrophoresis have permitted the isolation of specific bacterial messengers such as that for *E. coli* lipoprotein and those specific for the *E. coli lac* and *gal* operons. The latter two are polycistronic and have molecular weights of 1.75×10^6 [168] and 1.5×10^6 daltons [169] respectively (see Chapter 5). In general it appears that in bacteria, mRNA transcription is initiated at promoter sequences by DNA-dependent RNA polymerase as discussed in Chapter 10 and terminated at appropriate termination sequences. Means whereby these processes are regulated are discussed more fully in Chapter 13. Although the above mentioned messengers do not arise as a result of post-transcriptional processing of larger precursors, processing is clearly detectable in the production of the bacteriophage T7-specific 'early' mRNAs that appear in the initial stages of the infectious cycle in *E. coli*. Basically these arise from five cistrons arranged sequentially at the 5′-end of the bacteriophage T7 genome (accounting for 20 per cent of the total genome). These are initially transcribed together as one piece of RNA; however, this transcript is quickly cleaved in processes involving RNase III to yield five separate and distinctive monocistronic mRNAs [170–172]. This is illustrated in Fig. 11.16.

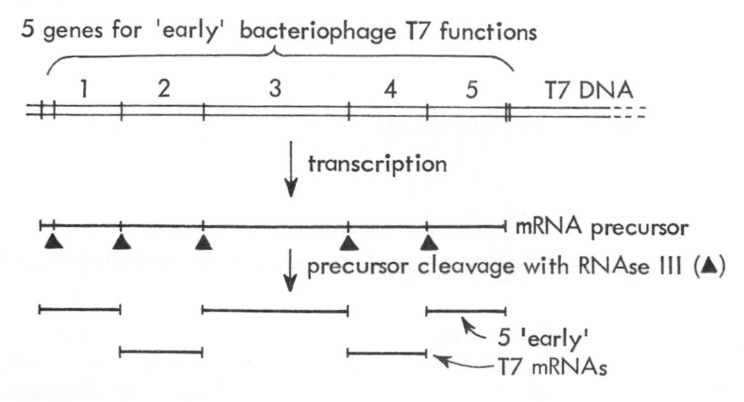

Fig. 11.16 Post-transcriptional processing of bacteriophage T7 early mRNAs.

In uninfected *E. coli*, recent data indicates that the transcription of the complex *rplJL-rpoBC* operon yields a high-molecular-weight RNA which is cleaved by RNase III [371]. This transcriptional unit codes for ribosomal proteins L10, L7/12 (see Chapter 12) and the RNA polymerase subunits β and β' (see Chapter 10), and the large transcript is cleaved some 200 nucleotides beyond the *rpl* genes [371].

11.6 EUKARYOTIC mRNA PRODUCTION

11.6.1 The globin mRNAs

When rabbit DNA, for example, is digested with a given restriction endonuclease (Chapters 3 and 6) about 10^6 different fragments are generated. The digested DNA can then be electrophoresed on an agarose gel and transferred to a nitrocellulose filter as described by Southern [173] (see Chapter 3). The fragments containing sequences complementary to β-globin mRNA can then be detected by hybridization with a ^{32}P-labelled cloned β-globin cDNA plasmid (see Chapter 14). This allows the physical mapping of specific restriction-enzyme cleavage sites in and around the rabbit β-globin gene directly in the genome itself [174] (Fig. 11.17). Detailed analysis of the physical map both in rabbit and mouse has shown that the β-globin gene does not exist as a contiguous stretch of DNA, but is present in three coding blocks (or exons) separated by both a large and a small intervening sequence (or intron) [175]. When a cloned 7 kb segment of mouse genomic DNA con-

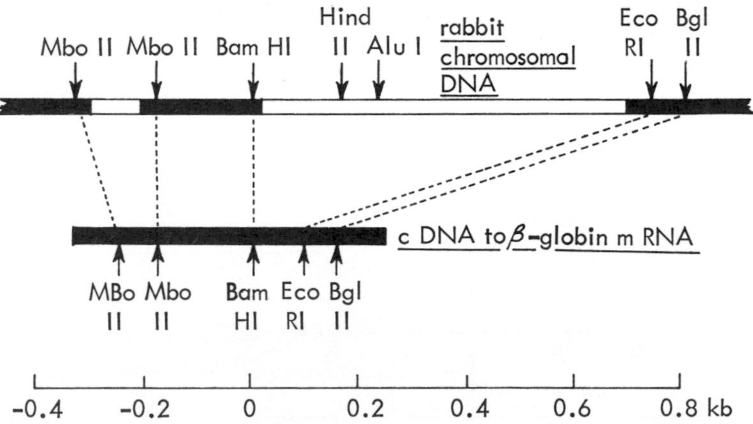

Fig. 11.17 A schematic comparison illustrating some of the restriction enzyme cleavage sites in cDNA to rabbit globin DNA and the rabbit chromosomal DNA containing the β-globin gene [302]. The distances given in kb are from the Bam HI site. The open regions represent intervening sequences.

taining the mouse β-globin gene was initially examined by the R-loop mapping technique [176, 177] using 9S globin mRNA further evidence for the large intervening sequence was obtained [178]. In this powerful electron-microscope technique (Fig. 11.18 and Plate V) the DNA duplex to be examined is partially denatured to allow a complementary strand (in this case globin mRNA) to hybridize and displace a single-strand or R-loop of DNA. In principle this permits a direct visualization of the globin gene region. However, in practice rather than a continuous R-loop within the 7 kb fragment *two* R-loop structures were initially readily detected. These were close to one another but were separated by a looped out double-stranded region for which there was no corresponding displaced single-stranded region. R-loops corresponding to the smaller intervening sequence are more difficult to detect. (Another means of detecting intervening sequences devised by Berk and Sharp [238] is shown in Fig. 11.19.) A detailed sequence examination of the large and small intervening sequences from mouse and rabbit β-globin genes showed them to be about equal length (large, 646 and about 580 bp; small 115 and 126 bp for mouse and rabbit respectively [24]). Moreover, they occur in precisely the same positions relative to the coding sequence (the large intervening sequence interrupts codons 104 and 105 and the small intervening sequence (codons 30 and 31). On the other hand the homologous intervening sequences show only a little sequence similarity and what similarity there is, is most evident at the junctions with coding sequences.

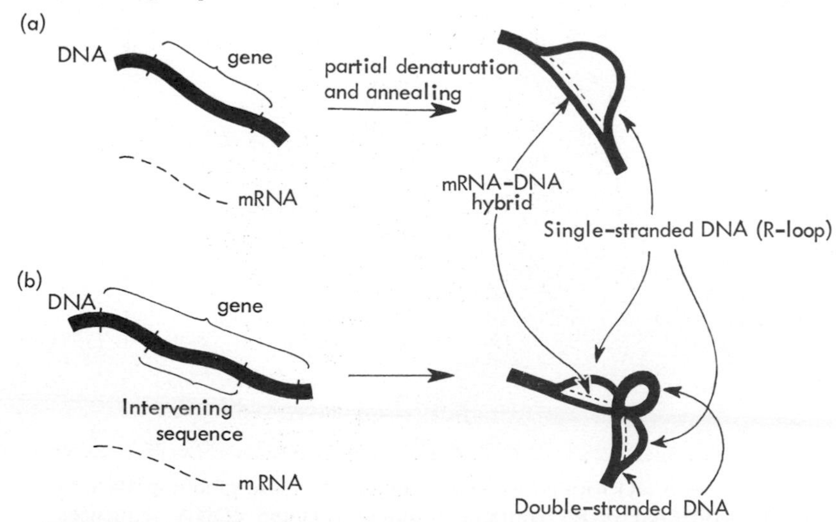

Fig. 11.18 A schematic indication of R-loop formation as applied to (a) a hypothetical gene without an intervening sequence and (b) a hypothetical gene with an intervening sequence.

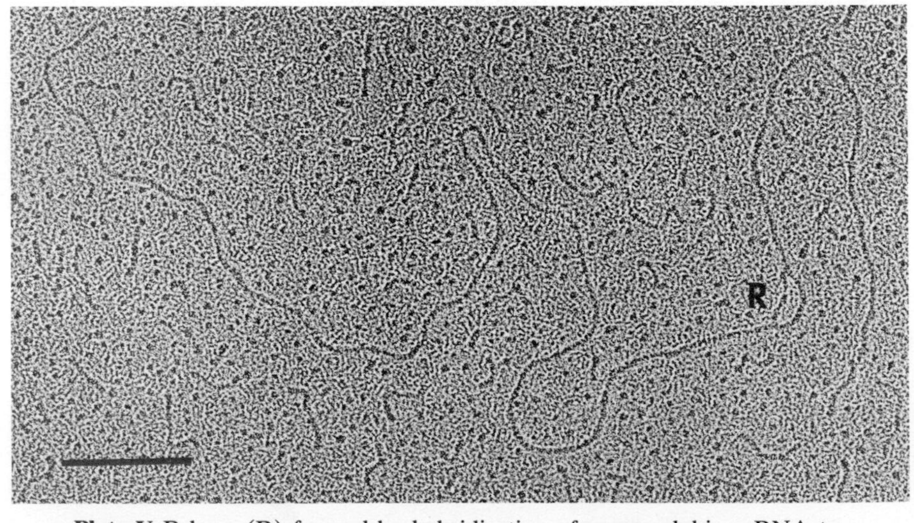

Plate V R-loop (R) formed by hybridization of mouse globin mRNA to HindIII—digested pCR1 containing mouse β-globin cDNA sequences inserted at the EcoRI site. The R-loop is stabilized by glyoxal which also extends the unhybridized RNA. Bar represents 0.25 μm. (By courtesy of Dr Lesley Coggins.)

In the mouse there is also another adult β-globin sequence (β^{min}) which would yield a product different by about nine amino acids. Again two intervening sequences split this gene into three coding blocks. The shorter intervening sequence is strongly preserved between the two mouse genes (β^{maj} and β^{min}) whereas the larger is less so [179] although the regions immediately adjacent to the coding regions are preserved in both cases. Intervening sequences have also been detected in chicken β-globin genes [181], human β, δ and γ genes [182].

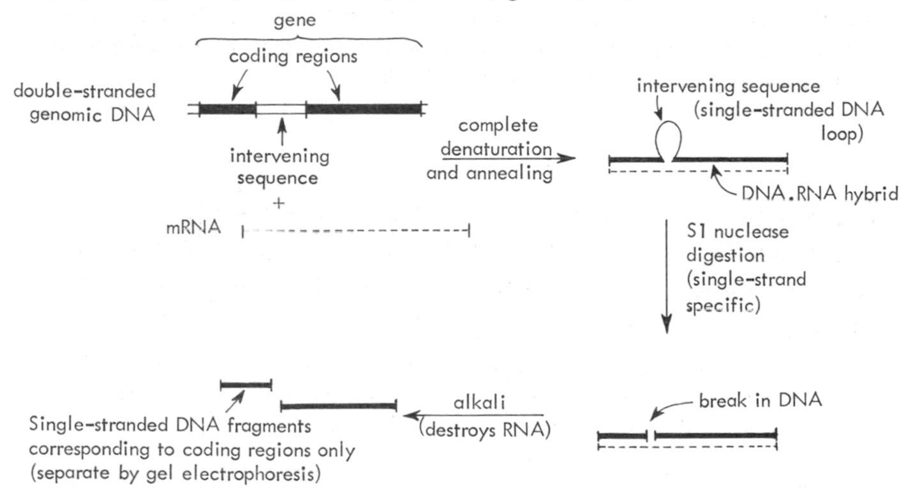

Fig. 11.19 A schematic illustration of an approach devised by Berk and Sharp to detect and locate an intervening sequence in a hypothetical segment of genomic DNA [238].

The α-globin genes of mouse appear to be interrupted by two intervening sequences which are, however, much shorter than those encountered in the β-genes (only 121 and 135 bp) [180]. In humans two α-genes are located 3.7 kb apart on chromosome 16 whereas the non-α-genes are closely linked 3.5 kb apart on chromosome 11 in a cluster with the arrangement γ^{G}-γ^{A}-δ-β [182, 226].

The question of transcription of globin genes has received most attention in the mouse system. The mouse β^{maj}-globin gene is initially transcribed in erythroid cells into a 15S molecule some three times the size of β-globin mRNA [183–185] which is processed to mature mRNA within 20 min [186]. Examination of this 15S transcript showed it to be 'capped' [187] and to have a 150 residue poly (A) 5′-terminus added post-transcriptionally [183, 186, 187]. In addition it contains sequences complementary to the intervening sequences as shown by electron microscopy and nuclease digestion of hybrids between 15S RNA and β-globin cDNA [188–190]. Since the 5′-terminal sequence of the 15S transcript and the mature mRNA map at the same position on the genomic DNA [191] it appears that the 15S transcript does not contain extra 5′-terminal sequences not found in β-globin mRNA. Thus the 15S precursor is converted into the mature mRNA without incurring any shortening at the 5′-end.

A question is, does the 5′-terminus of the 15S RNA represent the initiation of transcription? The 5′-terminal 'capped' regions of several adenovirus late mRNAs have sequences identical to a region of the adenovirus genome containing the late promoter [192] (see p. 356). Their structure contains some homology with the 'cap' region of mouse and rabbit β-globin genes [193] (Fig. 11.20).

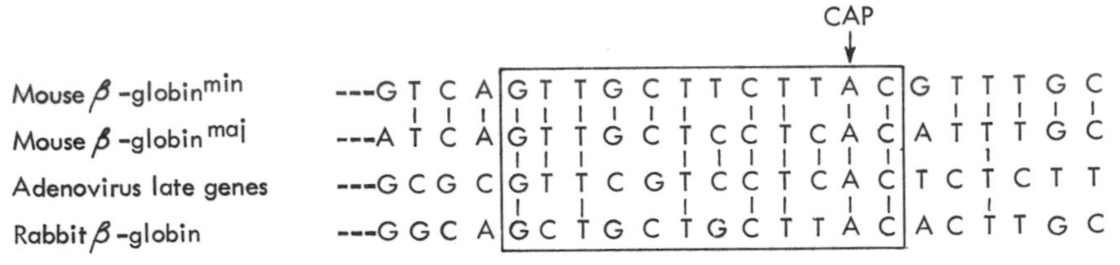

Fig. 11.20 Comparison of the regions surrounding the 'cap' and potential 'cap' sites. The boxed sequences contain homologies indicated by vertical lines. CAP↓ denotes the nucleotide corresponding to the penultimate nucleotide of the 'capped' mRNA.

In addition the 5′-flanking region of both mouse β-genes have a hepta-nucleotide TATAA$^{A}_{G}$G centered around 30 bp or three turns of the double helix upstream from the 'cap' site just described [179]. Molecular models have shown that this distance allows the TATAA and 'cap' sites each to be centred in the major groove and on the same face of the DNA-helix, thus conveniently located for interaction with say RNA polymerase [179]. A similar hexanucleotide is found identically placed with regard to the α-globin gene [180]. Interestingly, this hepta-nucleotide sequence also shares its first five nucleotides with the sequence described by Pribnow [377] for the proper binding of *E. coli* RNA polymerase to prokaryotic genes (the so called Pribnow box or hypothetical 'core' recognition region, R_c, as described in Chapter 10, and which precedes the transcription initiation site in prokaryotes by 5 to 7 nucleotides).

The β-globin gene structure has also been examined for potential regulatory regions towards the 3′-terminus (Fig. 11.21). The penta-nucleotide AATAA is located 18 bases before the poly(A) addition site in the β^{min} gene as is the case in the rabbit, human and mouse β^{maj} genes. In these latter three genes the poly(A) addition triplet seems to be TGC but in β^{min} it is TC. The reason for the uncertainty is that in both the rabbit and mouse genes the T(G)C sequence is followed by the dinucleotide AA so it is impossible to tell from these data whether the poly(A) is added to the C or to either of the A residues.

```
                                                     ↓ ↓ ↓
βmin----C T [A A T A A] A T G A T A T T T A C T C T   C A T C [T   C] A A T T C T G T---
βmaj----C A [A A T A A] A A A G C A T T T A   T G T T C A   C [T G C] A A T G A T G T----
                                                     ↑ ↑ ↑
```

Fig. 11.21 3′-untranslated regions of mouse β-globin genes. Ubiquitous pentanucleotide and possible poly(A) addition sequences are boxed. Arrows indicate possible poly(A) addition sites. The vertical lines indicate sequence homology.

At present, it is not known whether such sequences are involved in termination. A comparison of sequences at the 3′-ends of messengers transcribed by *E. coli* RNA polymerase has led to a model for transcription termination involving a GC-rich region capable of forming a hairpin followed by a run of U residues terminating the message. Although some of these features can be seen at regions corresponding to the 3′-ends of histone mRNAs (see Chapter 13) they do not occur in the case of many other mRNAs. The sequences immediately downstream from the last nucleotide coding for mRNA have now been examined in a number of genes and show little homology although they are often AT-rich and contain some runs of T residues of varying length. On the other hand, it is known from work with adenovirus that actual termination of late mRNA sequences occurs some distance downstream from the poly(A) addition site (p. 356). It is possible that the larger 27S non-polyadenylated globin containing RNAs that have been observed to occur with a short half-life in nuclei of certain erythroid cells [194–196] actually represent the primary transcript of the β-globin gene region, but which is quickly reduced in size to yield the well-studied 15S product which is then polyadenylated at the site discussed previously.

Regarding the processing events (see Fig. 11.22) that lead to the removal of the intervening sequences from the 15S precursor, intermediates have been detected in nuclei of erythroid cells which are consistent with the stepwise elimination of at least the larger intervening sequence by two or more cleavage-ligation ('splicing') reactions [190]. However, unlike the situation with tRNA, the enzyme (or enzymes) involved in these splicing events has not yet been detected. Thus without knowledge of the enzymes involved, it is difficult to comment on substrate recognition or mechanisms. Nevertheless, the availability of sequence information at the 'splice' point (Fig. 11.27)

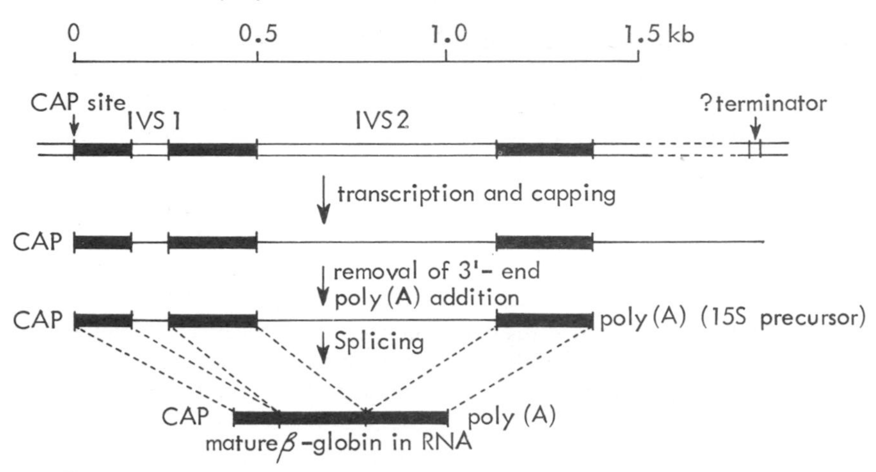

Fig. 11.22 The mouse β-globin gene: its transcription and the processing of transcription products to yield mRNA (IVS1 and IVS2 correspond to the small and large intervening sequences).

does put some limits on what the recognition sites might be. In the case of the globin genes, there is a striking divergence of sequence in the intervening sequences while sequences in the flanking regions are conserved. Whatever the mechanisms involved it is of interest that the complete β^{maj}-globin gene, including its intervening sequences and flanking sequences, has been cloned in the monkey virus, SV40, and the resulting recombinant used to infect cultured monkey kidney cells [198]. Despite the species and cell-type differences, the mouse globin signals for RNA splicing, poly(A) addition and translation are recognized, and substantial qualities of mouse β^{maj}-globin are produced. On the other hand processing of human β-globin mRNA precursor to mRNA is defective in three patients with β^+ thalassaemia. The genetic lesions may well be at splicing sites within intervening sequences in β-globin genes [374, 375].

Regarding the other globin genes, a precursor to mouse α-globin mRNA has been detected but it is somewhat smaller (850 nucleotides) which presumably reflects the smaller-sized intervening sequences [186, 197]. A precursor for human γ-globin on the other hand sediments at 16S [199].

11.6.2 Egg-white protein mRNAs

(a) *Ovalbumin.* The structure of the chicken ovalbumin gene turns out to be considerably more complex than that of the globin genes. Although the original restriction mapping suggested at least two

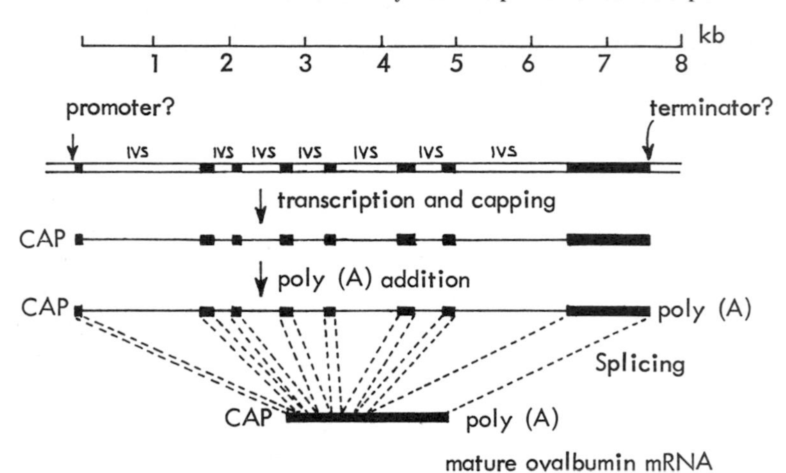

Fig. 11.23 The ovalbumin gene: its transcription and the processing of precursors to yield mature ovalbumin mRNA (IVS, intervening sequences).

intervening sequences, subsequent molecular cloning and sequence determination led to the conclusion that there are *seven* intervening sequences in the natural ovalbumin gene [200, 201] (Fig. 11.23), which is similarly organized in all chicken tissues examined. The minimal size of the transcriptional unit for ovalbumin is 7.8 kb. Approximately 30 bp upstream from the site coding for the 5′-end of the mRNA (which is assumed to be the 'cap' site) there is a sequence TATATAT which is strikingly reminiscent of the AT-rich sequence found at a similar position relative to the mouse β-globin genes (p. 350). Some additional homology with the globin genes is seen around 70 bp upstream from the presumed initiation/cap site and a model eukaryotic 'recognition' site for this particular region of GG$^{\mathrm{C}}_{\mathrm{T}}$CAATCT is proposed [202] in analogy to the prokaryotic recognition site (R_σ) discussed in Chapter 10. Transcription studied *in vitro* with calf thymus RNA polymerase II together with soluble factors from HeLa cells [353] support the view that such sequences are involved in the correct initiation processes [352]. However, the situation may be more complex. Ovalbumin is synthesized in *Xenopus* oocytes injected with a plasmid containing the ovalbumin gene but with the TATATAT site, the cap site, and first intervening sequence missing [363]. Although this suggests that these regions may not be absolutely required for transcription, accurate processing, transport and translation, the same may not be true for normal ovalbumin gene expression in the chicken oviduct.

The region corresponding to the 3′-end of the ovalbumin mRNA is shown in Fig. 11.24. Whilst no precise indication can be given at present

Fig. 11.24 3′-untranslated region of chick ovalbumin gene. Ubiquitous pentanucleotide is in box and possible poly(A) addition sites are indicated by the arrows.

regarding the actual site of transcription termination, it is likely to be close to the sequence shown in Fig. 11.24 as the maximum size of the nuclear transcripts containing ovalbumin gene sequences is no more than 7.8 kb [203]. To determine whether the intervening and structural sequences of the ovalbumin were transcribed into a large precursor that was processed to form mature mRNA, total oviduct nuclear RNA was first fractionated by size using agarose-gel electrophoresis in the presence of methyl mercury hydroxide [204] (p. 120) to eliminate secondary and tertiary structure. Then in a blotting procedure similar to that described by Southern [173] for the transfer of DNA fragments to cellulose nitrate, the size-fractionated nuclear RNA molecules were transferred to diazobenzyloxymethyl cellulose filter paper (p. 325) which can covalently bind single-stranded RNA molecules [205]. Once bound to this special cellulose the various size classes of oviduct nuclear RNA were then tested for their ability to hybridize to ^{32}P-labelled DNA probes corresponding to various parts of the ovalbumin gene. This revealed multiple species of RNA that are 1.3 to 4 times larger than ovalbumin mRNA and hybridize to both structural and intervening sequences of the ovalbumin gene. These results are consistent with the transcription of the entire ovalbumin gene into a long precursor followed by the excision of the intervening sequences to form the mature mRNA [206]. As was the case for the globin precursor, there is evidence that individual intervening sequences are not necessarily removed in a single step [202]. However, it seems that correct processing of the chicken precursors can be achieved in mouse cells as the natural chicken gene can be used to transform mouse cells [227].

A 46 kb region of the chicken genome containing the ovalbumin gene has been further analysed and shown to contain at least two other genes of unknown function. All three genes are orientated in the same direction and their expression in the chicken oviduct is under hormonal control. The three genes also show some sequence homologies, suggesting duplications have occurred in the ovalbumin gene region during evolution. [207].

(b) *Ovomucoid* is a major chicken egg-white protein whose synthesis is regulated by oestrogen and progesterone. The natural gene, which is 5.6 kb long and codes for an mRNA of 821 nucleotides, is split into at least eight segments by a minimum of *seven* intervening sequences of various sizes [208]. The shortest structural gene segment is as small as twenty nucleotides. All seven intervening sequences are located within the peptide-coding region of the gene. At 30 bp before the proposed initiation site is the heptanucleotide sequence TATATAT which, as already mentioned, is present in a similar location relative to the chicken ovalbumin gene. A search for high-molecular-weight precursors to the mature mRNA has revealed the presence in chicken oviduct nuclear RNA of multiple species of RNA which are 1.5 to 5 times larger than the messenger. By anology with ovalbumin RNA processing these data suggest that ovomucoid mRNA is derived from a primary transcript that contains intervening sequences [209].

(c) *Conalbumin* (or ovotransferrin) expression is also controlled at the transcriptional level by oestrogens and progesterone. The gene is one of the most baroque examples of the intervening sequence phenomenon. It contains a minimum of *sixteen* intervening sequences [210]! A possible promoter region (TATAAAA) occurs 25 bp from the 'cap' site as well as a possible eukaryotic recognition sequence GGACAAACA, 84 bp from the 'cap' site.

(d) *Lysozyme.* Chicken lysozyme gene also contains intervening sequences [211] and is about six times the length of the mature messenger.

11.6.3 Other eukaryotic mRNAs

Mammalian mRNA genes that have also been studied and which have intervening sequences include the non-allelic rat insulin genes (type II contains two intervening sequences, one 499 bp in the region encoding the connecting peptide of preproinsulin and the other 119 bp in the 5′-non-coding region of the mRNA from preproinsulin: type I only contains the smaller of these intervening sequences [212, 213]); human insulin gene (like rat insulin type II, it has two intervening sequences, 179 and 786 bp respectively [214]); human growth hormone gene (approx. 2.6 kb with three intervening sequences [215]); rat growth-hormone gene (approx. 3 kb with probably three intervening sequences [216]) rat serum-albumin gene (14.5 kb with thirteen intervening sequences [217]); mouse dihydrofolate reductase gene (42 kb with five intervening sequences [225]). In the case of the rat growth-hormone and serum-albumin genes high-molecular-weight nuclear RNA precursors have also been specifically identified [218, 219].

The δ-crystallin genes from chicks appear to be interrupted at least fourteen times [220]. A vitellogenin mRNA gene in *Xenopus* has twelve intervening sequences [221] and an analysis of putative mRNA precursors suggests that the order of splicing out of the different intervening sequences can vary.

Intervening sequences are also found in some insect genes. For example the fibroin gene of *Bombyx mori* contains a 970 bp intervening sequence as well as the putative promoter sequence TATAAAA the usual 30 bp from the initiation site [222].

In yeast, however, the genes for iso-1-cytochrome *c*, iso-2-cytochrome *c* and alcohol dehydrogenase appear to *lack* intervening sequences [223, 224, 347] although the actin gene *does* have a 304 bp intervening sequence [364, 376].

11.6.4 Virus messenger RNAs

RNA 'splicing' was in fact first discovered during studies on the specific mRNAs of DNA tumour viruses. However, the pattern of expression of the viral genes is extremely complex with a very large number of splicing events taking place. Only the major points will be discussed here, particularly where they demonstrate novel uses of RNA splicing.

Adenovirus 2 has a linear genome of 35 kb which after infection is transcribed in two phases, early and late [228]. In the late phase there is a single transcription start site at genome map position 16.45 [229]. However, although some of the 'late' genes are located thousands of nucleotides away (Fig. 11.25) each of the *fourteen* or so late mRNAs possess an identical 5′-sequence produced by the splicing on of *three* distinct segments of RNA (41, 71 and 88 nucleotides in length) originating from genome map positions 16.6, 19.6 and 26.6 respectively [230]. The precursor to *all* these fourteen mRNAs is a long primary transcript extending from map position 16.45 to 99.5 [231, 232]. The start cap site for transcription and the sequences 30 and 70 nucleotides

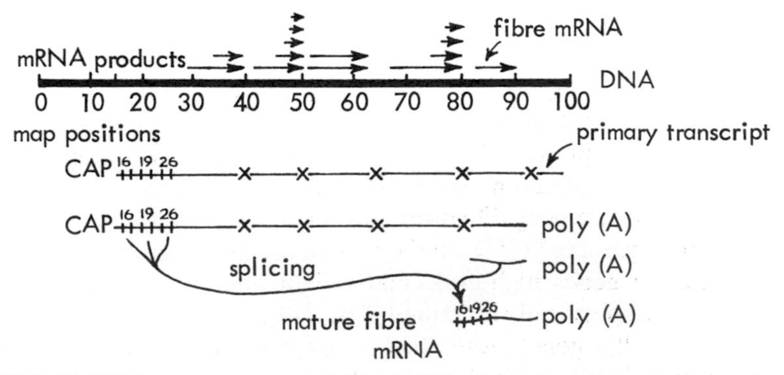

Fig. 11.25 The genome map positions of the late adenovirus 2 mRNAs (14 in 5 distinct groups). Below the genome is indicated the primary transcription product and cleavage sites (X) with possible processing steps involved in the production of just one of these mRNAs (fibre protein).

upstream from it have very distinct similarities to the corresponding regions already described for the globin and ovalbumin genes. The primary transcript is capped at the 5′-terminus and cleaved at a site corresponding to map position 92.5 at which poly(A) is added. Subsequent complex cleavage and 'splicing' events occurring in the nucleus, but possibly involving cytoplasmic factors [233], lead to the production of five distinct poly(A)-terminated RNAs. These in turn serve as the precursors to the mRNAs of the five groups that share 3′-sequences in common as well as having the common 5′-sequence ('leader') made up of the 16.6, 19.6 and 26.6 components described earlier (Fig. 11.25).

Regarding the 'early' mRNAs, they originate from four regions of the adenovirus genome. It seems that different mRNAs produced from single transcriptional units contain extensively overlapping sequences and differ from one another by the pattern in which sequences are spliced together with the removal of internal sequences [238] rather like the situation for the 'early' mRNAs produced on SV40 infection.

The small DNA viruses, SV40 and polyoma, contain covalently closed circular DNA genomes of around 5000 bp, which have been completely sequenced [239–241]. Again there are 'early' and 'late' phases of transcription. In the early phase a related group of proteins is synthesized called the 'transformation antigens'. In SV40 infections two are detected, large T (90 000 daltons) and small t (approx. 20 000 daltons) which has sequences in common with T [242]. In polyoma the situation is slightly more complex, there is a small t, a large T and a middle T antigen (see Chapter 4). The synthesis of the mRNAs for these transformation antigens demonstrates the power of RNA processing in the generation of more than one protein from the same sequence.

The SV40 large T antigen is synthesized from a 19S mRNA, but surprisingly the small t antigen is translated from a larger mRNA. Nevertheless, both mRNAs are the products of different splicing of a common primary transcript (Fig. 11.26). Both mRNAs have a common 5′-end but a larger segment of the primary transcript has been spliced out in the production of the large T mRNA. Sequence analysis of SV40 (and polyoma) reveals that there are translation stop codons (p. 406) in all reading frames in the region corresponding to genome map positions 54 to 60. Termination of translation in this region results in the synthesis of the small t antigen. When these stop codons are removed by splicing in the production of the mRNA for large T antigen, the synthesis of a much larger product is permitted but with the same *N*-terminal amino acid sequence (Fig. 11.26). Similar types of splicing allow the production of the appropriate mRNAs for the three T antigens from

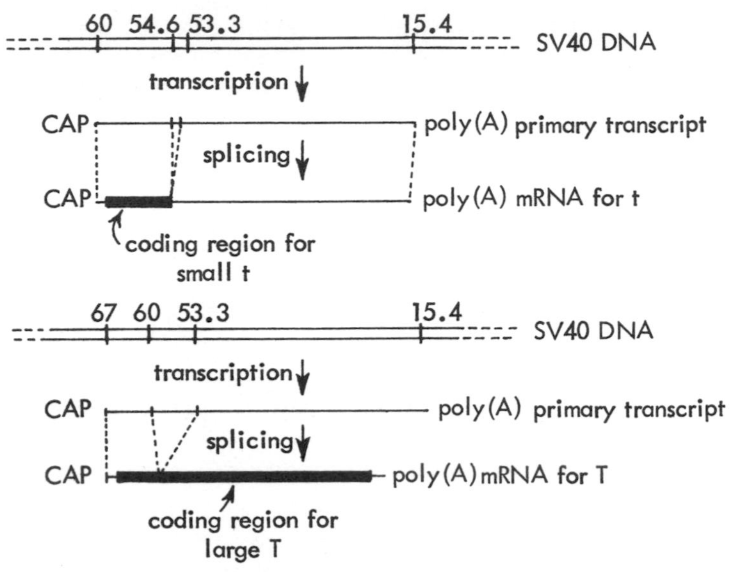

Fig. 11.26 Splicing pattern of SV40 early mRNAs (the numbers indicate approximate map positions on SV40 genome).

polyoma virus DNA [243, 244]. In the case of SV40, deletion mutants in the region of map positions 55–59 still permit the production of normal large T antigen [244]. Apparently the splicing takes place normally even when much of the intervening sequence is deleted, however these deletion mutants do not produce small t antigen as might be anticipated. Deletions in the coding sequences of SV40 do affect 'splicing' on the other hand especially when they are near 'splice' points. No stable mRNA is produced when 'splice' points are deleted [245].

Late in SV40 infection the other strand (L-strand) of the genome is used to direct the synthesis of three proteins, the major capsid protein VP1 and two minor proteins VP2 and VP3. VP1 is coded for by the region between map positions 95 and 16 but its mRNA contains a 5′-leader sequence of 203 nucleotides transcribed from a sequence about a thousand nucleotides away (position 72). The coding regions for VP2 and VP3 overlap with each other and part of the VP1 sequence [239, 240]. The mRNAs for VP2 and VP3 are also spliced and contain some of the 5′-leader sequences in common with VP1 mRNA. It is possible that different 5′-leader sequences regulate the level of mRNA utilization. The three mRNAs that encode the capsid proteins of polyoma virus are produced by the excision in the infected cell nucleus of different sequences from continuous transcripts of the L strand of the viral DNA [246].

Despite these complexities of post-transcriptional processing of eukaryotic viral mRNAs, it should be pointed out that the gene for polypeptide 1X which lies close to position 10 on the adenovirus 2 genome gives rise to an mRNA which unlike the mammalian mRNAs so far mentioned is *not* spliced [348].

11.7 POST-TRANSCRIPTIONAL PROCESSING AND mRNA FORMATION

11.7.1 Splicing

Sequence studies of the junctions between coding and intervening sequences (Fig. 11.27) have led to the proposal by Chambon [247] that all such junctions are related in sequence and may be derived from a prototype sequence of five to six nucleotides and that the 5′- and 3′-ends of intervening sequences are defined by the dinucleotides GU and AG respectively i.e. -AG ↓ GU...............AG ↓GU- (this rule, however, does not apply to the splice points in the tRNA genes [24] or in the ribosomal genes of *Tetrahymena* [113]). As sequences related to this prototype sequence occur elsewhere it is unlikely to be a sufficient signal for splicing specificity, it has been suggested that secondary structure

		Intervening sequence	
Ovalbumin			
	-----A A A U A A G	G U G A G C C ---(2)----- A U U A C A G	G U U G U U----
	------A G C U C A G	G U A C A G A ---(3)---- U A U U C A G	U G U G G C----
	-----C C U G C C A	G U A A G U U ---(4)---- U U U A C A G	G A A U A C-----
	-----A G A A A U G	G U A A G G U ---(5)----- C U U A A A G	G A A U U A-----
	-----G A C U G A G	G U A U A U G ---(6)----- G C U C C A G	G A A G A A-----
	-----U G A G C A G	G U A U G G C ---(7)---- C U U G C A G	C U U G A G----
Mouse β maj globin			
	-----U G G G C A G	G U U G G U A --small (1)-- U U U U U A G	G C U G C U-----
	-----C U U C A G G	G U G A C U C --large (2)-- C C C A C A G	C U C C U G-----
Rabbit β globin			
	------U G G G C A G	G U U G G U A --small (1)-- U U C U C A G	G C U G C U-----
	------C U U C A G G	G U G A G U U --large (2)-- C C U A C A G	U C U C C U-----
SV40 (T)			
	-----A A G U G A G	G U A U U U G --------- A U U U U A G	A U U C C A-----
SV40 (t)			
	-----C U A U A A G	G U A A A U G --------- A U U U U A G	A U U C C A-----

Fig. 11.27 Some sequences surrounding splice points [24].

in the primary transcript might bring into close proximity the ends of the coding regions to be spliced together. Tentative computer-based models of possible secondary structure for individual splicing events in SV40 [248] and adenovirus [249] have been proposed. However, an examination of the sequence of three ovalbumin-intervening sequences did not reveal any really stable structures which might facilitate the process [250]. Specific deletion mutants of SV40 suggest that correct splicing can occur even when large portions of intervening sequences have been removed. Using a mouse β-globin-SV40 DNA recombinant it appears that only 13 or fewer nucleotides of the coding sequence are required at the 5′-end of a specific splice point for correct splicing [251]. As previously mentioned it seems that some intervening sequences are not removed in a single step. The AGGU prototype sequence has been located at internal positions in some intervening sequences but not in others [250] and may be involved in preliminary excision events.

Unlike the eukaryotic tRNA situation no enzymatic splicing activity has yet been isolated for mRNA, although the reactions have been detected in isolated nuclei to which various soluble cytoplasmic and nuclear fractions have been added [233, 349]. Clearly endonucleases may be involved. A RNase III-like activity can be detected in mammalian nuclei (see Chapter 6) and a role for double-stranded structures in the processing of SV40 mRNAs is implied from studies with proflavine [359]. Whatever the mechanism, it appears to be complex and may well involve both nuclear and cytoplasmic factors [233]. Moreover, splicing of SV40 messengers seems to occur in differentiated mouse cells but not in undifferentiated cells derived from mouse teratocarcinoma [252]. A limitation to our understanding of the problem is the fact that the primary transcripts are immediately attached to certain structural proteins to form nuclear ribonucleoprotein particles (see later). Nevertheless, recent interest has focused on the detection [257] by specific antibodies, of six complexes each comprising a different small nuclear RNA (snRNA, see Chapter 5) but the same seven specific proteins. These complexes in turn seem to be associated with the above-mentioned ribonucleoprotein particles. Examination of the sequence of one of these small RNAs (U1A) [254] shows its 5′-terminal sequence to have some complementarity to the sequences found at splice points. It is hypothesized [253, 255, 350] that the 5′-terminal sequence of U1A RNA interacts with sequences at the splice and serves a template function in aligning the sequences to be spliced, possibly involving the enzymic activity of the proteins associated with the snRNA. A similar role is postulated for the small RNAs(VA), specifically coded for by adenovirus DNA, in the production of adenovirus mRNAs [365].

Whatever the mechanism of RNA splicing, it is a major and reliable mechanism in the expression of the eukaryotic gene. The cell can apparently quite happily cope with at least sixteen intervening sequences in the case of the conalbumin gene and remove all of them accurately to the precise base pair. Particularly noticeable in the case of conalbumin, ovalbumin, ovomucoid and lysozyme genes, all the coding regions are short with sizes ranging from 50 to 200 base pairs. Any attempt to explain the emergence and evolution of split genes must take into account the small size and high number of coding regions in these natural genes. This makes it difficult to accept without reservation the idea proposed by Crick [256] amongst others that these genes have evolved from already distinct regions each coding for a different structural domain, and that they were brought together in the genome by random shuffling. Although conalbumin mRNA is only 25 per cent longer than ovalbumin mRNA, the mRNA-coding portion of the natural conalbumin gene is in 17 pieces whereas the natural ovalbumin gene is in 8 pieces. On the other hand, in the case of immunoglobulin heavy chains, intervening sequences do seem to divide the gene into segments each encoding a separate structural domain [356, 357]. Additional data indicates a possible relationship between coding sequences and function in globin genes [358].

Whilst the polarity of an intervening sequence is critical for its function the species of origin and relative position appear less so [366]. Thus, intervening sequences may represent functional elements in the generation of stable mRNAs.

11.7.2 5′-terminal capping

A large proportion (40–60 per cent) of possible nuclear mRNA precursors are capped [258]. Mechanisms involved in the formation of capped structures in eukaryotic systems have been most extensively studied in the case of certain virus-specified mRNAs. In the case of vaccinia virus, reovirus and cytoplasmic polyhedrosis virus the capping reaction appears to involve the transfer of a guanosine monophosphate moiety from GTP to a 5′-diphosphate terminated mRNA followed by methylation at a 7 position of the guanine [259] (Fig. 11.28). However, in the case of vesicular stomatitis virus a different mechanism involving capping of a 5′-monophosphate appears to prevail [260].

The penultimate nucleotide at the 5′-terminus of capped viral mRNA appears to contain exclusively a purine base. Consequently it was felt highly probable that capping of eukaryotic viral mRNA species occurs at the sequence corresponding to the point of transcription initiation, a conclusion reinforced by the proximity of the promoter and cap site for

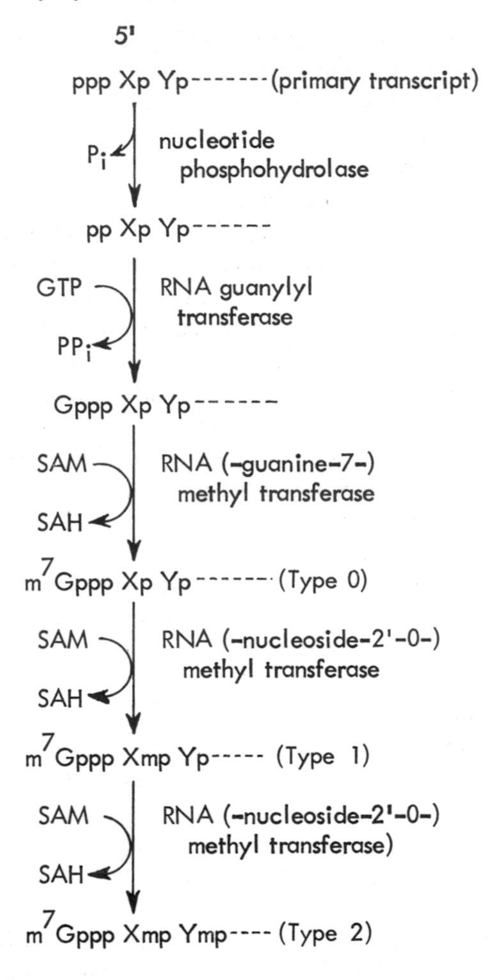

Fig. 11.28 Possible mechanism for the biogenesis of 5′-terminal cap structures (Cap 0; Cap 1; Cap 2) in eukaryotic mRNA precursors. SAM, S-adenosyl-L-methionine; SAH, S-adenosyl-L-homocysteine.

the 'late' adenovirus mRNAs (p. 356). Indeed as already pointed out in this chapter there is a remarkable homology between the cap and possible initiation sites of various viral and non-viral eukaryotic genes. Labelling experiments indicate that addition of 5′-cap structures to primary transcripts occurs very rapidly [261, 262] (even before poly(A) addition) in eukaryotic cells, probably by a mechanism similar to that described for vaccinia and reovirus mRNAs (Fig. 11.28). RNA guanylyl transferase has been detected in HeLa cell nuclei [263] and RNA (guanine-7-)-methyltransferase and RNA (nucleoside-2′-*O*-)-

methyltransferase have been purified from HeLa cell cytoplasm [264, 265], although such activities, at least in rat liver, may also associate with the nuclear ribonucleoprotein particles containing the mRNA precursors [266] (see later in this section).

Internal methylated sequences containing 6-methyladenine are also found in the large nuclear transcripts (two to six per molecule) [267] and it would appear that a large proportion of such sequences are conserved during the production of mRNA [268, 269].

11.7.3 3′-poly(A) addition

The inability to detect 'free' poly(A) in nuclei after very brief radioactive labelling [270] suggested that the poly(A) sequence synthesized in the nucleus is incorporated at the 3′-end of mRNA precursors in a series of sequential additions of AMP residues from substrate ATP, the reaction being carried out by poly(A) synthetase (Fig. 11.29). Addition of poly(A) had long been considered a post-transcriptional event, since it had been shown that cukaryotic DNA did not contain dT regions long enough to code for poly(A) [271]. In addition the process was insensitive to actinomycin D [270].

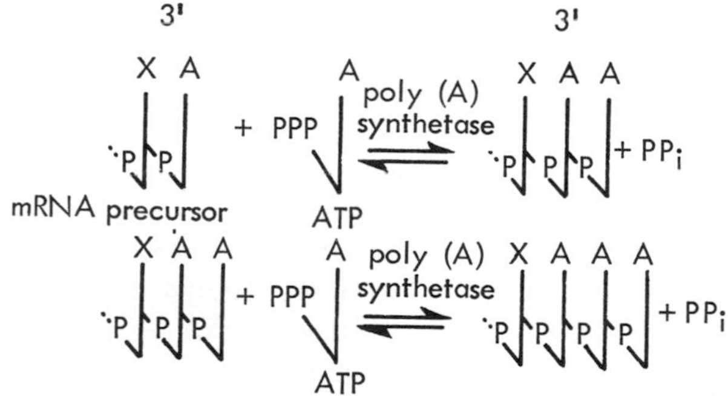

Fig. 11.29 The sequential addition of terminal units to a nuclear mRNA precursor to generate 3′-terminal poly(A) tracts. The precursor RNA here is being used as a primer rather than a template for the enzyme poly(A) synthetase which also requires a divalent cation such as Mn^{2+} or Mg^{2+}.

Addition of poly(A) appears to be generally an event which occurs relatively early after transcription. Some is added almost immediately after transcription, but some is also added some time after transcription [272], possibly to cleavage products of the primary transcript (see Fig. 11.25). Between 70 and 100 per cent of the nuclear poly(A) sequences present in the polyadenylated nuclear RNA transcribed from the adenovirus 2 late-transcription unit are transported to the cytoplasm [273].

Poly(A) synthetase activity (RNA terminal riboadenylate transferase EC 2.7.7.19) has been reported in a wide variety of eukaryotic cells [275, 276] and large-scale purification has been achieved in a few cases [277–279]. In several cell types it has been located in the cell nucleus although several workers have described cytoplasm-based activities [280]. In rat liver nuclei no detectable poly(A) synthetase was apparent in the nucleoli [281], whereas 'free' and chromatin bound forms could be distinguished which appear to exhibit different primer specificities [281, 282]. In addition, two poly(A) synthetase activities distinguishable on the basis of their ionic requirements have been reported in association with the nuclear ribonucleoprotein particles containing mRNA precursors [283].

11.7.4 mRNA precursors and their association with proteins

Evidence for an association of mRNA precursors with proteins in the nucleus has been obtained from a variety of ultrastructure studies. For example, electron-microscopic visualization has clearly indicated the presence of long ribonucleoprotein fibrils being transcribed from the DNA of the lateral loops of amphibian lampbrush chromosomes [285] as well as from dispersed HeLa cell chromatin [284–286] (see Plate VI).

These nuclear ribonucleoprotein complexes (protein to RNA ratio of about 4:1) have been isolated from eukaryotic nuclei either by extraction of intact nuclei with isotonic pH8 buffer [287] or after mechanical disruption of nuclei (e.g. sonication [289], French press [290], nitrogen cavitation [290]). The substructure of such complexes has been shown by electron microscopy to comprise a number of particles. 'Monoparticles' can be generated from such 'polyparticle' preparations by the controlled action of added RNase [290] (cf. the nucleosome structure of chromatin). Examination of purified 'monoparticles' indicates some size heterogeneity (200 Å–300 Å in diameter) [291–295].

It is generally believed that there are a few major polypeptides of about 30000 to 45000 daltons and several minor species from 45000 to 150000 daltons or more [288, 293, 294]. The smaller major polypeptide species (the 'core' polypeptides) have been suggested to function in the stabilization and packaging of the nuclear mRNA precursors [293], whereas the minor polypeptides may be contaminants or possible processing enzymes with only a transient association. Some tissue specificity has been demonstrated for the polypeptide components of the 'polyparticle' preparations isolated from disrupted nuclei [288]. However, the polypeptides of 'monoparticles' from a wide variety of nuclei are quite similar [292–294]. Polypeptides associated with the poly(A) segment of the nuclear mRNA precursors usually include two to three major

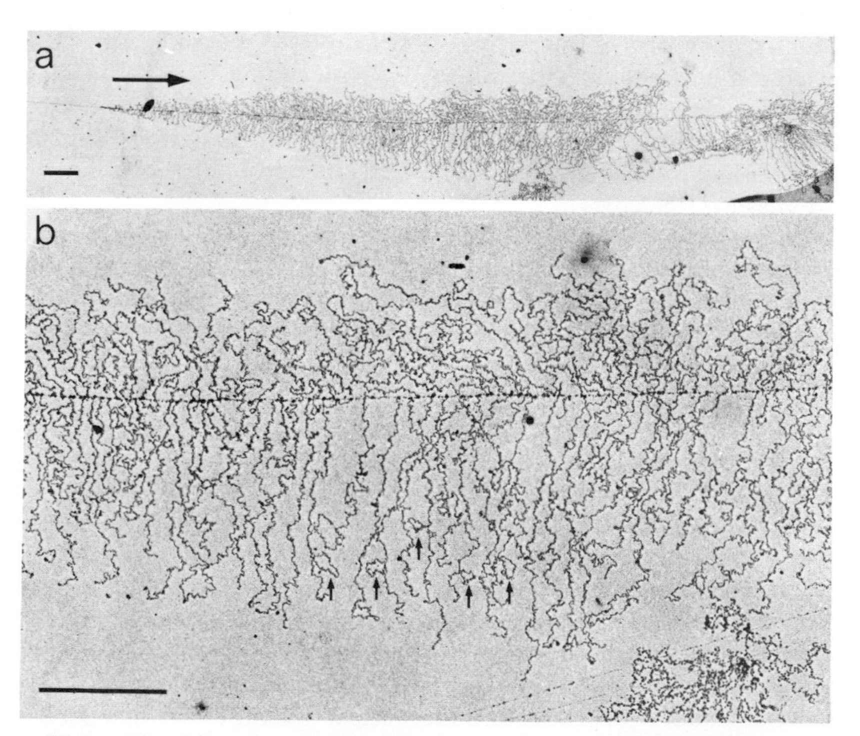

Plate VI Electron micrographs of spread transcriptionally-active chromatin from *Triturus oocytes.* (a) The structural features of a transcriptional unit showing the chromatin axis running from left to right and attached lateral ribonucleoprotein (RNP) fibrils forming a gradient of increasing lengths. The arrow indicates the direction of transcription. (b) Part of the same unit shown at higher magnification. Individual RNA polymerase molecules can be seen as densely stained particles along the chromatin axis and at the base of each RNP fibril. Secondary structures can be seen in the RNP fibrils themselves: these occur as hairpin-like configurations, or as loops (arrowed) which apparently can become detached from the main fibril as RNP circles. The bars indicate 1μm. (By courtesy of Dr U. Scheer and Dr J. Sommerville.)

species, especially a polypeptide of about 75000 daltons which is shared by the poly(A) sequence in the cytoplasmic complexes between mature mRNAs and protein [296, 297]. Recently it has been shown that this poly(A)-specific mRNA binding protein and the poly(A)-synthetase discussed in the previous section (p. 129) are antigenically related [298]. Thus poly(A) synthetase may well become associated with the newly synthesized mRNA precursors in the nucleus and catalyse the initial poly(A) addition reactions. The enzyme may then escort the mRNA through the nucleoplasm to the cytoplasm continuously catalysing the poly(A) extension reaction. Although the initial poly(A) addition may play a role in transport to mRNA to cytoplasm, chain extension may stabilize the mRNA by protecting it from degradation (see Chapter 13). Recent data from nuclease-digestion experiments, however, indicate that structural relationships of the poly(A)-binding protein, and the poly(A) regions may be different in nuclear RNA and cytoplasmic RNA [354].

Although several models have been proposed for the structure of the nuclear ribonucleoprotein particles [292, 299–301] the relationship of these particles to the associated small-nuclear RNA-containing protein complexes (mentioned in Section 11.7.1) remains unclear.

11.8 TRANSCRIPTION IN MITOCHONDRIA AND CHLOROPLASTS

The construction of both mitochondria and chloroplasts requires the contribution of two genetic systems: the nuclear system and a second system located in the organelle itself [303–306]. Although both systems carry out a basically similar sequence of reactions, DNA replication and transcription and the translation of mRNAs, the two systems do not have a single component in common. Nevertheless, the mitochondrial genetic system is dependent on the nuclear system for its functioning. Major enzyme complexes found in the inner mitochondrial membrane are synthesized part on mitochondrial ribosomes and part on cell sap ribosomes.

11.8.1 Mitochondria

The mitochondrial DNA (mtDNA) from a variety of organisms has now been examined and its molecular weight varies within a fairly narrow range, between 10×10^6 in animal tissues to 70×10^6 in higher plants (yeast is 49×10^6). With the exception of mtDNA from *Tetrahymena* and *Paramecium* circularity is the rule. Although, for example, mtDNA can account for up to 0.5 per cent of total human DNA, it appears that all mtDNA molecules in a single normal organism have an identical nucleotide sequence and there are no major gene repetitions. The complete 16 569 bp sequence is now available [307]. The genes for the 12S and 16S rRNAs, 22 tRNAs, cytochrome oxidase subunits I, II and III, ATPase subunit 6, cytochrome b and 8 other protein genes have been located. The two adjacent rRNA (16S and 12S) genes together with a number of tRNA genes scattered around the circle were found on one strand. The 12S rRNA gene and the tRNA^{Phe} gene are in fact joined end-to-end [335]. Other tRNA genes were found widely separated on the complementary strand (Fig. 11.30a). The sequences of the rRNA genes are homologous in some regions to eukaryotic and prokaryotic sequences, but nevertheless they are quite distinctive and indicate that human mtDNA did not originate from recognizable relatives of present-day organisms [367].

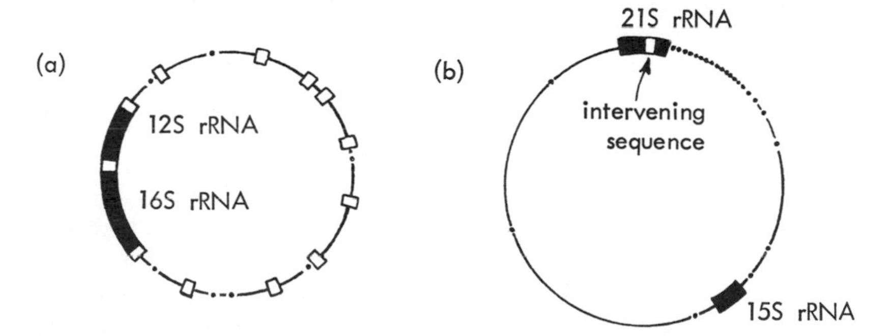

Fig. 11.30 Diagrammatic representation of some mitochondrial genes (a) human mitochondrial DNA, L-strand codes for 12S and 16S ribosomal RNAs and tRNAs marked □, the H-strand codes for tRNAs marked • (b) yeast mitochondrial DNA, all tRNAs are marked •. The same scale is not used for (a) and (b).

In yeast the mtDNA is about five times the size of human mtDNA. However, the size of animal mtDNA appears sufficient to code for all the known and putative genes. The extra DNA in the larger mtDNAs may merely serve a spacer function. Indeed restriction-mapping data show the genes for large (21S) and small (15S) ribosomal RNAs to be very far apart (25 000 bp) in yeast mtDNA [306] (Fig. 11.30b). Moreover, the gene for the 21S species is split and the intervening sequence varies in length in different yeast strains [306] and contains a long uninterrupted coding sequence able to specify a 235 amino acid polypeptide [370]. Processing requires two steps, excision of the intervening sequence and removal of the extra 3′-sequences [351]. The splice points show no homology with those so far found for nuclear, viral or chloroplast genes [368].

Although not all the regions coding for the mitochondrial proteins in yeast mtDNA (e.g. the three subunits of cytochrome *c* oxidase, cytochrome *b*, three subunits of the ATPase and a ribosome-associated protein) have been examined in detail, intervening sequences have also been found in the genes for cytochrome *b* and cytochrome oxidase subunit I (subunit II appears to have no intervening sequence in its gene [310]). Cytochrome *b* is a mosaic of at least five coding sequences and four intervening sequences. Moreover, certain mutations in these intervening sequences affect the synthesis of cytochrome oxidase subunit I [369]. The interruption of mRNA processing by such mutations has nevertheless allowed the deduction of a possible pathway for the maturation of cytochrome *b* mRNA [311] in which intervening sequences are sequentially spliced. Some of the intervening sequences appear to be released as approximately 1 kb *cicular* or *linear* RNA molecules [311]. Recent data indicate that the first splice removes the first intervening sequence so that translation reads through into the second intervening sequence. The product is an RNA maturase, a protein required to make the second splice which removes the second intervening sequence [379]. Unlike the mRNAs from human and *Xenopus* mitochondria, the yeast mitochondrial mRNAs appear to lack 3′-poly(A) terminii. However, circular RNAs in yeast mitochondria have been detected which may be active mRNAs, storage forms of mRNA or arise from cut and splice processes which generate mRNAs from long transcripts [312]. On the other hand the mRNAs in *Xenopus* mitochondria do not arise from split genes [308].

In human (HeLa) mitochondria, transcription is symmetrical resulting in complete transcription of both strands which are then processed to yield rRNAs, tRNAs and mRNAs which have 60–80 nucleotide-long 3′-poly(A) tails added post-transcriptionally [307] but appear to lack 5′-cap structures [309]. The processing involves precise endonucleolytic cleavages which often occur before and after tRNA sequences [378]. Such a tRNA 'punctuation' model does not appear to operate in the production of yeast mtDNA mRNAs. Also, there is no evidence for symmetrical transcription in yeast mitochondria.

11.8.2 Chloroplasts

Like mtDNA chloroplast DNA is also circular, but it is larger (around 10^8 daltons). Besides containing the genes for its two ribosomal RNAs and chloroplast tRNAs (20–40 per genome) there is the potential to code for more than 100 different proteins, however some of this DNA may well be 'spacer'. Proteins known to be coded for by chloroplast DNA include the large subunit of ribulose bisphosphate carboxylase, three components of the chloroplast coupling factor (CF1), cytochrome *f* and *b*559 and elongation factors T and G [304].

The sequence of chloroplast rRNA genes in *Zea mays* which are part of a 2000 bp inverted repeat sequence [313], has been examined and in the case of the 16S rRNA there is strong homology with the 16S rRNA species from *E. coli* [314]. Indeed a high-molecular-weight precursor (2.7×10^6) has been detected in spinach chloroplasts which appears to be processed along the lines encountered in *E. coli* [315]. On the other hand an intervening sequence has been detected in the 23S rRNA gene in *C. reinhardii* [316]. Two tRNA genes lie between the 16S and 23S ribosomal RNA genes in *Euglena* chloroplast DNA [373].

The mRNA for the large subunit of ribulose 1,5-bisphosphate carboxylase in *C. reinhardii* (1.2×10^3 daltons) appears to lack a poly(A) terminus but arises from a gene which lacks intervening sequences [317]. A single gene for this subunit is also detectable in maize chloroplasts [318].

11.9 RNA-DEPENDENT SYNTHESIS OF RNA

11.9.1 RNA-bacteriophages [319–324]

The single-stranded RNA genome of the small RNA bacteriophage (e.g. R17, MS2, f2, Qβ) is replicative in two stages. (1) The entering bacteriophage RNA acts as a messenger which, in conjunction with the ribosomes of the host *E. coli* cell, controls the formation of three bacteriophage-specific proteins (coat protein, a maturation protein and subunit II of the 'replicase'). (2) With the aid of the 'replicase' enzyme new bacteriophage RNA is synthesized. In Qβ infection the active 'replicase' enzyme comprises four subunits, the bacteriophage-coded subunit II (65000 daltons) mentioned above along with *three* subunits donated by the host which turn out to be the two protein-synthesis elongation factors, EF-Tu and EF-Ts, together with subunit I, the 70000 dalton 30S ribosomal protein, S1 (see Chapter 12).

The mechanism of replication is shown in broad outline in Fig. 11.31. The original (+) bacteriophage RNA strand forms an enzyme–RNA complex with active 'replicase', R. It is proposed that template recognition and initiation of RNA synthesis involves the interaction of the enzyme with one or two internal sites on the RNA molecule as well as with the 3′-end where transcription initiates (see also Chapter 13). The

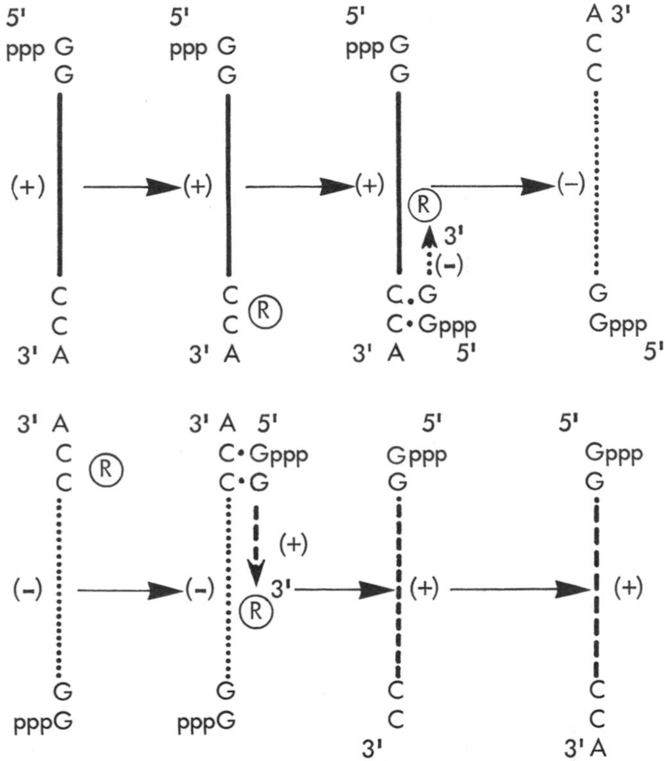

Fig. 11.31 Synthesis of Qβ bacteriophage RNA.

internal binding may help to position the enzyme in such a way that the 3′-end of the RNA is near the initiation site of the enzyme. Important in this recognition step is the subunit I.

The 'replicase' uses ribonucleoside 5′-triphosphates and synthesis is initiated (with the aid of an additional host factor) at the penultimate cytidine (C) residue rather than at the 3′-terminal adenosine (A), with the incorporation of a pppG residue as the first nucleotide at the 5′-end of the (−) strand. Synthesis precedes in the direction 5′→3′.

When the replicase has finished copying the plus strand into a full-length minus strand a termination event occurs followed either by enzymatic strand switching or by enzyme release and binding to a free, released minus strand. The termination event involves the addition of an A residue to the 3′-end of the minus strand. In the next stage, the replicase interacts with the bacteriophage (−) strand and using it as template catalyses the synthesis of the new (+) strand which grows in the direction 5′→3′. After completion of new (+), or progeny strands (broken lines) the enzyme adds a final adenosine residue, although this again is not specified by the template, to reform the -CCA terminal sequence.

11.9.2 RNA viruses of animal cells

(a) Plus-strand type

In the case of the picornaviruses, e.g. polio, encephalomyocarditis (EMC), and mengo viruses, and arboviruses, e.g. Semliki forest and Sinbis viruses, the general mechanism of virus replication is presumed to resemble that described for the small RNA bacteriophages [225, 326].

The isolation and partial purification of poliovirus specific RNA-dependent RNA polymerase (replicase) from poliovirus-infected HeLa cells has only recently been achieved [327]. The enzyme is able to initiate copying of poliovirus RNA without added primer, and requires a host factor which can be purified from a salt wash of ribosomes from uninfected cells [328].

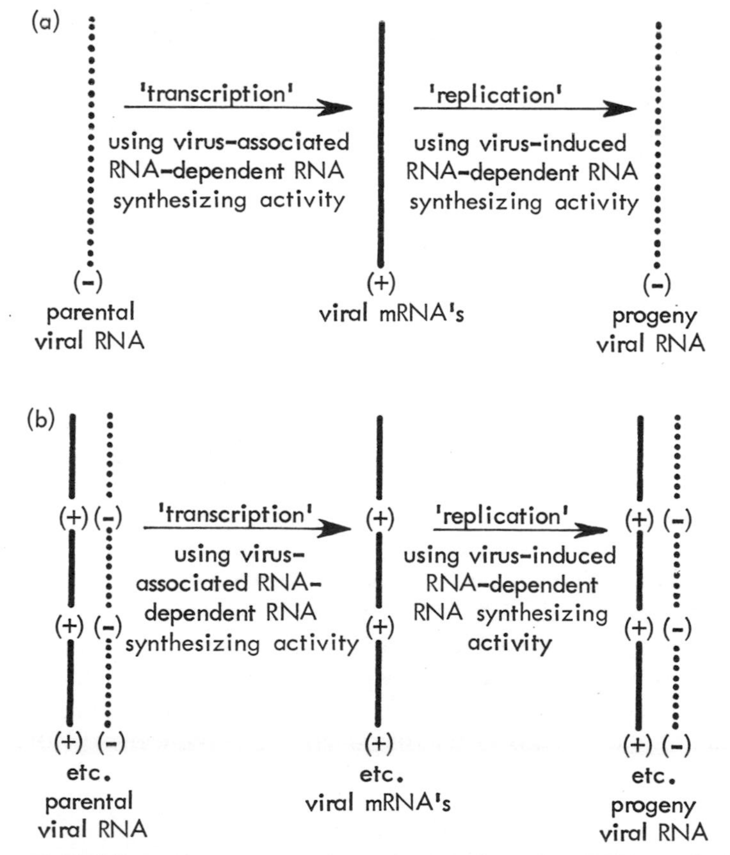

Fig. 11.32 Schematic representations of RNA-dependent RNA synthetic mechanisms operative in the replication of (a) minus-strand type viruses, (b) double-strand type viruses.

(b) Minus-strand type

The genome, however, in the case of the rhabdoviruses (e.g. vesicular stomatitis virus) and paramyxoviruses (e.g. Newcastle disease and Sendai viruses) is effectively of (−) complementarity. In infected cells this is first transcribed into translatable (+ type) messenger RNAs. This initial transcription is carried out by an RNA-dependent enzyme which is a structural component of the invading virus particle [329, 330, 334]. For genome replication, translation of these messengers is required to produce an RNA replicase activity which is thought to use the (+) type strands as templates (see Fig. 11.32a).

(c) Double-strand type

Another class of virus which contains an RNA-dependent RNA-transcribing enzyme as a structural element is the diplornavirus (e.g. reovirus) group [326, 331–334]. The virus enzyme carries out the initial transcription in the infected cell of the various segments of double-stranded RNA that comprise the genome, to produce (+) type messenger RNAs. Like the rhabdovirus and paramyxovirus situations described above, these mRNAs appear to lead to the induction of RNA replicase machinery which permits the production of double-stranded viral genomes using the viral mRNAs (+) as templates (see Fig. 11.32b).

11.9.3 Plant viruses and RNA-dependent RNA polymerases of plants

Many simple RNA viruses of plants only contain about half as much RNA as poliovirus, thus raising the question whether they specify a protein in the infected plant cell which, possibly in combination with host proteins, acts as a viral RNA replicase. Virus infection certainly causes considerable RNA-dependent RNA polymerase activity but it seems that such polymerase also occurs in the cytoplasmic fraction of *uninfected* plant cells [336–340]. The molecular weights of these RNA polymerases range from 140000 to 175000 daltons [340]. They require ribonucleoside 5′-triphosphates, Mg^{2+} a single-stranded RNA template and are resistant to actinomycin D, α-amanitin and rifampicin. Data are now accumulating suggesting a role in viral RNA replication [340].

REFERENCES

1 Hall, B.D. and Spiegelman, S. (1961), *Proc. Natl. Acad. Sci. USA,* **47,** 137.
2 Schildkraut, C.L., Marmur, J., Fresco, J. and Doty, P. (1961), *J. Biol. Chem.,* **236,** PC3.

3 Kinnell, D.E. (1971), *Prog. Nucleic Acid Res. Mol. Biol.*, **11,** 259.
4 Britten, R.J., Graham, D.E. and Neufeld, B.R. (1974), *Methods Enzymol.*, **29E,** 363.
5 Klein, W.H., Murphy, W., Attardi, G. Britten, R.J. and Davidson, E.H. (1974), *Proc. Natl. Acad. Sci. USA,* **71,** 1785.
6 Spradling, A., Penman, S., Campo, M.S. and Bishop, J.O. (1974), *Cell,* **3,** 23.
7 Davidson, E.R., Hough, B.R., Klein, W.H. and Britten, R.J. (1975), *Cell,* **4,** 217.
8 Bishop, J.O. (1972), *Gene Transcription in Reproductive Tissue* (ed. E. Diczfaluzy), Karolinska Symp. on Research Methods in Reproductive Endocrinology, p. 247.
9 Nygaard, A.P. and Hall, B.D. (1963), *Biochem. Biophys. Res. Commun.*, **12,** 98.
10 Gillespie, D. and Spiegelman, S. (1965), *J. Mol. Biol.,* **12,** 826.
11 Alwinc, J.C., Kemp, D.J. and Stark, G.R. (1977), *Proc. Natl. Acad. Sci. USA,* **74,** 5350.
12 Denhardt, D. (1966), *Biochem. Biophys. Res. Commun.*, **23,** 641.
13 Wetmur, J.G. (1976), *Annu. Rev. Biophys. Bioeng.*, **5,** 337.
14 McConaughy, B.L., Laird, C.D. and McCarthy, B.J. (1969), *Biochemistry,* **8,** 3289.
15 Hahn, W.E. and Laird, C.D. (1971), *Science,* **173,** 158.
16 Birnstiel, M.C., Sells, B.H. and Purdom, I.F. (1972), *J. Mol. Biol.*, **63,** 21.
17 Bishop, J.O., Morton, J.G., Rosbash, M. and Richardson, M. (1974), *Nature,* **250,** 199.
18 Birnie, G.D., MacPhail, E., Young, B.D., Getz, M.J. and Paul, J. (1974), *Cell Differ.*, **3,** 221.
19 Smith, J.D. (1976), *Prog. Nucleic Acid Res. Mol. Biol.*, **16,** 25.
20 Burdon, R.H. (1974), *Brookhaven Symp. Biol.*, **26,** 138.
21 Clarkson, S., Birnsteil, M. and Serra, V. (1973), *J. Mol. Biol.*, **79,** 391.
22 Altman, S. (1975), *Cell,* **4,** 21.
23 Altman, S., Bothwell, A.C.M. and Stark, B.C. (1974), *Brookhaven Symp. Biol.*, **26,** 12.
24 Abelson, J. (1979), *Annu. Rev. Biochem.*, **48,** 4035.
25 Stark, B., Kole, R., Bowman, E. and Altman, S. (1978), *Proc. Natl. Acad. Sci. USA,* **75,** 3717.
26 Bikoff, E.K., LaRue, B.F. and Gefter, M.L. (1975), *J. Biol. Chem.*, **250,** 6248.
27 Shimura, Y., Sakano, H. and Nagawa, F. (1978), *Eur. J. Biochem.*, **86,** 267.
28 Sakano, H. and Shimura, Y. (1978), *J. Mol. Biol.*, **123,** 287.
29 Schedl, P., Primakoff, P. and Roberts, J. (1974), *Brookhaven Symp. Biol.*, **26,** 53.
30 Beckmann, J.S., Johnson, P.F. and Abelson, J. (1977), *Science,* **196,** 205.
31 Cortese, R., Melton, D., Tranquilla, T. and Smith, J.D. (1978), *Nucleic Acids Res.*, **5,** 4593.
32 Hagenbuchle, O., Larson, D., Hall, G.I. and Sprague, K.U. (1979), *Cell,* **18,** 1217.

33 Garber, R.C. and Gage, P. (1979), *Cell,* **18,** 817.
34 Olson, M.V., Hall, B.D., Cameron, J.R. and Davis, R.W. (1979), *J. Mol. Biol.,* **127,** 285.
35 Yen, P.H., Sodja, A., Cohen, M., Conrad, S.E., Wu, M., Davidson, N. and Iglen, C. (1977), *Cell,* **11,** 763.
36 Schmidt, O., Mao, J.I., Silverman, S., Hovemann, B. and Soll, D. (1978), *Proc. Natl. Acad. Sci. USA,* **74,** 542.
37 Valenzuela, P., Venegas, A., Weinberg, F., Bishop, R. and Rutter, W.J. (1978), *Proc. Natl. Acad. Sci. USA,* **75,** 190.
38 Goodman, H.M., Olson, M.V. and Hall, B.D. (1977), *Proc. Natl. Acad. Sci. USA,* **74,** 5453.
39 Ogden, R.C., Beckman, J.S., Abelson, J., Kang, H.S., Soll, D. and Schmidt, O. (1979), *Cell,* **17,** 399.
40 Knapp, G., Beckman, J.S., Johnson, P.F., Fuhrman, S.A. and Abelson, J. (1978), *Cell,* **14,** 221.
41 O'Farrell, P.Z., Cordell, B., Valenzuela, P., Rutter, W.J. and Goodman, H.M. (1978), *Nature,* **274,** 438.
42 Melton, D.A., DeRoberts, E.M. and Cortese, R. (1980), *Nature,* **284,** 143.
43 Garber, R.L. and Gage, L.P. (1979), *Cell,* **18,** 817.
44 Garber, R.L. and Altman, S. (1979), *Cell,* **17,** 389.
45 Muller, F. and Clarkson, S.G. (1980), *Cell,* **19,** 345.
46 Kressman, A., Hofstetter, H., DiCapua, E., Grosschedl, H. and Birnstiel, M.L. (1979), *Nucleic Acids Res.,* **7,** 1749.
47 Berman, M.L. and Landy, A. (1979), *Proc. Natl. Acad. Sci. USA,* **76,** 4303.
48 Heidelberger, C., Harbers, E., Leibman, K.C., Takagi, Y. and Potter, V.R. (1956), *Biochim. Biophys. Acta,* **20,** 445.
49 Hoagland, M. (1960), in *The Nucleic Acids,* Vol. 3, p. 349 (eds E. Chargaff and J. N. Davidson), Academic Press, New York.
50 Zamecnik, P.C. (1960), *The Harvey Lectures,* **54,** 256.
51 Hecht, L.L., Stephenson, L. and Zamecnik, P.C. (1959), *Proc. Natl. Acad. Sci. USA,* **45,** 505.
52 Daniel, V. and Littauer, U.Z. (1963), *J. Biol. Chem.,* **238,** 2102.
53 Deutscher, M.P., Foulds, J., Morse, J.W. and Hilderman, R.H. (1975), *Brookhaven Symp. Biol.,* **26,** 124.
54 Deutscher, M.P. (1973), *Prog. Nucleic Acid Res. Mol. Biol.,* **13,** 51.
55 Nishimura, S. (1972), *Prog. Nucleic Acid Res. Mol. Biol.,* **12,** 49.
56 Soll, D. (1971), *Science,* **183,** 293.
57 Mandel, L.R. and Borek, E. (1963), *Biochemistry,* **2,** 555.
58 Hurwitz, J., Gold, M. and Anders, M. (1964), *J. Biol. Chem.,* **239,** 3462.
59 Tsutsui, E., Srinivasan, P.R. and Borek, E. (1966), *Proc. Natl. Acad. Sci. USA,* **56,** 1003.
60 Lane, B.G. and Tamaoki, T. (1969), *Biochim. Biophys. Acta,* **179,** 332.
61 Baguley, B.C. and Staehelin, M. (1968), *Biochemistry,* **7,** 45.
62 Rodeh, R., Geldman, M. and Littauer, U.Z. (1967), *Biochemistry,* **6,** 451.
63 Svensson, I., Bjork, G.K. and Lundahl, P. (1979), *Eur. J. Biochem.,* **9,** 216.
64 Agris, P.F., Sprermulli, L.M. and Brown, G.M. (1974), *Arch. Biochem. Biophys.,* **162,** 38.

65 Kraus, J. and Staehlin, M. (1974), *Nucleic Acids Res.,* **1,** 1455.
66 Kraus, J. and Staehlin, M. (1974), *Nucleic Acids Res.,* **1,** 1479.
67 Wierzbicka, H., Jakubowski, J. and Pawelkiewicz, J. (1975), *Nucleic Acids Res.,* **2,** 101.
68 Smolnar, N., Hellman, U. and Svensson, I. (1975), *Nucleic Acids Res.,* **2,** 993.
69 Smolnar, N. and Svensson, I. (1974), *Nucleic Acids Res.,* **1,** 707.
70 Gold, M., Hurwitz, J. and Anders, M. (1963), *Biochem. Biophys. Res Commun.,* **11,** 107.
71 Hacker, B. and Mandel, L.R. (1969), *Biochim. Biophys. Acta,* **190,** 38.
72 Silber, R., Goldstein, B., Berman, E., Decter, J. and Friend, C. (1967), *Cancer Res.,* **27,** 1264.
73 Borek, E. (1971), *Cancer Res.,* **31,** 596.
74 Mittleman, A., Hall, R.H., Yohn, D.S. and Grace, J.T. (1967), *Cancer Res.,* **27,** 1409.
75 Burdon, R.H. (1971), *Prog. Nucleic Acid Res. Mol. Biol.,* **11,** 33.
76 Wainfan, E., Srinivasan, P.R. and Borek, E. (1965), *Biochemistry,* **4,** 2845.
77 Wainfan, E., Srinivasan, P.R. and Borek, E. (1966), *J. Mol. Biol.,* **22,** 349.
78 Turkington, R.W. (1969), *J. Biol. Chem.,* **244,** 5140.
79 Van de Woude, G.F., Arlinghaus, R.B. and Polatnick, J. (1967), *Biochem. Biophys. Res. Commun.,* **29,** 483.
80 Kerr, S.J., Sharma, O.K. and Borek, E. (1971), *Cancer Res.,* **31,** 633.
81 Kerr, S.J. and Borek, E. (1973), *Adv. Enzyme Regul.,* **11,** 63.
82 Nomura, M., Morgan, E.A., Jaskunas, S.R. (1977), *Annu. Rev. Genetics,* **11,** 297.
83 Dunn, J.J. and Studier, F.W. (1973), *Proc. Natl. Acad. Sci. USA,* **70,** 3239.
84 Nickolaev, N., Silengo, L. and Schlessinger, D. (1973), *Proc. Natl. Acad. Sci. USA,* **70,** 3361.
85 Lund, E., Dahlberg, J.E. and Guthrie, C. (1979), in *Transfer RNA* (eds J. Abelson, D. Soll and P. Schimmel), Cold Spring Harbor Laboratory.
86 Young, R.A. and Steitz, J.A. (1978), *Proc. Natl. Acad. Sci. USA,* **75,** 3593.
87 Bram, R.J., Young, R.A. and Steitz, J.A. (1980), *Cell,* **19,** 393.
88 Dahlberg, A.E., Tokimatsu, H., Zahalak, M., Reynolds, F., Calvert, P.C., Rabson, A.B., Lund, E. and Dahlberg, J.E. (1978), *Proc. Natl. Acad. Sci. USA,* **75,** 3598.
89 Misra, T.K. and Apirion, D. (1979), *J. Biol. Chem.,* **254,** 11 154.
90 Sogin, M.L., Pace, N.R., Rosenberg, M., Weissman, S.M. (1976), *J. Biol. Chem.,* **251,** 3480.
91 Sogin, M.L., Pace, B. and Pace, N.R. (1977), *J. Biol. Chem.,* **252,** 1350.
92 Young, R.A. and Steitz, J.A. (1979), *Cell,* **17,** 225.
93 Gilbert, S.F., de Boer, H.A. and Nomura, M. (1979), *Cell,* **17,** 211.
94 Young, R.A. (1979), *J. Biol. Chem.,* **254,** 12 725.
95 Reeder, R. (1974), in *Ribosomes* (eds M. Nomura, A. Tissieres, and P. Lengyel), Cold Spring Harbor Press, New York, p. 489.
96 Birnstiel, M.L., Chipchase, M. and Spiers, J. (1971), *Prog. Nucleic Acid Res. Mol. Biol.,* **11,** 351.
97 Gall, J.G. (1974), *Proc. Natl. Acad. Sci. USA,* **71,** 3078.

98 Wellauer, P.K., Dawid, I.B., Brown, D.D. and Reeder, R.H. (1976), *J. Mol. Biol.*, **105**, 461.
99 Moss, T. and Birnstiel, M.L. (1979), *Nucleic Acids Res.*, **6**, 3733.
100 Solner-Webb, B. and Reeder, R.H. (1979), *Cell*, **18**, 485
101 Ritossa, F. (1976), in *The Genetics and Biology of* Drosophila (eds M. Ashburner and E. Novitsky), Academic Press, New York, p. 801.
102 Jordon, B.R. and Glover, D.M. (1977), *FEBS Lett.*, **78**, 271.
103 Wellauer, P.K., Dawid, I.B. and Tartof, K.D. (1978), *Cell*, **14**, 269.
104 Kidd, S.J. and Glover, D.M. (1980), *Cell*, **19**, 103.
105 Jolly, D.J. and Thomas, C.A. (1980), *Nucleic Acids Res.*, **8**, 67.
106 Long, E.O. and Dawid, I.B. (1979), *Cell*, **18**, 1185.
107 Planta, R.J., Meyernick, J.H. and Klootwijk, J. (1979), *FEBS 12th Meeting, Dresden*, **51**, *Gene Function*, p. 401.
108 Engberg, J. and Klenow, H. (1977), *Trends in Biochem. Sci.*, **2**, 183.
109 Din, N., Engberg, J., Kaffenberger, W. and Eckert, W.A. (1979), *Cell*, **18**, 525.
110 Wild, M.A. and Gall, I.E. (1979), *Cell*, **18**, 565.
111 Campbell, G.R., Littau, V.C., Melera, P.W., Allfrey, V.G. and Johnson, E.M. (1979), *Nucleic Acids Res.*, **6**, 1433.
112 Cech, T.R. and Rio, D.C. (1979), *Proc. Natl. Acad. Sci. USA*, **76**, 5051.
113 Wild, M.A. and Sommer, R. (1980), *Nature*, **283**, 693.
114 Zaug, A.J. and Cech, T.R. (1980), *Cell*, **19**, 331.
115 Meunier-Roturel, M., Cortedas, J., Macaya, C. and Bernardi, g. (1979), *Nucleic Acids Res.*, **6**, 2109.
116 Wellauer, P.K. and Dawid, I.B. (1979), *J. Mol. Biol.*, **128**, 289.
117 Krystal, M. and Arnheim, N. (1978), *J. Mol. Biol.*, **126**, 91.
118 Darnell, J.E. (1968), *Bact. Rev.*, **32**, 262.
119 Wellauer, P.K. and Dawid, I.B. (1973), *Proc. Natl. Acad. Sci. USA*, **70**, 2827.
120 Maden, B.E.H. (1971), *Prog. Biophys. Mol. Biol.*, **22**, 127.
121 Burdon, R.H. (1971), *Prog. Nucleic Acid Res. Mol. Biol.*, **11**, 33.
122 Ford, P.J. (1973), *Biochem. Soc. Symp.*, **37**, 69.
123 Starr, J. and Fefferman, R. (1964), *J. Biol. Chem.*, **239**, 3457.
124 Srinivasan, P.R., Nofal, S. and Sussman, C. (1964), *Biochem. Biophys. Res. Commun.*, **16**, 82.
125 Lane, B.G. (1965), *Biochemistry*, **4**, 212.
126 Wagner, E.K., Penman, S. and Ingram, V. (1967), *J. Mol. Biol.*, **29**, 271.
127 Maden, B.E.H. and Salim, M. (1974), *J. Mol. Biol.*, **88**, 133.
128 Greenberg, H. and Penman, S. (1966), *J. Mol. Biol.*, **21**, 523.
129 Vaughan, M.H., Soiero, R., Warner, J. and Darnell, J.E. (1976), *Proc. Natl. Acad. Sci. USA*, **58**, 1527.
130 Hesler, T.L., Davies, J.E. and Dahlberg, J.E. (1971), *Nature New Biol.*, **233**, 12.
131 Gordon, J. and Boman, H.G. (1967), *J. Mol. Biol.*, **9**, 638.
132 Hurwitz, J., Anders, M., Gold, M. and Smith, J. (1965), *J. Biol. Chem.*, **240**, 1256.
133 Nichols, J.L. and Lane, B.G. (1968), *Can. J. Biochem.*, **46**, 108.

134 Thammana, P. and Held, W.A. (1974), *Nature,* **251,** 682.
135 Culp, L.A. and Brown, G.M. (1970), *Arch. Biochem. Biophys.,* **137,** 222.
136 Brown, D.D. and Sugimoto, K. (1973), *Cold Spring Harbor Symp. Quant. Biol.,* **38,** 501.
137 Federoff, N. (1979), *Cell,* **16,** 697.
138 Federoff, N.V. and Brown, D.D. (1978), *Cell,* **13,** 701.
139 Jacq, C., Miller, J.R. and Brownlee, G.G. (1977), *Cell,* **12,** 109.
140 Rubin, G.M. and Hogness, D.S. (1975), *Cell,* **6,** 207.
141 Levis, R. (1978), *J. Mol. Biol.,* **122,** 279.
142 Sakonju, S., Bogenhagen, D.F. and Brown, D.D. (1980), *Cell,* **19,** 13.
143 Bogenhagen, D.F., Sakonju, S. and Brown, D.D. (1980), *Cell,* **19,** 27.
144 Kjeldgaard, N.O. and Gaussing, K. (1974), in *Ribosomes* (eds M. Nomura, A. Tissieres and P. Lengyl), Cold Spring Harbor Laboratory, New York, p. 369.
145 Nomura, M. (1976), *Cell,* **9,** 633.
146 Post, L.E., Strycharz, G.D., Nomura, M., Lewis, H. and Denis, P.P. (1979), *Proc. Natl. Acad. Sci. USA,* **76,** 1697.
147 Nikolaev, N., Birenbaum, M. and Schlessinger, D. (1975), *Biochim. Biophys. Acta,* **395,** 478.
148 Nomura, M. (1973), *Science,* **179,** 864.
149 Schlessinger, D. (1974), in *Ribosomes* (eds M. Nomura, A. Tissieres and P. Lengyl), Cold Spring Harbor Laboratory, New York, p. 393.
150 Nierhaus, K.H., Bordasch, K. and Homann, H.E. (1973), *J. Mol. Biol.,* **74,** 587.
151 Winicov, I. and Perry, R.P. (1974), *Brookhaven Symp. Biol.,* **26,** 201.
152 Hershey, A.D., Dixon, J. and Chase, M. (1953), *J. Gen. Physiol.,* **73,** 110.
153 Volkin, E. and Astrachan, L. (1956), *Virology,* **2,** 149, 433.
154 Bessman, M.J. (1963), *Molecular Genetics* (ed. J. H. Taylor), Academic Press, New York, Part I, p. 1.
155 Nomura, M., Hall, B.D. and Spiegelman, S. (1960), *J. Mol. Biol.,* **2,** 306.
156 Spiegelman, S., Hall, B.D. and Storck, R. (1961), *Proc. Natl. Acad. Sci. USA,* **47,** 1135.
157 Spiegelman, S. (1961), *Cold Spring Harbor Symp. Quant. Biol.,* **226,** 75.
158 Jacob, F. and Monod, J. (1961), *J. Mol. Biol.,* **3,** 318.
159 Brenner, S., Jacob, F. and Meselson, M. (1961), *Nature,* **190,** 576.
160 Gros, F., Hiatt, H., Gilbert, W., Kurland, C.G., Risebrough, R.W. and Watson, J.D. (1961), *Nature,* **190,** 581.
161 Hayashi, M. and Spiegelman, S. (1961), *Proc. Natl. Acad. Sci. USA,* **47,** 1564.
162 Bautz, E.K.F. and Hall, B.D. (1962), *Proc. Natl. Acad. Sci. USA,* **48,** 400.
163 Champe, S.P. and Benzer, S. (1962), *Proc. Natl. Acad. Sci. USA,* **48,** 332.
164 Marmur, J., Greenspan, C.M., Policek, E., Kahan, F.M., Levine, J. and Mandel, M. (1963), *Cold Spring Harbor Symp. Quant. Biol.,* **28,** 191.
165 Marmur, J. and Greenspan, C.M. (1963), *Science,* **142,** 387.
166 Tocchini-Valentini, G.P., Stodolsky, M., Aurisicchio, A., Sarnat, M., Fraziosi, F., Weiss, S.B. and Geiduschek, E.P. (1963), *Proc. Natl. Acad. Sci. USA,* **50,** 935.

167 Hayashi, M., Hayashi, M.N. and Spiegelman, S. (1963), *Science,* **140,** 1313.
168 Blundell, M. and Kennell, D. (1974), *J. Mol. Biol.,* **83,** 143.
169 Achord, D. and Kennell, D. (1974), *J. Mol. Biol.,* **90,** 581.
170 Dunn, J.J. and Studier, F.W. (1973), *Proc. Natl. Acad. Sci. USA,* **70,** 1559.
171 Dunn, J.J. and Studier, F.W. (1975), *Brookhaven Symp. Biol.,* **26,** 267.
172 Rosenberg, M., Kramer, R.A. and Steitz, J.A. (1974), *J. Mol. Biol.,* **89,** 777.
173 Southern, E.M. (1975), *J. Mol. Biol.,* **98,** 503.
174 Jeffreys, A.J. and Flavell, R.A. (1977), *Cell,* **12,** 1097.
175 Van den Berg, J., Van Ooyen, A., Mantei, N., Schambock, A., Grosveld, G., Flavell, R.A. and Weissmann, C. (1978), *Nature,* **276,** 37.
176 Thomas, M., White, R.C. and Davis, R.W. (1976), *Proc. Natl. Acad. Sci. USA,* **73,** 2294.
177 White, R.L. and Hogness, D.S. (1977), *Cell,* **10,** 177.
178 Tilgham, S.M., Tiemeier, D.C., Seidman, J.G., Peterlin, B.M., Sullivan, M., Maizel, J.V. and Leder, P. (1978), *Proc. Natl. Acad. Sci. USA,* **75,** 725.
179 Konkel, D.A., Maizel, J.V. and Leder, P. (1979), *Cell,* **18,** 865.
180 Nishioka, Y. and Leder, P. (1979), *Cell,* **18,** 875.
181 Richards, R.I., Shine, J., Ullrich, A., Wells, J.R.E. and Goodman, H.M. (1979), *Nucleic Acids Res.,* **7,** 1137.
182 Little, P.F.R., Flavell, R.A., Kooter, J.M., Annison, G. and Williamson, R. (1979), *Nature,* **278,** 227.
183 Curtis, P.J. and Weissmann, C. (1976), *J. Mol. Biol.,* **106,** 1061.
184 Kwan, S-P., Wood, T.G. and Lingrel, K.B. (1977), *Proc. Natl. Acad. Sci. USA,* **74,** 178.
185 Ross, J. (1976), *J. Mol. Biol.,* **106,** 403.
186 Ross, J. and Knecht, D.A. (1978), *J. Mol. Biol.,* **119,** 1.
187 Curtis, P.J., Mantei, N. and Weissmann, C. (1977), *Cold Spring Harbor Symp. Quant. Biol.,* **42,** 971.
188 Tilgham, S.M., Curtis, P.J., Tiemeier, D.C., Leder, P. and Weissmann, C. (1978), *Proc. Natl. Acad. Sci. USA,* **75,** 1309.
189 Kinniburgh, A.J., Mertz, J.E. and Ross, J. (1978), *Cell,* **14,** 681.
190 Kinniburgh, A.J. and Ross, J. (1979), *Cell,* **17,** 915.
191 Weaver, R.F. and Weissmann, C. (1979), *Nucleic Acids Res.,* **7,** 1175.
192 Ziff, E.B. and Evans, R.M. (1978), *Cell,* **15,** 1463.
193 Konkel, D.A., Tilgham, S.M. and Leder, P. (1978), *Cell,* **15,** 1125.
194 Bastos, R.N. and Aviv, H. (1977), *Cell,* **11,** 641.
195 Scherrer, K., Imaizumi-Scherrer, M-T., Reynaud, C.A. and Therwath, A. (1979), *Mol. Biol. Rep.,* **5,** 5.
196 Stair, R.K., Skoultchi, A.I. and Schafritz, D.A. (1977), *Cell,* **12,** 133.
197 Curtis, P.J., Mantei, N., Van den Berg, J. and Weissmann, C. (1977), *Proc. Natl. Acad. Sci. USA,* **74,** 3184.
198 Hamer, D.H. and Leder, P. (1979), *Nature,* **281,** 35.
199 Courtney, M. and Williamson, R. (1979), *Nucleic Acids Res.,* **7,** 1121.
200 Gannon, G., O'Hare, K., Perrin, F., Le Pennee, J.P., Benoist, C., Cochet, M., Breathnach, R., Royal, A., Garapin, A., Cani, B. and Chambon, P. (1979), *Nature,* **278,** 428.

201 Dugaiczyk, A., Woo, S.L.C., Colbert, D.A., Lai, E.C., Mace, M.L. and O'Malley, B.W. (1979), *Proc. Natl. Acad. Sci. USA,* **76,** 2253.
202 Benoist, C., O'Hare, K., Breathnach, R. and Chambon, P. (1980), *Nucleic Acids Res.,* **8,** 127.
203 Roop, D.R., Tsai, M-J. and O'Malley, B.W. (1980), *Cell,* **19,** 63.
204 Bailey, J.M. and Davidson, N. (1976), *Anal. Biochem.,* **70,** 75.
205 Alwine, J.C., Kemp, D.J. and Stark, G.R. (1977), *Proc. Natl. Acad. Sci. USA,* **74,** 5350.
206 Roop, D.R., Nordstrom, J.L., Tsai, S.Y., Tsai, M-J. and O'Malley, B.W. (1978), *Cell,* **15,** 671.
207 Royal, A., Garapin, A., Cami, B., Perrin, F., Mandel, J.L., Le Meur, M., Bregegegre, F., Gannon, F., Le Pennec, J.P. Chambon, P. and Kourilsky, P. (1979), *Nature,* **279,** 125.
208 Lai, E.C., Stein, J.P., Catterall, J.F., Woo, S.L.C., Mace, M.L., Means, A.R. and O'Malley, B.W. (1979), *Cell,* **18,** 829.
209 Nordstrom, J.L., Roop, D.R., Tsai, M-J. and O'Malley, B.W. (1979), *Nature,* **278,** 328.
210 Cochet, M., Gannon, F., Hen, R., Marateaux, L., Perrin, F. and Chambon, P. (1979), *Nature,* **282,** 567.
211 Nguyen-Huu, M.G., Stratmann, M., Groner, G., Wurtz, T., Land, H., Giesecke, K., Sippel, A.E. and Schutz, G. (1979), *Proc. Natl. Acad. Sci. USA,* **76,** 76.
212 Lomedico, P., Rosenthal, N. Estratiadis, A., Gilbert, W., Kolodner, R. and Tizard, R. (1979), *Cell,* **18,** 545.
213 Cordell, B., Bell, G., Tischer, E., De Noto, F.M., Ullrich, A., Pictet, R., Rutter, W.J. and Goodman, H.M. (1979), *Cell,* **18,** 533–543.
214 Bell, G.T., Picet, R.L., Rutter, W.I., Cordell, B., Tischer, E. and Goodman, H.M. (1980), *Nature,* **284,** 26.
215 Fiddes, J.C., Seeburg, P.H., De Noto, F.M., Hallewell,R.A., Baxter, J.D. and Goodman, H.M. (1979), *Proc. Natl. Acad. Sci. USA,* **76,** 4294.
216 Soreq, H., Harpold, M., Evans, R., Darnell, J.E. and Bancroft, F.C. (1979), *Nucleic Acids Res.,* **6,** 2471–2482.
217 Sargent, T.D., Wu, J.R., Sala-Trepat, J.M., Wallace, R.B., Reyes, A.A. and Bonner, J. (1979), *Proc. Natl. Acad. Sci. USA,* **76,** 3256.
218 Harpold, M.M., Dolner, P.R., Evans, R., Bancroft, F.C. and Darnell, J.E. (1979), *Nucleic Acids Res.,* **6,** 3133.
219 Strair, R.K., Yap, S.H., Nadel-Ginard, B. and Shafritz, D.A. (1978), *J. Biol. Chem.,* **253,** 1328.
220 Bhat, S.P., Jones, R.E., Sullivan, M.A. and Piatrigorsky, J. (198), *Nature,* **284,** 234.
221 Ryffel, G.U., Wyler, T., Muellener, D.B. and Weber, R. (1980), *Cell,* **19,** 53.
222 Tsujimoto, Y., Suzuki, Y. (1979), *Cell,* **18,** 591.
223 Smith, M., Leung, D.W., Gillam, S. and Astell, C.R. (1979), *Cell,* **16,** 753.
224 Williamson, U.H., Bennetzen, J. and Young, E.T. (1980), *Nature,* **283,** 214.

225 Nunberg, J., Kaufman, R.J., Chang, A.C.Y., Cohen, S.N. and Schimke, R.T. (1980), *Cell,* **19,** 355.
226 Sanders-Haigh, L., Anderson, E.F. and Francke, U. (1980), *Nature,* **283,** 683.
227 Lai, E.C., Woo, S.L.C., Bordelon-Riser, M.E., Frazer, T.H. and O'Malley, B.W. (1980), *Proc. Natl. Acad. Sci. USA,* **77,** 244.
228 Flint, J. (1977), *Cell,* **10,** 153.
229 Weill, P.A., Luse, D.S., Segall, J. and Roeder, R.G. (1979), *Cell,* **18,** 469.
230 Zain, S., Sambrook, J., Roberts, R.J., Keller, W., Fried, M. and Dunn, A.R. (1979), *Cell,* **16,** 851.
231 Fraser, N.W., Nevins, J.R., Ziff, E. and Darnell, J.E. (1979), *J. Mol. Biol.,* **129,** 643.
232 Nevins, J.R. (1979), *J. Mol. Biol.,* **130,** 493.
233 Blanchard, J-M., Weber, J., Jelinek, W. and Darnell, J.E. (1978), *Proc. Natl. Acad. Sci. USA,* **75,** 5344.
238 Berk, A.J. and Sharp, P.A. (1978), *Cell,* **14,** 695.
239 Reddy, V.B., Thimmapaya, B., Dhar, R., Subramanian, K.M., Zain, B.S., Pan, J., Ghosh, P.K., Celma, M.L. and Weissman, S.M. (1978), *Science,* **200,** 494.
240 Fiers, W., Conteras, R., Haegeman, G., Rogiers, R., Van de Voorde, A., Van Heuverswyn, H., Van Herreweghe, J., Volkaert, G. and Ysebert, M. (1978), *Nature,* **273,** 113–20.
241 Soeda, E., Arrand, J.R., Smolar, N., Walsh, J.E. and Griffin, B.E. (1980), *Nature,* **283,** 445.
242 Crawford, L.V. (1980), *Trends in Biochem. Sci.,* **5,** 39.
243 Rigby, P. (1979), *Nature,* **282,** 781.
244 Shenk, T.E., Carbon, J. and Berg, P. (1976), *J. Virol.,* **18,** 664.
245 Lai, C-J. and Khoury, G. (1979), *Proc. Natl. Acad. Sci. USA,* **76,** 71.
246 Piper, P., Wardale, J. and Crew, F. (1979), *Nature,* **282,** 686.
247 Breathbach, R., Benoist, C., O'Hare, K., Gannon, F. and Chambon, P. (1978), *Proc. Natl. Acad. Sci. USA,* **75,** 4853.
248 Reddy, V., Ghosh, P., Lebowitz, P., Piatak, M. and Weissman, S. (1979), *J. Virol.,* **30,** 279.
249 Zain, S., Gingeras, T.R., Bullock, P., Wong, G. and Gelinas, R.E. (1979), *J. Mol. Biol.,* **135,** 413.
250 Benoist, C., O'Hare, K., Breathnach, R. and Chambon , C. (1980), *Nucleic Acids Res.,* **8,** 127.
251 Hamer, D. and Leder, P. (1979), *Cell,* **17,** 737.
252 Segal, S., Levine, A.J. and Khoury, G. (1979), *Nature,* **280,** 335.
253 Lerner, M.R., Boyle, J.A., Mount, S.M., Wolin, S.L. and Steitz, J.A. (1980), *Nature,* **283,** 220.
254 Reddy, R., Ro-Choi, T.S., Henning, D. and Busch, H. (1974), *J. Biol. Chem.,* **249,** 6486.
255 Roberts, R.J. (1980), *Nature,* **283,** 132.
256 Crick, F.H.C. (1979), *Science,* **204,** 264.
257 Lerner, M.R. and Steitz, J.A. (1979), *Proc. Natl. Acad. Sci. USA,* **76,** 5495.
258 Perry, R.P., Kelley, D.E., Fridirici, K.H. and Rottman, F.M. (1975), *Cell,* **6,** 13.

259 Shatkin, A.J. (1976), *Cell,* **9,** 645.
260 Abraham, G., Rhodes, D.P. and Banerjee, A.K. (1975), *Cell,* **5,** 51.
261 Salditt-Georgieff, M., Harpold, M., Chen-Kiang, S. and Darnell, J.E. (1980), *Cell,* **19,** 69.
262 Schibler, U. and Perry, R.P. (1976), *Cell,* **9,** 121.
263 Wei, C-M and Moss, B. (1977), *Proc. Natl. Acad. Sci. USA,* **74,** 3758.
264 Ensinger, M.J. and Moss, B. (1976), *J. Biol. Chem.,* **251,** 5283.
265 Keith, J.M., Ensinger, M.J. and Moss, B. (1978), *J. Biol. Chem.,* **253,** 5033.
266 Bajszar, G., Szabo, G., Simoncsits, A. and Molnar, J. (1978), *Mol. Biol. Rep.,* **4,** 93.
267 Salditt-Georgieff, M., Jelinek, W., Darnell, J.E., Furiuchi, Y., Morgan, M. and Shatkin, A. (1976), *Cell,* **7,** 227.
268 Schibler, U., Marcu, K.B. and Perry, R.P. (1977), *J. Mol. Biol,* **115,** 695.
269 Chen-Kiang, S., Nevins, J.R. and Darnell, J.E. (1979), *J. Mol. Biol.,* **135,** 733.
270 Jelinek, W., Adesnik, M., Salditt, M., Sheiness, D., Wall, R., Molloy, G.R., Philipson, L. and Darnell, J.E. (1973), *J. Mol. Biol.,* **75,** 515.
271 Shenkin, A. and Burdon, R.H. (1974), *J. Mol. Biol.,* **85,** 19.
272 Derman, E. and Darnell, J.E. (1974), *Cell,* **3,** 255.
273 Nevins, J.R. and Darnell, J.E. (1978), *Cell,* **15,** 1477.
274 Puckett, L. and Darnell, J.E. (1977), *J. Cell. Physiol.,* **90,** 521.
275 Edmonds, M. and Winters, M.A. (1976), *Prog. Nucleic Acid Res. Mol. Biol.,* **17,** 149.
276 Jacob, S.T. and Rose, K. (1978), *Methods in Cancer Res.,* **14,** 191.
277 Winters, M.A. and Edmonds, M. (1973), *J. Biol. Chem.,* **248,** 4756.
278 Tsiapalis, C.M., Dorson, J.W. and Bollum, F.J. (1975), *J. Biol. Chem.,* **250,** 4486.
279 Rose, K.M. and Jacob, S.T. (1976), *Eur. J. Biochem.,* **67,** 11.
280 Mans, R. and Stern, G. (1974), *Life Sci.,* **14,** 437.
281 Jacob, S.T., Roe, F.J. and Rose, K.M. (1976), *Biochem. J.,* **153,** 733.
282 Rose, K.M., Bell, L.E. and Jacob, S.T. (1977), *Nature,* **267,** 178.
283 Niessing, J. and Sekeris, C.E. (1973), *Nature New Biol.,* **243,** 9.
284 Miller, O.L. and Hamkalo, B.A. (1972), *Int. Rev. Cytol.,* **33,** 1.
285 Miller, O.L. and Baaken, A.H. (1972), *Karolinska Symp. Res. Methods Reproduc. Endocrinol.,* **5,** 155.
286 Scott, S.E.M. and Sommerville, J. (1974), *Nature,* **250,** 680.
287 Samarina, O.P., Krichevskaya, A.A. and Georgiev, G.P. (1966), *Nature,* **210,** 1319.
288 Pederson, T. (1974), *J. Mol. Biol.,* **83,** 163.
289 Faiferman, I. and Pogo, A.O. (1975), *Biochemistry,* **14,** 3808.
290 Georgiev, G.P. and Samarina, O.P. (1971), *Adv. Cell Biol.,* **2,** 47.
291 Lukanidin, E.M., Zalmanzon, E.S., Komaromi, L., Samarina, O.P. and Georgiev, G.P. (1972), *Nature New Biol.,* **238,** 193.
292 Martin, T., Billings, P., Pullman, J., Stevens, B. and Kinniburgh, A. (1978), *Cold Spring Harbor Symp. Quant. Biol.,* **42,** 899.
293 Beyer, A.L., Christensen, M.E., Walker, B.W. and Le Stourgeon, W.M. (1977), *Cell,* **11,** 127.

294 Karn, J., Vidali, G., Boffa, L.C. and Allfrey, V.G. (1977), *J. Biol. Chem.*, **252,** 7307.
295 Stevenin, J., Devilliers, G. and Jacob, M. (1976), *Mol. Biol. Rep.*, **2,** 241.
296 Kish, V.M. and Pederson, T. (1975), *J. Mol. Biol.*, **95,** 227.
297 Molnar, J. and Samarina, O.P. (1975), *Mol. Biol. Rep.*, **2,** 1.
298 Rose, K.M., Jacob, S.T. and Kumar, A. (1979), *Nature,* **279,** 260.
299 Samarina, O.P., Aitkhozina, N.A. and Besson, J. (1973), *Mol. Biol. Rep.*, **1,** 193.
300 Malcolm, D.B. and Sommerville, J. (1977), *J. Cell Sci.*, **24,** 143.
301 Sekeris, C.E. and Neissing, J. (1975), *Biochem. Biophys. Res. Commun.*, **62,** 642.
302 Van den Berg, J., Van Ooyen, A., Mantei, N., Schambock, A., Grosveld, G., Flavell, R.A. and Weissmann, C. (1978), *Nature,* **276,** 37.
303 Borst, P. (1977), *Trends in Biochem. Sci.*, **2,** 31.
304 Ciferri, O. (1978), *Trends in Biochem. Sci.*, **3,** 256.
305 Tzagoloff, A., Macino, G. and Sebald, W. (1979), *Annu. Rev. Biochem.*, **48,** 419.
306 Borst, P. and Grivell, L.A. (1978), *Cell,* **15,** 705.
307 Anderson, S., Bankier, A.T., Barrell, B.G., de Bruijn, M.H.L., Coulson, A.R., Drouin, J., Eperon, I.C., Nierlich, D.P., Roe, B.A., Sanger, F., Schreier, P.H., Smith, A.J.H., Staden, R., Young, I.G. (1981), *Nature,* **290,** 457.
308 Rastli, E. and Dawid, I.B. (1979), *Cell,* **18,** 501.
309 Grohman, K., Amalric, F., Crews, S. and Attardi, G. (1978), *Nucleic Acids Res.*, **5,** 637.
310 Coruzzi, G. and Tzagoloff, A. (1979), *J. Biol. Chem.*, **254,** 9324.
311 Halbreich, A., Pajot, P., Foucher, M., Grandchamp, C. and Slonimiski, P. (1980), *Cell,* **19,** 321.
312 Arnberg, A.C., Van Ommen, G-J.B., Grivell, L.A., Van Bruggen, E.F.J. and Borst, P. (1980), *Cell,* **19,** 313.
313 Bedbrook, J.R., Kolodner, R. and Borograd, L. (1977), *Cell,* **11,** 739.
314 Schwarz, Z.S. and Kossel, H. (1980), *Nature,* **283,** 739.
315 Hartley, M.R. and Heard, C. (1979), *Eur. J. Biochem.*, **96,** 301.
316 Allet, B. and Rochaix, J.D. (1979), *Gene,* **18,** 55–60.
317 Malnoe, P., Rochaix, J-D., Chua, N.H. and Spahr, P.F. (1979), *J. Mol. Biol.*, **133,** 417.
318 Bedbrook, J.R., Coen, D.M., Beaton, A.R., Bogorad, L. and Rich, A. (1979), *J. Biol. Chem.*, **354,** 905.
319 Weissmann, C., Billeter, M.A., Goodman, H.M., Hindley, J. and Weber, M. (1973), *Annu. Rev. Biochem.*, **42,** 303.
320 Hindley, J. (1973), *Br. Med. Bull.*, **29,** 236.
321 Federoff, N. (1975), in *RNA phages* (ed. N. Zinder), Cold Spring Harbor Harbor N.Y., Cold Spring Laboratory, p. 235.
322 Kaman, R.I. (1975), *ibid.*, p. 203.
323 Kuo, C-H., Eoyang, L. and August, J.T. (1975), *ibid.*, p. 259.
324 Blumenthal, T. and Carmichael, G.C. (1979), *Annu. Rev. Biochem.*, **48,** 525.
325 Baltimore, D. (1971), *Bact. Rev.*, **35,** 235.
326 Shatkin, A.J. (1974), *Annu. Rev. Biochem.*, **43,** 643.

327 Dascupta, A., Baron, M.H. and Baltimore, D. (1979), *Proc. Natl. Acad. Sci. USA,* **76,** 2679.
328 Dascupta, A., Zabel, P. and Baltimore, D. (1980), *Cell,* **19,** 423.
329 Szilagy, J.F. and Uryuayev, L. (1973), *J. Virol.,* **11,** 279.
330 Perlman, S.M. and Huang, A.S. (1975), *Virus Research* (ed. C.F. Fox and W.S. Robinson), Academic Press, New York, p. 105.
331 Joklik, W.K. (1975), *Virus Research* (ed. C.F. Fox and W.S. Robinson), Academic Press, New York. p. 105.
332 Shatkin, A.J. and Sipe, J.D. (1968), *Proc. Natl. Acad. Sci. USA,* **61,** 1462.
333 Smith, R.E., Zweernick, H.J. and Joklik, W.K. (1969), *Virology,* **39,** 79.
334 Luria, S.E., Darnell, J.E., Baltimore, D. and Campbell, A. (1979), in *General Virology,* 3rd edn, John Wiley and Sons, New York and Chichester, U.S.
335 Crews, S. and Attardi, G. (1980), *Cell,* **19,** 775.
336 Astier-Manifacier, S. and Cornuet, P. (1971), *Biochim. Biophys. Acta,* **232.**
337 Duda, C.T., Zaitlin, M. and Siegel, A. (1973), *Biochim. Biophys. Acta,* **319,** 62.
339 Fraenkel-Conrat, H. (1976), *Virology,* **172,** 23.
340 Fraenkel-Conrat, H. (1979), *Trends in Biochem. Sci.,* **4,** 184.
341 Wilkes, P.R., Birnie, G.D. and Paul, J. (1979), *Nucleic Acids Res.,* **6,** 2193.
342 Kleinman, L., Birnie, G.D., Young, B.D. and Paul, J. (1977), *Biochemistry,* **16,** 1218.
343 Selker, E. and Yanofsky, C. (1980), *Nucleic Acids Res.,* **8,** 1033.
344 Engelke, D.R., Ng, S-Y., Shastry, B.S. and Roeder, R.G. (1980), *Cell,* **19,** 717.
345 Smith, G.P. (1976), *Science,* **191,** 528.
346 Tschudi, C. and Pirrotta, V. (1980), *Nucleic Acids Res.,* **8,** 441.
347 Montgomery, D.L., Leung, D.W., Smith, M., Shalit, P., Faye, G. and Hall, B.D. (1980), *Proc. Natl. Acad. Sci. USA,* **77,** 541.
348 Alestrom, P., Akusjarvi, G., Perricaudet, M., Mathews, M.B., Klessig, D.F. and Pettersson, U. (1980), *Cell,* **19,** 671.
349 Hamoda, H., Igarashi, T. and Murumatsu, M. (1980), *Nucleic Acids Res.,* **8,** 587.
350 Rodgers, J. and Wall, R. (1980), *Proc. Natl. Acad. Sci. USA,* **77,** 1877.
351 Merten, S., Synenki, R.M., Locker, J., Christianson, T. and Rabinowitz, M. (1980), *Proc. Natl. Acad. Sci. USA,* **77,** 1417.
352 Wasylyk, B., Kedinger, C., Corden, J., Brison, O. and Chambon, P. (1980), *Nature,* **285,** 367.
353 Wu, G.U. (1978), *Proc. Natl. Acad. Sci. USA,* **75,** 2175.
354 Baer, B.W. and Kronberg, R.D. (1980), *Proc. Natl. Acad. Sci. USA,* **77.**
355 Cudny, H. and Deutscher, M. (1980), *Proc. Natl. Acad. Sci. USA,* **77,** 837.
356 Gough, N.M., Kemp, D.J., Tyler, B.M., Adams, J.M. and Cory, S. (1980), *Proc. Natl. Acad. Sci. USA,* **77,** 554.
357 Sakano, H., Rogers, J.H., Hupi, K., Brack, C., Traunecker, A., Maki, R., Wall, R. and Tonegawa, S. (1979), *Nature,* **271,** 627.
358 Eaton, W.A. (1980), *Nature,* **284,** 183.
359 Chui, N.H., Bruszewski, W.B. and Salzman, N.P. (1980), *Nucleic Acids Res.,* **8,** 153.

360 Long, E.O. and Dawid, I.B. (1980), *J. Mol. Biol.,* **138,** 837.
361 Kaback, D.B. and Davidson, N. (1980), *J. Mol. Biol.,* **138,** 745.
362 Gubler, U., Wyler, T., Seebeck, T. and Braun, R. (1980), *Nucleic Acids Res.,* **8,** 2647.
363 Wickens, M.P., Woo, S., O'Malley, B.W. and Gurdon, J.B. (1980), *Nature,* **285,** 628.
364 Galliwitz, D. and Sures, I. (1980), *Proc. Natl. Acad. Sci. USA,* **77,** 2546.
365 Mathews, M.B. (1980), *Nature,* **285,** 575.
366 Guss, P. and Khoury, G. (1980), *Nature,* **286,** 634.
367 Eperon, I.C., Anderson, S. and Nierlich, D.P. (1980), *Nature,* **286,** 460.
368 Bos, J.L., Osinga, K.A., Van der Horst, G., Hecht, N.B., Tabak, H.F., Van Ommen, G-J.B. and Borst, P. (1980), *Cell,* **20,** 207.
369 Van Ommen, G-J.B., Boer, P.H., Groot, G.S.P., Haan, M.D., Roosendaal, E., Grivell, L.A., Haid, A. and Schweyen, R.J. (1980), *Cell,* **20,** 173.
370 Dujon, B. (1980), *Cell,* **20,** 185.
371 Barry, G., Squires, C. and Squires, C.L. (1980), *Proc. Natl. Acad. Sci. USA,* **77,** 3331.
372 Defranco, D., Schmidt, O. and Soll, D. (1980), *Proc. Natl. Acad. Sci. USA,* **77,** 3365.
373 Graf, L., Kossel, H. and Stutz, E. (1980), *Nature,* **286,** 908.
374 Marquat, L.E., Kinniburgh, A.J., Beach, L.R., Honig, G.R., Lazerson, J., Ershler, W.B. and Ross, J. (1980), *Proc. Natl. Acad. Sci. USA,* **77,** 4287.
375 Kantor, J.A., Turner, P.H. and Nieuhuis, A.W. (1980), *Cell,* **21,** 179.
376 Ng, R. and Abelson, J. (1980), *Proc. Natl. Acad. Sci. USA,* **77,** 3912.
377 Pribnow, D. (1979), in *Biological Regulation and Development,* Vol. 1, p. 219 (ed. R.F. Goldberger), Plenum Press, New York.
378 Ojala, D., Montoya, J. and Attardi, G. (1981), *Nature,* **290,** 470.
379 Lazowska, J., Jacq, C., Slonimski, P.P. (1980), *Cell,* **22,** 333.

The biological function of RNA: protein synthesis

12

12.1 RNA AND THE EXPRESSION OF GENETIC INFORMATION

The first indication that RNA might in some way be involved in protein synthesis came from the early experiments of Caspersson [1] using spectrophotometric methods, of Brachet [2] using histochemical techniques, and of Davidson [3] using chemical methods, all of whom showed that RNA was particularly abundant in cells engaged in the synthesis of protein either for growth or for secretion. It is now known that the three types of RNA, messenger RNA, transfer RNA and ribosomal RNA, are involved in expressing as protein the biological information contained in the DNA of the cell.

The amino acid sequences in proteins are determined by the sequences of bases in the discrete regions of the cellular DNA called genes. The first proof of colinearity of gene and protein came from an analysis of mutants of the head protein of bacteriophage T4 [4] and the A protein of tryptophan synthetase of *E. coli* [5]. In the process of *duplication* or *replication* (Chapter 8) exact copies of the DNA are made for hereditary transmission. In the process of *transcription* (Chapter 10) the genetic information is transferred from the DNA to the complementary or messenger RNA. Finally, on the ribosome, the genetic information is *translated* from the four-letter language of the mRNA into the twenty-letter language of the proteins in the process of protein synthesis. The topic of translation will be discussed here with particular emphasis on the part played by the nucleic acids. More detailed treatment of this subject may be found elsewhere [6–14].

12.2 THE CODON AS A NUCLEOTIDE TRIPLET

In proteins, twenty different kinds of amino acid are commonly found, whereas only four main kinds of base occur in the nucleic acids. The genetic code describes how a sequence derived from twenty or more units is determined by a sequence derived from four units of a different type (for reviews see [15–20]).

Since there are only four kinds of base but twenty kinds of amino acid, the correspondence cannot be a simple 1:1 relationship between bases and amino acids. Nor are there sufficient combinations of two bases (4^2, i.e. 16) to account for twenty amino acids. It had been suspected, therefore, that each amino acid would be determined by a sequence of at least three bases, which would give sixty-four combinations (4^3)—more than adequate for the coding of the twenty amino acids. Crick and his colleagues [16, 21, 22] produced fairly clear-cut evidence that the *triplet* theory was correct and that what they termed the *codon* is a sequence of three nucleotides. Their experiments were carried out on the A and B cistrons of the rII locus of bacteriophage T4 in which, as Benzer had shown by careful genetic mapping, one particular region of the DNA determines whether or not the bacteriophage can attack strain K of *E. coli*; and they used proflavine (Section 9.2.4) to bring about either the insertion of an additional base into the DNA sequence or the deletion of a single base.

If we assume that the sequence of bases in a portion of DNA is as shown in the top line of Fig. 12.1, and that the message is read in groups of three from left to right starting at the first C, then the insertion by the mutagen of another base N in the second triplet from the left will upset the reading of all triplets to the right of the point of insertion (Fig. 12.1, second line). The mutant so produced will be seriously defective and will not infect strain K. However, if a further mutation can now be produced which brings about the removal of the third A from the left, the fourth, fifth and subsequent triplets will read correctly and only the second and third triplets will be faulty (Fig. 12.1, third line). Only two amino acids, corresponding to these two triplets, will be 'wrong', and if

```
         CAT|CAT|CAT|CAT|CAT|CAT| ---------
   +1    CAT|CAN|TCA|TCA|TCA|TCA| T--------
+1 -1    CAT|CAN|TCT|CAT|CAT|CAT| ---------
   +2    CAT|NCA|NTC|ATC|ATC|ATC| AT-------
   +3    CAN|TNC|ANT|CAT|CAT|CAT| CAT---
```

Fig. 12.1 Hypothetical sequence of bases in a DNA strand showing genetic message in triplets. Addition of one base (second line) or two bases (fourth line) makes the code unreadable, but it can be restored if one base is added and another removed nearby (third line). The message is still readable if three bases are inserted (last line).

the presence of these two amino acids does not affect the structure of the protein significantly the bacteriophage will behave normally and will infect strain K. In practice it was in fact found that bacteriophages with an insertion and a deletion close together behave normally, whereas the chances of normal behaviour diminishes as the distance between the insertion and the deletion increases.

It is, moreover, possible to combine mutants in other ways. When two (+) mutations are combined, the recombinants are defective (Fig. 12.1, fourth line) but three (+) or three (−) mutations behave normally and infect strain K (Fig. 12.1, bottom line).

These results can best be interpreted by assuming that coding takes place by consecutive triplets in the nucleic acid. The insertion of one or two bases at any point will so alter the sequence of triplets as to make the code unreadable, whereas if three bases are added (or if one base is added and another is deleted) the sequence of triplets is restored after the first two changes, and the original message on the DNA can be interpreted as before.

12.3 THE ADAPTOR FUNCTION OF tRNA

It was difficult in molecular terms to envisage a direct interaction between a particular triplet of nucleic acid bases and a specific amino acid. Crick proposed that specific interaction with the triplet codon required an *adaptor* triplet or oligonucleotide, to which the amino acid was attached [23]. The existence and nature of the adaptor molecule was established by Hoagland with the discovery that a previously neglected class of low-molecular-weight RNA, transfer RNA, could accept amino acids and transfer them to growing polypeptide chains [24, 25]. As already described in Chapter 5, the tRNA molecule is considerably larger than the adaptor Crick envisaged. It consists of about 75 nucleotides, at the 3′ end of which an amino acid may be covalently attached. A loop or bend near the centre of the chain contains a triplet of bases (anticodon) that can link with a corresponding codon on the messenger RNA by complementary base-pairing (see Fig. 5.5). The genetic code is thus, in effect, determined by which aminoacyl-tRNAs can recognize which triplet codons.

12.4 CODON ASSIGNMENTS

12.4.1 Amino acid codons

The problem of assigning triplets of bases to each of the 20 amino acids was attacked in several ways.

(a) *The use of biosynthetic messengers*

The first approach utilized protein-synthesizing systems prepared from cell-free extracts of *E. coli* [26]. Such extracts contain ribosomes, tRNAs, aminoacyl-tRNA synthetases and other enzymes, and, in the crude state, also DNA and messenger RNA. When ATP is added together with GTP, Mg^{2+}, K^+, and amino acids, the amino acids are readily incorporated into an acid-insoluble protein product and the incorporation process can be followed by using amino acids labelled with ^{14}C. When the DNA in such extracts is destroyed by DNase, protein synthesis ceases after the messenger RNA has been depleted, but can be restored by adding RNA fractions from various sources. These latter were found to include synthetic polynucleotides produced by the action of polynucleotide phosphorylase (see Section 10.4). This led Nirenberg and his co-workers to examine which amino acids were incorporated into polypeptide when different synthetic polynucleotides were added.

In 1961, Nirenberg and Matthaei [27] observed that, when the synthetic polymer poly(U) was added to the cell-free system with mixtures of 20 amino acids, only one amino acid in each mixture being radioactive, the only amino acid to be incorporated into acid-insoluble protein-like material was phenylalanine, and the product was polyphenylalanine. When taken with the evidence [21] that the codon is a triplet, this established that phenylalanine may be coded for by UUU, the first codon to be deciphered. Similarly, poly(A) and poly(C) were found to direct the synthesis of polylysine and polyproline, respectively. Nirenberg and his colleagues, and Ochoa and his colleagues, extended this approach from homopolynucleotides to heteropolynucleotides. Such heteropolynucleotides could be synthesized with various defined base compositions. Thus, although they had random sequences, the statistical incidence of different triplet codons could be calculated, and correlated with the relative incorporation of different ^{14}C-labelled amino acids [28, 29].

Such use of random heteropolynucleotides could never do more than indicate the base composition of the codons specifying different amino acids. Khorana and his colleagues were subsequently able to attack this problem more directly with copolymers of defined sequence, prepared by an elegant combination of organic chemical and enzymic syntheses. The enzyme DNA polymerase I (see Section 8.4.2) can be used to extend a short double-helical oligodeoxyribonucleotide, e.g. poly d(A–C).poly d(T–G), previously prepared by chemical means. This synthetic DNA is then transcribed using RNA polymerase (see Chapter 10), to yield a specific polyribonucleotide poly(U–G) containing U and G in a repeating sequence. This can then be used in a cell-free protein-synthesizing system in order to determine codon assignments (Table 12.1) [30, 31].

Table 12.1 Amino acid incorporations stimulated by mRNA containing repeating nucleotide sequences [31].

Messenger	Amino acids incorporated	Messenger	Amino acids incorporated
Repeating dinucleotides		*Repeating trinucleotides*	
Poly(U-C)	Ser-Leu	Poly(G-U-A)	Val, Ser
Poly(A-G)	Arg-Glu	Poly(U-A-C)	Tyr, Leu, Ile, Ser
Poly(U-G)	Val-Cys	Poly(A-U-C)	Ile, Ser, His
Poly(A-C)	Thr-His	Poly(G-A-U)	Met, Asp
Repeating trinucleotides		*Repeating tetranucleotides*	
Poly(U-U-C)	Phe, Ser, Leu	Poly(U-A-U-C)	Tyr, Leu, Ile Ser
Poly(A-A-G)	Lys, Glu, Arg	Poly(G-A-U-A)	none
Poly(U-U-G)	Cys, Leu, Val	Poly(U-U-A-C)	Leu, Thr, Tyr
Poly(C-A-A)	Gln, Thr, Asn	Poly(G-U-A-A)	none

For example, a $(A\text{-}C)_n$ sequence will be read as ACA-CAC-ACA-CAC-A . . . and will yield a polypeptide containing two amino acids, alternating those coded by ACA and CAC. In fact, the amino acids incorporated are threonine and histidine when poly(A-C) is used as messenger. Which of these codons corresponds to which amino acid? In this case the results of the translation of the polytrinucleotide, poly(C-A-A), provides the answer. From this sequence three homopolymers should be coded corresponding to the triplets CAA, AAC, and ACA. This is because the starting point is not clearly defined and the message may be read in any of the three forms:

. . C A A C A A C A A C A . .
. . C A A C A A C A A C A . .
. . C A A C A A C A A C A . .

In practice, the amino acids incorporated are glutamine, threonine and asparagine. As poly(C-A-A) only specifies one triplet, ACA, that is also found in poly(A-C), above, this triplet must code for threonine, the only amino acid incorporated in both cases. Thus the other triplet, CAC, specified by poly(A-C), must correspond to histidine.

In other cases the fact that more than one triplet codon may specify a single amino acid prevents unambiguous assignments being made from these results alone. However, by taking into account the previous data obtained with random heteropolynucleotides, and the results of mutation experiments [see (c) below], the amino acids specified by about half the codons were established [30, 31].

(b) *The ribosome binding technique*

At about the same time as Khorana and his colleagues were performing the studies just described, Nirenberg and Leder were employing a different, and less ambiguous, approach to the problem. Instead of translating synthetic messengers, they made use of the specific codon–anticodon interaction between oligonucleotide triplets and aminoacyl-tRNA that they found could occur on ribosomes *in vitro* [32]. When a mixture containing ribosomes, aminoacyl-tRNA and a triplet is allowed to react under suitable ionic conditions, and is then poured onto a nitrocellulose filter, the free tRNA (and triplet) passes through the filter, whereas the ribosomes—and any bound tRNA—are retained on it. By using a series of 20 different amino acid mixtures each containing one ^{14}C-labelled amino acid, it is possible to identify the amino acid corresponding to a triplet by means of the radioactivity absorbed by the filter. For example, the trinucleotide GUU retains valyl-tRNA whereas UGU and UUG do not [33]. All 64 possible triplets were synthesized and tested, and more than 50 of them gave unambiguous results [34].

(c) *The use of mutations*

The cell-free techniques, described in (a) and (b) above, established the amino acid assignments shown in Table 12.2. Confirmatory evidence was obtained from genetic mutations, and this was of especial value as it derived from intact cells.

The principle involved in the use of base-substitution mutants may be illustrated by the results obtained with the aid of artificially induced mutants of tobacco mosaic virus (TMV) [35–37]. When TMV RNA is treated with nitrous acid two changes are brought about: (i) cytosine is deaminated to uracil (C→U), and (ii) adenine is deaminated to hypoxanthine, which is equivalent to guanine in coding (A→G) [35, 36]. When HNO_2-treated TMV RNA is used to infect tobacco plants, mutants may be produced in which single amino acids in the viral protein are replaced by different amino acids at certain positions in the polypeptide chain in such a way that the replacements can be correlated with the changes A→G or C→U (Fig. 12.2). For example, leucine may be replaced by phenylalanine corresponding to the change CUU→UUU, or alanine may be replaced by glycine in accordance with the change GCA→GCG. Similar evidence is obtained from the mutations affecting the A protein of tryptophan synthetase [38], and from the different varieties of human haemoglobin [19].

Table 12.2 The genetic code.

5′-OH terminal base	Middle base				3′-OH terminal base
	U	C	A	G	
U	Phe	Ser	Tyr	Cys	U
	Phe	Ser	Tyr	Cys	C
	Leu	Ser	STOP	STOP	A
	Leu	Ser	STOP	Trp	G
C	Leu	Pro	His	Arg	U
	Leu	Pro	His	Arg	C
	Leu	Pro	Gln	Arg	A
	Leu	Pro	Gln	Arg	G
A	Ile	Thr	Asn	Ser	U
	Ile	Thr	Asn	Ser	C
	Ile	Thr	Lys	Arg	A
	Met*	Thr	Lys	Arg	G
G	Val	Ala	Asp	Gly	U
	Val	Ala	Asp	Gly	C
	Val	Ala	Glu	Gly	A
	Val†	Ala	Glu	Gly	G

STOP = Chain termination codon
* Also usual initiation codon for fMet
† Also occasionally initiation codon for fMet

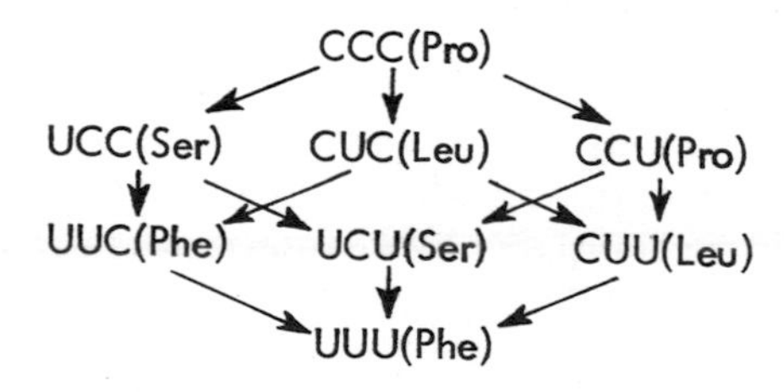

Fig. 12.2 Steps by which the triplet CCC which codes for proline may be changed by deamination to the triplet UUU which codes for phenylalanine. The amino acids corresponding to each triplet are shown on the right of the codon.

Protein sequence determination was also performed on '*frame shift*' mutants of the type that had been used to establish the triplet nature of the genetic code (Fig. 12.1). The results were indeed consistent with the mutations having been caused by the insertion or deletion of a base producing a shift in the reading frame of the genetic message, according to the predicted genetic code [39].

(d) *Evidence from bacteriophage RNAs with regions of known sequence*
Final confirmation of the genetic code *in vivo* has come from determination of the sequences of mRNAs. The first such analysis was undertaken on the small RNA bacteriophages (see Section 13.3), the genome of which is also the mRNA. In 1969 Sanger and his colleagues published the nucleotide sequence of a segment of RNA from bacteriophage R17 which correlated exactly with the amino acids known to be in positions 81–99 of the coat protein [40]. Subsequently Fiers and his colleagues determined the complete sequence of the related bacteriophage MS2, and confirmed all the amino acid assignments of the genetic code [41].

12.4.2 Termination codons

The amino acid assignments derived by the methods described in Section 12.4.1 (a), (b) and (c) accounted for 61 of the possible 64 triplets. The remaining three, UAA, UAG, and UGA, designated STOP in Table 12.2, do not code for any amino acid, but are signals for polypeptide chain termination.

The existence of specific termination signals was indicated by genetic studies with a class of mutants termed '*nonsense*', to distinguish them from the '*missense*' mutants in which one amino acid was replaced by another [42]. The identities of the termination codons responsible for two classes of the mutants ('amber' = UAG, and 'ochre' = UAA) were deduced from mutagenesis experiments analogous to those described in 12.4.1 (c) above, but employing hydroxylamine and 2-aminopurine as mutagens [43]. A third class of 'nonsense' mutant (sometimes called 'opal') was subsequently discovered and shown to correspond to the termination codon UGA [44, 45]. The designation of codons UAG and UAA as terminators explains why neither poly(G-A-U-A) nor poly (G-U-A-A) could be translated by Khorana and his colleagues (Table 12.1). The termination codon assignments have also been confirmed by the nucleotide sequences of mRNAs.

Although tRNA is *not* involved in decoding the termination triplets (see Section 12.9.3, below), certain mutant tRNAs have been found to recognize them. These occur in the 'suppressor' strains of *E. coli*, so called for their ability to suppress particular classes of 'nonsense' mutants. Nucleotide-sequence analysis of such a mutant of a tyrosine tRNA from an amber suppressor strain showed that its anticodon is changed from 3′ AUG 5′ to 3′ AUC 5′. Thus it is able to insert tyrosine into a polypeptide chain in response to the termination codon UAG, rather than to the tyrosine codon UAC [46].

12.4.3 Initiation codons

It was originally thought that no specific codon served to signal the start of a polypeptide chain. This was because all 64 triplets were otherwise accounted for, and no common triplet was required for synthetic polynucleotides to be translated *in vitro* (Table 12.1). However, in 1964 Marcker and Sanger discovered that, in *E. coli*, methionyl-tRNA could be formylated in the α-amino position of the methionine residue [47]. This tRNA, *N*-formylmethionyl-tRNA, was thus only capable of forming a peptide bond with its α-carboxyl group. When this was considered with earlier data showing that methionine occupied the amino-terminal position of about 45% of *E. coli* proteins [48], it raised the possibility that bacterial proteins might all be initiated at a codon specifying formylmethionine, the formyl group and, sometimes, the methionyl residue being subsequently removed.

Direct evidence for this came from the use of cell-free systems to translate the RNA of bacteriophages R17 and f2. (This translation of natural mRNA requires a much lower Mg^{2+} concentration than the promiscuous translation of artificial polynucleotides.) The coat proteins of the bacteriophages synthesized *in vivo* have amino-terminal sequences commencing Ala.Ser. . . . However, coat proteins were synthesized *in vitro* with amino-terminal sequences commencing fMet.Ala.Ser. . . . [49, 50]. Presumably the cell-free system lacked the deformylase [51, 52] and aminopeptidase [52] activities that were subsequently demonstrated to be present in the intact cell.

The tRNA that inserts the initiating fMet into polypeptide chains, $tRNA_f^{Met}$, has a different nucleotide sequence from the tRNA that inserts Met internally: $tRNA_m^{Met}$ [53, 54]. Both species accept methionine, but only the Met-$tRNA_f$ can then be formylated by a *transformylase* enzyme that has N^{10}-formyl-tetrahydrofolic acid as a cofactor [55]. In the ribosome-binding assay fMet–tRNA will recognize the triplets, AUG, GUG and UUG [55, 56]. Both AUG and, to a much lesser extent, GUG have been found as natural initiation codons in those mRNAs, the nucleotide sequences of which have so far been determined. There is also an instance of reinitiation at an internal UUG codon in an mRNA for the *lac* repressor protein, in which there is an 'amber' mutation that causes premature termination [57].

12.5 THE DEGENERACY OF THE GENETIC CODE

Since many of the 20 amino acids are coded by more than one triplet (Table 12.2), the code is said to be degenerate.

Triplets coding for the same amino acid are not distributed at random, but are grouped together so that they generally share the same 5′ and middle base. This has the consequence that mutations producing a change in the base at the 3′ position of the codon often have no effect on the amino acid specified. Furthermore, the different amino acids are segregated to a considerable extent on the basis of chemical similarity (hydrophobicity, hydrophilicity, acidity and basicity). Thus a mutation in the 5′ base of any of the six leucine codons would give a codon specifying another hydrophobic amino acid. Such a change might not impair the function of a particular globular protein, if the altered amino acid merely performed a structural role in the hydrophobic core of this protein. It has therefore been argued that the specific arrangement of codons in the genetic code serves to reduce the potentially harmful effect of possible mutations.

One might have expected that each degenerate codon would require its own tRNA with a corresponding anticodon. Some such discrete *iso-accepting* tRNAs, recognizing the same amino acid, were found. However Crick argued that there might be a degree of latitude in the complementary base pairing between the base in the 3′ position of the codon and that in the 5′ position of the anticodon [58]. This '*wobble*', as he called it, would allow the two pairs of 3′ codon bases that are grouped together in degenerate codons, U and C, and A and G, to be recognized by a common 5′ anticodon base. From stereochemical considerations he suggested that the 5′ anticodon base, G, (but not A) might be able to pair with *both* U or C in the 5′ position of the codon; and that the 5′ anticodon base, U, (but not C) might be able to pair with *both* A or G in the 5′ position of the codon. Moreover, it had already been found that several tRNA anticodons contained the nucleoside inosine (which pairs only with C in a double-helix), and he suggested that this might be able to pair with A, U or C in the 3′ position of the mRNA codon. This is illustrated for some known anticodons in Fig. 12.3. The hypothesis requires, however, that inosine be excluded from the 5′ position of the anticodon of tRNAs such as those for Phe, Tyr, Cys, His, Asn, Ser or Asp, to prevent misreading of those adjacent codons with A in the 3′ position.

Direct evidence supporting the 'wobble' hypothesis came from ribosome-binding experiments with purified individual tRNA species of known base sequence [59, 60]. All tRNA species characterized subsequently have been found to have codon recognition patterns consistent with the 'wobble' hypothesis and with the restriction on inosine usage mentioned above.

Crick proposed the 'wobble' hypothesis at a time when the genetic code was not quite completely deciphered. It has subsequently been necessary to place a further restriction on allowed bases in one anticodon: the Ile-tRNA reading the codon AUA requires to have I in the 5′ position, as a U in that position would allow wobble pairing with the methionine codon AUG. The two currently available sequences of Ile-tRNAs are consistent with this restriction [62]. Another point should be made regarding the reading of the codon AUG. Whereas the other amino acid codons always have the normal strict base-pairing in their 5′ and middle positions, the initiator, fMet-tRNA, has a 'wobble' pairing between its 3′ U [53] and the 5′ position of codons AUG, GUG, and even UUG. A possible explanation for this is provided by experiments which indicate that the anticodon loop of $tRNA_f^{Met}$ has a different conformation from that of other tRNAs [61, 273].

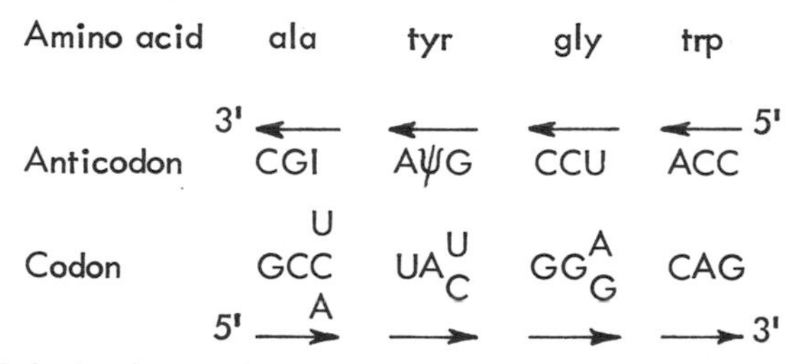

Fig. 12.3 Codon interactions of certain known tRNA anticodons predicted by the 'wobble' hypothesis (see text). The codons are written in the conventional direction (5′→3′) while the anticodons are written backwards (3′←5′) to show the base pairings in an antiparallel direction. No anticodon is known with a base A in the 5′ position, which, according to the hypothesis would pair only with a base U in the 3′ position of the codon.

Whilst discussing the *degenerate* tRNAs, which recognize more than one codon, it should be mentioned that there are, so-called, redundant, iso-accepting tRNAs, sharing exactly the same anticodon, but differing at other points in their nucleotide sequences. Examples are $tRNA_1^{Tyr}$ and $tRNA_2^{Tyr}$ from *E. coli,* and $tRNA^{Ser1}$ and $tRNA^{Ser2}$ from yeast (see Fig. 5.4). These differ only in the bases at a very few positions [62], suggesting that they arose by gene duplication and subsequent mutation. Such gene duplication (which in other cases may produce genes coding for identical tRNAs) is assumed to help protect the organism against catastrophic mutation in its tRNAs.

Now that whole or partial sequences of an appreciable number of mRNAs (or their corresponding genes) are known, one can ask what is the relative usage of the different degenerate codons for particular amino acids. It appears that for certain codons the usage is non-random, and that the pattern of preferred codons is different in prokaryotes and in eukaryotes [63]. For example, there seems to be a preference for G and C over A and U in the 3′ position of eukaryotic codons. For initiation, usage of AUG is heavily favoured over GUG, which is found at the initiation site of only three of about 80 prokaryotic mRNAs [64]. In eukaryotes no initiator codon other than AUG has yet been found in about 40 sequences. In this respect it is interesting that a study of mutants of yeast iso-1-cytochrome *c* suggested that GUG cannot replace AUG as an initiator codon in that protein [65]. All three termination codons have been found in mRNA sequences from both prokaryotes and eukaryotes, but it is not yet clear whether usage is random or not.

12.6 THE UNIVERSALITY OF THE GENETIC CODE

Experiments with tRNAs from eukaryotes suggested that they recognized the same triplet codons as the corresponding tRNAs from prokaryotes [66], and the identity of the codes in the two orders has been confirmed by the correspondence of the amino acid sequences of a number of eukaryotic cytoplasmic proteins with the nucleotide sequences of their mRNAs. [The only minor difference (see Section 12.10.1) is that the initiator tRNA specified by AUG in eukaryotic cytoplasm carries a methionine that is not formylated.]

Studies with mitochondrial systems, however, indicate that certain mitochondrial mRNA codons have different assignments from those in cytoplasmic mRNAs. This was first evident from the finding that the amino acid at position 46 of yeast mitochondrial ATPase subunit 9 was threonine, whereas the mitochondrial DNA sequence indicated the corresponding mRNA codon to be CUA, which normally codes for leucine [67, 68], and it appears that all the codons of the class CUN are read as threonine rather than leucine in yeast mitochondria [302]. Human and bovine [303] and *N. crassa* [304] mitochondria do not have this peculiarity, the class of codons, CUN, specifying leucine. However, they share with yeast mitochondria the use of the codon UGA for tryptophan, rather than for termination [69, 70, 302–304]. Moreover, in mammalian (but not in yeast) mitochondria AUA may specify methionine, rather than isoleucine. A fundamental difference in the way mitochondrial tRNAs can interact with mRNA codons underlies these altered codon assignments. This arises from the fact that mitochondria contain only about 24 different species of tRNA; less than the minimum 31 required by the 'wobble' hypothesis [58] to read the cytoplasmic genetic code. In fact the restrictions imposed by the 'wobble' hypothesis are violated in mitochondria and most of the block of four codons specifying a single amino acid are read by a single tRNA with an unmodified U in the 5′ anticodon (wobble) position [302–304]. Unique features of mitochondrial tRNA or ribosome structure presumably facilitate this greater latitude in U 'wobble'.

12.7 AMINOACYLATION OF tRNA

Before a tRNA molecule can act as an adaptor by interacting with its corresponding anticodon in the decoding process, it must first be 'charged' with its cognate amino acid [8]. The enzymes responsible for this process are called *aminoacyl-tRNA synthetases* (*amino acid-tRNA ligases,* EC 6.1.1.) and catalyse the reaction illustrated in Fig. 12.4.

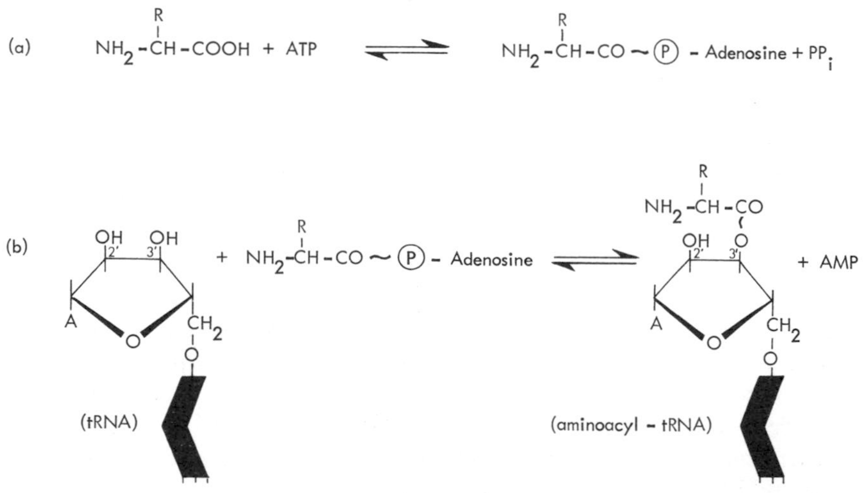

(a) $NH_2-CH(R)-COOH + ATP \rightleftharpoons NH_2-CH(R)-CO\sim(P) - Adenosine + PP_i$

(b) (tRNA) + $NH_2-CH(R)-CO\sim(P)$ - Adenosine $\rightleftharpoons$ (aminoacyl - tRNA) + AMP

Fig. 12.4 The activation of amino acids and their attachment to tRNA. The reaction occurs in two stages, (a) and (b), both of which are catalysed by the same enzyme, and to which the intermediate is bound (see text). The symbol ~ represents a bond with a relatively high standard free energy of hydrolysis.

The reaction occurs in two stages [25, 71], in the first of which the amino acid is *activated* by ATP to form an aminoacyl adenylate:

$$ATP + \text{amino acid}_1 + E_1 \rightleftharpoons (\text{amino acid}_1 - AMP)E_1 + PP_i$$

The aminoacyl–adenylate complex then reacts with the terminal adenosine moiety of the appropriate tRNA to form an aminoacyl-tRNA:

$$(\text{amino acid}_1\text{–AMP})E_1 + \text{tRNA}_1 \rightleftharpoons \text{tRNA}_1\text{–amino acid}_1 + \text{AMP} + E_1$$

Although the amino acid is located at the 3′-OH of the ribose of the terminal adenosine moiety of tRNA during peptide-bond formation, the initial point of attachment can be the 2′-OH, the 3′-OH, or either, depending on the amino acid [72]. After attachment, rapid migration between the two positions is possible [73]. The aminoacyl–ester linkage has a relatively high free energy of hydrolysis, derived from the ATP hydrolysed during its activation. This is important as it provides the necessary energy for the subsequent peptide-bond formation to occur [74].

Despite the fact that multiple species of tRNA (isoaccepting tRNAs) exist for a single amino acid (see Section 12.5), there appears to be only one aminoacyl-tRNA synthetase for each amno acid. Even the different initiating and elongating methionyl-tRNAs are recognized by the same enzyme. At first sight it appears that different aminoacyl-tRNA synthetases have very different structures, some being large single-chain enzymes, some multi-chain enzymes of similar subunit size, and some multi-chain enzymes of different subunit size [75]. Despite this apparent structural dissimilarity, all the synthetases may in fact have a number of functionally similar subunits or domains with a mass of about 30000 daltons [76].

It has already been asserted that the specificity of decoding resides in the complementary base pairing between the anticodon of the tRNA and the codon of the mRNA. Indeed, once the amino acid has been enzymically attached to the tRNA, it may be converted chemically into a different amino acid without altering the codon–anticodon interaction. This was shown in the classic experiment in which Cys-tRNA^{Cys} was reduced with Raney nickel to give Ala-tRNA^{Cys}, which then incorporated alanine in response to codons for cysteine in a cell-free system [77].

The maintenance of the accurate decoding conferred by the specific codon–anticodon interaction is absolutely dependent on the aminoacyl-tRNA synthetases charging the tRNAs with their correct cognate amino acids. This requires the enzymes to be specifically able to recognize both amino acids and nucleic acids.

There are several pairs of amino acids differing in structure by no more than a single methyl group, and this poses a real problem in discrimination for the synthetases. Thus, it was observed that valine bound appreciably to Ile-tRNA synthetase [78]. There is good reason to believe that a 'proof-reading' or 'editing' mechanism exists in those synthetases for which there are inappropriate isosteric or smaller amino acids with which the tRNA may be mischarged. This is thought to involve a hydrolytic site on the enzyme, close to but distinct from the acylation site, to discharge such inappropriate aminoacyl-tRNAs 79.

The initial binding of the similarly shaped tRNAs to the synthetases is also non-specific, and recognition would appear to involve molecular interactions at a number of points along the inside angle of the 'L' in the three-dimensional structure of the tRNA, these sometimes extending from anticodon to acceptor stem. Structural features of tRNAs (and even of certain viral genomic RNAs) necessary for recognition by the appropriate synthetase have been studied by a variety of methods. However the features identified differ with different synthetases, and there appears to be no single common recognition site (reviewed in [75, 80, 81]).

12.8 GENERAL ASPECTS OF POLYPEPTIDE FORMATION

The interaction of the aminoacylated tRNA with the mRNA takes place on the ribosomes where successive codons are read in an ordered manner and the amino acids linked together to form a polypeptide chain. The direction of growth of the polypeptide chain is from the amino terminus to the carboxyl terminus [82], and the direction of reading of the mRNA is $5' \rightarrow 3'$ [83, 39]. As this latter is also the direction in which the mRNA is synthesized, it is possible, in prokaryotes, for ribosomes to start translating a mRNA before its transcription is complete [84, 85]. The rates of protein synthesis in prokaryotes and eukaryotes appear to be quite similar. It has been reported that this is 15 amino acids/s for β-galactosidase in *E. coli* [86], and 7 amino acids/s for globin chains in rabbit reticulocytes [87]. The formation of peptide bonds is catalysed by an enzymic activity, *peptidyl transferase,* which is an integral part of the larger ribosomal subunit [88], and there are two sites on the ribosome at which the reacting molecules of tRNA are bound. One of these sites, the A-site, is occupied by successive molecules of aminoacyl-tRNA, whereas the other site, the P-site, is occupied prior to peptide bond formation by the tRNA carrying the growing polypeptide chain, peptidyl-tRNA. There is considerable evidence that these sites are physically distinct [89] and that no more than two molecules of tRNA can be bound by a ribosome at any one time [90].

Figure 12.5 is a schematic representation of the ribosome just before the aminoacyl–ester bond of the peptidyl-tRNA is broken and the polypeptide chain transferred to the α-amino group of the aminoacyl-tRNA at the A-site. The events involved in the reading of successive codons are described in detail below (Section 12.9) and involve relative movement of the 5′ portion of the mRNA and the ribosome away from one another. The overall length of the mRNA and the rate of initiation are usually such that the initiation codon can attach another ribosome before the first one completes its polypeptide chain. In fact several ribosomes are normally found on a given molecule of mRNA, translating different parts of it simultaneously, and such groups of ribosomes are termed *polyribosomes* or *polysomes* (see Plate VII). These may be visualized by electron microscopy [91], and polysomes containing different numbers of ribosomes may be resolved by sucrose-density-gradient centrifugation [92]. The size of the polysomes increases

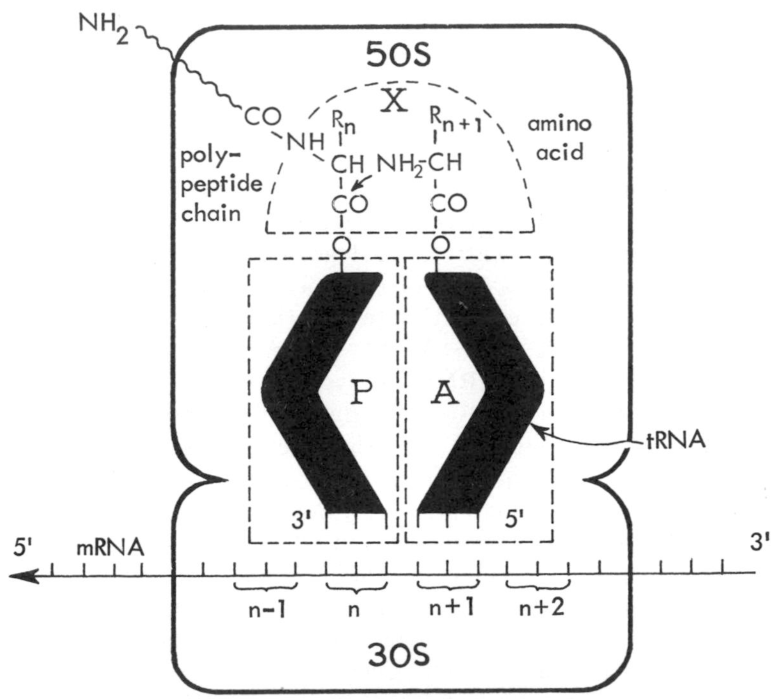

Fig. 12.5 Diagrammatic representation of a prokaryotic ribosome. Two tRNA molecules are bound to the ribosome in response to the mRNA codons designated n and $n+1$. The tRNA bearing the growing polypeptide chain is occupying the peptidyl site (rectangular area, marked P), and the tRNA bearing an amino acid is occupying the aminoacyl site (rectangular area, marked A). The peptidyl transferase centre where the peptide bond formation is catalysed is represented by the semicircular area, marked X.

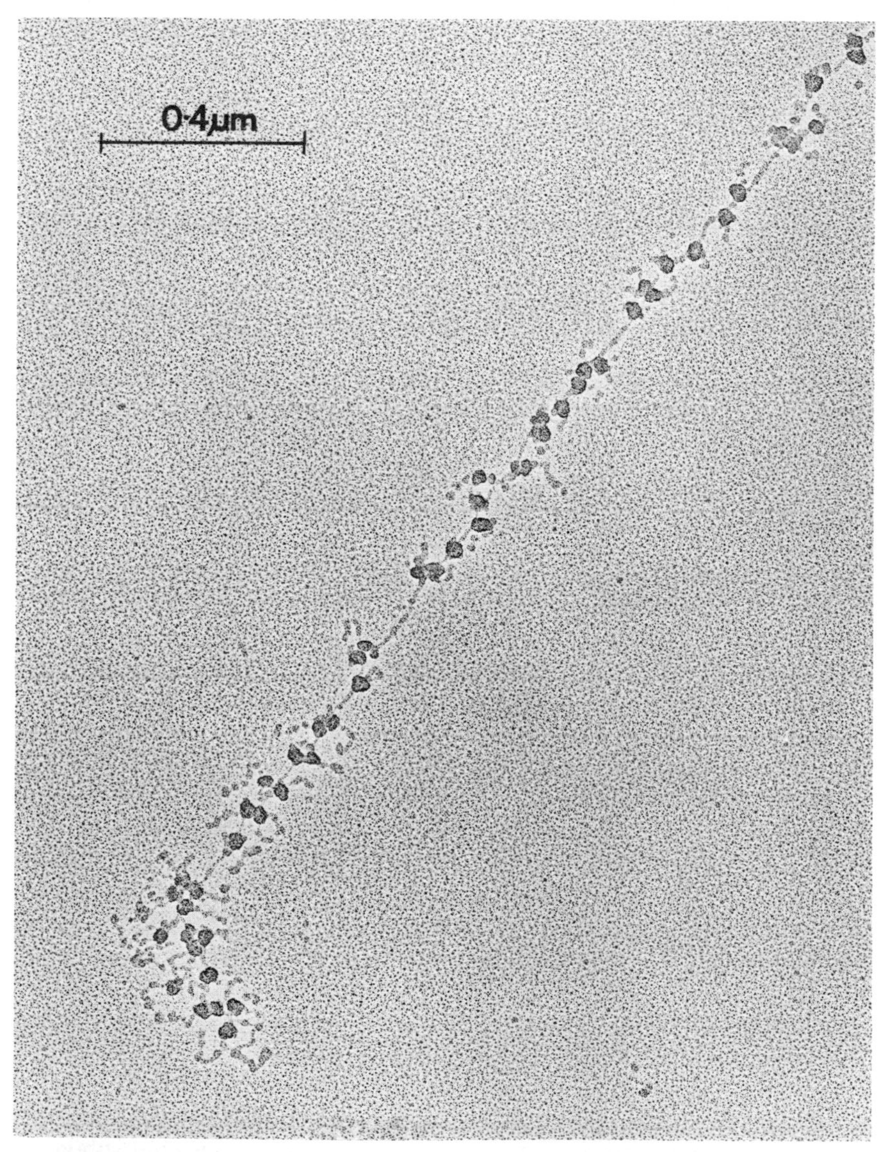

Plate VII Electron micrograph showing the translation of silk fibroin mRNA on polysomes. The extended fibrous fibroin molecules can be seen emerging from the ribosomes (dark irregular particles). The length of the fibroin molecules increases from the top right to the bottom left of the frame, indicating that this is the $5' \rightarrow 3'$ direction along the mRNA. (Courtesy of Dr Steven L. McKnight and Dr Oscar L. Miller, Jr.)

with the size of the mRNA: polysomes synthesizing haemoglobin β-chains (molecular weight 16 000 approx.) contain four to five ribosomes [93], whereas those synthesizing myosin heavy chains (molecular weight, 200 000 approx.) contain about 50–60 ribosomes [94].

The ribosome is only responsible for synthesizing a single polypeptide chain comprising no more than the 20 genetically coded amino acids. After peptide-bond formation, amino acids may form additional inter- or intra-molecular covalent bonds, be subject to specific proteolysis, or be chemically modified in any of a large number of different possible ways [95].

12.9 THE EVENTS ON THE BACTERIAL RIBOSOME

The following discussion uses the uniform nomenclature for bacterial protein synthesis factors. The relationship of this to the various older nomenclatures is given in [96].

12.9.1 Chain initiation [97–100]

In polypeptide chain initiation fMet–tRNA is bound to the initiation codon of the mRNA on the 30S ribosomal subunit [101] and the resulting 30S initiation complex then reacts with the 50S ribosomal subunit to give a 70S initiation complex. This process requires GTP and the initiation factors, IF-1, IF-2 and IF-3 [102–104], and in polycistronic mRNAs can occur independently at several different initiation sites [105].

In bacterial-cell extracts the initiation factors are found associated with the 30S subunit, from which they have been extracted and purified [97, 106]. As already mentioned (Section 12.4.3), the requirement for AUG and factors in initiation was only seen *in vitro* when salt-washed ribosomes and a suitably low concentration of Mg^{2+} (usually about 5 mM) were used [107, 108]. The exact sequence of events in initiation and the precise roles of all the factors is still uncertain, partly because of the concerted manner in which they act. For this reason, the initiation scheme presented in Fig. 12.6 avoids, for the most part, giving more detail than there is general agreement upon.

The initiating 30S subunit most probably has bound to it IF-3 and IF-1 (Fig. 12.6a) which are involved in generating free ribosomal subunits after polypeptide-chain termination (see Section 12.9.3). Their role in initiation is distinct from this latter as they are required for the formation of a 30S initiation complex (Fig. 12.6b) even when 50S subunits are not present [109]. IF-3 (molecular weight 22 000) is primarily involved in binding mRNA to the ribosome [110]. Although it is needed for

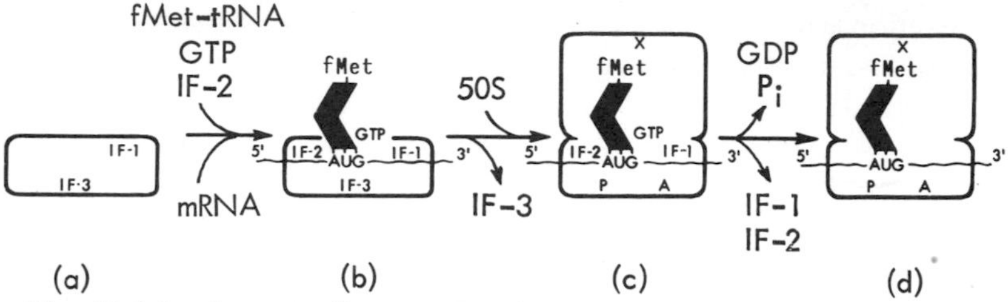

Fig. 12.6 A schematic diagram of prokaryotic polypeptide chain initiation. The ribosome and tRNA are represented as in Fig. 12.5.

translation of natural mRNAs, there is no absolute requirement for IF-3 in either the AUG-dependent ribosome binding of fMet–tRNA or the translation of artificial polynucleotides such as $AUGA_n$ [111]. This suggests that, either directly or indirectly, IF-3 facilitates the recognition of the untranslated 'leader' sequences that precede the initiating AUGs of natural mRNAs. Because polycistronic mRNAs, such as the coliphage RNAs, have separate initiation sites for each cistron, some of these sites are necessarily far from the 5′ end of the mRNA, which cannot therefore be the basis of recognition. As is discussed further in Section 12.11, the ability of the leader sequence of nucleotides to base-pair with the 3′ end of 16S rRNA is thought to be the major factor determining the recognition of an initiator AUG codon, rather than one meant to specify an internal methionine. As well as this positive recognition, mRNA secondary structure may exert a negative control, by preventing initiation at certain internal AUG codons [112].

The primary role of IF-2 (molecular weight, 118000) is to bind fMet–tRNA to the ribosome in a reaction that requires GTP and is stimulated by the other initiation factors, especially IF-1 [113]. As it is possible *in vitro* to bind either mRNA or fMet–tRNA independently to 30S ribosomal subunits, it is not certain which reaction occurs first *in vivo*. Nor is it clear whether the unstable ternary complex between IF-2, fMet–tRNA and GTP, found *in vitro* [114, 115], actually occurs free of the ribosome *in vivo,* or whether such a complex is found only on the 30S subunit [116, 117]. Undoubtedly interaction between IF-2 and fMet–tRNA occurs at some point, and the factor must recognize some specific structural feature of the initiator tRNA as it will not interact with aminoacyl-tRNAs, including Met–$tRNA_m$ [115].

Once a 30S initiation complex, containing mRNA and fMet–tRNA, has been formed (Fig. 12.6b) the 50S ribosomal subunit can associate with it (Fig. 12.6c) causing the release of IF-3 [110]. The non-hydrolysable analogue of GTP, 5′-guanylylmethylene diphosphonate, has been used to show that hydrolysis of GTP is not required for this step, but GTP hydrolysis is required for the fMET–tRNA to become available for reaction with puromycin (an analogue of aminoacyl-tRNA, binding in the A site: see Section 12.13) [118], at which stage IF-2 and (probably) IF-1 are released [119]. As GTP hydrolysis does not cause relative movement of mRNA and the ribosome, fMet–tRNA must be bound directly at the P-site 120, 121. The role of GTP hydrolysis, which is discussed in more detail below (Section 12.12), cannot, therefore, be for movement of the fMet–tRNA from A-site to P-site.

It will be evident from the foregoing that although IF-1 (molecular weight, 9000) is absolutely required for initiation, its role is not clearly defined. The fact that it cycles on and off the ribosome during initiation established that it is indeed an initiation factor, rather then a loosely bound ribosomal protein such as S1 (see Section 12.12). Lower-molecular-weight sub-species of IF-2 and IF-3 have been described, but, although these have functional activity, they are almost certainly the result of proteolysis of the parent protein, and have no separate role in initiation [122, 123].

12.9.2 Chain elongation [89, 124–126]

In polypeptide-chain elongation an aminoacyl-tRNA binds to the A-site of the ribosome and reacts with the peptidyl-tRNA (Fig. 12.5) or fMet–tRNA (Fig. 12.7) in the P-site, accepting the growing polypeptide chain. The tRNA is then moved across to the P-site (translocated), with concomitant movement of the mRNA and expulsion of the deacylated tRNA, in order to make the A-site available for the next aminoacyl-tRNA [127–129]. Elongation requires three soluble factors, EF-Tu, EF-Ts, EF-G [130], and the hydrolysis of two molecules of GTP.

Elongation factor EF-Tu (molecular weight 42 000–48 000 [131, 132]) is responsible for the ribosomal binding of the aminoacyl-tRNA corresponding to the mRNA codon in the A-site (arbitrarily designated as Ala in Fig. 12.7), prior to which it forms a soluble ternary complex with the tRNA and GTP 133, 134. All aminoacyl-tRNAs will form this complex, but fMet–tRNA will not 134. The non-hydrolysable analogue of GTP, 5′-guanylylmethylene diphosphonate, will allow the aminoacyl-tRNA to bind to the 70S ribosome, but the GTP must be hydrolysed before peptide bond formation can occur [127]. The GTP hydrolysis is not required for the peptidyl transferase reaction itself (see below) and its possible role is discussed in Section 12.12. The EF-Tu and GDP are released from the ribosome as a complex. In this form the EF-Tu cannot react with GTP or aminoacyl-tRNA, and it is the function of EF-Ts (molecular weight, 31 000–34 000 [132, 135]) to displace GDP from the EF-Tu · GDP complex [136, 137]. This results in the formation of an

EF-Tu · EF-Ts complex from which the EF-Tu · GTP complex can be regenerated (Fig. 12.7).

The structure of EF-Tu has been studied by X-ray crystallography and it is of interest that it has overall dimensions similar to those of tRNA, despite its greater molecular weight [138]. Both EF-Tu and EF-Ts may also be involved in cellular RNA synthesis (see Chapter 13): certainly they constitute two of the three host subunits of Qβ replicase [139]. The abundance of EF-Tu in bacterial cells (some 5 per cent of total cell protein), and its excess over EF-Ts and ribosomes, have provoked suggestions of yet further possible functions for this protein [140]. The fact that there are two separate genes coding for EF-Tu in *E. coli* may be significant in this respect [141].

The aminoacyl-tRNA bound in the A-site (Fig. 12.7b) can now be

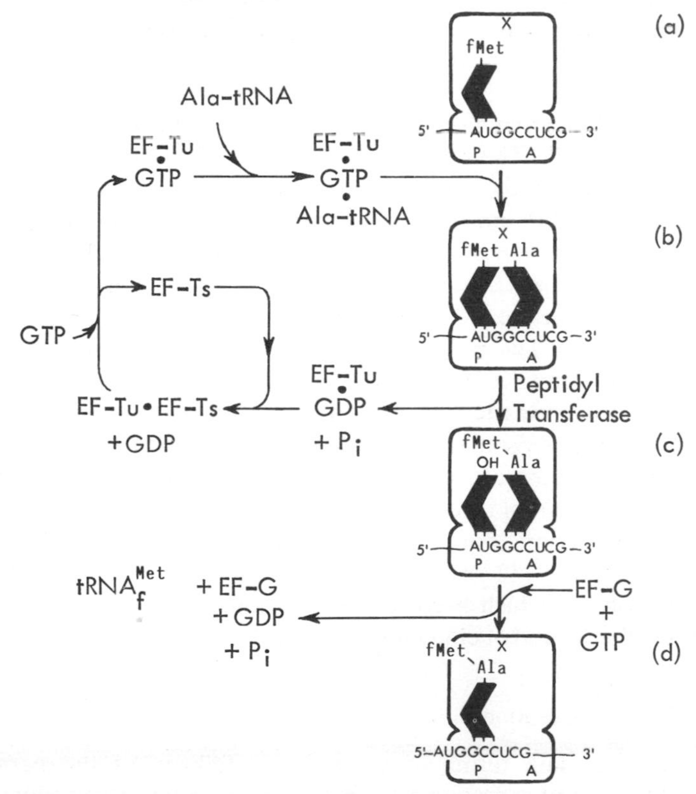

Fig. 12.7 A schematic diagram of prokaryotic polypeptide chain elongation. For convenience the 70S initiation complex of Fig. 12.6d has been taken as the starting point, (a), although the scheme is equally true for 70S ribosomes bearing any peptidyl-tRNA in the P-site (e.g. d.). Likewise the designations of the mRNA triplet in the A-site as coding for Ala, and the third triplet for Ser, are purely arbitrary.

linked to the carboxyl group of the fMet or nascent peptide, through the catalytic activity of the *peptidyl transferase* centre of the 50S ribosomal subunit [142, 88]. As already mentioned, the thermodynamic free energy for peptide-bond formation comes from the hydrolysis of the 'energy-rich' acyl–ester bond of the fMet-tRNA, which in its turn is derived from the ATP hydrolysed during aminoacylation. The fact that GTP and supernatant factors are not required for the transpeptidation was perhaps most convincingly confirmed when it was discovered that, in the presence of ethanol (about 50 per cent), a 3′ hexanucleotide fragment of fMet-tRNA could react with puromycin on the isolated 50S ribosomal subunit [143].

Extension of the 'fragment reaction' to even smaller oligonucleotide fragments has shown that CCA-fMet is the smallest species that can occupy the P-site of the peptidyl transferase [144], and other experiments with peptidyl-tRNA showed that puromycin can be replaced by CA-Gly at the A-site [145].

The translocation of the peptidyl-tRNA from the A-site to the P-site requires the elongation factor EF-G (molecular weight 72000–84000 [146, 147]) and GTP. This reaction has been shown to allow movement of the peptidyl end of the tRNA so that it becomes reactive towards puromycin [127, 148], movement of the mRNA relative to the ribosome [120, 121], and ejection of the deacylated-tRNA from the P-site [149], The reaction requires hydrolysis of the GTP, the non-hydrolysable analogue being inactive [127, 128], although it will allow EF-G to bind to ribosomes [146]. The molecular mechanism underlying this process is perhaps the most intriguing and the least understood aspect of protein biosynthesis. It is clearly possible that a large part of the structural complexity of the ribosome, including even the division of the ribosome into subunits, may be a consequence of the need for this specific and concerted movement of macromolecules.

After translocation (Fig. 12.7d), one cycle of elongation has been completed (cf. Fig. 12.7a). The vacant A-site now contains a new mRNA codon, to which a corresponding aminoacyl-tRNA can bind, starting another round of elongation.

12.9.3 Chain termination [150–152]

In polypeptide-chain termination the ester linkage of the peptidyl-tRNA is hydrolysed in response to one of three termination codons (see Section 12.4.2) in a reaction involving two of the three release factors, RF-1, RF-2 and RF-3 [153–156]. The deacylated tRNA and the mRNA are expelled from the ribosome in the presence of release factor, RRF, and EF-G [157], liberating free ribosomal subunits. The subunits will associate to form 70S ribosomes unless prevented from doing so by IF-3 and IF-1 (see Section 12.9.1).

In contrast with the other 61 codons, the three specific terminators are not read by tRNAs. This was shown by using RNA from an amber mutant of bacteriophage f2 that directs the synthesis of the *N*-terminal hexapeptide of the coat protein *in vitro*. Only the six appropriate aminoacyl-tRNAs and supernatant proteins were required for release of the peptide [158]. This same system was subsequently used with purified elongation factors and release factors to show that release of the peptide required the amber termination codon (UAG) to be at the P-site of the ribosome [159]. To study conveniently the factor requirements for termination at all three codons, an assay was developed in which the termination codons could direct the release of fMet-tRNA, previously bound to ribosomes in the presence of the triplet AUG [160]. This led to the resolution [155] of two release factors, RF-1 (molecular weight 44000 [161]) and RF-2 (molecular weight 47000 [161]), of different codon specificities:

RF-1 for UAA or UAG
RF-2 for UAA or UGA

The third release factor, RF-3 is not codon specific and has no release activity in the absence of the other factors [156]. It stimulates the release of polypeptide promoted by the other factors and seems to stimulate both binding and release of these latter from the ribosome [162]. Its action is stimulated by GTP, but a requirement for GTP hydrolysis during termination in prokaryotes is not firmly established as GDP can replace GTP in the reaction *in vitro*.

There are quite strong grounds for thinking that the actual hydrolysis of the peptidyl ester linkage is catalysed by the peptidyl transferase centre of the ribosome, the reaction specificity of which has been modified by the binding of the release factors. This was suggested by the finding that the peptidyl transferase would catalyse the formation of an ester link to fMet-tRNA or its hexanucleotide fragment (see Section 12.9.2), if certain alcohols were presented to the ribosome instead of aminoacyl-tRNA [163]. If the hydroxyl group of an alcohol could replace the α-amino group of an aminoacyl-tRNA as a reactive nucleophile, it was possible that the hydroxyl group of water might do likewise. This suggestion was supported by the fact that a number of antibiotics (e.g. sparsomycin and chloramphenicol, see Section 12.13) and ionic conditions known to inhibit the peptidyl transferase reaction were found also to inhibit the termination reaction *in vitro* [155, 164].

After the release of the peptide, the mRNA and deacylated tRNA are still attached to the ribosome (Fig. 12.8b) and must be removed before

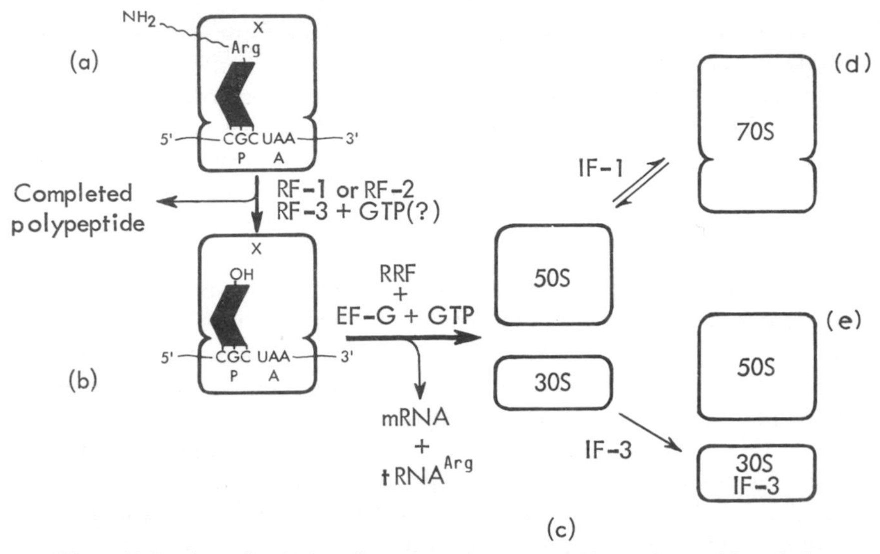

Fig. 12.8 A schematic diagram of prokaryotic polypeptide chain termination. The amino acid designation in the P-site is purely arbitrary. Other possible termination codons in the A-site are UAG and UGA (see text).

subunits can be regenerated for another round of protein synthesis. This process has been studied rather indirectly by assaying the release of ribosomes from polysomes after termination by puromycin [157], or using an amber coat protein mutant of bacteriophage R17 (cf. f2 above [165]). A clear requirement was observed for GTP, EF-G and ribosome release factor, RRF (molecular weight, 18000 [157]). It has also been shown that RRF is required for synthesis of β-galactosidase in a coupled DNA-dependent transcription and translation system *in vitro* [166]. Although it might be expected that the primary role of EF-G in this process would be the expulsion of deacylated tRNA, there appear to be no data bearing on this question.

The released ribosomes can be in the form of 70S ribosomes or 30S and 50S subunits. It was originally thought that 70S ribosomes were released first and subsequently converted into subunits by a dissociation factor [167]. Although such an activity was identified [168], and later shown to be IF-3 [169], it seems most likely that it operates as an *anti-association factor,* preventing the 50S subunit from associating with the 30S · IF-3 complex [170]. Inactive 70S ribosomes do accumulate in cells, especially when inhibition of initiation results in a relative excess of 30S subunits over IF-3 [167]. To regenerate subunits when conditions improve there must be an equilibrium between 70S ribosomes and ribosomal subunits. Initiation factor IF-1 is thought to accelerate this reaction in both directions, without altering the position of its equilibrium [171]. Thus IF-1 will only stimulate dissociation if IF-3 is available to prevent the subunits reassociating [172].

12.10 THE EVENTS ON THE EUKARYOTIC RIBOSOME

Eukaryotic ribosomes catalyse essentially the same process as prokaryotic ribosomes. Although the details of eukaryotic protein synthesis are less-well understood, it would seem that the differences from prokaryotic protein synthesis are relatively minor for elongation and termination, but much greater for initiation. The following discussion is almost completely confined to cytoplasmic protein synthesis. For details of what little is known about the distinct protein-synthesizing systems of mitochondria and chloroplasts, the reader is directed elsewhere [173–175].

12.10.1 Chain initiation

The first way in which it was realized that eukaryotic initiation differed from that in prokaryotes was in the initiating amino acid. Thus, although fMet-tRNA could be detected in yeast and rat liver mitochondria (in which it is now known to be the initiating amino acid) it was not found in the corresponding cytoplasms [176]. Nevertheless two distinct species of methionine tRNA were isolated from the cytoplasm: $tRNA_m^{Met}$ and $tRNA_f^{Met}$, the latter so designated because *in vitro* the $Met\text{-}tRNA_f$ could be formylated using *E. coli* transformylase [177]. Smith and colleagues [178, 179] used the Krebs II ascites tumour cell-free system of Mathews and Korner [180] to translate artificial polynucleotides, and obtained evidence that the methionine from the $Met\text{-}tRNA_f$ was preferentially incorporated at the amino-terminus of the peptide products. Shortly afterwards it was demonstrated that methionine is transiently present at the amino-termini of rabbit α- and β-globins [181, 182] and trout protamine [183] synthesized *in vivo*.

The more fundamental difference in eukaryotic initiation is the greater number of factors involved, at least eight [184–188] having been obtained in a pure state by a sufficient number of different workers to be designated by a common nomenclature [189]. A scheme indicating the stages at which they are thought to act in the initiation of haemoglobin synthesis is shown in Fig. 12.9. Certain points should be noted. The $Met\text{-}tRNA_f$ has been clearly shown to bind to the 40S subunit before the mRNA in both an unfractionated reticulocyte lysate [190] and with purified components [191]. Moreover, it is quite clear in eukaryotes that the initiator tRNA forms a stable ternary complex with eIF-2 and GTP

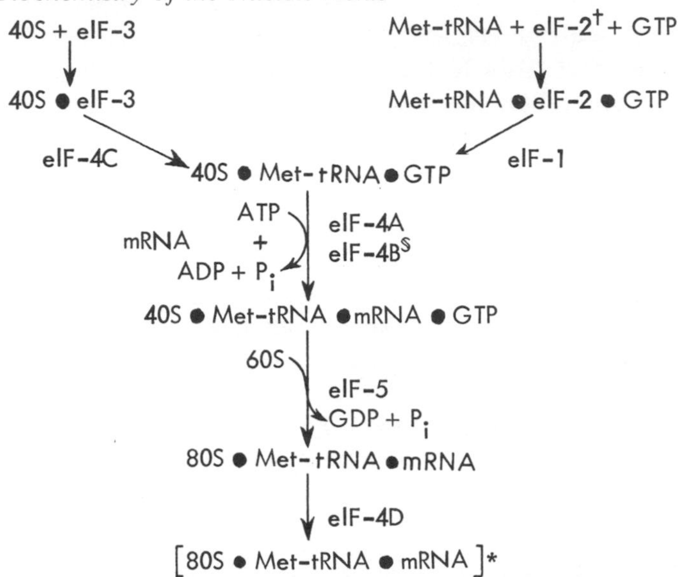

Fig. 12.9 A schematic diagram of eukaryotic polypeptide chain initiation. (After [187, 305, 306]). *This complex, unlike that before eIF-4D action, can react with the second aminoacyl-tRNA. †A factor designated co-eIF2-A may also act here [197]. §A factor designated 'cap-binding factor' may also act here [196].

[191–193]. The hydrolysis of ATP is necessary for the mRNA-binding reaction [194, 185], and there are several factors (eIF-1, -3, -4A, -4B and -4C) that may be involved in this step. One of these, eIF-3, is extremely complex, consisting of at least eight separate polypeptide chains with an aggregate molecular weight of about 700 000 [184, 195]. This abundance of factors may, to some extent, be related to the peculiar 5′-terminal structures of eukaryotic mRNAs (see Chapter 5), and to the possibility of translational control of protein synthesis (see Chapter 13). Other initiation factors that may be needed have not been included in this scheme. However, there is very persuasive evidence for the involvement of a protein responsible for the 'cap'-binding activity previously thought to reside in other factors, especially eIF-4B [196], and of at least one [197] of the various proteins described as being able to stimulate the activity of eIF-2 [197–200] (see also Chapter 13). It was first found in eukaryotic cell-free systems that polyamines, partially replacing Mg^{2+}, considerably enhance initiation [201], similar results subsequently being obtained with prokaryotic systems.

12.10.2 Chain elongation

Eukaryotic elongation factors were first described by Schweet and coworkers [202]. Initially it appeared that there was only a single aminoacyl-tRNA binding factor, EF-1, in eukaryotes, corresponding to EF-Tu + EF-Ts in prokaryotes [203]. However, the extreme tendency of EF-1 preparations to aggregate [204], seems initially to have obscured the fact that the factor that resembles EF-Tu and forms a ternary complex with aminoacyl-tRNA and GTP (now called EF-1_α, molecular weight 50000 [14]) was associated with another activity [205, 206] (now called EF-1_β, molecular weight, 26000 [207]). This factor appears to have properties analogous to the prokaryotic EF-Ts [207, 208]. One interesting feature of EF-1_α is that, unlike EF-Tu, it can form a ternary complex with GTP and the initiator tRNA [209]. The reason that the unformylated Met-$tRNA_f$ in eukaryotes is not in fact used for internal insertion of methionine residues [178] is unclear.

The elongation factor involved in the translocation reaction in eukaryotes, EF-2, has been purified [210, 211], and seems closely analogous to EF-G in prokaryotes. One point of interest is its specific inactivation by diphtheria toxin, which transfers ADP-ribose from NAD^+ to an unusual modified histidine residue in EF-2 [212–214, 307].

Eukaryotic, like prokaryotic, ribosomes possess an intrinsic peptidyl transferase activity, which has also been studied by using the 'fragment reaction' [215]. Although the eukaryotic peptidyl transferase is inhibited by certain antibiotics (e.g. sparsomycin) that inhibit prokaryotic peptidyl transferase, it is resistant to the action of others (e.g. chloramphenicol), and hence must differ somewhat in its structure.

12.10.3 Chain termination

A single factor, RF, was found to catalyse the release of the completed polypeptide chain from eukaryotic ribosomes [216]. This appears to recognize all three termination codons, UAA, UAG and UGA, although it is necessary to use tetranucleotides to assay for this in the fMet-tRNA release reaction [217]. In contrast to the prokaryotic factors, the eukaryotic release factor shows a clear requirement for GTP hydrolysis for its action *in vitro*.

Although one imagines that there is an eukaryotic factor analogous to TRF, none has so far been described. Eukaryotic ribosomes have been shown to be liberated from polysomes as subunits, and a pool of inactive 80S monomers is present in eukaryotic cells [218, 219]. One of the initiation factors, eIF-3, has an anti-association activity *in vitro* [195, 220], and in this respect may be analogous to prokaryotic factor IF-3.

12.11 THE ROLE OF RIBOSOMAL RNA IN PROTEIN SYNTHESIS

The structure of rRNA has already been described (Chapter 5). What function does this have in the ribosome? It had for a long time been assumed that the role of rRNA was confined to organizing the more functionally important proteins both in the assembly and final structure of the ribosome. Indeed, certain ribosomal proteins were shown to bind specifically to regions of the rRNA (six to the 16S rRNA, ten to the 28S rRNA and three to the 5S rRNA [221]), and some of these appear to participate in the initial stages of the assembly of the ribosomal subparticles *in vivo* [222]. More recent work with less denatured RNA and protein preparations has significantly extended the list of proteins known to interact with rRNA [223], and it may very well be that, in the assembled ribosome, most proteins are in contact with rRNA.

Such a structural role for rRNA does not preclude an additional, more active, role of base-pairing with the RNAs that bind to the ribosome. The best-documented interaction of this type is between prokaryotic mRNAs and 16S rRNA. Shine and Dalgarno [224] determined the sequence of the 3′-end of *E. coli* 16S rRNA and observed a polypyrimidine sequence complementary to various parts of the untranslated polypurine sequences that had been found to precede the initiator AUG codons of the coat protein, A protein, and replicase genes of R17 and related bacteriophages (see Fig. 5.10). They therefore suggested that this polypyrimidine region of 16S rRNA might promote the interaction of the 30S subunit with the region of the mRNA containing the initiator AUG. This suggestion was important, because it offered a possible answer to the question of why 30S ribosomes bind to the initiator AUG codon rather than one specifying on internal methionine (see Section 12.9.1). As the sequences of more *E. coli* mRNAs (both phage and bacterial) were determined, complementarity to the 16S rRNA continued to be found, as illustrated for a selection of mRNAs in Table 12.3. The complementarity varies between 3 and 9 bases, and is even found in those cases where the preinitiation sequence is actually the 3′-end of a preceding cistron. A table of about 80 mRNAs from *E. coli*, and their complementarity to rRNA, is to be found in [64], which also reviews this topic.

What is the evidence that this potential base pairing actually occurs and is the basis of the selection of the correct initiation codon? The occurrence of this base pairing has been shown by partial nuclease treatment of a 70S initiation complex, which allowed the isolation of a hybrid formed between 50 nucleotides at the 3′-end of 16S rRNA and 30 nucleotides from around the initiation site of the A protein cistron of bacteriophage R17 [225]. Furthermore, an oligonucleotide complementary to the 3′-terminus of 16S rRNA blocked the formation of an initiation complex with bacteriophage Qβ RNA, but not with AUG [226]. The role of this interaction in the selection of the correct initiation codon has been elegantly demonstrated in a study of the

Table 12.3 Complementarity between pre-initiation regions of *E. coli* mRNAs and 16S rRNA.

16S rRNA	3′ $_{HO}$A U U C C U C C A U A G . . . . 5′
MS2 coat	5′ . . . . . . U C A A C C G G A G U U U G A A G C *A U G* . . . 3′
MS2 replicase	5′ . . . . . . C A A A C A U G A G G A U U A C C C *A U G* . . . 3′
MS2 A protein	5′ . . . . U C C U A G G A G G U U U G A C C U *G U G* . . 3′
λ Cro	5′ . . A U G U A C U A A G G A G G U U G U *A U G* . . . . 3′
gal E	5′ . . . A G C C U A A U G G A G C G A A U U *A U G* . . . . . . 3′
β-lactamase	5′ . U A U U G A A A A A G G A A G A G U *A U G* . . . . 3′
Lipoprotein	5′ . . . . A U C U A G A G G G U A U U A A U A *A U G* . 3′
Ribosomal protein S12	5′ . . . . A A A A C C A G G A G C U A U U U A *A U G* . . . . 3′
RNA polymerase *β*	5′ . . . . A G C G A G C U G A G G A A C C C U *A U G* . . . . 3′
trp E	5′ . . . . C A A A A U U A G A G A A U A A C A *A U G* . . . . 3′

(Regions of complementarity are underlined, and the initiation codons are italicized. Original references to the sequences are given in [64].)

translation of bacteriophage T7 mRNA with mutations in its preinitiation, or 'Shine and Dalgarno', sequence [227].

Although a potential for such an mRNA–rRNA interaction is clearly a necessary condition for initiation, it is not a sufficient condition, and the actual occurrence of such an interaction is influenced by other factors such as secondary structure (see Sections 12.9.1 and 13.3.1).

The foregoing applies only to *E. coli* and probably other prokaryotic mRNAs. It is notable that although there is strong conservation of much of the 3′-end of the rRNA of the small subunit in *E. coli* and eukaryotes, the key CCUCC sequence involved in the 'Shine and Dalgarno' interaction (see Table 12.3) is absent from 18S rRNA [228]. The preinitiation sequences of eukaryotic mRNAs show no consistent pattern of complementarity to 18S rRNA and it is clear that eukaryotes have some quite different method of selecting the correct initiation codon. This problem is discussed by Kozak [229] who suggests that ribosomes may bind to the 5′-end of monocistronic eukaryotic mRNAs and advance towards the 3′-end of the mRNA, stopping at the first AUG codon.

More speculative is the possibility that base-pairing occurs between the 3′-ends of 16S rRNA and 23S rRNA, and is involved in the formation of 70S monomers from 30S and 50S ribosomal subunits [230]. There is complementarity between these RNA sequences, but only if the 16S rRNA adopts a secondary structure different from that which it is thought to adopt for the 'Shine and Dalgarno' base-pairing with mRNA [225]. However, the mutual exclusion of these two patterns of base pairing suggests a molecular basis for the dual function of IF-3 as anti-association factor and mRNA-binding factor (see Section 12.9). The interaction of IF-3 with the 30S subunit might cause the secondary structure of the 3′-end of 16S rRNA to change from that in which only interaction with 23S rRNA is possible, to that in which only interaction with mRNA is possible. Although IF-3 appears to bind to a region of the 30S subunit near the 3′-end of 16S rRNA that also appears to be near the 3′-end of 23S rRNA [225], direct evidence in support of this interesting suggestion is at present lacking.

Complementarity was also noted between an invariant sequence CGAAC within prokaryotic 5S rRNA, and the conserved sequence GTψCR of prokaryotic tRNAs, or the GUψCR of eukaryotic tRNAs. This led to the suggestion that there might be interaction between these when aminoacyl-tRNA binds to the ribosome [231]. Indirect evidence supporting this possibility came from the finding that the oligonucleotide TψCG inhibits non-enzymic ribosomal binding of aminoacyl tRNA [232]. Binding of Met-$tRNA_f$ to 80S ribosomes was not inhibited [233], suggesting interaction only at the A-site, and it is interesting in this respect that the eukaryotic initiator tRNA contains AUCG rather than TψCG [234]. Indeed, it has been suggested that fMet-tRNA may interact with 23S rRNA, rather than 5S rRNA [300]. It must be pointed out, however, that in the three-dimensional structure of yeast $tRNA^{Phe}$ (Fig. 2.7) the TψC loop is hydrogen bonded to the dihydrouridine loop. Thus the proposed interaction can only occur if the conformation of aminoacyl-tRNA on the ribosome differs appreciably from this. In fact there is an increasing body of physical evidence that the codon–anticodon interaction does cause the GTψCR sequence to become exposed [235]. Also consistent with this is the fact that *in vitro* the oligonucleotide TψCG can substitute for the appropriate deacylated tRNA in the stringent-factor dependent synthesis of guanosine nucleotides (see Section 13.2) which normally requires codon recognition at the A-site of the ribosome [236].

The interaction of the TψC loop of the tRNA with 5S rRNA cannot yet be considered proven, especially as the relevant area of the 5S rRNA may also be masked [237]. However, the general idea of a codon–anticodon interaction promoting ribosomal binding of other regions of the tRNA is attractive because the codon–anticodon interaction alone cannot account for the energy required to bind the tRNA to the ribosome [238].

12.12 THE ROLE OF RIBOSOMAL PROTEINS IN PROTEIN SYNTHESIS

Before considering the function of the protein components of the ribosome it is necessary to describe these in rather more detail than given in Chapter 5. In *E. coli* the 30S ribosomal subunit has 21 distinct proteins, designated S1–S21, one copy of each being present per 30S subunit, except for S6, which appears to be present in two copies per subunit [239]. It was originally thought that there were 34 distinct proteins, L1–L34, on the 50S ribosomal subunit. However, L8 is in fact a complex of L7/L12 and L10, and L26 is identical to S20, there being an average 0.2 copies of L26 and 0.8 copies of S20 per 70S ribosome. Thus it is best regarded that there are 32 distinct proteins on the 50S subunit. All but two of these proteins are completely dissimilar serologically and in amino acid sequence, and are present as single copies. The exceptions are L7 and L12, L7 being the *N*-acetylated form of L12, and the two proteins together are present in a total of four copies per 50S subunit. Apart from proteins S1, S6, L7 and L12, the ribosomal proteins are all chemically basic, with molecular weights in the range 9000 to 35000.

The total amino acid sequences of all of the ribosomal proteins have now been determined. It is worth mentioning that physical and immunochemical studies indicate that several of the ribosomal proteins have elongated rather than globular structures. Much of our knowledge of the structure of ribosomal proteins is a result of the work of Wittmann and his colleagues, who have reviewed this topic in detail [240–243].

One powerful tool for the study of ribosomal proteins has been the reassembly of active subparticles from their individual components *in vitro*. This technique was pioneered by Nomura and his colleagues, first with the 30S subunit of *E. coli* [244], and later with the 50S subunit of *B. stearothermophilus* [245]. More recently conditions have been found for the reassembly of the 50S subunit of *E. coli* [246]. Omission of single proteins from the mixture used to reassemble 30S subunits demonstrated that, although almost all of them (we now believe all of them) were required for the production of particles with biological activity, omission of only S4, S7, S8, S16 or S19 prevented the assembly of a particle sedimenting at about 30S [247]. Thus one function of these proteins is clearly in the assembly of the subparticle. Whether they are also directly involved in the function of the assembled 30S particle is less clear, although in the case of S16, there is evidence to suggest that this may not be so [248].

It has generally been difficult to assign specific functions to those proteins not implicated in the assembly of the ribosomal subparticles. Omission of a given protein in the reconstitution assay usually causes a loss of activity in all the different partial reactions of protein biosynthesis (exceptions are discussed later). This is consistent with a model of the ribosome in which there is extensive co-operativity between proteins to form the different catalytic and substrate-binding sites on the ribosome. The question of which proteins are physically located at which sites on the ribosome has been extensively studed by affinity labelling, and the results obtained by this and other techniques have been reviewed [249–251]. These results can be used to build up a picture of the proteins present in various functional ribosomal domains, especially when related to the data on relative protein proximities. The latter have been obtained by a variety of different techniques, including assembly mapping [252–254], protein–protein cross-linking [255], generation of fragments with ribonuclease [223], singlet energy transfer between fluorescently labelled proteins [256], and low-angle neutron scattering [257]. It should be possible to relate all these data to the position of the proteins in the overall topography of the 70S ribosome (Plate VIII), as determined by electron microscopy of subunits linked by antibodies to individual ribosomal proteins [242, 258, 259]. Although this technique is potentially very powerful, it contains pitfalls because of the difficulty both in obtaining completely mono-specific anti-sera and in

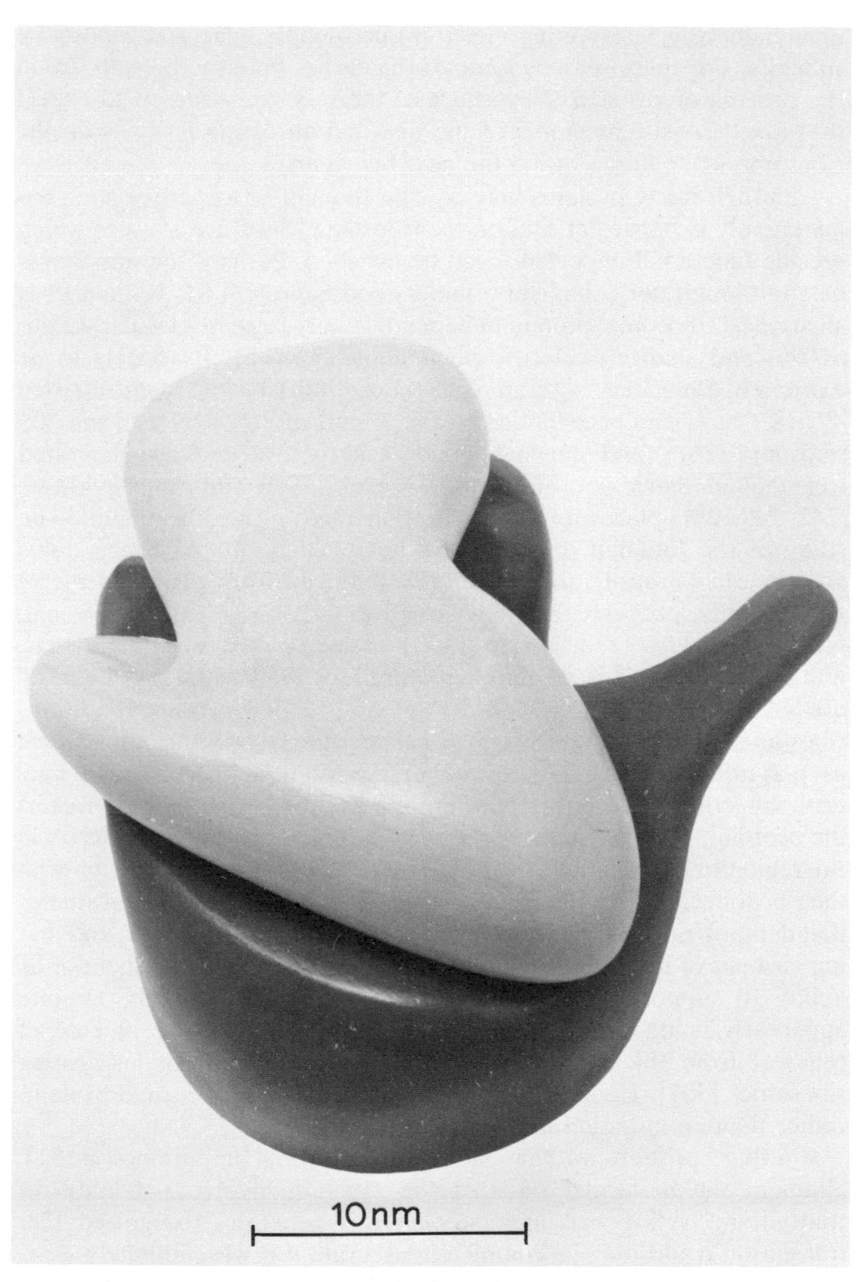

Plate VIII Model of the 70S ribosome of *E. coli* based on the electron-microscopic studies of Lake and co-workers. The 30S subunit is light, the 50S subunit is dark. (Original photograph of a model provided by courtesy of Dr James A. Lake.)

unambiguously interpreting electron-microscopic images of ribosomal subunits. For these reasons some of the earlier published results are in the process of revision. Nevertheless, there is every reason to expect that a self-consistent model of the distribution of the proteins on the ribosome will emerge within the next few years.

Although many proteins may best be thought of as participating co-operatively in particular sites on the ribosome, there are a few to which specific functional importance can be ascribed. Perhaps the one that is best (although not completely) understood is protein S1. Protein S1 is an atypical ribosomal protein in being relatively large (molecular weight 61000) and acidic (isoelectric point about 5.0–5.5). It appears to be extremely elongated, with an axial ratio of 10:1 having been reported [260]. It has been cross-linked to the 3′-end of 16S rRNA in the 30S ribosome [261], and shown to bind *in vitro* to a nuclease-generated fragment of the 3′-end of 16S rRNA [262]. This and other evidence [242, 249, 251] places it at the 'initiation site' on the 30S subunit. The study of the function of protein S1 has been facilitated because this protein can be easily and specifically removed from the ribosome *in vitro*. (This property may be related to other, extra-ribosomal, functions *in vivo*, see Section 13.2.) In this way it was found that, although protein S1 is absolutely required for the translation of natural mRNAs (bacteriophage MS2 RNA etc.) under certain conditions ribosomes lacking protein S1 can translate artificial polynucleotides such as polyuridylic acid [263]. However, translation of denatured bacteriophage RNA did not require protein S1, suggesting that the role of the protein might be to 'melt' the secondary structure of the mRNA in the region of its initiation site, hence facilitating correct interaction with the ribosome [264]. Support for this idea has come from physical studies that demonstrate that protein S1 is indeed able to unfold RNA, and that impairment of its ability to do this is paralleled by an impairment of its ability to support protein synthesis on the ribosome [265]. Despite apparently being only required for initiation, and despite its ease of removal from the 30S subunit, protein S1 is present on elongating ribosomes [301]. Hence it must be considered as a ribosomal protein, rather than an initiation factor.

Another protein with a particular functional importance is S12, although the molecular basis of the function involved, 'fidelity' of translation, is not well understood. It is generally recognized that translation could only be completely accurate if it were infinitely slow, and so a balance between speed and fidelity is necessary. Protein S12 was first implicated in 'fidelity' by the results of studies of mutant ribosomes from *E. coli,* resistant to the drug streptomycin, which inhibits protein biosynthesis at suitable concentrations *in vitro* (see Section

12.13). By performing mixed reconstitution experiments, with one protein from the mutant and the rest (and the 16S rRNA) from the wild type, it was established that protein S12 was the protein altered in the streptomycin-resistant mutant [266]. Streptomycin, at lower concentrations, causes misreading of the genetic message, and it was also noted that in ribosomes reconstituted with protein S12 from the streptomycin-resistant bacteria the extent of misreading was much decreased. Decreased misreading was also observed with ribosomes lacking S12 entirely (although in this case the rate of protein synthesis was lower) [266]. A related, but somewhat different, observation regarding protein S12 concerns the translation of coliphage R17 RNA. It is known that ribosomes from *B. stearothermophilus* will only translate the A-protein cistron of this RNA, and not the coat protein and replicase cistrons recognized by the ribosomes of *E. coli* (see also Section 13.3.1) [267]. Ribsomes were reconstituted from proteins of the two types of bacterium and protein S12 was found to be partially responsible for the cistron selection. The source of 16S rRNA was the other important factor [268], although this cannot be explained in terms of 'Shine and Dalgarno' base pairing to the different 3′-terminal sequences of the 16S rRNAs. These and other data clearly implicate protein S12 in an important direct or indirect interaction with the mRNA, although the precise purpose of this remains unclear.

Single component omission experiments have been less successful in identifying individual proteins responsible for the catalytic activities associated with ribosomes. However, in combination with various other techniques the results have helped to establish a group of proteins situated near to, and required for, the *peptidyl transferase* activity [251]. It has, in fact, been suggested that protein L16 is the peptidyl transferase [269], but conclusive evidence is not yet to hand.

Although our understanding of the *GTPase* activities associated with the ribosome is similarly incomplete, these activities are clearly fundamental to ribosome function. They will be discussed in this section because they appear to involve the proteins L7/L12. These multi-copy acidic proteins (see above) have been implicated in the EF-G-dependent GTPase reaction by the results of experiments involving their selective removal from the ribosome [270], their chemical cross-linking to EF-G [271], and the demonstration that only antisera to these proteins prevent formation of an EF-G·GDP complex [272]. There are reasons for thinking that there is only a single site on the ribosome that is associated with the GTPase reactions of the different factors [274–276], and it is, therefore, interesting that a requirement for proteins L7/L12 has been demonstrated in each of initiation, elongation and termination [277]. Proteins L7/L12 themselves are not, however, thought to contain the GTPase activity, nor is the protein L11 [278], even though it has also been implicated in GTP hydrolysis [279]. Although a complex of 5S rRNA and ribosomal proteins L5, L18 and L25 has been found to have GTPase activity [277, 280], a GTPase activity has also been described for EF–Tu in the absence of ribosomes [281] (see Section 12.13).

Despite this uncertainty, proteins L7/L12 do appear to form the ribosomal binding site for the aforementioned factors. Moreover, the formal resemblance of these proteins to the acidic, multi-copy, nucleotide triphosphate-associated actin and myosin, accorded well with the early speculation of a similarity between the mechanisms of translocation and muscle contraction [282]. The attraction of the original conception of this analogy was initially diminished when it was found that GTP hydrolysis was also needed during binding of initiator- and aminoacyl-tRNAs to the ribosome (see Section 12.9). One way of explaining the latter is in terms of movement of the tRNA from some kind of entry site to the P- or A-site (e.g. the I-site proposed for fMet-tRNA [97], or the R-site proposed for aminoacyl-tRNA [259]). However, the emergence of evidence for a single GTPase site on the ribosome, to which EF-Tu and EF-G could not bind simultaneously, led to the suggestion of a unitary role for the GTP hydrolysis in expelling the factors from the ribosome after they had fulfilled their functions [12, 14]. The presence of the factor on the ribosome was regarded as preventing the next reaction from occurring. The role of EF-G envisaged in this model would be to prime the ribosome for subsequent translocation which would *not*, in itself, require GTP. It is interesting in this regard that ribosomes can be made to perform a factor- and energy-independent translocation *in vitro*, albeit at rather a low rate [283, 284].

The foregoing discussion has concentrated solely on prokaryotic ribosomes. Much less is known about the functions of the proteins of eukaryotic ribosomes (reviewed in [285–287]). Although one would expect structural similarities to complement the functional similarities of peptidyl transferase and GTPase (at the very least), apparently only one of the prokaryotic ribosomal proteins, L7/L12, has a counterpart on eukaryotic ribosomes [277, 285]. The eukaryotic proteins corresponding to *E. coli* L7/L12 are modified by the phosphorylation of certain serine residues [287–289]. However, the functional significance of this modification is no more apparent than that of the *N*-acetylation found in bacterial L7.

12.13 INHIBITORS OF PROTEIN SYNTHESIS

Many antibiotics act by inhibiting bacterial protein biosynthesis, and both these and certain specific inhibitors of eukaryotic protein biosynthesis have been useful in studying the details of these processes. Some of these will be discussed briefly here, extensive accounts being available elsewhere [290–294].

Puromycin (Fig. 12.10) is a nucleoside derivative which closely resembles the 3′-terminal nucleoside residue of a loaded tRNA molecule [295]. It competes with aminoacyl-tRNA molecules in its capacity to serve as an acceptor for the peptidyl group of peptidyl-tRNA during protein synthesis on both prokaryotic and eukaryotic ribosomes. The consequence is that synthesis of complete proteins is prevented and, instead, peptides are produced which bear a puromycin residue covalently bonded to the carboxy-terminal group. The uses of puromycin to study both the ribosomal sites at which tRNA is bound, and the peptidyl transferase reaction, have already been described (see Section 12.9).

Streptomycin is a specific inhibitor of prokaryotic protein biosynthesis and also causes misreading of the mRNA. Poly(U), for instance, codes not only for phenylalanine but, in the presence of streptomycin, for serine, isoleucine and leucine also. The antibiotic binds to the 30S

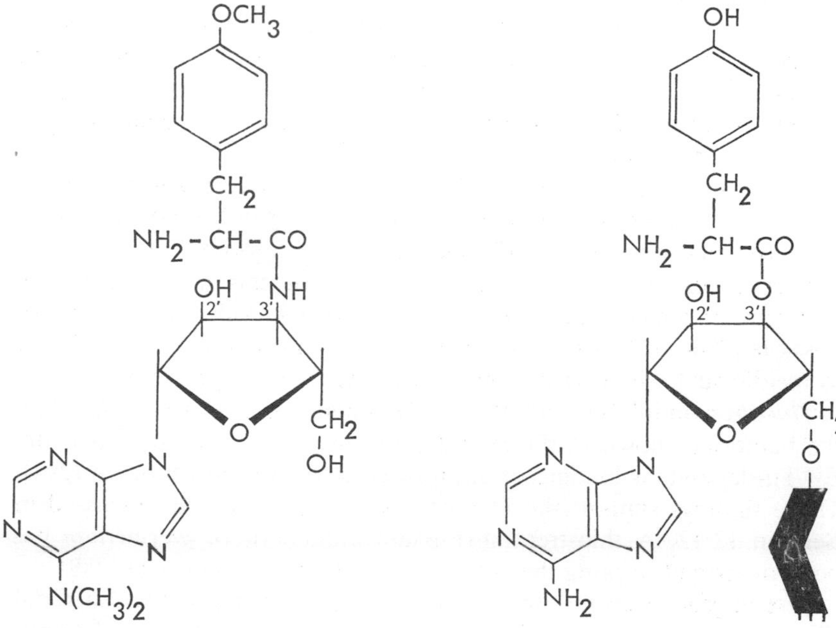

OCH_3 CH_2 NH_2 - CH - CO OH NH 2′ 3′ O CH_2 OH N N N N $N(CH_3)_2$

OH CH_2 NH_2 - CH - CO OH O 2′ 3′ O CH_2 O N N N N NH_2

Puromycin Tyrosyl - tRNA

Fig. 12.10 The structure of puromycin compared with that of the terminal adenosine residue of Tyr-tRNA. Compare also Fig 12.4 and Fig. 12.5.

ribosomal subunit, and the use of streptomycin-resistant mutants in studying ribosomal protein S12 has been described in Section 12.12.

Tetracycline binds to the small subunit of prokaryotic ribosomes and blocks the binding of aminoacyl-tRNA to the A-site. It has been used to distinguish between tRNA bound in the two sites [296].

Chloramphenicol binds to the large bacterial subunit and inhibits the peptidyl transferase reaction. Its use in studying the possible role of the peptidyl transferase in the termination reaction has already been mentioned (Section 12.9.3). It has been possible to prepare suitably modified derivatives for use as affinity labels to identify proteins at the peptidyl transferase centre (see Section 12.12). Chloramphenicol (and other specific inhibitors of prokaryotic ribosomal function, such as streptomycin and erythromycin) also inhibit mitochondrial and chloroplast protein biosynthesis [173–175]. This suggests that organelle protein biosynthesis is more similar to that in prokaryotes than in the cytoplasm of eukaryotes, and this conclusion is supported by the use of fMet-tRNA for initiation in organelles (see Section 12.10.1).

Sparsomycin, which also blocks peptidyl transfer at some stage, does so with both bacterial and eukaryotic ribosomes. For example, its specific inhibition of elongation allowed it to be used to retard the rate of eukaryotic protein biosynthesis sufficiently to allow short nascent peptides, still bearing the amino-terminal methionine residue, to be detected [181].

Erythromycin affects prokaryotic ribosomes and is thought to act at the translocation step.

Fusidic acid inhibits translocation in both prokaryotic and eukaryotic systems by preventing the release of EF-G (or EF-2)·GDP complexes from the ribosome after GTP hydrolysis (see Section 12.9.2). It was important in demonstrating the existence of this complex, in the presence of which binding of aminoacyl-tRNA with EF-Tu was not possible [274–276]. A mutant form of EF-G was found in cells resistant to fusidic acid, allowing the gene for EF-G to be mapped [297].

Thiostrepton inhibits both the EF-G-dependent hydrolysis of GTP on the bacterial ribosome, the IF-2-dependent GTPase reaction, and the EF-Tu-dependent binding of aminoacyl-tRNA to the ribosome [294]. These findings support the idea of a single GTPase site, discussed in Section 12.12. A thiostrepton-resistant mutant of *B. megaterium* has been described, lacking the equivalent of *E. coli* protein L11 [278].

Kirromycin prevents the release of EF-Tu from the bacterial ribosome after GTP hydrolysis. In the absence of ribosomes, the antibiotic has been found to cause an EF-Tu-dependent hydrolysis of GTP, suggesting that the GTPase site is on the factor rather than the ribosome [281] (but see also Section 12.12).

Kasugamycin specifically inhibits the initiation of prokaryotic protein biosynthesis. Kasugamycin-resistant mutants of *E. coli* are remarkable in having ribosomes in which the 16S rRNA is altered. The sequence $m^6_2Am^6_2A$, starting 24 residues from the 3′-end, is unmethylated in the mutant [298]. The proximity of this to the first nine or so residues that can participate in 'Shine and Dalgarno' base pairing (Section 12.11) is worth pointing out.

Pactamycin, at suitable concentrations, specifically inhibits both eukaryotic and prokaryotic initiation. It was used in determining the gene order of proteins derived from the single polypeptide translated from picornaviral RNAs [299].

Cycloheximide is a specific inhibitor of both initiation and elongation of protein synthesis in the cytoplasm of eukaryotes but not in their mitochondria. Because it prevents ribosomes 'running off' polysomes it has also been used to 'freeze' the polysome profile of cells immediately before their isolation. Cycloheximide-resistant mutants with altered 60S ribosomal subunits have been described. Cycloheximide is much used to inhibit protein synthesis in intact cells because its effects are reversible.

Diphtheria toxin specifically inactivates eukaryotic EF-2, as mentioned in Section 12.10. This is the basis of its toxicity *in vivo.*

REFERENCES

1 Caspersson, T. (1942), *Naturwissenschaften,* **29,** 33.
2 Brachet, J. (1950), *Chemical Embryology,* Interscience, New York.
3 Davidson, J.N. and Waymouth, C. (1944), *Biochem. J.,* **38,** 39.
4 Sarabhai, A.S., Stretton, A.O.W., Brenner, S. and Bolle, A. (1964), *Nature,* **201,** 13.
5 Yanofski, C., Carlton, B.C., Guest, J.R., Helinski, D.R. and Henning, V. (1964), *Proc. Natl. Acad. Sci. USA,* **51,** 266.
6 Crick, F.H.C. (1958), *Symp. Soc. Exp. Biol.,* **12,** 139, Cambridge University Press.
7 Watson, J.D. (1964), *Bull. Soc. Chim. Biol.,* **46,** 1399.
8 Novelli, G.D. (1967), *Annu. Rev. Biochem.,* **36,** 449.
9 Lengyel, P. and Söll, D. (1969), *Bact. Rev.,* **33,** 264.
10 Lucas-Lenard, J. and Lipmann, F. (1971), *Annu. Rev. Biochem.,* **40,** 409.
11 Haselkorn, R. and Rothman-Denes, L.B. (1973), *Annu. Rev. Biochem.,* **42,** 397.
12 Lengyel, P. (1974), in *Ribosomes* (ed. M. Nomura, A. Tissières and P. Lengyel), Cold Spring Harbor Laboratory Monograph Series, p. 13.
13 Weissbach, H. and Pestka, S. (1977), *Molecular Mechanisms of Protein Biosynthesis,* Academic Press, New York.

14 Weissbach, H. (1979), in *Ribosomes: Structure, Function and Genetics* (ed. G. Chambliss, G.R. Craven, J. Davies, K. Davis, L. Kahan and M. Nomura), University Park Press, Baltimore, p. 377.
15 Crick, F.H.C. (1963), *Prog. Nucl. Acid Res.*, **1**, 163.
16 Crick, F.H.C. (1967), *Proc. R. Soc. B*, **167**, 331.
17 Woese, C.R. (1967), *The Genetic Code, The Molecular Basis for Genetic Expression*, Harper and Row, New York.
18 Ycas, M. (1969), *The Biological Code*, North-Holland Publishing Co., Amsterdam.
19 Jukes, T.H. and Gatlin, L. (1971), *Prog. Nucl. Acid Res. Mol. Biol.*, **11**, 303.
20 Jukes, T.H. (1977), *Comprehensive Biochemistry* (ed. M. Florkin and E.H. Stotz), Elsevier, Amsterdam. Vol. 24, p. 235.
21 Crick, F.H.C., Barnett, L., Brenner, S. and Watts-Tobin, R.J. (1961), *Nature*, **192**, 1227.
22 Crick, F.H.C. (1966), *Cold Spring Harbor Symp. Quant. Biol.*, **31**, 3.
23 Crick, F.H.C., Griffith, J.S. and Orgel, L.E. (1957), *Proc. Natl. Acad. Sci. USA*, **43**, 416.
24 Hoagland, M.B., Zamecnik, P.C. and Stephenson, M.L. (1957), *Biochim. Biophys. Acta*, **24**, 215.
25 Hoagland, M.B., Stephenson, M.L., Scott, J.F., Hecht, L.I. and Zamecnik, P.C. (1958), *J. Biol. Chem.*, **231**, 241.
26 Matthaei, J.H. and Nirenberg, M.W. (1961), *Proc. Natl. Acad. Sci. USA*, **47**, 1580.
27 Nirenberg, M.W. and Matthai, J.H. (1961), *Proc. Natl. Acad. Sci. USA*, **47**, 1588.
28 Nirenberg, M.W., Matthai, J.H., Jones, O.W., Martin, R.G. and Barondes, S.H. (1963), *Fed. Proc. Fed. Am. Soc. Exp. Biol.*, **22**, 55.
29 Speyer, J.S., Lengyel, P., Basilio, C., Wahba, A.J., Gardner, R.S. and Ochoa, S. (1963), *Cold Spring Harbor Symp. Quant. Biol.*, **28**, 559.
30 Khorana, H.G. (1965), *Fed. Proc. Fed. Am. Soc. Exp. Biol.*, **24**, 1473.
31 Khorana, H.G., Büchi, H., Ghosh, H., Gupta, N., Jacob, T.M., Kössel, H., Morgan, R., Narang, S.A., Ohtsuka, E. and Wells, R.D. (1966), *Cold Spring Harbor Symp. Quant. Biol.*, **31**, 39.
32 Nirenberg, M. and Leder, P. (1964), *Science*, **145**, 1399.
33 Leder, P. and Nirenberg, M. (1964), *Proc. Natl. Acad. Sci. USA*, **52**, 420.
34 Nirenberg, M., Caskey, C.T., Marshall, R., Brimacombe, R., Kelly, D., Doctor, B., Hatfield, D., Levin, J., Rottman, F., Pestka, S., Wilcox, F. and Anderson, W.F. (1966), *Cold Spring Harbor Symp. Quant. Biol.*, **31**, 11.
35 Wittmann, H.G. (1962), *Z. Vererbungslehre*, **93**, 491.
36 Wittmann, H.G. and Wittmann-Liebold, B. (1963), *Cold Spring Harbor Symp. Quant. Biol.*, **28**, 589.
37 Fraenkel-Conrat, H. (1964), *Sci. Amer.*, **211**(4), 46.
38 Yanofsky, C. (1967), *Sci. Amer.*, **216**(5), 80.
39 Terzaghi, E., Okada, Y., Streisinger, G., Emrich, J., Inouye, M. and Tsugita, A. (1966), *Proc. Natl. Acad. Sci. USA*, **56**, 500.
40 Adams, J.M., Jeppesen, P.G.N., Sanger, F. and Barrell, B.G. (1969), *Nature*, **223**, 1009.

41 Fiers, W., Contreras, R., Duerinck, F., Hageman, G., Iserentant, D., Merregaert, J., Min Jou, W., Molemans, F., Raeymaekers, A., Van den Berghe, A., Volckaert, G. and Ysebaert, M. (1976), *Nature,* **260,** 500.
42 Benzer, S. and Champe, S.P. (1962), *Proc. Natl. Acad. Sci. USA,* **48,** 1114.
43 Brenner, S., Stretton, A.O.W. and Kaplan, S. (1965), *Nature,* **206,** 994.
44 Brenner, S., Barnett, L., Katz, E.R. and Crick, F.H.C. (1967), *Nature,* **213,** 449.
45 Sambrook, J.F., Fan, D.P. and Brenner, S. (1967), *Nature,* **214,** 452.
46 Goodman, H.M., Abelson, J., Landy, A., Brenner, S. and Smith, J.D. (1968), *Nature,* **217,** 1019.
47 Marcker, K.A. and Sanger, F. (1964), *J. Mol. Biol.,* **8,** 835.
48 Waller, J.P. (1963), *J. Mol. Biol.,* **10,** 319.
49 Adams, J.M. and Capecchi, M.R. (1966), *Proc. Natl. Acad. Sci. USA,* **55,** 147.
50 Webster, R.E., Engelhardt, D.L. and Zinder, N.S. (1966), *Proc. Natl. Acad. Sci. USA,* **55,** 155.
51 Adams, J.M. (1968), *J. Mol. Biol.,* **34,** 571.
52 Takeda, M. and Webster, R.E. (1968), *Proc. Natl. Acad. Sci. USA,* **60,** 1487.
53 Dube, S.K., Marcker, K.A., Clark, B.F.C. and Cory, S. (1968), *Nature,* **218,** 232.
54 Cory, S., Marcker, K.A., Dube., S.K. and Clark, B.F.C. (1968), *Nature,* **220,** 1039.
55 Marcker, K.A., Clark, B.F.C. and Anderson, J.S. (1966), *Cold Spring Harbor Symp. Quant. Biol.,* **31,** 279.
56 Ghosh, H.P., Söll, D. and Khorana, H.G. (1967), *J. Mol. Biol.,* **25,** 275.
57 Files, J.G., Weber, K., Coulondre, C. and Miller, J.H. (1975), *J. Mol. Biol.,* **95,** 327.
58 Crick, F.H.C. (1966), *J. Mol. Biol.,* **19,** 548.
59 Söll, D., Cherayil, J.D. and Bock, R.M. (1967), *J. Mol. Biol.,* **29,** 97.
60 Söll, D. and RajBhandary, U.L. (1967), *J. Mol. Biol.,* **29,** 113.
61 Wrede, P., Woo, N.H. and Rich, A. (1979), *Proc. Natl. Acad. Sci. USA,* **76,** 3289.
62 Gauss, D.H., Grüter, F. and Sprinzl, M. (1979), *Nucl. Acids Res.,* **6,** r1.
63 Grantham, R., Gautier, C. and Gouy, M. (1980), *Nucl. Acids Res.,* **8,** 1893.
64 Steitz, J.A. (1980), in *Ribosomes: Structure, Function and Genetics* (ed. G. Chambliss, G.R. Craven, J. Davies, K. Davis, L. Kahan and M. Nomura), University Park Press, Baltimore, p. 479.
65 Stewart, J.W., Sherman, F., Shipman, N.A. and Jackson, M. (1971), *J. Biol. Chem.,* **246,** 7429.
66 Marshall, R.E., Caskey, C.T. and Nirenberg, M. (1967), *Science,* **155,** 820.
67 Hensgens, L.A.M., Grivell, L.A., Borst, P. and Bos, J.L. (1979), *Proc. Natl. Acad. Sci. USA,* **76,** 1663.
68 Li, M. and Tzagoloff, A. (1979), *Cell,* **18,** 47.
69 Barrell, B.G., Bankier, A.T. and Drouin, J. (1979), *Nature,* **282,** 189.
70 Fox, T.D. (1979), *Proc. Natl. Acad. Sci. USA,* **76,** 6534.
71 Lagerkvist, U., Rymo, L. and Waldenström, J. (1966), *J. Biol. Chem.,* **241,** 5391.
72 Julius, D.J., Fraser, T.H. and Rich, A. (1979), *Biochemistry,* **18,** 604.

73 Griffin, B.E., Jarman, M., Reese, C.B., Sulston, J.E. and Trentham, D.R. (1966), *Biochemistry,* **5,** 3638.
74 Zachau, H.G. and Feldman, H. (1965), *Prog. Nucl. Acid Res. Mol. Biol.,* **4,** 217.
75 Schimmel, P.R. and Söll, D. (1979), *Annu. Rev. Biochem.,* **48,** 601.
76 Hartley, B.S. (1979), in *Transfer RNA: Structure, Properties and Recognition* (ed. P.R. Schimmel, D. Söll and J.N. Abelson), Cold Spring Harbor Laboratory Monograph Series, p. 223.
77 Chapeville, F., Lipmann, F., von Ehrenstein, G., Weisblum, B., Ray, Jr., W. J. and Benzer, S. (1962), *Proc. Natl. Acad. Sci. USA,* **48,** 1086.
78 Baldwin, A.N. and Berg, P. (1966), *J. Biol. Chem.,* **241,** 839.
79 Fersht, A.R. (1979), in *Transfer RNA: Structure, Properties and Recognition,* (ed. P.R. Schimmel, D. Söll and J.N. Abelson), Cold Spring Harbor Laboratory Monograph Series, p. 247.
80 Ofengand, J. (1977), in *Molecular Mechanisms of Protein Biosynthesis* (ed. H. Weissbach and S. Pestka), Academic Press, New York, p. 7.
81 Goddard, J.P. (1977), *Proc. Biophys. Molec. Biol.,* **32,** 233.
82 Dintzis, H.M. (1961), *Proc. Natl. Acad. Sci. USA,* **47,** 247.
83 Thach, R.E., Cecere, M.A., Sundrararajan, T.A. and Doty, P. (1965), *Proc. Natl. Acad. Sci. USA,* **54,** 1167.
84 Byrne, R., Levin, J.G., Bladen, H.A. and Nirenberg, M.W. (1964), *Proc. Natl. Acad. Sci. USA,* **52,** 140.
85 Miller, O.L,Jr., Hamkalo, B.A. and Thomas, C.A. (1970), *Science,* **169,** 392.
86 Lacroute, F. and Stent, G. (1968), *J. Mol. Biol.,* **35,** 165.
87 Knopf, P.M. and Lamfrom, H. (1965), *Biochim. Biophys, Acta,* **95,** 398.
88 Maden, B.E.H., Traut, R.R. and Monro, R.E. (1968), *J. Mol. Biol.,* **35,** 333.
89 Harris, R.J. and Pestka, S. (1977), in *Molecular Mechanisms of Protein Biosynthesis* (ed. H. Weissbach and S. Pestka), Academic Press, New York, p. 413.
90 Roufa, D.J., Skogerson, F.E. and Leder, P. (1970), *Nature,* **227,** 567.
91 Warner, J., Knopf, P.M. and Rich, A. (1963), *Proc. Natl. Acad. Sci. USA,* **49,** 122.
92 Wettstein, F.O., Staehelin, T. and Noll, H. (1963), *Nature,* **197,** 430.
93 Lodish, H.F. and Jacobsen, M. (1972), *J. Biol. Chem.,* **247,** 3622.
94 Heywood, S.M., Dowben, R.M. and Rich, A. (1967), *Proc. Natl. Acad. Sci. USA,* **57,** 1002.
95 Uy, R. and Wold, F. (1977), *Science,* **198,** 890.
96 Caskey, T., Leder, P., Moldave, K. and Schlessinger, D. (1972), *Science,* **176,** 195.
97 Ochoa, S. and Mazumder, R. (1974), in *The Enzymes,* Vol. 10, p. 1 (ed. P.D. Boyer), Academic Press, New York.
98 Grunberg-Manago, M. and Gros, F. (1977), *Prog. Nucleic Acid Res. Mol. Biol.,* **20,** 209.
99 Revel, M. (1977), in *Molecular Mechanisms of Protein Biosynthesis* (ed. H. Weissbach and S. Pestka), Academic Press, New York, p. 245.
100 Grunberg-Manago, M. (1980), in *Ribosomes: Structure, Function and Genetics* (ed. G. Chambliss, G.R. Craven, J. Davies, K. Davis, L. Kahan and M. Nomura), University Park Press, Baltimore, p. 445.

101 Guthrie, C. and Nomura, M. (1968), *Nature,* **219,** 232.
102 Stanley, W.M., Salas, M., Wabha, A.J. and Ochoa, S. (1966), *Proc. Natl. Acad. Sci. USA,* **56,** 290.
103 Iwasaki, K., Sabol, S., Wabha, A.J. and Ochoa, S. (1968), *Arch. Biochem. Biophys.,* **125,** 542.
104 Revel, M., Lelong, J.C., Brawerman, G. and Gros, F. (1968), *Nature,* **219,** 1016.
105 Steitz, J.A. (1969), *Nature,* **224,** 957.
106 Hershey, J.W.B., Yanov, J., Johnston, K. and Fakunding, J.L. (1977), *Arch. Biochem. Biophys.,* **182,** 626.
107 Nakamoto, T. and Kolakofsky, D. (1966), *Proc. Natl. Acad. Sci. USA,* **55,** 606.
108 Lucas-Lenard, J. and Lipmann, F. (1967), *Proc. Natl. Acad. Sci. USA,* **57,** 1050.
109 Sabol, S., Sillero, M.A.G., Iwasaki, K. and Ochoa, S. (1970), *Nature,* **228,** 1269.
110 Vermeer, C., Van Alphen, W., Van Knippenberg, P.H. and Bosch, L. (1973), *Eur. J. Biochem.,* **40,** 295.
111 Salas, M., Hille, M.B., Last, J.A., Wabha, A.J. and Ochoa, S. (1967), *Proc. Natl. Acad. Sci. USA,* **57,** 387.
112 Steege, D.A. (1977), *Proc. Natl. Acad. Sci. USA,* **74,** 4163.
113 Chae, Y.-B., Mazumder, R. and Ochoa, S. (1969), *Proc. Natl. Acad. Sci. USA,* **63,** 828.
114 Rudland, P.S., Whybrow, W.A. and Clark, B.F.C. (1971), *Nature New Biol.,* **231,** 76.
115 Lockwood, A.H., Chakraborty, P.R. and Maitra, V. (1971), *Proc. Natl. Acad. Sci. USA,* **68,** 3122.
116 Groner, Y. and Revel, M. (1971), *Eur. J. Biochem.,* **22,** 144.
117 Groner, Y. and Revel, M. (1973), *J. Mol. Biol.,* **74,** 407.
118 Anderson, J.S., Dahlberg, J.E., Bretscher, M.S., Revel, M. and Clark, B.F.C. (1967), *Nature,* **216**, 1072.
119 Benne, R., Ebes, F. and Voorma, H.O. (1973), *Eur. J. Biochem.,* **38,** 265.
120 Thach, S.S. and Thach, R.E. (1971), *Proc. Natl. Acad. Sci. USA,* **68,** 1791.
121 Kuechler, E. (1971), *Nature New Biol.,* **234,** 216.
122 Eskin, B., Treadwell, B., Redfield, B., Spears, C., Kung, H.-F. and Weissbach, H. (1978), *Arch. Biochem. Biophys.,* **189,** 531.
123 Suryanarayana, T. and Subramanian, A.R. (1977), *FEBS Lett.,* **79,** 264.
124 Lucas-Lenard, J. and Beres, L. (1974), in *The Enzymes,* Vol. 10, p. 53 (ed. P.D. Boyer), Academic Press, New York.
125 Miller, D.L. and Weissbach, H. (1977), in *Molecular Mechanisms for Protein Biosynthesis* (ed. H. Weissbach and S. Pestka), Academic Press, New York, p. 324.
126 Bermek, E. (1978), *Prog. Nucleic Acid Res. Mol. Biol.,* **21,** 63.
127 Haenni, A.-L. and Lucas-Lenard, J. (1968), *Proc. Natl. Acad. Sci. USA,* **61,** 1363.
128 Erbe, R.W., Nau, M.M. and Leder, P. (1969), *J. Mol. Biol.,* **39,** 441.
129 Gupta, S.L., Waterson, J., Sopori, M.L., Weissman, S. and Lengyel, P. (1971), *Biochemistry,* **10,** 4410.

130 Lucas-Lenard, J. and Lipman, F. (1966), *Proc. Natl. Acad. USA,* **55,** 1562.
131 Miller, D.L. and Weissbach, H. (1970), *Arch. Biochem. Biophys.,* **141,** 26.
132 Arai, K.I., Kawakita, M., Kaziro, Y., Kondo, T. and Ui, N. (1973), *J. Biochem. Tokyo,* **73,** 1095.
133 Ravel, J.M., Shorey, R.L. and Shive, W. (1967), *Biochem. Biophys. Res. Commun.,* **29,** 68.
134 Ono, Y., Skoultchi, A., Klein, A. and Lengyel, P. (1968), *Nature,* **220,** 1304.
135 Hachmann, I., Miller, D.L. and Weissbach, H. (1971), *Arch. Biochem. Biophys.,* **147,** 457.
136 Weissbach, H., Redfield, B. and Brot, N. (1971), *Arch. Biochem. Biophys.,* **144,** 224.
137 Beaud, G. and Lengyel, P. (1971), *Biochemistry,* **10,** 4899.
138 Morikawa, K., LaCour, T.F.M., Nyborg, J., Rasmussen, K.M., Miller, D.L. and Clark, B.F.C. (1978), *J. Mol. Biol.,* **125,** 325.
139 Blumenthal, T., Landers, T.A and Weber, K. (1972), *Proc. Natl. Acad. Sci. USA,* **69,** 1313.
140 Jacobson, G.R. and Rosenbusch, J.P. (1976), *Nature,* **261,** 23.
141 Jaskunas, S.R., Lindahl, L., Nomura, M. and Burgess, R.R. (1976), *Nature,* **257,** 458.
142 Traut, R.R. and Monro, R.E. (1964), *J. Mol. Biol.,* **10,** 63.
143 Monro, R.E. (1967), *J. Mol. Biol.,* **26,** 147.
144 Monro, R.E., Černa, J. and Marcker, K.A. (1968), *Proc. Natl. Acad. Sci. USA,* **61,** 1042.
145 Rychlík, I., Chládek, S. and Žemlička, J. (1967), *Biochim. Biophys. Acta,* **138,** 640.
146 Parmeggiani, A. and Gottchalk, E.M. (1969), *Cold Spring Harbor Symp. Quant. Biol.,* **34,** 377.
147 Leder, P., Skogerson, L.E. and Nau, M.M. (1969), *Proc. Natl. Acad. Sci. USA,* **62,** 454.
148 Brot, N., Ertel, R. and Weissbach, H. (1968), *Biochem. Biophys. Res. Commun.,* **31,** 563.
149 Lucas-Lenard, J. and Haenni, A-L. (1969), *Proc. Natl. Acad. Sci. USA,* **63,** 93.
150 Caskey, C.T. (1973), *Adv. Protein Chem.,* **27,** 243.
151 Tate, W.P. and Caskey, C.T. (1974), in *The Enzymes,* Vol. 10, p. 87 (ed. P.D. Boyer), Academic Press, New York.
152 Caskey, C.T. (1977), in *Molecular Mechanisms of Protein Biosynthesis* (ed. H. Weissbach and S. Pestka), Academic Press, New York, p. 443.
153 Ganoza, M.C. (1966), *Cold Spring Harbor Symp. Quant. Biol.,* **31,** 273.
154 Capecchi, M.R. (1967), *Proc. Natl. Acad. Sci. USA,* **58,** 1144.
155 Scolnick, E., Tompkins, R., Caskey, T. and Nirenberg, M. (1968), *Proc. Natl. Acad. Sci. USA,* **61,** 768.
156 Milman, G., Goldstein, J., Scolnick, E. and Caskey, T. (1969), *Proc. Natl. Acad. Sci. USA,* **63,** 183.
157 Hirashima, A. and Kaji, A. (1972), *Biochemistry,* **11,** 4037.
158 Bretscher, M.S. (1968), *J. Mol. Biol.,* **34,** 131.
159 Capecchi, M.R. and Klein, H.A. (1969), *Cold Spring Harbor Symp. Quant. Biol.,* **34,** 469.

160 Caskey, C.T., Tomkins, R., Scolnick, E., Caryk, T. and Nirenberg, M. (1968), *Science,* **162,** 135.
161 Klein, H.A. and Capecchi, M.R. (1971), *J. Biol. Chem.,* **246,** 1055.
162 Goldstein, J.L. and Caskey, C.T. (1970), *Proc. Natl. Acad. Sci. USA,* **67,** 537.
163 Fahnestock, S., Neumann, H., Shashoua, V. and Rich, A. (1970), *Biochemistry,* **9,** 2477.
164 Vogel, Z., Zamir, A. and Elson, D. (1969), *Biochemistry,* **8,** 5161.
165 Ogawa, K. and Kaji, A. (1975), *Eur. J. Biochem.,* **58,** 411.
166 Kung, H.-F., Treadwell, B.V., Spears, C., Tai, P.-C. and Weissbach, H. (1977), *Proc. Natl. Acad. Sci. USA,* **74,** 3217.
167 Davis, B.D. (1971), *Nature,* **231,** 153.
168 Subramanian, A.R., Ron, E.Z. and Davis, B.D. (1968), *Proc. Natl. Acad. Sci. USA,* **61,** 761.
169 Subramanian, A.R. and Davis, B.D. (1970), *Nature,* **228,** 1273.
170 Kaempfer, R. (1972), *J. Mol. Biol.,* **71,** 583.
171 Naaktgeboren, N., Roobol, K. and Voorma, H.O. (1977), *Eur. J. Biochem.,* **72,** 49.
172 Miall, S.H. and Tamaoki, T. (1972), *Biochemistry,* **11,** 4826.
173 Boyton, J.E., Gillham, N.W. and Lambowitz, A.M. (1980), in *Ribosomes: Structure, Function and Genetics* (ed. G. Chambliss, G.R. Craven, J. Davies, K. Davis, L. Kahan and M. Nomura), University Park Press, Baltimore, p. 903.
174 Bogorad, L. and Weil, J.H. (1977), *Nucleic Acid and Protein Synthesis in Plants,* Plenum, New York.
175 Gillham, N.W. (1978), *Organelle Heredity,* Academic Press, New York.
176 Smith, A.E. and Marcker, K.A. (1968), *J. Mol. Biol.,* **38,** 241.
177 Caskcy, C.T., Redfield, B. and Weissbach, H. (1967), *Arch. Biochem. Biophys.,* **120,** 119.
178 Smith, A.E. and Marcker, K.A. (1970), *Nature,* **226,** 607.
179 Brown, J.C. and Smith, A.E. (1970), *Nature,* **226,** 610.
180 Mathews, M.B. and Korner, A. (1970), *Eur. J. Biochem.,* **17,** 328.
181 Jackson, R. and Hunter, T. (1970), *Nature,* **227,** 672.
182 Wilson, D.B. and Dintzis, H.M. (1970), *Proc. Natl. Acad. Sci. USA,* **66,** 1282.
183 Wigle, D.T. and Dixon, G.H. (1970), *Nature,* **227,** 676.
184 Schreier, M.H., Erni, B. and Staehelin, T. (1977), *J. Mol. Biol.,* **116,** 727.
185 Trachsel, H., Schreier, M.H., Erni, B. and Staehelin, T. (1977), *J. Mol. Biol.,* **116,** 755.
186 Safer, B. and Anderson, W.F. (1978), *CRC Crit. Rev. Biochem.,* **5,** 261.
187 Benne, R. and Hershey, J.W.B. (1978), *J. Biol. Chem.,* **253,** 3078.
188 Thomas, A., Goumans, H., Amesz, H., Benne, R. and Voorma, H.O. (1979), *Eur. J. Biochem.,* **98,** 329.
189 Anderson, W.F., Bosch, L., Cohn, W.E., Lodish, H., Merrick, W.C., Weissbach, H., Wittmann, H.G. and Wool, I.G. (1977), *FEBS Lett.,* **76,** 1.
190 Darnbrough, C., Legon, S., Hunt, T. and Jackson, R. (1973), *J. Mol. Biol.,* **76,** 379.

191 Schreier, M.H. and Staehelin, T. (1973), *Nature New Biol.*, **242,** 35.
192 Chen, Y.C., Woodley, C.L., Bose, K.K. and Gupta, N.K. (1972), *Biochem. Biophys. Res. Commun.*, **48,** 1.
193 Benne, R., Amesz, H., Hershey, J.W.B. and Voorma, H.O. (1979), *J. Biol. Chem.*, **254,** 3201.
194 Marcus, A. (1970), *J. Biol. Chem.*, **245,** 955.
195 Trachsel, H. and Staehelin, T. (1979), *Biochim. Biophys. Acta,* **565,** 305.
196 Sonenberg, N., Rupprecht, K.M., Hecht, S.M. and Shatkin, A.J. (1979), *Proc. Natl. Acad. Sci. USA,* **76,** 4345.
197 Ghosh-Dastidar, P., Yaghami, B., Das, A., Das, H.K. and Gupta, N.K. (1980), *J. Biol. Chem.*, **255,** 365.
198 Dasgupta, A., Majumder, A., George, A.D. and Gupta, N.K. (1976), *Biochem. Biophys. Res. Commun.*, **71,** 1234.
199 De Haro, C. and Ochoa, S. (1978), *Proc. Natl. Acad. Sci. USA,* **75,** 2713.
200 Ranu, R.S. and London, I.M. (1979), *Proc. Natl. Acad. Sci. USA,* **76,** 1079.
201 Konecki, D., Kramer, G., Pinphanichakarn, P. and Hardesty, B. (1975), *Arch. Biochem. Biophys.*, **169,** 192.
202 Arlinghaus, R., Shaeffer, J. and Schweet, R. (1964), *Proc. Natl. Acad. Sci. USA,* **51,** 1291.
203 McKeehan, W.L. and Hardesty, B. (1969), *J. Biol. Chem.*, **244,** 4330.
204 Schneir, M. and Moldave, K. (1968), *Biochim. Biophys. Acta,* **166,** 58.
205 Iwasaki, K., Mizumoto, K., Tanaka, M. and Kaziro, Y. (1973), *J. Biochem. Tokyo,* **74,** 849.
206 Prather, N., Ravel, J.M., Hardesty, B. and Shive, W. (1974), *Biochem. Biophys, Res. Commun.*, **57,** 578.
207 Slobin, L.I. and Möller, W. (1978), *Eur. J. Biochem.*, **84,** 69.
208 Nagata, S., Motoyoshi, K. and Iwasaki, K. (1976), *Biochem. Biophys. Res. Commun.*, **71,** 933.
209 Richter, D. and Lipmann, F. (1970), *Nature,* **227,** 1212.
210 Galasinski, W. and Moldave, K. (1969), *J. Biol. Chem.*, **244,** 6527.
211 Raeburn, S., Collins, J.F., Moon, H.M. and Maxwell, E.S. (1975), *J. Biol. Chem.*, **250,** 720.
212 Collier, R.J. (1967), *J. Mol. Biol.*, **25,** 83.
213 Honjo, T., Nishizuka, Y. and Hayaishi, O. (1968), *J. Biol. Chem.*, **243,** 3553.
214 Collier, R.J. (1975), *Bacteriol. Rev.*, **39,** 54.
215 Neth, R., Monro, R.E., Heller, G., Battaner, E. and Vázquez, D. (1970), *FEBS Lett.*, **6,** 198.
216 Goldstein, J.L., Beaudet, A.L. and Caskey, C.T. (1970), *Proc. Natl. Acad. Sci. USA,* **67,** 99.
217 Beudet, A.L. and Caskey, C.T. (1971), *Proc. Natl. Acad. Sci. USA,* **68,** 619.
218 Hogan, B.L. and Korner, A. (1968), *Biochim. Biophys. Acta,* **169,** 139.
219 Kaempfer, R. (1969), *Nature,* **222,** 950.
220 Thompson, H.A., Sadnik, I., Scheinbuks, J. and Moldave, K. (1977), *Biochemistry,* **16,** 2221.
221 Zimmerman, R.A. (1974), in *Ribosomes* (ed. M. Nomura, A. Tissières and P. Lengyel), Cold Spring Harbor Monograph Series, p. 225.
222 Nashimoto, H., Held, W., Kaltschmidt, E. and Nomura, M. (1971), *J. Mol. Biol.*, **62,** 121.

223 Brimacombe, R., Stöffler, G. and Wittmann, H.G. (1978), *Annu. Rev. Biochem.*, **47**, 217.
224 Shine, J. and Dalgarno, L. (1974), *Proc. Natl. Acad. Sci. USA,* **71**, 1342.
225 Steitz, J.A. and Jakes, K. (1975), *Proc. Natl. Acad. Sci. USA,* **72**, 4734.
226 Taniguchi, T. and Weissmann, C. (1978), *Nature,* **275**, 770.
227 Dunn, J.J., Buzash-Pollert, E. and Studier, F.W. (1978), *Proc. Natl. Acad. Sci. USA,* **75**, 2741.
228 Hagenbüchle, O., Santer, M., Steitz, J.A. and Mans, R.J. (1978), *Cell,* **13**, 551.
229 Kozak, M. (1979), *Cell,* **15**, 1109.
230 Van Duin, J., Kurland, C.G., Dondow, J., Grunberg-Manago, M., Barnlant, C. and Ebel, J.P. (1976), *FEBS Lett.*, **62**, 111.
231 Brownlee, G.G., Sanger, F. and Barrell, B.G. (1968), *J. Mol. Biol.,* **34**, 379.
232 Ofengand, J. and Henes, C. (1969), *J. Biol. Chem.,* **244**, 6241.
233 Grummt, F., Grummt, I., Gross, H.J., Sprinzl, M., Richter, D. and Erdmann, V.A. (1974), *FEBS Lett.*, **42**, 15.
234 Simsek, U. and RajBhandary, U.L. (1972), *Biochem. Biophys. Res. Commun.*, **49**, 508.
235 Möller, A., Manderschied, U., Lipecky, R., Bertram, S., Schmitt, M. and Gassen, H.G. (1979), in *Transfer RNA: Structure, Properties and Recognition* (ed. P.R. Schimmel, D. Söll and J.N. Abelson), Cold Spring Harbor Monograph Series, p. 459.
236 Richter, D., Erdmann, V. and Sprinzl, M. (1974), *Proc. Natl. Acad. Sci. USA,* **71**, 3226.
237 Larrinua, I. and Delitas, N. (1979), *Proc. Natl. Acad. Sci. USA,* **76**, 4400.
238 Kurland, C.G., Rigler, R., Ehrenberg, M. and Blomberg, C. (1975), *Proc. Natl. Acad. Sci. USA,* **72**, 4248.
239 Subramanian, A.R. (1980), *J. Biol. Chem.,* **255**, 6941.
240 Wittmann, H.G. (1974), in *Ribosomes* (ed. M. Nomura. A. Tissières and P. Lengyel), Cold Spring Harbor Monograph Series, p. 93.
241 Wittmann, H.G. and Wittmann-Liebold, B. (1974), in *Ribosomes* (ed. M. Nomura, A. Tissières and P. Lengyel), Cold Spring Harbor Monograph Series, p. 115.
242 Stöffler, G. and Wittmann, H.G. (1977), in *Molecular Mechanisms of Protein Biosynthesis* (ed. H. Weissbach and S. Pestka), Academic Press, New York, p. 117.
243 Wittmann, H.G., Littlechild, J.A. and Wittmann-Liebold, B. (1980), in *Ribsomes: Structure, Function and Genetics,* (ed. G. Chambliss, G.R. Craven, J. Davies, K. Davis, L. Kahan and M. Nomura), University Park Press, Baltimore, p. 51.
244 Traub, P. and Nomura, M. (1968), *Proc. Natl. Acad. Sci. USA,* **59**, 777.
245 Nomura, M. and Erdmann, V.A. (1970), *Nature,* **228**, 144.
246 Dohme, F. and Nierhaus, K.H. (1976), *J. Mol. Biol.,* **107**, 585.
247 Held, W.A. and Nomura, M. (1973), *Biochemistry,* **12**, 3273.
248 Nomura, M. and Held, W.A. (1974), in *Ribosomes* (ed. M. Nomura, A. Tissières and P. Lengyel), Cold Spring Harbor Monograph Series, p. 193.

249 Pongs. O., Nierhaus, K.H., Erdmann, V.A. and Wittmann, H.G. (1974), *FEBS Lett.*, **40**, S28.
250 Kuechler, E. and Ofengand, J. (1979), in *Transfer RNA: Structure, Properties and Recognition* (ed. P.R. Schimmel, D. Söll and J.N. Abelson), Cold Spring Harbor Monograph Series, p. 413.
251 Cooperman, B. (1980), in *Ribosomes: Structure, Function and Genetics* (ed. G. Chambliss, G.R. Craven, J. Davies, K. Davis, L. Kahan and M. Nomura), University Park Press, Baltimore, p. 531.
252 Held, W.A., Ballou, B., Mizushima, S. and Nomura, M. (1974), *J. Biol. Chem.*, **249**, 3103.
253 Laughrea, M. and Moore, P.B. (1978), *J. Mol. Biol.*, **122**, 109.
254 Roth, H.E. and Nierhaus, K.H. (1980), *Eur. J. Biochem.*, **103**, 95.
255 Traut, R.R., Lambert, J.M., Boileau, G. and Kenny, J.W. (1980), in *Ribsomes: Structure, Function and Genetics* (ed. G. Chambliss, G.R. Craven, J. Davies, K. Davis, L. Kahan and M. Nomura), University Park Press, Baltimore, p. 89.
256 Huang, K.-H., Fairclough, R.H. and Cantor, C.R. (1975), *J. Mol. Biol.*, **97**, 443.
257 Moore, P.B. (1980), in *Ribosomes: Structure, Function and Genetics* (ed. G. Chambliss, G.R. Craven, J. Davies, K. Davis, L. Kahan and M. Nomura), University Park Press, Baltimore, p. 171.
258 Stöffler, G., Bald, R., Kastner, B., Lührmann, R., Stöffler-Meiliche, M., Tischendorf, G. and Tesche, B. (1980), in *Ribosomes: Structure, Function and Genetics* (ed. G. Chambliss, G.R. Craven, J. Davies, K. Davis, L. Kahan and M. Nomura), University Park Press, Baltimore, p. 171.
259 Lake, J. (1980), in *Ribosomes: Structure, Function and Genetics* (ed. G. Chambliss, G.R. Craven, J. Davies, K. Davis, L. Kahan and M. Nomura), University Park Press, Baltimore, p. 111.
260 Laughrea, M. and Moore, P.B. (1977), *J. Mol. Biol.*, **112**, 399.
261 Kenner, R.A. (1973), *Biochem. Biophys. Res. Commun.*, **51**, 932.
262 Dahlberg, A.E. and Dahlberg, J.E. (1975), *Proc. Natl. Acad. Sci. USA*, **72**, 2940.
263 Van Dieijen, G., Van der Laken, C.J., Van Knippenberg, P.H. and Van Duin, J. (1975), *J. Mol. Biol.*, **93**, 351.
264 Van Dieijen, G., Van Knippenberg, P.H. and Van Duin, J. (1976), *Eur. J. Biochem.*, **64**, 511.
265 Thomas, J.O., Boublik, M., Szer, W. and Subramanian, A.R. (1979), *Eur. J. Biochem.*, **102**, 309.
266 Ozaki, M., Mizushima, S. and Nomura, M. (1969), *Nature*, **222**, 333.
267 Lodish, H.F. (1969), *Nature*, **224**, 867.
268 Held, W.A., Gette, W.R. and Nomura, M. (1974), *Biochemistry*, **13**, 2115.
269 Nierhaus, K.D. (1980), in *Ribosomes: Structure, Function and Genetics* (ed. G. Chambliss, G. R. Craven, J. Davies, K. Davis, L. Kahan and M. Nomura), University Park Press, Baltimore, p. 267.
270 Kischa, K., Möller, W. and Stöffler, G. (1971), *Nature New Biol.*, **233**, 62.
271 Acharya, A.S., Moore, P.B. and Richards, F.M. (1973), *Biochemistry*, **12**, 3108.

272 Highland, J.H., Bodley, J.W., Gordon, J., Hasenbank, R. and Stöffler, G. (1973), *Proc. Natl. Acad. Sci. USA,* **70,** 147.
273 Woo, N.H., Roe, B.A. and Rich, A. (1980), *Nature,* **286,** 346.
274 Richman, N. and Bodley, J.W. (1972), *Proc. Natl. Acad. Sci. USA,* **69,** 686.
275 Cabrer, B., Vázquez, D. and Modellel, J. (1972), *Proc. Natl. Acad. Sci. USA,* **69,** 733.
276 Miller, D.L. (1972), *Proc. Natl. Acad. Sci. USA,* **69,** 752.
277 Möller, W. (1974), in *Ribosomes* (ed. M. Nomura, A. Tissières and P. Lengyel), Cold Spring Harbor Monograph Series, p. 711.
278 Stark, M. and Cundliffe, E. (1979), *J. Mol. Biol.,* **134,** 767.
279 Maassen, J.A. and Möller, W. (1978), *J. Biol. Chem.,* **253,** 2777.
280 Erdmann, V.A. (1976), *Prog. Nucleic Acid Res. Mol. Biol.,* **18,** 45.
281 Sander, G., Ivell, R., Crechet, J.-B. and Parmeggiani, A. (1980), *Biochemistry,* **19,** 865.
282 Lipmann, F. (1969), *Science,* **164,** 1024.
283 Pestka, S. (1969), *J. Biol. Chem.,* **244,** 1533.
284 Gavrilova, L.T. and Spirin, A.S. (1971), *FEBS Lett.,* **17,** 324.
285 Wool, I.G. and Stöffler, G. (1974), in *Ribosomes* (ed. M. Nomura, A. Tissières and P. Lengyel), Cold Spring Harbor Monograph series, p. 417.
286 Bielka, H and Stahl, J. (1978), *International Review of Biochemistry. Amino Acid and Protein Biosynthesis II,* Vol. 18, p. 79 (ed. H. R. V. Arnstein), University Park Press, Baltimore.
287 Wool, I.G. (1979), *Annu. Rev. Biochem.,* **48,** 719.
288 Mathieson, A.T., Möller, W., Amons, R. and Yaguchi, M. (1980), in *Ribosomes: Structure, Function and Genetics* (ed. G. Chambliss, G. R. Craven, J. Davies, K. Davis, L. Kahan and M. Nomura), University Park Press, Baltimore, p. 297.
289 Leader, D.P. (1980), in *Molecular Aspects of Cellular Regulation,* Vol. 1, *Protein Phosphorylation in Regulation,* p. 203 (ed. P. Cohen), Elsevier/North Holland, Amsterdam.
290 Pestka, S. (1971), *Annu. Rev. Microbiol.,* **25,** 487.
291 Vázquez, D. (1974), *FEBS Lett.,* **40,** S63.
292 Pestka, S. (1977), in *Molecular Mechanisms of Protein Biosynthesis* (ed. H. Weissbach and S. Pestka), Academic Press, New York, p. 467.
293 Vázquez, D. (1979), *Antibiotic Inhibitors of Protein Biosynthesis,* Springer-Verlag, Berlin.
294 Cundliffe, E. (1980), in *Ribosomes: Structure, Function and Genetics* (ed. G. Chambliss, G. R. Craven, J. Davies, K. Davis, L. Kahan and M. Nomura), University Park Press, Baltimore, p. 555.
295 Nathans, D. (1964), *Proc. Natl. Acad. Sci. USA,* **51,** 585.
296 Bodley, J.W. and Zieve, F.T. (1969), *Biochem. Biophys. Res. Commun.,* **36,** 463.
297 Tanaka, M., Kawano, G. and Kinoshita, T. (1971), *Biochem. Biophys. Res. Commun.,* **42,** 564.
298 Helser, T.L., Davies, J.E. and Dahlberg, J.E. (1971), *Nature New Biol.,* **233,** 12.
299 Summers, D.F. and Maizel, J.V. (1971), *Proc. Natl. Acad. Sci. USA,* **68,** 2852.

300 Dahlberg, J.E., Kintner, C. and Lund, E. (1978), *Proc. Natl. Acad. Sci. USA,* **75,** 1071.

301 Van Knippenberg, P.H., Hooykaas, P.J.J. and Van Duin, J. (1974), *FEBS Lett.,* **41,** 323.

302 Bonitz, G., Berlani, R., Coruzzi, G., Li, M., Macino, G., Nobrega, F.G., Nobrega, M.P., Thalenfeld, B.E. and Tzagoloff, A. (1980), *Proc. Natl. Acad. Sci. USA,* **77,** 3167

303 Barrell, B.G., Anderson, S., Bankier, A.T., de Bruijn, M.H.L., Chen, E., Coulson, A.R., Drouin, J., Eperon, I.C., Nierlich, D.P., Roe, B.A., Sanger, F., Schreier, P.H., Smith, A.J.H., Staden, R. and Young, I.G. (1980), *Proc. Natl. Acad. Sci. USA,* **77,** 3164.

304 Heckman, J.E., Sarnoff, J., Alzner-DeWeerd, B., Yin, S. and RajBhandary, U.L. (1980), *Proc. Natl. Acad. Sci. USA,* **77,** 3159.

305 Thomas A., Spaan, W., Van Steeg, H., Voorma, H.O. and Benne, R. (1980), *FEBS Lett.,* **116,** 67.

306 Goumans, H., Thomas, A., Verhoeven, A., Voorma, H.O. and Benne, R. (1980), *Biochim. Biophys. Acta,* **608,** 39.

307 Van Ness, B.G., Howard, J.B. and Bodley, J.W. (1980), *J. Biol. Chem.,* **255,** 10710.

Nucleic acids and the regulation of protein synthesis

13

So far no attempt has been made to explain why cells do not continually make all the proteins which they are capable of producing. Various calculations suggest that the *E. coli* genome has, on the basis of size consideration, the potential to code for around 2000 average size proteins. On the other hand, it is estimated that the likely number of enzymes needed to provide the necessary metabolites is probably only of the order of one-third of that number. Furthermore, although it appears that each *E. coli* bacterium contains a total of roughly 10^7 protein molecules, a great variation appears to exist in the numbers of each protein type. For instance, some may be represented 500 000 times and some only once or twice [1]. Indeed, some enzymes are produced only when the need for their activity arises. Thus, there may be mechanisms for ensuring and controlling the selective utilization of the genome for the synthesis of proteins to cope with the spectrum of cellular requirements. How is this regulation of gene function achieved?

As was mentioned in Chapter 12, the actual synthesis of protein results from mRNA translation, a final step in the overall process of gene expression; the initial step being the transcription of RNA molecules from specific regions, or genes, of the DNA genome. Both translation and transcription are subject to a variety of molecular regulatory mechanisms which can potentially influence the level of a particular protein within a cell.

13.1 REGULATION OF TRANSCRIPTION

A wide variety of molecular mechanisms are now known whereby the production of various RNAs within the cell can be regulated.

13.1.1 RNA chain initiation and its regulation

Normally nucleotide units are added to growing RNA chains at a rate of 40–50 nucleotides per second at 37°C in *E. coli.* This rate, however, can vary with temperature and with other environmental changes, but under normal conditions the amount of RNA made in a bacterium is limited not so much by the rate of growth of RNA chains as by their rate of initiation. This varies quite considerably for individual types of molecule. Ribosomal RNA molecules, for instance, are required in fairly large amounts and can be initiated in *E. coli* at the rate of one molecule per second. On the other hand, a gene coding for a protein present in very small amounts may be transcribed as infrequently as once every bacterial generation [2]. Having initiated an RNA chain, the RNA polymerase moves away from the promoter site transcribing the adjacent genetic material and leaving the initiation site open to a second polymerase molecule. The frequency of these initiations will determine the proximity of RNA polymerase molecules on the genomic sequence in question [2]. In the case of the ribosomal genes this may be as close together as is sterically possible. How is the frequency of initiation regulated?

(a) *Operons and regulatory molecules*

In bacterial cells a cluster of functionally related genes that is regulated and transcribed as a unit is known as an *operon* (Fig. 13.1) [3–6]. Operons are usually transcribed into polycistronic mRNA [7]. This reflects the fact that each operon has only one promoter site at which RNA polymerase binds prior to initiating transcription [8, 9]. A regulatory region located between the promoter and the structural genes provides a means whereby the cell can regulate the frequency with which RNA polymerase molecules are allowed to transcribe the operon.

Regulatory regions are of two types [10]. One of these types are *operator* sequences. These overlap the RNA polymerase binding sites (p. 307) at the promoter and can be pictured as regions that are normally 'open' allowing RNA polymerase molecules past at a high frequency and are modulated by being progressively 'closed' [10]. *Initiator* genes on the other hand are regions that are normally 'closed' thus restricting the passage of RNA polymerase molecules and are modulated by being progressively 'opened'. In other cases the mechanism for altering the frequency with which RNA polymerase can transcribe the operon involves a specific regulatory protein and usually a specific small molecule for each operon [10, 11].

Most regulatory proteins are bifunctional. They can interact with specific sites in the genome, their interaction at such sites influencing the

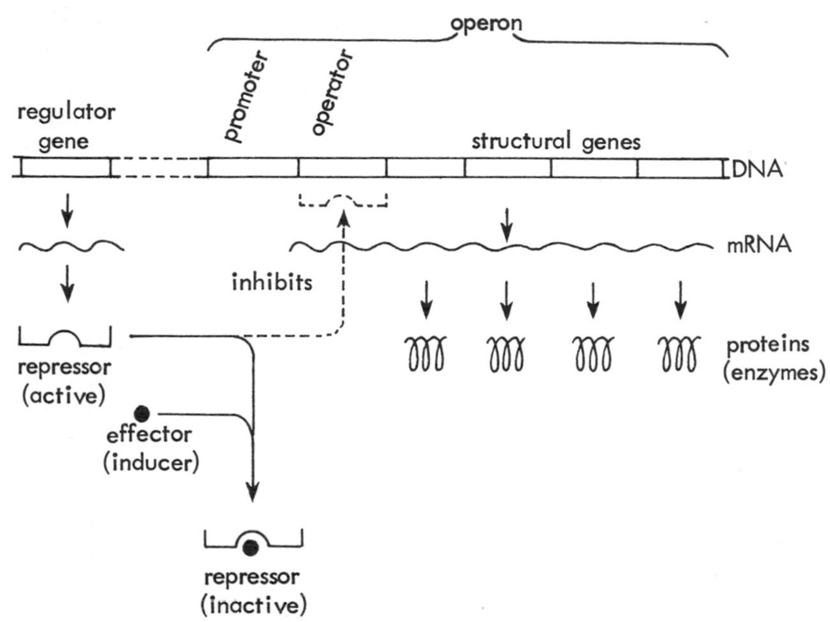

Fig. 13.1 A schematic representation of a negatively controlled inducible operon.

transcription of adjacent genes. In addition they possess the capacity to interact with small molecules. For an *operator* gene the regulatory protein is known as a *repressor.* An *activator* is the regulatory protein that binds to an *initiator* (or *positive activator*) gene. In the case of repressible operons the regulatory protein has no activity by itself and is called an *aporepressor.* Only when it binds an appropriate small molecule known as the *corepressor* does it assume the properties of a *repressor.* Usually the corepressor is the end-product of a biosynthetic pathway, or a molecule closely related to the end-product. In the case of inducible operons the regulatory protein is the active *repressor.* A small molecule known as the *inducer* (or *effector*) binds to the *repressor* and thereby either renders it unable to bind to the operator gene (in negative control) or imparts to it a new ability to bind to the initiator gene (in positive control), i.e. a *repressor* is converted into an *initiator.* Generally *inducers* are substrates of a catabolic pathway or a molecule closely related to the substrate.

(b) *Induction*

A schematic representation of a negatively controlled inducible system is shown in Fig. 13.1. It shows a regulatory gene for the synthesis of repressor, a promoter region and an operator region as well as a number of structural genes that make up an operon. Perhaps the best studied example of such a system is the *lac* operon of *E. coli* [3, 12, 13]. When glucose is depleted and lactose is present in the medium, lactose enters the cell and is converted into β-allolactose and this compound acts as inducer [14] of the *lac* operon thus increasing the cellular level of the emzymes involved in lactose metabolism. As mentioned above, an *operator* is the binding sequence for a specific repressor. It overlaps the RNA polymerase binding site at the promoter [15–18]. When bound to the *operator,* a *repressor* sterically excludes the RNA polymerase molecule from its normal crucial promoter contacts (p. 307); thus blocking promoter recognition [19–21]. *Repressor* binding occurs with native *operator* DNA in the native configuration [22] and in the bound state the *repressor* makes strong sequence-coupled interactions with the *operator* DNA and exhibits very low equilibrium binding constants (approximately 10^{-12}M) [23–25]. In the case of the *E. coli* lac operon the *operator* sequence not only overlaps the binding site for RNA polymerase [27], it contains a large palindrome [17] (Fig. 13.2). Studies

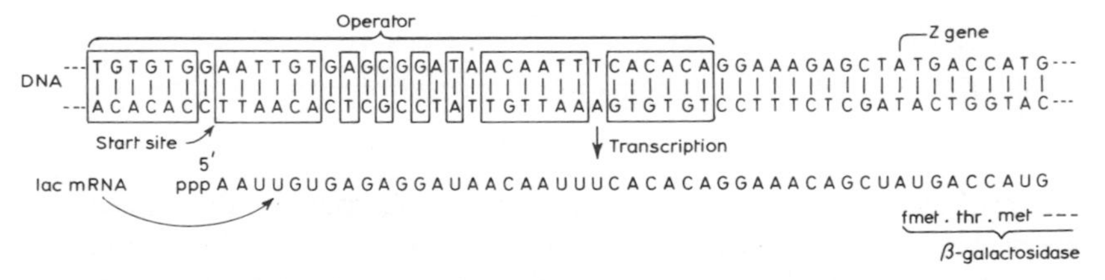

Fig. 13.2 The lac operator region and the *z* gene showing the nature of the 5′-end of the *lac* mRNA.

using chemical probes show that the *lac repressor* (a symmetric tetramer [26, 28], makes several non-symmetric contacts with *operator* DNA [29]. *In vivo* repressor-operator binding is mediated by the *inducer* (or *effector*) molecule allolactose, a by-product of lactose catabolism [14]. This effect, however, can be mimicked by the analogue, isopropyl-thio-β-D-galactoside (IPTG) [28]. IPTG can attach to the operator-bound *repressor* and drive it off the operator sequence [23, 24, 28]. Another well-studied repressor is the coliphage λ cI repressor [30] which mediates the choice between lytic growth of the bacteriophage or lysogenic growth that involves integration of λ DNA into the host genome ('the prophage state'). λ-*repressor* controls the expression of all λ lytic growth genes by interacting with specific operator sequences at two particular promoters p_L and p_R [30]. The operator sequences overlap the promoters, and as in the *lac* system described earlier, the bound repressor excludes the host RNA polymerase [21, 31]. Each operator is actually a compound repressor-binding site and the functional repressor is an oligomer [30, 32]. The mechanism of derepression (induction of integrated prophage) is not known but appears to involve the destruction of the repressor protein [33] (recA protein, may be involved). Thus induction in this case may well be atypical and not involve an interaction of a small effector molecule with the repressor.

In other inducible systems the operon is regulated by positive control; an example being the arabinose operon of *E. coli* [34, 332]. Whilst the general arrangement of genes and of the products they specify in such systems is quite similar to negatively controlled systems the crucial difference is that when the *repressor* interacts with the *inducer*, it not only cannot bind to the *operator*, but it becomes a positive *activator* binding to an *initiator* region on the genome. A highly schematic representation of a positively controlled inducible system is given in Fig. 13.3.

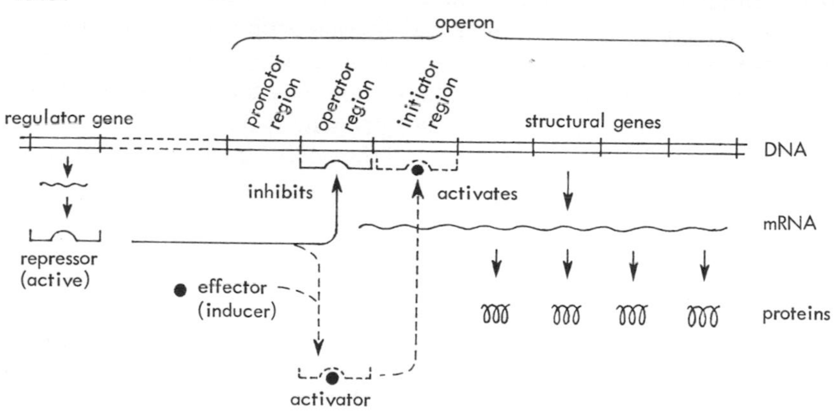

Fig. 13.3 A schematic representation of a hypothetical positively controlled inducible operon.

Activator proteins appear to function by stabilizing the recognition complex between polymerase and certain *promoters* to ensure promoter melting. Specific binding sequences (*initiator* regions) for *activator* proteins can exist near *promoters,* adjacent to the sequences covered by RNA polymerase but without blocking its access to whatever promoter contacts are available. It is likely that protein–protein interactions between activator and incoming RNA polymerase stabilize the recognition complex, thereby ensuring the transition to RS complex (p. 307).

A particularly well-studied protein with some of these properties is the catabolite activator protein (CAP) of *E. coli.* Cyclic AMP binds to CAP inducing a conformational change [39, 40] which enables CAP to bind to regions close to several promoters including the *lac* promoter [35–38]. CAP probably binds to the DNA as a dimer [41] and since the CAP-binding sequence is a palindrome [18], each CAP monomer (22 500 daltons) recognizes equivalent sequence information. CAP binding increases *lac* transcription about 50-fold [36, 37] probably as a result of increasing the stability of the RNA polymerase–*lac* promoter recognition complex brought about by CAP interaction with the RNA polymerase [186].

(c) *Repression*

A schematic model for a hypothetical negatively repressible system is shown in Fig. 13.4. The regulatory gene specifies a protein *(apo repressor)* preventing transcription when combined with, say, the end-product of a metabolic pathway acting a *corepressor.* The pathway for the synthesis of the amino acid tryptophan is among a number of systems regulated in this fashion, tryptophan acting as the *corepressor.* Thus when the organism grows in the presence of exogenous tryptophan the intracellular level of tryptophan maintains the repression of the tryptophan operon. In the absence of exogenous tryptophan, however, there is insufficient active repressor and the operon becomes derepressed [42].

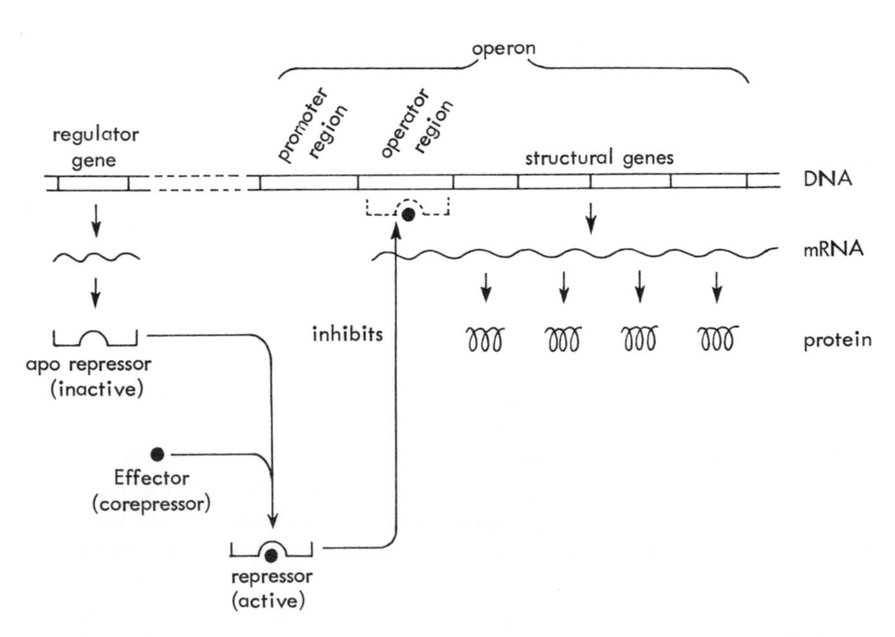

Fig. 13.4 A schematic representation of a hypothetical negatively controlled repressible system.

(d) *Autogenous regulation*
Autogenous regulation represents a variation on the above themes of induction and repression. The product of a structural gene can act as a *repressor* or *activator* to regulate the expression of the operon [43]. Thus an enzyme, or a structural protein, can have an additional role as a regulatory protein.

(e) *Divergent transcription from a common regulatory region*
In an *E. coli* operon such as *lac* or *gal,* transcription initiated at a promoter site proceeds in a fixed direction. In some gene clusters such as the arginine biosynthetic gene *arg ECBH* and the biotin genes *bio ABFCD,* although all genes of the cluster are under coordinate control, transcription starts from a position internal to the cluster and proceeds both leftward and rightward from it [44, 45]. Evidence that transcription of all biotin genes emanates from a common control region comes from a study of operator and promoter regions [46].

(f) *Overlapping transcription from different regulatory regions*
Certain situations are known where although there is a physical overlap of two oppositely orientated transcripts, they are subject to diverse controls and are formed at different times. For example following λ bacteriophage infection of *E. coli* the λ *cro* gene is transcribed rightward from promoter p_R as part of the cro cI O P Q transcript [47]. During the lysogenic state, under the influence of bacteriophage-specified proteins gpcII and gpcIII, transcription proceeds leftward through this region [48]. The functional significance of this overlap is at present unknown.

13.1.2 Regulation of transcriptional termination

Interest in the mechanism of transcriptional termination was stimulated by the finding that positive regulation of bacterial transcription can be imposed at this level as well as at the initiation steps just discussed.

Bacteriophage λ is able to suppress normal transcriptional termination through the formation of an *antiterminator.* At a molecular level one of the best studied examples is the N function in λ bacteriophage transcription. N-protein [49, 50] probably acts by binding to the RNA polymerase when at certain promoter sites (p_L and p_R) and subsequently prevents the normal functioning of rho factor at terminators t_{L1} and t_{L2}, probably as a result of excluding rho from the ternary complex. In the tryptophan (*trp*) operon of *E. coli,* transcription initiation at the promoter is regulated in a negatively repressible system regulated by the intracellular levels of tryptophan [20, 42]. In addition, at a site termed the *attenuator*, which precedes the structural genes of the operon, transcription can either be terminated to give a 140-residue 'leader' transcript, or be allowed to proceed into the structural genes [51]. Termination at the *attenuator* is regulated by the levels of charged and uncharged $tRNA^{Trp}$ [52, 53]. Most likely the charged $tRNA^{Trp}$ is required for the translation of that particular part of the 'leader' sequence that codes for an unusual 14-residue peptide containing two adjacent tryptophan residues [54, 55]. *In vitro* the terminated 240 nucleotide 'leader' transcript shows extensive secondary structure [56] and studies with mutants have indicated that the capacity to form secondary structure is essential for the regulation of transcriptional termination at the *attenuator*. There appear to be *at least* two hairpin loop structures possible, which include the nucleotides at the 3′-terminus of the leader RNA. One of these structures resembles a transcription terminator hairpin structure (see Chapter 10) whilst the other, and possibly the preferred structure, is much larger. These structures at least are mutually exclusive since the same nucleotides form part of one half stem of each hairpin. It is proposed [57] that under normal conditions where tryptophan is plentiful, ribosomes translate the leader RNA behind the RNA polymerase and cover some of the nucleotides needed to form the large hairpin stem. On the other hand they would not cover certain nucleotides included in the terminator hairpin, since they would have stopped at a nonsense codon some distance beforehand. The transcription terminator hairpin can thus form, and brake the progress of transcription. Actual termination of transcription occurs in the presence of rho factor (see Chapter 10). In situations where tryptophan is deficient, charged $tRNA^{Trp}$ will be limiting and ribosomes are likely to pause or stall at the two adjacent tryptophan codons in the leader RNA sequence. The large hairpin is unlikely to be disrupted so that the terminator hairpin does not form, and thus the RNA polymerase can transcribe through the attenuator into the *trp* genes. Similar models have been developed to explain attenuation in the leucine and histidine operons of *Salmonella typhimurium* [58, 59, 109] and in the *ilv* GEDA operon in *E. coli* [304, 305].

Premature termination of transcription has been noted in certain eukaryotic systems and it is possible that the phenomenon of transcriptional attenuation also exists in eukaryotes [333–336].

13.1.3 Structural modifications to RNA polymerases

RNA polymerase can be modified to change the promoter sequences that it recognizes [60, 61]. This can result in a change in the transcription programme of the cells involved. A temporal sequence of such modifications constitutes a programme of transcriptional development.

In bacteria the sigma factor (p. 306) is the determinant that directs the polymerase to promoter sites. Replacement of the sigma factor directs the polymerase to utilize different promoters [60, 61].

During infection of *B. subtilis* with bacteriophage SP01, transcription occurs in three stages, *early, middle* or *late.* The product of an *early* SP01 gene (p28) [62], a sigma-like protein, replaces the normal sigma factor in the host-cell RNA polymerase such that it now transcribes *middle* genes preferentially [60, 63, 64]. Two *middle*-gene products (p33 and p34) combine to act as further sigma-like activities complexing with the host RNA polymerase to elicit the specific transcription of SP01 *late* genes [60, 62]. On the other hand under certain unfavourable conditions *B. subtilis* undergoes a reversible change to a spore-forming state. This is accompanied by an alteration in transcription pattern involving a change in promoter selectivity probably again directed by a sigma-like factor [60].

During infection of *E. coli* with T4 bacteriophage, host RNA synthesis rapidly stops and a well-defined series of changes in the pattern of T4 DNA transcription (e.g. 'early' to 'late' RNA) occurs, which may result from changes observed in the polypeptides that make up the RNA polymerase subunit structure. For example, within four minutes of infection, the two α-subunits are enzymically modified. The modification involves the *covalent* attachment of an adenosine diphosphoribose unit to each subunit. The attachment appears to be through its terminal ribose to a guanido nitrogen of a specific arginine residue [65], the donor being NAD^+. This is in addition to the binding of several bacteriophage-coded proteins including the products of T4 genes 33 and 55 [66, 67]. Although their actual mode of action is not yet understood it is probable that these factors regulate the specificity of the host RNA polymerase.

13.1.4 The complexity of RNA polymerases

When the bacteriophage T7 infects its host *E. coli,* the RNA polymerase of the host transcribes a portion of the invading genome corresponding to the 'early' genes (see Chapter 11). The product of one of these genes in an mRNA which, when translated by the host's protein synthetic machinery, yields a new and simple RNA polymerase (a single polypeptide chain of mol. wt about 110 000) [68] which is not affected by rifampicin. The T7 specified RNA polymerase can only initiate RNA synthesis at promoter sites on T7 DNA other than those at which the host's RNA polymerase can initiate. In this way the T7 specified polymerase, with a different initiation specificity, simply synthesizes a collection of RNAs, known as the 'late' mRNAs, from the remainder of the T7 genome [68]. A similar type of RNA polymerase is induced on bacteriophage T3 infection of *E. coli* [68]. The reasons for the apparent simplicity of these bacteriophage-coded polymerases is not understood. It may be that, in some way, complexity of polymerase structure reflects the organism's requirement for higher degrees of control. The host, *E. coli,* is capable of rapid adaptation to different growth conditions, and perhaps this is reflected in the greater complexity of its polymerase, the initiation specificity of which can be altered by interaction with regulatory proteins. A similar situation is known to prevail in *B. subtilis* which can alter its growth pattern quite dramatically (e.g. spore formation as already mentioned). On the other hand, some bacteria like the *halophilic* group have a fairly stable growth pattern and appear to have simple type polymerases (two polypeptide chains each of 25000 daltons) [69].

13.1.5 The variety of RNA polymerases

A striking difference between the transcriptional equipment of eukaryotic cells and prokaryotic cells is the presence in the latter of more than one variety of polymerase. From Chapter 10 it will be remembered that so far there have been detected three distinguishable classes of complex type polymerases in eukaryotic nuclei. As was discussed in that chapter it would appear that initiation specificity (i.e. for rRNA, tRNA, mRNA etc.) may lie not only in the complexity of the polymerases but also in their interaction with nuclear components as well as with the template DNA.

Another question is whether the levels of the various polymerases can regulate the levels of the various RNAs produced. In higher eukaryotes, qualitative changes in RNA synthesis can be induced by hormone treatments, nutritional deficiencies, etc. The levels of activity of the various purified polymerases have been shown to vary considerably. Type II polymerase is high in rapidly growing cells [93] (up to 50000 molecules per cell) but is low in non-dividing cells [70] (as low as 800 molecules per cell). Type I polymerase is more problematic to determine owing to its metabolic turnover (half-life 1.5 hours [71]). In the wide range of differentiative or reversible metabolic transitions so far studied, significant changes in the levels of RNA polymerase activity have been detected. It should be emphasized, however, that in no instance is it yet known whether those changes are casually related to the onset of such transitions. A dramatic change in polymerase level accompanies the initiation of embryonic development [72, 73]. In reversible changes initiated, for instance, by hormones, nutrients, etc., modest changes in polymerase I and sometimes in polymerase II activity have been observed. Steroid hormones are thought to bind with specific cytoplasmic receptors. The complex of steroid and receptor in turn is believed to interact with the chromosome components to induce the appropriate response. Steroid hormone induction is accompanied by changes in polymerase activity [74–78]. Type I increases within two hours, for example, of oestrogen treatment, at which time a major product is the 45S rRNA precursor [82]. Subsequently, the activity of type II increases and specific mRNAs such as ovalbumin mRNA can be detected within three hours. Results of several experiments suggest that these increased levels of polymerase activity may be due to modulation of activity through interaction with hormone–receptor complexes [79, 80] rather than *de novo* synthesis. However, a crucial point is that these changes all occur subsequent to the production of an oestrogen-induced protein which appears within 30 minutes [77, 81]. The synthesis of the mRNA for this protein may occur within 15 minutes after hormone administration and may be the *primary* biosynthetic response to oestrogen.

13.1.6 Multiple genes

Another potential means of increasing the number of transcripts of a particular region of the genome is to have the sequences in question represented a number of times in the genome. This, as already mentioned in Chapter 11, is the case for tRNA and ribosomal RNA genes (including those for 5S RNA). The actual number of copies of these genes seems to increase with the complexity of the organism. However, mechanisms exist to regulate the transcription of these multiple copies. Both *Drosophila* and *Xenopus* appear to be able to tolerate a loss of more than half the normal ribosomal DNA sequences before the rate of ribosomal RNA falls [83]. At the molecular level there is some evidence that increases in the rate of rRNA synthesis occur by progressive activation of additional transcription units [84]. As mentioned in Chapter 11 the 'spacer' DNA encountered between the multiple eukaryotic ribosomal genes is free to diverge both in length and sequence. It then seems likely that the multiple ribosomal genes themselves experience mutagenic events, yet they retain a remarkable degee of sequence similarity. Whilst the actual mechanisms involved in maintaining this impressive homogeneity remain unknown, it is possible that grossly mutant sequences are repaired before they have a chance to exert a lethal effect (the 'correction theory'). Alternatively a few faulty ribosomes might be lethal because many essential mRNAs bound on those defective ribosomes would remain untranslated or incorrectly translated thus impairing the growth and development of the organism ('the selection theory') [85].

Also recurring in multiple copies are the genes for the histones [86, 87]. In sea urchins the genes for all the five major histone species comprise some 0.7 per cent of the genome and are arranged in a colinear fashion in tandemly repeated gene clusters (Fig. 13.5). There appear to be no intervening sequences within the mRNA coding regions. In sea urchins at least all the histone genes share the same polarity (i.e. only one of the DNA strands serves as the template for the synthesis of all five histone mRNAs) and the same relative 3′–5′ order of the five genes on the transcribed strand. The topology of the histone gene repeat units of several sea urchin species that are separated by millions of years of evolution is remarkably well conserved. Nevertheless the AT-rich spacer regions between the genes in *Psammachinus miliaris* are smaller than those found in *Strongylocentrotus purpuratus* which in turn are smaller than those found in *Lytechinus pictus* [87]. In *L. pictus* there are two types of non-allelic histone gene repeat units with similar overall topology, but with some differences in the sequences of the spacer regions between the genes [87].

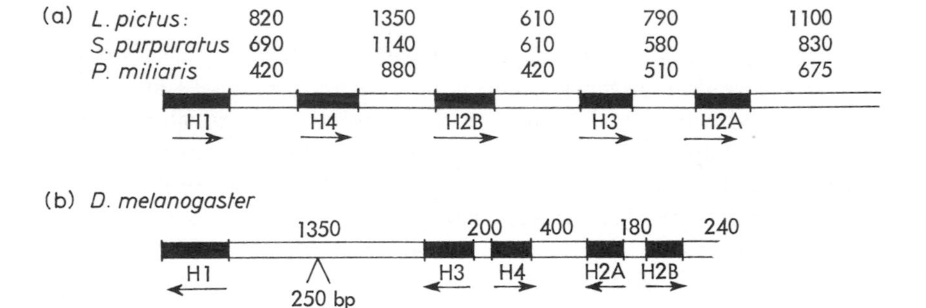

Fig. 13.5 Schematic diagram of histone gene repeat units from (a) three sea urchin species and (b) *Drosophila melanogaster*. Solid regions represent the histone genes and the open regions 'spacer' sequences and the numbers indicate the lengths of the spacer regions (in base pairs). Arrows indicate the direction in which the various genes are transcribed.

In *Drosophila* there are about 100 copies of a histone gene repeat unit (Fig. 13.5) at region 39 D-E of chromosome 2. However, these repeat units can be divided into two classes, the only apparent difference between them being a 250 base pair extra sequence (Fig. 13.5) in the spacer region (the longer repeat unit being present three times more often than the shorter). The two classes are arranged in tandem, shorter ones sometimes adjacent to the longer type. An important feature of the

Drosophila histone genes is that they are not all transcribed from the same strand. H4 and H2B are transcribed from one strand whereas H1, H3 and H2B from the other [87, 88] (Fig. 13.5).

The arrangement of histone genes in other eukaryotes is so far not known. However, in *Xenopus,* mouse and human, it is clear that the genes exist in multiple copies (10- to 50-fold reiteration). On the other hand a cloned 8 kb fragment of DNA from yeast contains only H2A and H2B genes so it is unlikely that an organization similar to that in *Drosophila* and sea urchin exists in yeast [87].

The regulation of histone-gene transcription is a matter of some controversy. Histone protein synthesis is closely coupled to DNA replication in somatic cells [89] (although this is not the case in oogenesis or early development where both histones and histone mRNAs are stored for use after fertilization). One hypothesis is that the S-phase replication regulation of histone messenger RNA activity is centred on a transcriptional level control mechanism involving non-histone proteins as regulatory molecules [89]. On the other hand post-transcriptional processing of histone mRNA precursors may be a key feature of histone–DNA coupling [90]. The results of searches for the presence of nuclear high-molecular-weight precursors have been conflicting when cleavage stages of sea urchin embryos have been examined. However, possible precursors are more obvious at the gastrula stage [91]. From *Paracentrotus lividus* embryos evidence has been obtained for four or five histone gene sequences linked in a high-molecular-weight RNA [309]. It is possible that processing of histone gene transcripts at early stages may be very much more rapid than at later stages. Alternatively early and late histone genes may be transcribed and processed in different ways.

Examination of some of the sequences [92–95] immediately preceding and following the protein-coding regions has not only revealed evolutionary conservation between sea urchin species, but also the fact that these homologies occur in topologically analogous positions relative to the individual coding regions. It is likely that these conserved sequences represent essential regulatory signals, for initiation [92–95] and termination [96]. For instance the 5′-termini of the five histone mRNAs coincide with a heptanucleotide PyPyATTCPu in the genome [183, 184, 306]. This is preceded by the sequence TATAAATA some 20–25 nucleotide pairs upstream from the 5′-termini of H2A, H2B and H3 mRNAs (H1 and H4 have AT-rich regions at analogous points [183]). Injection of genomic fragments that include the sea urchin H2A gene and its 5′-flanking regions into *Xenopus* oocyte nuclei indicates that the TATAATA or 'Hogness box' region does indeed have a regulatory role [185]. Deletion of this sequence did not abolish transcription of the gene, but a number of novel 5′-mRNA termini were generated. It may possibly function as a 'selector' sequence for gene transcription. Deletion of the heptanucleotide (or 5′-cap sequence) mentioned above, also did not abolish gene transcription, but again generated other mRNA 5′-termini [185].

With regard to the conserved sequence blocks downstream from the sea urchin histone coding sequences [306] they show some of the features of typical prokaryotic terminator or attenuator sequences as well as with RNA polymerase III terminator sequences. Most striking is a GC-rich inverted repeat sequence followed by a relatively AT-rich region. Although the mode of production of the histone mRNA 3′-termini is not known, the fact that they map within, or only a few nucleotides downstream from this feature, is consistent with a termination model.

Although histone proteins fall into five major classes, recent evidence has pointed out sequence heterogeneities in some classes which are now thought to reflect functional differences. For example H2A protein found at early cleavage stages in sea urchins has a methionine residue at position 50 whereas late H2A protein does not [97]. The appearance of stage-specific histone proteins suggests that different sets of histone genes are uniquely active at different stage of development. This has been confirmed by studies on histone mRNAs at different stages of sea urchin development. Each of the five major histone mRNAs synthesized late in development differs from its earlier counterpart by a least 10 per cent divergence in sequence [91].

Besides the histone genes and the V-region immunoglobulin genes (which will be discussed later in this chapter) certain other protein coding genes are known to exist in multiple copies. The keratins for example are a family of homologous peptide chains for which there are 25 to 35 different mRNAs arising from some 100–200 genes containing unique and repetitive elements [98, 99]. In *Dictyostelium* there are at least 15–20 actin genes [100, 101] with some sequence heterogeneity and an average spacing of 7 kb. Multiple dispersed genes for actin are also found in *Drosophila* [102, 103]. Vitellogenin, the yolk-protein precursor is derived, in *Xenopus laevis,* from large 6.3 kb mRNAs which originate from two main groups of DNA regions differing by about 20 per cent in their nucleotide sequence. Each of these main groups of genes contain two subgroups which differ in sequence by about 5 per cent, but all four groups are present in a single animal [104].

The chorion (egg shell) of the silk moth *(Antherea polyphemus)* consists of more than a hundred different proteins encoded by linked multigene families mapping at three neighbouring clusters on a single chromosome. Within the two major size classes of proteins (A or B) individual components show extensive sequence homology [105]. Remarkably the genes for the A and B classes of protein are arranged in *alternating sequence* and *divergent orientation* [106]!

After heat-shock treatment of *Drosophila* cells the appearance of a particular 70 000 dalton protein is readily observed. Each gene for this protein is part of a 3 kb common unit made up of elements x, y and z. The mRNA coding region is restricted to a portion of the largest element, z. The remainder of z and both x and y are located upstream of this region. The precise arrangement and the number of copies of this basic unit are subject to variation at both the 87A and 87C cytogenetic loci [107]. Two of the genes are found as an inverted repeat [108].

13.1.7 Gene amplification

During oogenesis in *Xenophus laevis,* the ribosomal RNA genes are amplified well above their chromosomal level. The first and preferential amplification of certain of the tandem array of ribosomal repeat units occurs in both sexes when only 9 to 16 primordial germ cells are present and the resulting 10- to 40-fold increase in ribosomal RNA gene number is maintained until meiosis begins [110]. At meiosis the sexes differ. In the male, all detectable amplified ribosomal DNA is lost, but in the female there is a transient reduction in ribosomal DNA followed by a dramatic second wave of amplification to about 2500 times the normal chromosomal level of ribosomal genes [111, 112]. This second wave of amplification is believed to involve a rolling-circle replication mechanism (see Chapter 8) giving rise to extrachromosomal circles of DNA [113, 114].

Another example of gene amplification in eukaryotic cells is the dramatic multiplication of the genes encoding dihydrofolate reductase in cultured mammalian cells resistant to the anticancer drug methotrexate (see Chapter 7). Cancer cells seem to become resistant to the drug by manufacturing extra molecules of the enzyme faster than the drug can inhibit their growth. When cultured animal cells are grown in the presence of low levels of methotrexate, cells are selected with low level of reductase gene amplification. On subsequent stepwise increments in drug concentration, cells are selected with progressively increasing reductase gene number. Cells with up to a thousand copies are encountered [115–117] and it appears that a very large repeating unit is involved in the amplification of the reductase gene [118], although the mechanism is not clear. In addition, cells resistant to normally lethal concentrations of *N*-(phosphoacetyl)-L-aspartate have been found to contain hundred-fold increases in the activity of aspartate transcarbamylase. This also is accompanied by some amplification. In this case it is the aspartate transcarbamylase genes [115].

Although amplification of certain repetitive DNA sequences has been observed during chicken cartilage and neural retina differentiation [119] it is not known whether such amplification events have a general significance in differentiation processes. Amplification of genes for chorion proteins has been detected during oogenesis in *Drosophila melanogaster* [307].

13.1.8 Gene sequence rearrangement

A commonly held view until recent times was that chromosomes were conservative structures in which a precise amount of genetic information was arranged in a definite sequential order. Normally this order was preserved when information is exchanged between chromosomes and guaranteed by recombination enzymes that function only with paired sectors of homologous DNA. However, it appears, as already mentioned in Chapter 4, that processes such as inversion, duplication and translation, often involving recombination between non-homologous chromosomal regions, can alter the sequence of genetic information.

(a) *Insertion sequences*

Studies on the *gal* and *lac* operons of *E. coli* revealed an unusual group of mutants that were caused by the insertion of a segment of DNA *into* the operon [20]. When integrated in a gene of an operon a DNA sequence of this kind not only abolishes the function of the gene but also leads to a reduction in the activities of all other genes downstream in the direction of transcription. Such inserted elements of DNA are known as insertion sequences [IS] (see also Chapter 4) and fall into non-homologous classes, such as IS1, IS2, IS3, IS4 and IS5 [121, 123]. They range in size from 800 to 1400 base pairs and are thus different from any known bacteriophage. Moreover they appear to be normal constituents of the chromosomes and plasmids of gram-negative bacteria [22]. For instance, the chromosomes of *E. coli* contain roughly eight copies of IS1, five copies of IS2 and three copies of IS3. IS1 has also been detected by molecular hybridization in *Salmonella typhimurium* and *Citrobacter freundii,* as well as in the coliphage P1. IS2 and IS3 occur on the F episome and on several drug resistance episomes [122].

The effects of IS elements have been shown to be at the level of transcription, possibly as a result of containing a combination of nonsense and terminator sequences which lead to rho-factor-dependent termination of transcription [124]. Revertants of IS-induced mutations can be detected and are believed to arise by the exact excision of the IS element. The enzymes involved in integration and excision have not yet been identified but they are independent of known recombination enzymes, since both events occur in bacteria where normal homologous recombination has been eliminated by mutation (recA$^-$). The process of excision and reintegration of the element at another site is termed transposition [122].

Transposons are elements closely related to IS sequences and are also lengths of DNA but which include several genes. In the case of the translocatable drug-resistance elements they range from 4500 base pairs (Tn1, Tn2 and Tn3) to 20000 base pairs (Tn4). Heteroduplex analysis has revealed that many of the transposons are bordered at each end by identical sequences in either reverse or direct orientation with respect to one another (Fig. 13.6). Tn5 (which carries kanamycin resistance) and Tn10 (tetracycline resistance) contain long inverted repeat sequences at their termini (2450 and 1400 base pairs respectively), Tn1, Tn2, Tn3 and Tn4 (ampicillin resistance) have inverted repeat sequences at their ends of only 140 base pairs, whilst Tn9 (chloramphenicol resistance) has direct repeat sequences of around 750 base pairs. Closer examination has revealed that the repeated sequences at the end of Tn9 and Tn10 are homologous to insertion sequences IS3 and IS1 respectively [122] which may cooperate to mediate the translocation of the intervening genetic material [122].

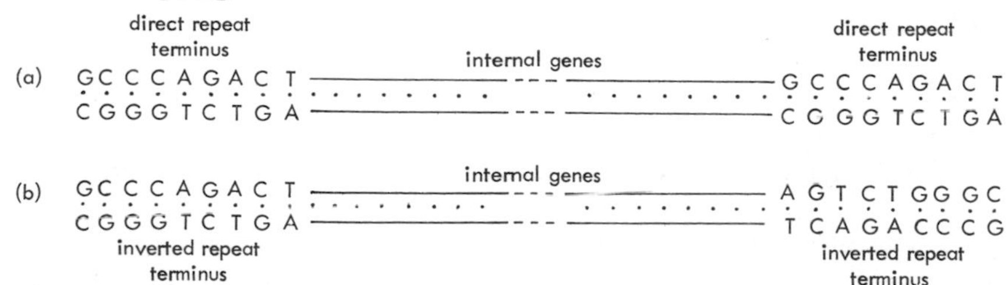

Fig. 13.6 Diagram of hypothetical transposon-like structures with (a) direct repeat sequences, or (b) inverted repeat sequences, at their terminii.

Transposons have been shown to move from the R (drug resistance) plasmid in which they normally reside to other plasmids (e.g. F episome), to the bacterial chromosome and to various bacteriophages [125, 126]. Such transposition, as is the case for IS elements, occurs in hosts which are recA$^-$, indicating that the enzymes involved are independent of normal recombination functions. Indeed in the case of Tn2 it is clear that the 4600 bp transposon itself encodes functions that affects its own transposition as well as a β-lactamase gene that renders host bacterial cells resistant to the antibiotics penicillin and ampicillin. Recombinant DNA techniques have revealed that the region known to influence the actual transposition events specifies a peptide of 21000 daltons that represses transposition of Tn3 and acts at the level of transcription to regulate its own synthesis. Subsequent studies showed that in mutants which failed to synthesize this repressor, there was a marked synthesis of a 100000 dalton peptide ('transposonase') required for the transposition of Tn3 and that its gene is transcribed in a divergent direction from the adjacent repressor gene [127, 128] (Fig. 13.7). The site of insertion and orientation of Tn3 are at least partly determined by the primary nucleotide sequence of the recipient genome and may result from the combined effects of AT-richness plus homology of the recipient genome with around 18 bp of the short terminal inverted repeat sequences of Tn3 [129]. Tn5 is a 5400 bp piece of DNA which encodes resistance to kanamycin as well as neomycin. As mentioned, it has long 150 bp inverted repeats which flank a central region 2500 bp in length. Surprisingly the inverted repeat sequences have different functional properties. They differ with respect to RNA polymerase binding, the promotion of neomycin resistance, the polypeptides they code for and their overall role in the transposition process [130]. The molecular consequence of transposition at the sequence level was first elucidated for IS1, and was the direct repetition of a nine base-pair sequence on either side of the inserted element [312]. Similarly disposed repeated sequences have now been detected for all elements so far analysed although the length of repeat can be different (3, 4, 5, 9 and 11 base pair repeats have been reported [312]).

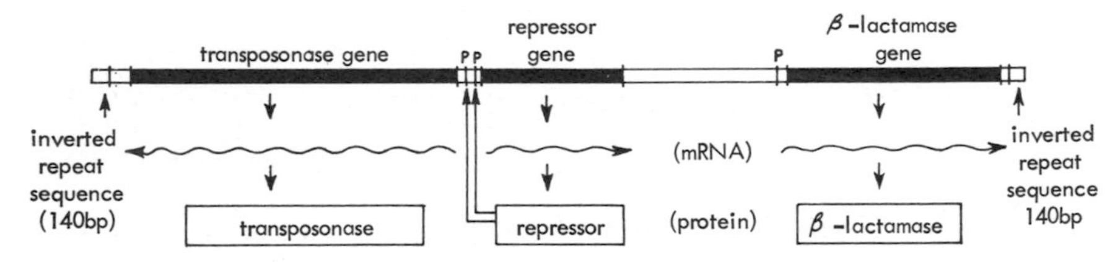

Fig. 13.7 Schematic diagram showing approximate location of Tn3 genes, the direction of their transcription and their regulation. A repressor protein acts to regulate transcription from the promoters (p) of the transposonase and repressor genes which are bidirectionally transcribed (see the text).

The effects of transposable genetic elements extend beyond their ability to join together unrelated DNA segments and move genes around among such segments. More frequently they may be involved in promoting both the rearrangement of genetic information on chromosomes and the deletion of genetic information. For example an additional feature of the translocation processes involving Tn10 are various deletions and inversions which originate at the internal termini of the 1400 base-pair terminal inverted repeat sequences [131, 132]. The deletions appear to involve the removal of a single contiguous piece of DNA beginning at one such terminus. The mechanism of inversion on the other hand is complex [131].

A transposable element that can also exist in an infectious virus is the bacteriophage Mu [120]. In the bacteriophage particle the Mu DNA is sandwiched between two short segments of bacterial DNA it has picked up from a bacterial chromosome. When the Mu bacteriophage infects a new cell it sheds the old bacterial DNA and is transposed to a site in the new host chromosome. Mu can catalyse a remarkable series of chromosome rearrangements. These include fusion of two separate and independently replicating DNA molecules, the transposition of segments of the bacterial chromosome to plasmids, the deletion of DNA and the inversion of segments of the chromosome [121, 126].

(b) *Translocation elements in eukaryotes*

The genetic properties of 'controlling elements' in maize have similarities to the properties of translocatable elements in bacteria [133, 134]. Genetic phenomena resembling controlling elements have also been reported in *Drosophila* [135].

It has been postulated that middle repetitive DNA in higher eukaryotes could be composed of IS-like sequences and thus have a regulatory role. Using recombinant DNA, two specific families of repeated sequences have been revealed in *Drosophila* called 'copia' and '412' respectively [136]. These are each present in about 35 copies dispersed throughout the genome, but of particular interest is the fact that they are each bordered by short terminal direct repeat sequences as encountered in bacterial transposon structures. Recent data indicate that such sequences can be dispersed to different genomic sites during culture of certain cell lines (10^{-3}–10^{-4} transpositions per element per generation) [137]. Additionally, comparison of different strains of *Drosophila melanogaster* indicate that '412' and 'copia' are capable of evolutionary rapid transposition to new chromosomal sites [138]. Considerable evidence is now accumulating for a substantial number of other mobile dispersed genes in *Drosophila* [187] and also for transposable elements different from 'copia' which contain *inverted* terminal repeats and are *heterogeneously* constructed [313].

Families of specific repeated sequences have also been isolated from yeast. One of these, Tyl, consists of a 5.6 kb repeat sequence including a direct repeat sequence of 0.25 kb at each end. About 35 copies of Tyl occur dispersed per haploid genome and specific sequence alterations involving the transposition of Tyl elements have been found [139]. For instance the his4-912 mutation results from insertion of Tyl into the 5′-non-coding region of the his4 gene of yeast [310, 311]. A duplication of five base pairs of wild type his4 DNA flanks the inserted element. As mentioned above, the creation of flanking duplications is a characteristic of all prokaryotic transposable elements [312] and suggests that the mechanism of transposition is similar between at least one eukaryote and the prokaryotes. In other genetic experiments transposition is also believed to be responsible for the differentiation of yeast mating types. Structural studies are consistent with the view that interconversion of mating type is the result of the replacement of one 'cassette' of DNA at the single mating type locus with another 'cassette' containing DNA coding for the opposite mating type allele [140, 141]. 'Cassettes' are envisaged as being stored as repeated sequences or silent copies elsewhere in the genome [140].

Another type of sequence recently shown to resemble that of a bacterial transposable element are the retrovirus proviruses. *Direct* repeats are found at both ends of the integrated viral DNA at the junctions with cellular DNA [314, 333, 338, 339].

(c) *Immunoglobulin genes*

Another example of a rearrangement of DNA sequences which affects gene expression concerns the regions encoding the immunoglobulins in higher organisms. Each immunoglobulin molecule (IgG) in mouse for example consists of four polypeptide chains, two identical light chains (λ or κ) and two identical heavy chains (from classes μ, γ, ϵ or α). The light chains are each composed of two regions namely the *N*-terminal variable (V_L) region and a *C*-terminal constant region (C_L) [142]. From experiments where fragments from restriction endonuclease digests of total DNA from embryos or antibody-producing cells are hybridized with mRNAs for various light chains it is clear that the coding regions for the C_L and V_L regions are some distance apart from one another in the germ line DNA but brought closer during the differentiation processes leading to the antibody-producing cells [143]. Subsequent

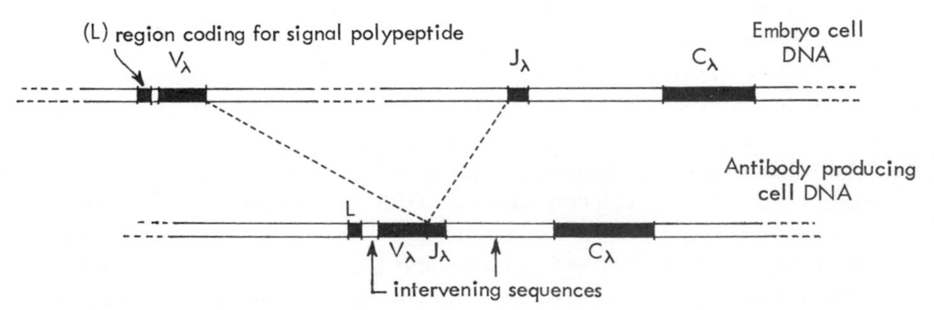

Fig. 13.8 Arrangements of mouse λ chain gene sequences in embryos and λ chain-producing plasma cells.

experiments in which the nucleotide sequence of various cloned genomic fragments were analysed [144–147] have demonstrated for example that in the germ line (Fig. 13.8), a certain λ-chain V region (V_λ) coding sequence has an intervening sequence of 93bp in the region coding for the hydrophobic 'signal' peptide (p. 469) some four codons after the initiator codon (p. 402). In addition the coding sequence only extends to a position equivalent to the 98th amino acid of the V_λ-region. The coding sequence for the remaining 14 amino acids of the V_λ-region are found some considerable distance away as a segment called the 'J_λ-region' which in turn is 1250 bp from the sequence encoding the C_λ-region. Thus among the many processes giving rise to the antibody-producing cell line there is a DNA recombination event(s) (Fig. 13.8) between the J_λ-region and V_λ-region to produce a 'gene' for the light-chain with two intervening sequences, one in the 'signal' peptide and the other a sequence of 1250 base pairs separating the J_λ and C_λ regions. Whilst it is clear that a DNA rearrangement is required for the expression of immunoglobulin genes the mechanism is not yet clear, although there are data supporting recombination involving deletion of DNA sequences between V and J regions [148]. Mapping of cloned immunoglobulin κ-light chains has also established that variable and constant regions are separated by intervening sequences in κ-chain secreting cells [146, 149–151], the removal of these intervening sequences occurring at the RNA level as one of the final steps in the post-transcriptional processing of an mRNA precursor [153]. Pulse–chase and ultraviolet mapping data indicate κ-chain mRNA to be derived from a 10.6 kb primary transcript, whose transcription appears to be initiated some 5.8 kb upstream for the V_κ-region coding sequences in the genome [152, 153]. Thus a definitive view of the post-transcriptional processing events must await a fuller understanding of the natural light-chain gene. However, abnormal V–J joining results in the absence of specific splice signals and ensuing abnormal splicing reactions yield a mutant light chain [329, 330].

In the case of the heavy-chain genes, the constant region (C_H) is divided into five classes (C_μ, C_δ, C_γ, C_ϵ and C_α) each of which associates with a variable region (V_H) at a certain stage of B-cell differentiation. C_μ seems to be the heavy chain class expressed initially. However, heavy chain gene expression involves the switch which is manifest by the predominance of the IgM (μ-chain class) in the primary immunological response and by the predominance of IgG (γ-chain class) in the secondary response. At the terminal stage of this pathway there can also be a switch to IgA (α-chain class) production. At the DNA level this would seem to involve a V_H gene first being expressed with one C_H gene (C_μ) then being 'switched' from this C_H gene to a different one (C_γ) and possibly later to C_α. Various mechanisms for this have been proposed, but recent data [154–156] support two distinct types of DNA rearrangement [315, 316]. Integration between a V_H gene and the C gene with subsequent switching facilitated by possible deletion [161, 188, 315, 316] of DNA between the integrated V_H gene and the next C_H gene (Fig. 13.9) from the expressed allelic chromosome [337]. (Detailed studies of the C_μ region itself have shown it to contain three intervening sequences separating regions coding for structural domains of the constant region [158].)

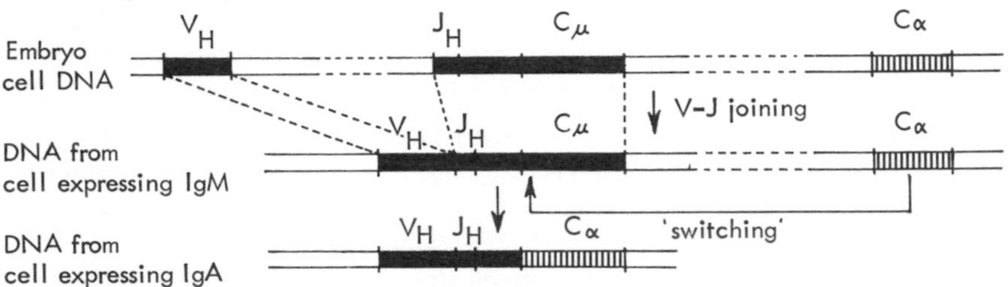

Fig. 13.9 A simplified diagram illustrating two types of DNA rearrangement involving V to J joining and C_H 'switching' in immunoglobulin heavy chain gene formation.

The extensive repertoire of immunoglobulins can be accounted for in two ways: multiplicity of germline variable region genes (V_L and V_H) and somatic divrsification of V-region genes [145]. Nucleic acid hydridization and DNA sequence data indicate that variability may arise somatically by mutation of a germline V_λ gene. In contrast, for κ-chains the pattern of variability is extensive and can be explained in part by somatic diversification but also by multiple germline genes [157].

Rearrangement of DNA sequences has also been shown to be involved in the expression of genes for variant surface antigens in pathogenic African trypanosomes which evade the immune system of their mammalian hosts by sequential expression of alternatives of cell-surface glycoproteins [159, 160].

13.1.9 Pseudogenes

Recently a particular DNA sequence has been isolated from the mouse genome which is closely related to the normal α-globin gene. However, it lacks the two intervening sequences found in the normal productive gene [317, 318]. Because of the similarities this sequence has been termed an α-globin-related *pseudogene*. It has been found to be almost identical in two different mouse strains [317, 318].

Several other globin-related sequences have also been isolated [319, 320]. In rabbit a gene designated β_2 has been found 5′ to the productive β-gene. It has some sequence homology to the adult and embryonic

β-globin sequences. In the human β- and α-like sequences occur on the 5′-side of the productive β- and α-globin genes respectively. The α-like sequence in the human [321] differs in several ways from the productive α-globin gene that would make it incapable of coding for functional globin polypeptides. On the other hand it does have intervening sequences which, however, have an organization which make it unlikely that they could be spliced out.

It would appear therefore that *pseudogenes*, either lacking, or having unusual intervening sequences are associated with productive α- and β-type globin genes, and it has been postulated that they might be involved in the control of expression of the productive globin genes [317]. Preliminary experiments indicate that these pseudogenes are not transcribed in erythroid tissues at least [317].

Regarding the evolution of the α-pseudogene it is possible that it could be a present day descendant of an old duplicate of a primitive globin gene which never had intervening sequences. A simple explanation for its evolutionary persistence is that it serves some important function. Alternatively the α-pseudogene could be a more recent duplication of an α-globin gene with both intervening sequences. However, the precise removal of the intervening sequences would require a mechanism analogous to the 'splicing' mechanism described for the removal of intervening sequences from RNA molecules. Recent data provide some support for this latter view. Chickens, rats and humans have a preproinsulin gene with two intervening sequences in homologous positions [317, 322, 323]. However, rats have an additional functional preproinsulin gene containing only the smaller first intervening sequence [322]. It is postulated that the two rat genes arose from a duplication after which the larger intervening sequence was precisely eliminated [317].

Another important gene which has related sequences located upstream from its 5′-end is ovalbumin. Analysis of large cloned DNA fragments overlapping the chicken ovalbumin gene has revealed the existence of two other genes termed X and Y [324]. All three genes are split into eight pieces [325] and are orientated in the same direction (5′-X-Y-ovalbumin-3′) and share some closely related sequences and are expressed in chick oviduct under steroid hormone control. It is suggested that gene duplications have occurred in this region of the chicken genome. Although the intervening sequences have diverged widely there is considerable conservation in the coding regions [325].

13.1.10 Transcription and post-transcriptional processing in eukaryotes

It is clear from a variety of molecular hybridization experiments that not all of the eukaryotic genome is transcribed. For example, in the rat only

10.9 per cent of single-copy DNA sequences are transcribed in the liver [161]. In kidney the value is 5.3 per cent, in spleen 4.8 per cent, in thymus 4.6 per cent and in brain 15.6 per cent [162]. Mixing experiments with RNAs from these tissues indicated that a proportion of the transcripts were common to each tissue. Indeed a general view from these and other experiments was that various tissues of an organism all express a common set of genes and the differences between them were due to the expression of genes specific for the further characteristics of the tissue. For instance, in muscle development the proportion of the mouse genome transcribed increases [163]. Sequences are transcribed in differentiated cells that were not transcribed in their progenitors. Using a specific globin cDNA probe a search was carried out for globin mRNA sequences in the steady-state RNA population of embryonic chicken haematocytoblasts, fibroblasts or muscle cells [164]. Failure to detect any globin RNA sequences strengthened the view that globin genes are expressed only in cells of an erythropoetic lineage and are thus regulated at the level of transcription by some means or other. However, other groups subsequently detected globin sequences albeit at very low levels in a number of non-erythroid cell lines and tissues [165, 166]. Nevertheless in labelling experiments it was possible to show that at least the rate of transcription of globin genes increased very shortly after dimethyl sulphoxide induction of erythroleukaemic cells and could account for the observed increase in the abundance of globin mRNA after induction [167]. A problem with all these studies is that a particular RNA sequence which is rapidly degraded by post-transcriptional processing events in the nucleus might easily go undetected, or appear to occur at very low levels.

Control of ovalbumin gene expression by oestrogen on chick oviduct also seems to be exerted at the transcriptional level from the results of labelling experiments. After withdrawal of diethylstilboestrol there was a depletion of high-molecular-weight ovalbumin mRNA precursors. Re-administration of the hormone induced a rapid accumulation of ovalbumin mRNA sequences in the nuclei [168]. The ovomucoid gene has also been found to be regulated by oestrogen in the chick oviduct system in much the same way, the evidence for control at the transcriptional level being the rapid accumulation in the nuclei of gene transcripts after hormone treatment [169]. In male *Xenopus,* liver vitellogenin mRNA sequences are not readily detectable [170]. However, administration of oestradiol provokes transcription of the genes [171].

The effect of prolactin on rat mammary gland in organ culture is also to increase the rate of transcription, this time of the casein genes (by two- to four-fold). However, in addition it was found that the half-life of the casein gene transcripts is *also* increased (17- to 25-fold) [172]. Thus hormonal control of casein gene expression is not an all-or-none process occurring at the transcriptional level. It also clearly involves post-transcriptional mechanisms. Evidence has also been gathered to show that selective stabilization of ovalbumin and conalbumin mRNA sequences may occur after oestrogen administration to chick oviduct [173, 174]. The control of chick ovalbumin and conalbumin gene expression may thus be analogous to that of the rat casein gene.

Whilst dimethyl sulphoxide treatment of Friend murine erythroleukaemic cells leads to a tenfold increase in globin sequences in the nuclear RNA, the increase in globin sequences on polysomes is 25–50-fold [175]. Chicken erythroblasts transformed with avian erythroblastosis virus cease to synthesize haemoglobin and globin mRNA disappears from the cytoplasm. However, globin sequences remain in the nucleus and the globin genes continue to be transcribed [176]. During maturation of chicken erythroid cells globin genes are transcribed and globin sequences are present in the nucleus but during the first 24 hours of culture none of these sequences appear in the cytoplasm. A considerable amount of globin mRNA then moves into the cytoplasm indicating a specific post-transcriptional control mechanism [177]. In summary, therefore, globin gene expression can be controlled both at the transcriptional and post-transcriptional level.

In reality the differences between some tissues and cell types may be quantitative rather than qualitative. For instance, in male *Xenopus* liver in which the protein vitellogenin is never normally produced there may be a very low level of transcription of the genes with the transcripts being degraded very rapidly by post-transcriptional mechanisms. In fact, processing reactions could well be taking place before transcription is complete so that a complete transcript never exists. Given the appropriate hormonal stimulus transcription can be increased by several orders of magnitude and the gene transcripts become more readily detectable. The processes of transcription and post-transcriptional processing may well be interdependent in eukaryotes, making distinction difficult. There is, however, clear evidence from sea urchins that post-transcriptional processing can lead to qualitative control of gene expression. Blastula and pluteus stage embryos transcribe the same 40 per cent of the genome as do adults, however, the homology between the polysomal RNA of these tissues is very limited [178, 179]. In other words sequences present as mRNA in one cell type but not another are nevertheless present in the nuclei of both cell types. Large changes in mRNA population do not necessarily imply any changes in DNA transcription.

Comparable situations do not seem to occur in higher eukaryotes where it is more likely that post-transcriptional processing is both quantitative and qualitative. In mammals, study of nuclear RNA populations from various tissues has shown that not all the sequences are held in common. Indeed in the rat, as already mentioned, large differences exist in the proportion of the genome transcribed into nuclear RNA in different tissues [162]. On the other hand when polysomal RNA populations are compared the amount of sequence homology is surprisingly high [180]. A total of 55 per cent of the mRNA sequences of mouse embryo, mouse brain and mouse liver are common [181] although tissue differences in mRNA abundances may be marked [182].

A general feature of the eukaryotic genome appears to be the lack of extensive clustering of genes equivalent to the situation in bacteria. Functional clustering appears to have been lost during the evolution of eukaryotes so that any models for control must provide a means to regulate together genes localized at different sites in the genome. Some eleven years ago the notion that a coordinate regulatory system of animal genomes is encoded in networks of repetitive sequence relationships was put forward in a model proposed by Britten and Davidson [340]. It seemed reasonable, that regardless of the level of regulatory interactions, the coordinate control of sets of functionally related structural genes during development and differentiation might involve the participation of some of the repetitive sequence elements which occur extensively in eukaryotic DNA (see Chapter 4). Such repeat sequences are known to be interspersed with single-copy DNA [341], but it is only recently that their relationship to particular transcription units has been studied. In the case of the human β-type globin genes for example, five repetitive sequences have been found in the intergenic sequences which flank three of these genes [342]. However, as already mentioned, current evidence suggests that the extent of variation in the transcription of structural genes in eukaryotes is far more limited than originally assumed. Davidson and Britten [343] have subsequently proposed that nuclear RNA includes continuously synthesized RNA copies of the structural genes regions of the genome and that regulatory interactions occur between these 'copies' and complementary repetitive sequence transcripts by the formation of RNA–RNA duplexes. They propose a model for regulation of gene expression in which it is the formation of repetitive RNA–RNA duplexes which controls the production of mRNA in a *post-transcriptional* process. Regulatory DNA regions would be transcribed into 'control RNAs'. However, direct evidence for this model is not yet available.

13.1.11 Alternative post-transcription processing pathways

The possibility of developmentally programmed post-transcriptional processing of RNA must also be considered. Recently it has been demonstrated that *two* mRNAs can be produced from a *single* immunoglobulin heavy chain gene by alternative RNA processing pathways. Two separate 3′-terminal sequences for μ-chain RNA are encoded in the genome, one that specifies an amino acid sequence appropriate for membranes and a second for secretion [326–328]. At different stages of immunocyte development different μ-chain mRNAs predominate. A situation with some similarities exists in sea urchins. Four sizes of mRNA appear to arise from a single gene expressed in early embryos. These RNAs constitute a set of alternatively partially overlapping transcripts from the same genomic region [344]. In the mouse a *single* α-amylase gene specifies *two* different tissue specific mRNAs. It is possible that the salivary gland and liver α-amylase mRNAs arise from differential processing [345].

13.2 CO-ORDINATE REGULATION OF TRANSCRIPTION AND TRANSLATION

When a bacterium, such as *E. coli*, is deprived of amino acids that are required for growth, not only is protein synthesis decreased, but so also is RNA synthesis [189]. This phenomenon is known as the 'stringent response' (see [190, 191] for reviews) and affects the synthesis of different types of RNA selectively. Specific inhibition of the synthesis of rRNA and tRNA, but not of the bulk of mRNA, was first to be recognized. Later it was found that the synthesis of certain mRNAs (e.g. those coding for ribosomal proteins) is also inhibited, although that of others (e.g. *lac* mRNA) is in fact stimulated. It is clearly advantageous for the cell to be able to avoid wasting valuable resources in synthesizing more ribosomes etc., which a deficiency of amino acids prevents being utilized. Likewise it is of benefit for the cell to retain the ability to synthesize essential proteins with the diminished supply of amino acids it can provide for itself through protein breakdown. Studies of the mechanism of the stringent response have been particularly assisted by two findings. One was that the lack of a single amino acid, or even of a functional aminoacyl-tRNA synthetase, would provoke the response. This latter was most clearly demonstrated with bacterial mutants having a temperature-sensitive aminoacyl-tRNA synthetase [192]. The other finding was that bacterial mutants could be obtained that did not show the stringent response. These mutant strains are known as 'relaxed' strains, and the initial strains examined all had mutations that mapped at what is now termed the *rel*A locus [193]. The breakthrough in understanding the initial events in the stringent response came with the discovery by Cashel and Gallant that when stringent, but not relaxed, strains of *E. coli* were starved of amino acids two unusual nucleotides (or 'magic spots') appeared [194]. These were subsequently found to be guanosine 5′-diphosphate,3′-diphosphate (ppGpp) and guanosine 5′-triphosphate,3′-diphosphate (pppGpp) [195]. A third nucleotide, guanosine 5′-diphosphate,3′monophosphate has subsequently been found [196]. The presence of these nucleotides (particularly ppGpp) generally correlates very well with inhibition of RNA synthesis and, despite certain exceptions to this correlation [191], there is no doubt that these molecules play a key role in stringent control. The fact that the absence of an active aminoacyl-tRNA synthetase could provoke the stringent response had lead to the suspicion that deacylated tRNA might be the inducing molecule. Evidence supporting this came when it was found that deacylated tRNA stimulated the synthesis of ppGpp and pppGpp in an 'idling' reaction occurring on ribosomes that had been isolated from cells undergoing the stringent response [197]. The reaction by which pppGpp is synthesized utilizes both ATP and GTP (see Fig. 13.10) and requires that a codon cognate to the deacylated tRNA be present in the A-site of the ribosome. The reaction is catalysed by a protein known as 'stringent factor' (molecular weight, 75 000), which has been shown to be the product of the *rel*A gene [197]. It is perhaps worth mentioning that ribosomal protein L11, the absence of which from the ribosomes of *B. megaterium* does not prevent protein synthesis occurring (see Section 12.13), is nevertheless required for the stringent response [198].

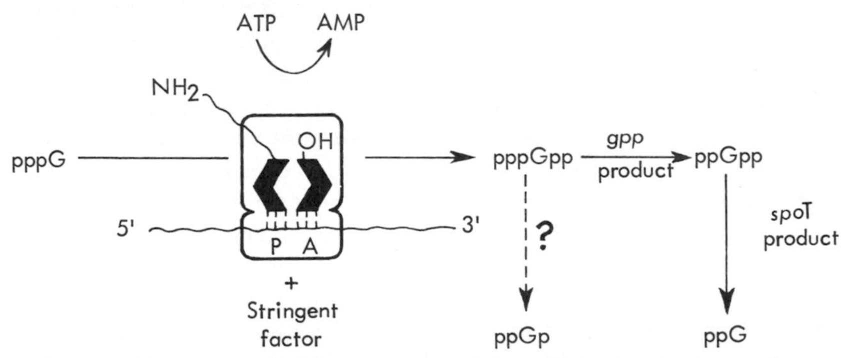

Fig. 13.10 Pathways for the synthesis and degradation of guanine nucleotides involved in 'stringent control'.

Although ribosomes can utilize GDP to synthesize ppGpp *in vitro,* it is thought that the reaction *in vivo* mainly gives rise to pppGpp, which is converted into ppGpp by the gene product of the *gpp* locus [191]. This latter nucleotide is further degraded to ppG (GDP) in a reaction catalysed by the gene product of the *spo*T locus [199] (Fig. 13.10). The control of degradation of ppGpp, although incompletely understood, is of some interest. One reason for this is that accumulation of ppGpp (but not pppGpp) also occurs when bacterial cells are deprived of an energy source, and, as this can occur in *rel*A$^-$ strains, must involve a non-ribosomal mechanism. Although there may also be a non-ribosomal pathway for the synthesis of ppGpp, there is evidence for slow ribosomal synthesis of pppGpp (and hence ppGpp) in the absence of the stringent factor, and it is thought that one mechanism by which such ppGpp accumulates in these conditions is by inhibition of its degradation. It should be added that the stringent response also results in the inhibition of other metabolic processes, such as fat synthesis, that are more directly related to energy charge. However, the discussion here is confined to the mechanism of the control of RNA synthesis.

Many studies have investigated the effects of ppGpp on RNA synthesis *in vitro*. In some cases there is good correlation between the effects *in vitro* and the stringent response *in vivo*. This is especially so for mRNAs; the synthesis of some (e.g. those for ribosomal proteins) being inhibited, wheres that of others (e.g. those for *lac* mRNA and *his* mRNA) is stimulated by the nucleotide [191]. In the case of rRNA and tRNA the situation is less clear; some workers finding specific inhibition of rRNA and tRNA synthesis, whereas others find only non-specific effects [191]. To try to explain this discrepancy Travers has studied the synthesis of rRNA and tRNA *in vitro* and found that this depends critically on the conditions of ionic strength and temperature employed. He regards these latter as determining whether the rRNA and tRNA promoters are 'open' or 'closed', and suggests that *in vivo* the state of the promoters may be influenced by proteins acting either directly on the DNA, or on the RNA polymerase [200]. One such protein, Ψ, suggested to affect RNA polymerase, was originally found by him to specifically stimulate rRNA synthesis, an effect that was antagonized by ppGpp [201]. Factor Ψ is, in fact, EF-Tu·EF-Ts (see Section 12.9.2). It certainly does participate (together with ribosomal protein S1, which has also been reported to affect transcription [202, 203]) in the RNA-dependent RNA synthesis of *E. coli* bacteriophages, such as Qβ, R17 and MS2, which is, incidentally, also subject to stringent control [204]. Clearly if it were established that EF-Tu·EF-Ts and ribosomal protein S1 (see Section 12.12) do participate in *E. coli* RNA synthesis, this would also provide a coarse mechanism for co-ordinating RNA and protein synthesis. Indeed comparable coarse positive control of RNA polymerase activity by fMet-tRNA and IF-2 has been proposed [205, 308]. However, it must be said that the failure to demonstrate a direct interaction between EF-Tu·EF-Ts and *E. coli* RNA polymerase, and the fact that the effects are only seen under certain conditions *in vitro*, has prevented these ideas being generally accepted. Nevertheless, it is clear that ppGpp, either directly or indirectly, must be responsible for the specific inhibition of rRNA and tRNA transcription during the stringent response.

More recently evidence has been obtained that ppGpp has an additional role during amino acid starvation. This is to reduce translational errors, which are clearly more likely to occur if a particular aminoacyl-tRNA is in short supply [191]. Further study of this phenomenon may help to increase our understanding of translational fidelity (see Section 12.12).

Finally it should be emphasized that ppGpp does not occur in higher eukaryotes [206], in which RNA and protein synthesis appear to be less tightly coupled, at least in the short term [207].

13.3 REGULATION OF TRANSLATION

The regulation of protein synthesis in bacteria predominantly involves the modulation of transcription, as described in Section 13.1. Although there may be cases of modulation of the translation of bacterial mRNAs (for example in the co-ordinate synthesis of ribosomal proteins [208]), the prokaryotic mRNAs, the translational control of which has been studied in most detail, are the stable *E. coli* bacteriophage RNAs. Discussion will be restricted to these in the hope of illustrating principles which may apply more widely (see Section 13.3.1). There is more evidence for quantitative and, in certain cases, qualitative control of the translation of the longer-lived eukaryotic mRNAs (see [209–212] for reviews), although our knowledge of the mechanisms involved here is limited. Only the better understood examples will be discussed, therefore, together with a consideration of certain intrinsic features of mRNAs that may influence the life of the mRNA or the subcellular location of its products.

13.3.1 Control of the translation of prokaryotic mRNA in bacteriophages

The *E. coli* bacteriophages to be considered are those in the group which includes R17, MS2, and f2, and the related, but serologically distinct, bacteriophage Qβ. As discussed in Chapter 5 (see Fig. 5.10) the single-stranded RNAs of these bacteriophages also act as mRNAs and have three clearly defined cistrons. These code for a, so-called, A protein (which is present in one copy per virion, and one function of which is to ensure proper encapsidation of the viral RNA), a coat protein (180 copies of which are present per virion), and a subunit of the viral replicase (the other three subunits of which are EF-Tu, EF-Ts, and ribosomal protein S1) [213, 214]. Different amounts of these three proteins are synthesized during bacteriophage infection of *E. coli in vivo,* about 10–20 times as much coat protein as the other two proteins having been found at the end of infection [215]. There are, in addition, temporal differences in the synthesis of the proteins: synthesis of the replicase subunit and A-protein apparently predominating early in the replicative cycle, whereas that of coat protein predominates at later times and continues after synthesis of the replicase subunit and A-protein has declined.

Studies of the mechanisms underlying these phenomena have utilized the fact that the bacteriophage RNAs can be translated in an *E. coli* cell-free system *in vitro* [216], and have been assisted by knowledge of partial sequences of many of the RNAs and, ultimately, of the complete sequence of MS2 RNA [217]. The ratio of the amounts of the three proteins synthesized *in vitro* from f2 RNA is roughly comparable to that of those synthesized *in vivo* (coat protein : replicase subunit : A protein = 20:5:1 [216]). However, it should be noted that the temporal differences in translation *in vivo* are not mimicked during translation *in vitro.*

The determination of the gene order shown in Fig. 5.10, and the demonstration of independent initiation at the initiation codons of each cistron [218], disposed of early ideas that the cause of differential translation was analogous to that in certain polar mutants of polycistronic mRNAs [219]. Nor were the sequences of the pre-initiation regions of the cistrons consistent with the relative extents of translation being governed by the extent of 'Shine and Dalgarno' base-pairing with 16S rRNA (see Table 12.3). Furthermore, the claims that the differential translation was a result of differential efficiencies of cistron-specific sub-species of IF-3 [220, 221] have not been substantiated by purification of different sub-species of the factor [209].

Instead it appears that secondary structure exerts a major influence over the quantitative control of initiation at the three sites. Lodish showed that destruction of the secondary structure of bacteriophage f2 RNA by mild formaldehyde treatment or by heating led to different relative amounts of the products translated from the RNA. The synthesis of replicase subunit was increased to the level of coat protein, and the synthesis of A-protein was also increased substantially [222]. Furthermore, when isolated oligonucleotides containing the initiation sites for the three cistrons were compared, ribosomal binding was greatest to the site for the A-protein [223].

The proposed secondary structure for MS2 RNA (Fig. 5.13) shows the initiation site of the replicase subunit hydrogen-bonded to part of the coat protein cistron, the initiation site of the latter being well exposed [224]. This is consistent with the idea that part of the coat protein cistron must be translated before there can be initiation of the synthesis of replicase protein. This idea had been based on the fact that an amber mutation at amino acid position 6 of the bacteriophage f2 coat protein (see Section 12.9.3) severely repressed expression of the cistron for the replicase subunit, but that this effect was abolished if the bacteriophage RNA were treated with formaldehyde [222]. In contrast, amber mutations at the 50th, 54th and 70th amino acid did not have this effect (see [224]). It can be seen from inspection of Fig. 5.13 that translation of only the first six codons of the coat protein cistron of the related bacteriophage MS2 would not break the putative hydrogen bonds to the replicase subunit initiation site, but that translation to the 50th or subsequent mutant termination codons would.

It has been found that the A-protein cistron is translated *in vitro* more efficiently on the nascent RNA of replicative intermediates, and it has been suggested that translation from the mature RNA *in vivo* may be completely prevented by the secondary structure [225]. Although secondary structure is clearly a major factor in determining the extent of translation of the three cistrons, the fact that ribosomes from *B. stearothermophilus* will translate the A-protein cistron of the mature RNA of these *E. coli* bacteriophages (see Section 12.12) suggests that our understanding of this phenomenon may still be incomplete.

Although the temporal regulation of the translation of the three cistrons is not mimicked during simple translation of the bacteriophage RNA *in vitro,* experiments *in vitro* have provided some suggestions of how the temporal regulation may operate. Addition of coat protein inhibits translation of the replicase subunit cistron by binding specifically to the initiation site [226, 227], offering a persuasive explanation of the later decline in synthesis of the replicase subunit *in vivo,* as the coat protein builds up. If it is true that translation of the A protein can only occur on the replicative intermediate [225], the later decline in the synthesis of this protein *in vivo* can be explained quite well by the concomitant decline in RNA synthesis [215]. It is more difficult to see how the translation of replicase subunit can predominate over that of coat protein early in infection if the translation of coat protein is a pre-requisite for that of replicase subunit, as discussed above. A possible solution to this dilemma is afforded by the suggestion of an alternative, non-hydrogen-bonded, structure for the initation site of the replicase subunit cistron [224]. It has been found that ribosomal protein S1 inhibits the synthesis of coat protein [228], and such an effect might explain the initial delay in the synthesis of the latter, if there were translation of replicase from the alternative RNA structure. This inhibition of coat-protein translation would later be relieved as the newly synthesized replicase subunit associated with protein S1, which constitutes one of the host-specified subunits of the replicase [229].

An alternative explanation can be envisaged for the inhibition of coat protein synthesis by ribosomal protein S1 *in vitro*. This is, that it reflects an inhibition which occurs *in vivo* with protein S1 integrated into the replicase. The purpose of this would be to prevent ribosomes starting along the coat protein in the 5′→3′ direction as these would prevent the movement of the replicase in the 3′→5′ direction (see Chapter 11). It has been suggested that the replicase would simultaneously attach to the 3′ end of the RNA, which would start being replicated, but that the replicase would still remain attached to the coat-protein initiation region until a considerable proportion of the transcription had occurred [230]. This is illustrated in Fig. 13.11. Binding sites for ribosomal protein S1 have, in fact, been identified, both near the 3′ terminus and near the start of the coat-protein cistron of Qβ RNA [231, 232], supporting this idea. An inhibitory role for protein S1, alone, still cannot be excluded however.

Another aspect of control in the RNA bacteriophages appears to be restricted to bacteriophage Qβ. This bacteriophage actually synthesizes minor amounts of a fourth protein, thought to be composed of the coat protein and the product of translation of the subsequent intercistronic region and beyond, as a result of 'read-through' of the single UGA termination codon [233]. Although bacteriophages R17 and MS2 terminate their coat proteins with a double stop, UAA UAG, (Fig. 5.13), and no single UGA terminator is found in the other RNA bacteriophage cistrons, single UGA terminators do occur in cellular mRNAs. As these are regarded as 'leaky' it has been suggested that such 'read-through' plays a wider role, in eukaryotes as well as prokaryotes [234].

More recently it has been found that bacteriophages f2 and MS2 code for a fourth protein required for host-cell lysis [301–303]. This is initiated about 50 bases from the end of the coat protein cistron, but *out of phase* with the reading frame of the latter, and involves the translation of the intercistronic region and about 140 bases of the replicase cistron before finishing at an out-of-phase termination codon. This viral translation strategy recalls that of ØX174 (see Section 4.5), although in the latter case it is not certain whether a single mRNA is translated in different reading frames: separate mRNAs appear to be transcribed from at least one of the pairs of overlapping ØX174 genes (A and B). The amount of lysis protein translated from MS2 RNA *in vivo* is only very small. However, the initiation region of this cistron appears to be exposed in the secondary structure (Fig. 5.13), and contains a potential 'Shine and Dalgarno' sequence of four bases. Perhaps there are further conditions

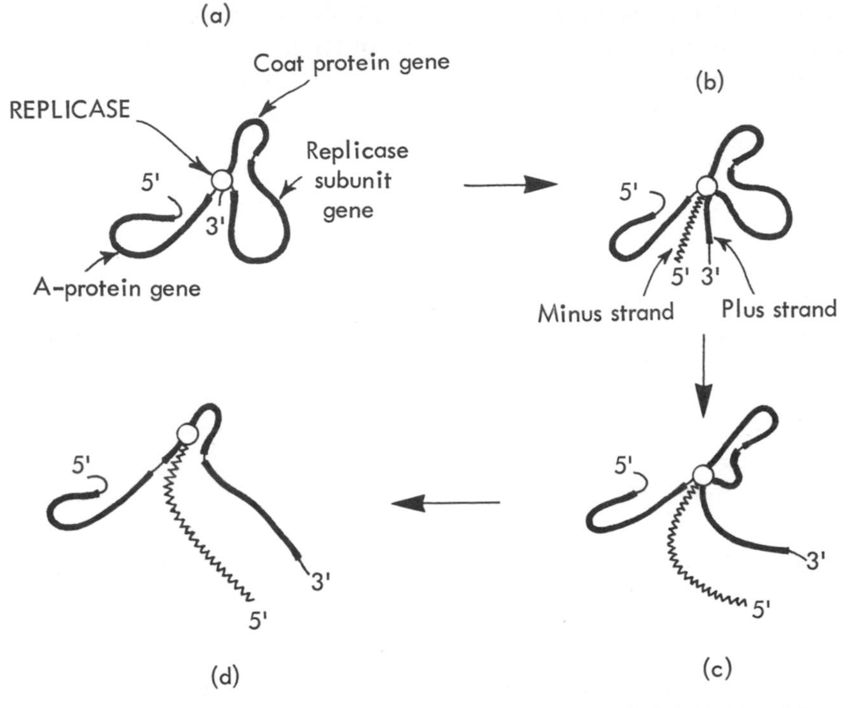

Fig. 13.11 Model for initiation of the replication of Qβ RNA with simultaneous inhibition of initiation of the translation of the coat-protein cistron. (a) The binding of the replicase near the 5′ end of the coat-protein cistron prevents further ribosomes attaching. (b) When all the ribosomes have run off the coat protein and replicase cistrons, the 3′ end of the Qβ RNA can start to be replicated by the replicase, still also attached to the coat-protein cistron. (c) Replication has advanced further into the replicase cistron. Initiation of its translation is prevented because of the secondary structure adopted when the coat protein is not being translated (see the text). (d) Replication has proceeded as far as the coat-protein cistron and at some stage (arbitarily represented as that shown) the replicase will detach from the coat-protein initiation site so that it can travel into the A-protein cistron. (Translation of the A-protein cistron is prevented by its secondary structure—see the text.)

that must be fulfilled before a mRNA sequence can act as an efficient initiation site.

13.3.2 Control of the cellular location of the products of translation

Certain proteins synthesized by cells of both eukaryotes and prokaryotes are destined for secretion or, in the case of prokaryotes, sequestration in the periplasmic space. The ribosomes synthesizing such proteins (unlike the majority of those synthesizing proteins to be retained intracellularly) are located on the membranes of the rough endoplasmic reticulum (in eukaryotes) or on the inner cell membrane (in prokaryotes). The proteins are extruded through the membrane as they are synthesized and, in the case of eukaryotes, pass from the

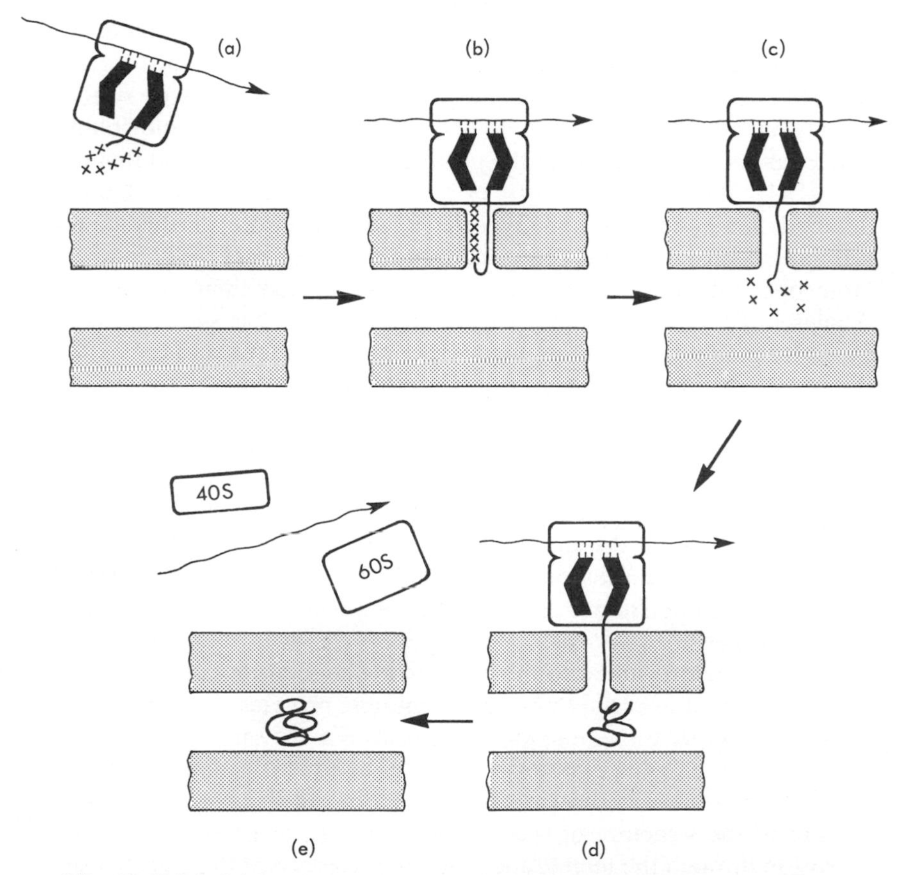

Fig. 13.12 Model for the segregation of secretory proteins into the lumen of the endoplasmic reticulum. (a) A 'free' ribosome initiates the synthesis of a secretory protein bearing a signal peptide (xxxxxx) at its *N*-terminus. (b) This hydrophobic peptide attaches to the membrane and passes through towards the lumen, probably as shown, the ribosome attaching to the membrane *via* its 60S subunit. (c) At some time before completion of the chain the signal peptide is cleaved and probably degraded. (d) The growing polypeptide continues to pass through the membrane. (e) On completion the protein passes along towards the Golgi apparatus and secretory vacuoles, and the ribosomal subunits may be released.

cisternae of the endoplasmic reticulum, *via* the Golgi apparatus, to secretory vacuoles (see [235–237] for reviews). The basis of this segregation is, at least in part, a sequence in the mRNA coding for a hydrophobic peptide, the '*signal peptide*', which by its affinity for the hydrophobic membrane leads the protein across the latter (reviewed in [238–240]).

In most cases the 'signal peptide' is an extension of the amino-terminus of the mature protein, and is excised after the protein has passed through the membrane (Fig. 13.12). The first evidence for this was obtained by Milstein and co-workers who used a cell-free system, lacking membranes, to translate the mRNA for an immunoglobulin light chain *in vitro,* and found the product to be larger than that found *in vivo* [241]. Blobel and co-workers were responsible for formulating a detailed model (the 'signal hypothesis') for the role of the amino-terminal extensions in the segregative process [242], and for obtaining a large body of evidence that, in general, supported their predictions. Analysis of the structure of many 'signal peptides' has shown them to range from about 15 to 30 amino acid residues in length, and to be quite disparate in sequence, apart from a general hydrophobicity and the presence of a small amino acid at the position adjacent to the cleavage site. The 'signal peptide' need not, however, be either at the amino-terminus or be cleaved subsequently, for in the case of the secreted oviduct protein, ovalbumin, the signal peptide lies within an internal structural region of the molecule [243]. It is for this reason that the nascent protein is represented in Fig. 13.12 as threading itself through the membrane starting from an internal point rather than from its amino-terminus.

Signal peptides, similar to those occurring on secretory proteins, have also been found in certain nascent membrane proteins [244]. However, these proteins are thought to lodge themselves in the same membrane on which they are being synthesized, this subsequently 'flowing' to the plasma membrane in the case of eukaryotic cells. Thus, some other feature of the structure of these proteins must prevent their complete extrusion through the membrane. Blobel has discussed this problem and that of the direction of proteins to other subcellular locations, and has proposed the general name of 'topogenic sequences' for the various different signals controlling such direction [245].

13.3.3 Control of mRNA lifespan

Although it is not inconceivable that differential rates of breakdown of prokaryotic mRNAs might regulate the amounts of their products [246, 247], only for eukaryotic mRNAs are there sufficient experimental data to justify discussion of this type of regulation. Here there is evidence that an intrinsic feature of many eukaryotic mRNAs, the 3′-poly(A) sequence (see Section 5.7), may influence their degradation, the loss of this or its reduction to a certain minimum size leaving them vulnerable to exonuclease action.

This proposition is supported by the following evidence. The poly(A) sequences of cytoplasmic mRNAs become progressively shorter with time [248–250]; histone mRNAs, which lack poly(A), have shorter half-lives than most eukaryotic mRNAs [251]; and deadenylated globin mRNA is translated less efficiently than normal globin mRNA when injected into *Xenopus* oocytes [252], but not when added to cell-free systems that only sustain translation for short periods [253]. The exponential decay of poly(A)-containing mRNA implies that the susceptibility of such mRNA to degradation does not increase with decreasing size [251]. Rather, there is evidence that when the size of the poly(A) segment falls below a certain threshold value the mRNA suddenly becomes extremely susceptible to degradation [252]. Because of our ignorance about the mechanism by which eukaryotic mRNAs are degraded we can only speculate on the means by which poly(A) exerts its apparent effect. However, it is worth mentioning the fact that at least one protein has been described that is associated with the poly(A) regions of polysomal mRNA [254], and one possibility is that this protein might protect the mRNA from degradation.

Cytoplasmic extension of the poly(A) regions of mRNAs can occur under certain circumstances, although it has not yet been established whether the half-life of the mRNA is altered thereby [299]. A different mechanism, however, must account for the documented increase in the half-life of 26S myosin mRNA that occurs after fusion of myoblasts, for in this case the size of the poly(A) regions is not altered [300].

13.3.4 Quantitative control of translation

There are a number of circumstances in which the rate of protein synthesis in eukaryotic cells alters in a way that appears to affect the translation of all mRNAs to a similar extent (see Section 13.3.6). However, the only cell where there is any understanding of the mechanism involved is the reticulocyte, the protein synthesis of which is inhibited under various conditions that all appear to affect the same component of initiation [255–257].

Most intensively studied has been the regulatory role of haem, the absence of which from rabbit reticulocyte lysates cause rapid inhibition of protein synthesis. (Haem is rapidly converted in aqueous solution into an oxidized form, haemin, which is what is actually used in such studies.) Although one can rationalize this effect in terms of the co-ordination of the synthesis of the predominant reticulocyte protein, globin, with the availability of its prosthetic group, haem, it must be emphasized that, in fact, the synthesis of all reticulocyte proteins is similarly affected [258]. Early work established that it was the initiation of protein synthesis that was inhibited in cells deprived of haemin, and that the reaction involved was the binding of Met-tRNA$_f$ to the 40S ribosomal subunit [259], the reaction that requires eIF-2 (See Fig. 12.9). At the same time the actual inhibitor was partially purified from the lysates, and found to have properties consistent with it being a cytoplasmic protein, pre-existing in an inactive form in normal cells [260].

This inhibitor (known as HCR, haem-controlled repressor; or HRI, haem-regulated inhibitor) was subsequently found to be a cyclic AMP-independent protein phosphokinase that specifically phosphorylates the 35000 dalton subunit of eIF-2 [261]. Although the circumstantial evidence that this phosphorylation is the cause of the inhibition is overwhelming, it has not been possible to detect any effect of phosphorylation on the ability of purified eIF-2 to bind Met-tRNA to 40S ribosomal subunits. There are reports, however, of inhibition of the binding of Met-tRNA$_f$ using cruder preparations of phosphorylated eIF-2, in which it is claimed there are factors (not shown in Fig. 12.9) that stimulate the binding reaction. It has therefore been concluded that the phosphorylation of eIF-2 prevents the enhancement of this reaction by these initiation factors (see [262]). The situation will, however, remain unclear until the factors present in these preparations are properly characterized, and their role in the initiation process established.

Although this type of explanation may in fact come to provide a satisfactory resolution of the difficulty, it should be remarked that not under every circumstance does the phosphorylation state of eIF-2 correlate well with the protein-synthetic activity of the lysate [256]. This probably indicates that the regulation has additional levels of complexity, and in this respect it may be mentioned that an anti-inhibitor protein has been described [263]. At present our knowledge of the mechanism and intervening steps in the activation of the inhibitor by haemin is also deficient, as is our understanding of the dephosphorylation of eIF-2 and its control.

Other stimuli can also provoke the phosphorylation of eIF-2 and the inhibition of protein synthesis in reticulocyte lysates. These include low concentrations of double-stranded RNA (see also Section 13.3.5), oxidized glutathione, and even the elevated pressure associated with ultracentrifugation. In the case of double-stranded RNA, a different protein kinase inhibitor is activated [261]. The fact that haem is not the only possible stimulus for the phosphorylation of eIF-2 suggests that this phosphorylation may be a more general control mechanism, and an apparently similar type of regulation has been reported to occur in rat liver [264]. It should also be mentioned that under certain conditions the extent of phosphorylation of another component of the 40S initiation complex varies in concert with changes in the protein-synthetic activity of the cell. This is a basic ribosomal protein, designated (eukaryotic) S6, which can accept up to five phosphoryl residues *in vivo*. In this case the phosphorylation of the protein is greatest in rapidly growing cells, but, as yet, no causal link has been demonstrated between the phosphorylation and the rate of protein synthesis [265].

13.3.5 Qualitative control of translation

In addition to the co-ordinate regulation of the translation of the mRNAs for all the proteins in a cell, discussed in Sections 13.3.4 and 13.3.6, it is necessary to consider the possibility of the selective stimulation or inhibition of the translation of mRNAs for specific proteins. The definite examples of this type of control that can be found where the metabolism of a cell is dramatically altered by virus infection or by interferon action will be discussed separately from the more speculative question of discrimination between different cellular mRNAs.

(a) *Interferon*

Animal cells infected with viruses produce a glycoprotein called interferon, which acts on adjacent uninfected cells, enabling them to resist infection by inhibiting the replicative cycle of the virus [266]. Interferons are species-specific but are effective against a wide spectrum of animal viruses. It appears probable that the transcription as well as the translation of viral mRNAs is specifically inhibited in interferon-treated cells, but only the control of latter process will be considered here.

The synthesis of interferon by cells may also be induced with double-stranded RNA, which is present during the replication of RNA viruses, and is probably involved in viral-induced production of interferon. When double-stranded RNA is added to lysates of interferon-treated cells there is an inhibition of protein synthesis which is much greater than that provoked by the interferon treatment or the double-stranded RNA alone. One consequence of this treatment is the activation of a protein kinase that phosphorylates eIF-2 and another protein of unknown function [267]. [In contrast to reticulocytes (see Section 13.3.4) this phosphorylation of eIF-2 occurs hardly at all when the lysates are treated with double-stranded RNA alone.] A second consequence is the activation of an enzyme that catalyses the synthesis of the trinucleotide, pppA2′p5′A2′p5′A [268]. This nucleotide activates a ribonuclease that degrades mRNA [269], and a phosphodiesterase that degrades the CCA end of tRNAs [345].

Although these translational inhibitions are of undoubted importance, it must be stressed that they are not specific for viral mRNAs, at least in the lysates *in vitro*. This could, of course, mean that neither of them is responsible for the specific inhibition *in vivo*. However, in the face of the diversity of viruses the replication of which is inhibited in interferon-treated cells, some workers have considered whether a specific inhibition could be produced by a basically non-specific mechanism. One suggestion is that, if viral mRNAs have a higher affinity for ribosomes than host mRNAs, any general inhibitor that is most effective against high-affinity mRNAs will, therefore, have a greater effect on viral protein synthesis [266]. A theoretical analysis suggested that inhibition of elongation (but not of initiation) would have this type of effect [209, 270], and a study of the effects of inhibitors of elongation on the translation of some cellular and viral mRNAs is consistent with this possibility [271]. Nevertheless it is not obvious how the inhibitors of protein synthesis so far identified in interferon-treated cells could specifically inhibit elongation. An alternative suggestion is that the inhibitors might be produced at the sites of viral mRNA in the cells, and their operation largely restricted to these sites because of rapid inactivation as they diffuse away.

(b) *Viral inhibition of host protein synthesis*

Many, but not all, eukaryotic viruses specifically inhibit protein synthesis in their hosts. Clearly a number of different mechanisms may be involved, and for one type of virus, at least, a non-general mechanism has been found. This is for poliovirus, which is unusual in that its mRNA lacks the 5′ 'cap' structure (see Section 5.7) [272]. The translation of its mRNA does not, therefore, require the 'cap'-binding initiation factor (Section 12.10.1) that 'capped' mRNAs require. It has been found that poliovirus infection results in loss of cellular 'cap'-binding activity [273], suggesting a simple way by which the viral mRNA would be favoured.

Nevertheless, some other viruses might share a common mechanism for inhibition of host protein synthesis, perhaps because of structural features that confer on them the high affinity for ribosomes, suggested above. Such high affinities may be mediated *via* initiation factors, for it has been shown that encephalomyocarditis viral RNA will outcompete host mRNA for eIF-4B [274]. Another suggestion is that viral mRNAs are more efficient under altered ionic conditions resulting from damage to the cell membrane during infection [275]. Such a mechanism may operate for certain viruses but does not, however, seem generally to hold true.

(c) *Differential translation of cellular mRNAs*
The idea that differentiated cells differ in their ability to translate different mRNAs has been raised most consistently in relation to the translation of myosin mRNA in cell-free systems from muscle and non-muscle cells. Heywood and co-workers find myosin mRNA to be less-well translated in cell-free systems from rabbit reticulocytes than in those from chick muscle, with the converse true for globin mRNA. Recently they have ascribed the difference to a difference in the specificities of the multi-component eIF-3 from the two sources [276]. However, because of the complexity of this factor, the precedents for its contamination with other activities such as 'cap'-binding factor, and the ease with which a wide spectrum of mRNAs has been translated in cell-free systems from reticulocytes and wheat germ, one is hesitant in accepting the existence of cell-specific initiation factors.

An alternative possibility is that the relative amounts of the various initiation factors differ in different cells, and that this could differentially affect translation of mRNAs having different affinities or requirements for the factors. Although there is no evidence for the former proposition, it is known that mRNAs do differ in their requirement for eIF-4A [277]. Furthermore, it has been demonstrated that β-globin mRNA is translated more efficiently than α-globin mRNA both *in vivo* [278] and *in vitro* [279]. (It appears that more α-globin mRNA is present in reticulocyte cells to maintain production of similar amounts of both chains [280].) *In vitro* it has been found that this difference reflects different rates of initiation [279], and it has been reported that addition of eIF-4A and eIF-4B to reticulocyte cell-free system synthesizing more β- than α-globin, can relieve this imbalance [281]. Thus, β-globin mRNA may have a greater affinity than α-globin mRNA for these factors (although, in the case of eIF-4B, the possibility of contaminating 'cap'-binding protein must again be borne in mind). Whether this effect is related to the different cap 1 : cap 2 ratios reported for α- and β-globin mRNA [282] is a matter for speculation.

13.3.6 Untranslated cytoplasmic mRNA

Although our knowledge of the mechanisms involved in the quantitative control of translation is limited to a few systems, there are several others the mechanism of which is unclear, but in which there is good evidence that reversible control of translation does operate. These include inhibition of protein synthesis in mitotic cells [283], in cells deprived of nutrients [284] or serum [285], and in certain tissues deprived of particular anabolic hormones [286–289]. The reversibility of these effects implies that during the inhibition the cells contain untranslated mRNA.

This must also be true of unfertilized oocytes and certain dormant or developing cells that translate pre-existing mRNA when activated. Examples of these latter are sea urchin eggs [290], cysts of *A. salina* [291] and pre-fusion myoblasts [300].

It would appear that the untranslated mRNA is not associated with ribosomes, but is complexed to certain proteins in the form of messenger ribonuclear protein (mRNP) particles of sedimentation coefficient ranging from about 20 to 120S [291–294]. Such mRNP particles (sometimes termed 'informasomes') are also found, and in fact were first described, in normal cells [295, 296]. It is difficult to decide which of the protein components of these mRNPs are intrinsic and which contaminants. However, it does appear that the free cytoplasmic mRNPs lack a protein present in the mRNPs that can be derived from polyribosomes by treatment with EDTA [297]. This protein is the 76000 dalton protein which binds to the poly(A) region of the mRNA (see Section 13.3.3). It has, in fact, been reported that a small uridine-rich RNA species is associated with muscle mRNPs and represses their translation, perhaps by binding to their poly(A) regions [298]. This question, and the molecular mechanisms by which mRNA enters and leaves mRNPs, require further study.

REFERENCES

1 Stent, G.S. (1971), *Molecular Genetics,* Freeman, San Francisco.

2 Travers, A. (1974), *Biochemistry of Nucleic Acids,* MTP. International Review of Science Biochemistry Series 1, Vol. 6 (ed. K. Burton), Butterworths, London.

3 Jacob, F. and Monod, J. (1961), *J. Mol. Biol.,* **3,** 318.

4 Ames, B.N. and Hartman, P. (1963), *Cold Spring Harbour Symp. Quant.* **28,** 349.

5 Buttin, G. (1963), *J. Mol. Biol.,* **7,** 164.

6 Yanofsky, C. (1971), *J. Am. Med. Ass.,* **218,** 1026.

7 Martin, R.G. (1963), *Cold Spring Harbor Symp. Quant. Biol.,* **28,** 357.

8 Jacob, F., Ullman, A. and Monod, J. (1964), *C.R. Acad. Sci.,* **258,** 3125.

9 Epstein, W. and Beckwith, J. (1968), *Annu. Rev. Biochem.,* **37,** 411.

10 Goldberger, R.F. (1979), in *Biological Regulation and Development* Vol. I, Plenum Press, New York and London, p. 1.

11 Barkley, M.D. and Bourgeois, S. (1978), in *The Operon* (eds. J.H. Miller and S.S. Reznikoff), Cold Spring Harbor Laboratory, p. 177.

12 Beckwith, J. (1978), *ibid.,* p. 11.

13 Miller, J.H. (1978), *ibid.,* p. 13.

14 Jobe, A, and Bourgeois, S. (1972), *J. Mol. Biol.,* **75,** 303.

15 Bennet, G.N., Schweingruber, M.E., Brown, K.D., Squires, C. and Yanofsky, C. (1976), *Proc. Natl. Acad. Sci. USA,* **73,** 235.

16 Maniatis, T., Ptashne, M., Backman, K., Kleid, D., Flashman, S., Jeffrey, A. and Maurer, R. (1975), *Cell,* **5,** 109.
17 Gilbert, W. and Maxam, A. (1973), *Proc. Natl. Acad. Sci. USA,* **70,** 3581.
18 Dickson, R.C., Abelson, J., Barnes, W.M. and Reznikoff, W.S. (1975), *Science,* **187,** 27.
19 Majors, J. (1975), *Proc. Natl. Acad. Sci. USA,* **72,** 4394.
20 Squires, C., Lee, F. and Yanofsky, C. (1975), *J. Mol. Biol.,* **92,** 93.
21 Meyer, B., Kleid, D. and Ptashne, M. (1975), *Proc. Natl. Acad. Sci. USA,* **72,** 4785.
22 Wang, J., Barkley, M. and Bourgeois, S. (1974), *Nature,* **251,** 247.
23 Riggs, A.D., Suzuki, H. and Bourgeois, S. (1970), *J. Mol. Biol.,* **48,** 67.
24 Riggs, A.D., Newby, R.F. and Bourgeois, S. (1970), *J. Mol. Biol.,* **51,** 303.
25 Rose, K. and Yanofsky, C. (1974), *Proc. Natl. Acad. Sci. USA,* **71,** 3134.
26 Beyreuther, K. (1978), in *The Operon* (eds. J.H. Miller, and W.S. Reznikoff), Cold Spring Harbor Laboratory, p. 123.
27 Reznikoff, W.S. and Abelson, J.N. (1978), *ibid.,* p. 221.
28 Gilbert, W. and Muller-Hill, B. (1970), in *The Lactose Operon* (eds. J.R. Beckwith and D. Zisper), Cold Spring Harbor Laboratory, p. 93.
29 Gilbert, W., Maxam, A. and Mirzabekov, A. (1976), in *Control of Ribosome Synthesis* (eds. N.O. Kjelgaard and O. Maaloe), Alfred Benzon Symposium IX, Munksgaard, Copenhagen, p. 139.
30 Ptashne, M., Backman, K., Humayun, M.Z., Jeffrey, A., Maurer, R., Meyer, B. and Sawer, R.T. (1976), *Science,* **194,** 156.
31 Walz, A. and Pirrotta, V. (1975), *Nature,* **254,** 118.
32 Brack, C. and Pirrotta, V. (1975), *J. Mol. Biol.,* **96,** 139.
33 Craig, N.L. and Roberts, J.W. (1980), *Nature,* **283,** 26.
34 Englesberg, E. and Wilcox, G. (1974), *Annu. Rev. Genetics,* **8,** 219.
35 Zubay, G., Schwartz, D. and Beckwith, J. (1970), *Proc. Natl. Acad. Sci. USA,* **66,** 104.
36 de Crombrugghe, B., Chen, B., Anderson, W., Nissley, S., Gottesman, M., Pastan, J. and Perlman, R. (1971), *Nature New Biol.,* **230,** 37.
37 Majors, J. (1975), *Nature,* **256,** 672.
38 Mitra, S., Zubay, G. and Landy, A. (1975), *Biochem. Biophys. Res. Commun.,* **67,** 857.
39 Krakow, J.S. and Pastan, I. (1973), *Proc. Natl. Acad. Sci. USA,* **70,** 2529.
40 Wu, F. Y-H, Nath, K., and Wu C-W (1974), *Biochemistry,* **13,** 2567.
41 Anderson, W., Schneider, A., Emmer, M., Perlman, R. and Pastan, I. (1971), *J. Biol. Chem.,* **246,** 5929.
42 Yanofsky, C. (1976), in *Molecular Mechanisms in the control of gene expression* (eds. D. Nierlich and W.J. Rutter), Academic Press, New York, p. 75.
43 Savageau, M.A. (1979), in *Biological Regulation and Development* (ed. R.F. Golderger), Plenum Press, New York and London, p. 57.
44 Elseviers, D., Cunin, R., Glandsdorff, N., Baumberg, S. and Ashcroft, E. (1972), *Mol. Gen. Genet.,* **117,** 349.
45 Guha, A., Saturen, Y. and Szybalski, W. (1971), *J. Mol. Biol.,* **56,** 53.
46 Ketner, G. and Campbell, A. (1975), *J. Mol. Biol.,* **96,** 13.

47 Campbell, A. (1979), in *Biological Regulation and Development* Vol. I (ed. R.F. Golderger), Plenum Press, New York and London, p. 19.
48 Hayes, S. and Szybalski, W. (1971), *Mol. Gen. Genet.*, **7,** 289.
49 Roberts, J.W. (1976), in *RNA polymerase* (eds. R. Losick and M. Chamberlin), Cold Spring Harbor Laboratory, p. 247.
50 Herkowitz, I. (1974), *Annu. Rev. Genetics,* **7,** 389.
51 Bertrand, K., Squires, C. and Yanofsky, C. (1976), *J. Mol. Biol.*, **103,** 319.
52 Morse, D.E. and Morse, A.N.C. (1976), *J. Mol. Biol.*, **103,** 209.
53 Yanofsky, C. and Soll, L. (1977), *J. Mol. Biol.*, **113,** 663.
54 Zurawski, G., Elseviers, D., Stauffer, G.V. and Yanofsky, C. (1978), *Proc. Natl. Acad. Sci. USA,* **75,** 5988.
55 Lee, F., Bertrand, K., Bennett, G. and Yanofsky, C. (1978), *J. Mol. Biol.*, **121,** 193.
56 Lee, F. and Yanofsky, C. (1977), *Proc. Natl. Acad. Sci. USA,* **74,** 4365.
57 Oxender, D.L., Zurawski, G. and Yanofsky, C. (1979), *Proc. Natl. Acad. Sci. USA,* **76,** 5524.
58 Gemmill, R.M., Wessler, S.R., Kerr, E.B. and Calvo, J.M. (1979), *Proc. Natl. Acad. Sci. USA,* **76,** 4941.
59 Johnston, H.M., Barnes, W.M., Chumley, F.G., Bossi, L. and Roth, J.R. (1980), *Proc. Natl. Acad. Sci. USA,* **77,** 508.
60 Losick, R. and Pero, J. (1976), in *RNA polymerase* (eds. R. Losick and M. Chamberlin), Cold Spring Harbor Laboratory, p. 227.
61 Rabussay, D. and Geiduschek, E.P. (1976), *Comp. Virol.*, **8,** 1.
62 Fujita, D.J., Ohlsson-Wilhelm, B.M. and Geiduschek, E.P. (1971), *J. Mol. Biol.*, **57,** 301.
63 Duffy, J.J. and Geiduschek, E.P. (1975), *J. Biol. Chem.*, **250,** 4530.
64 Duffy, J.J. and Geiduschek, E.P. (1977), *Cell,* **8,** 595.
65 Goff, C.G. (1974), *J. Biol. Chem.*, **249,** 6181.
66 Ratner, D. (1974), *J. Mol. Biol.*, **89,** 803.
67 Rabussay, D. and Geiduschek, E.P. (1977), *Proc. Natl. Acad. Sci. USA,* **74,** 5305.
68 Bautz, E.K.F. (1976), in *RNA polymerase* (eds. R. Losick and M. Chamberlin), Cold Spring Harbor Laboratory, p. 273.
69 Louis, B.G. and Fitt, P.S. (1971), *FEBS Lett.*, **14,** 143.
70 Keller, W. and Goor, R. (1970), *Cold Spring Harbor Symp. Quant. Biol.*, **35,** 671.
71 Yu, F.L. and Feigelson, P. (1972), *Proc. Natl. Acad. Sci. USA,* **69,** 2833.
72 Roeder, R.G. (1972), *Molecular Genetics and Developmental Biology* (ed. M. Sussman), Prentice-Hall, New Jersey, p. 163.
73 Rutter, W.J., Morris, P.W., Goldberg, M., Paula, M. and Morris, R.W. (1973), *The Biochemistry of Gene Expression in Higher Organisms* (eds. J.K. Pollak and J.W. Lee), Australian and New Zealand Book Co., Sydney, p. 89.
74 Mainwaring, W.I.F., Mangan, F.R. and Peterkin, B.M. (1971), *Biochem. J.*, **123,** 619.
75 Glasser, S.R., Chytil, F. and Spelsberg, T.C. (1972), *Biochem. J.*, **130,** 947.
76 Yu, F.L. and Feigelson, P. (1971), *Proc. Natl. Acad. Sci. USA,* **68,** 2177.
77 Baulieu, E.E., Wira, C.R., Milgrom, E. and Raynaud-Jammet, C. (1972),

Gene Transcription in Reproductive Tissue (ed. E. Diczfaluzy), Karolinska Symp. on Research Methods in Reproductive Endocrinology.
78 Borthwick, N.M. and Smellie, R.M.S. (1975), *Biochem. J.*, **147**, 101.
79 Mueller, G.C., Vonderhaar, B., Kim, U.H. and Le Mahieu, M. (1972), *Recent Prog. Horm. Res.*, **28**, 1.
80 Mohla, S., De Sombre, E.R. and Jensen, E.V. (1972), *Biochem. Biophys. Res. Commun.*, **46**, 661.
81 Barnea, A. and Gorski, J. (1970), *Biochemistry*, **9**, 1899.
82 Knowler, J.T. and Smellie, R.M.S. (1971), *Biochem. J.*, **125**, 605.
83 Krider, H.M. and Plant, W. (1972), *J. Cell Sci.*, **11**, 675.
84 McKnight, S.L. and Miller, O.L. (1976), *Cell*, **8**, 305.
85 Goodenough, U. (1978), *Genetics*, 2nd Edn, Holt, Rinehart and Winston.
86 Birnstiel, M.L., Kressman, A., Schaffner, W., Portmann, R. and Busslinger, M. (1978), *Philos. Trans. R. Soc. London Ser. B*, **283**, 319.
87 Kedes, L.H. (1979), *Annu. Rev. Biochem.*, **48**, 837.
88 Lifton, R.P., Goldberg, M.L., Karp, R.W. and Hogness, D.S. (1977), *Cold Spring Harbor Symp. Quant. Biol.*, **42**, 1047.
89 Stein, G.S., Stein, J.L., Kleinsmith, L.J., Jansing, R.L., Park, W.D. and Thomson, J.A. (1977), *Biochem. Soc. Symp.*, **42**, 137.
90 Melli, M., Spinelli, G., Wyssling, H. and Arnold, E. (1977), *Cell*, **11**, 651.
91 Kunkel, N.S. and Weinberg, E.S. (1978), *Cell*, **14**, 313.
92 Sures, I., Maxam, A., Cohn, R.H. and Kedes, L.H. (1976), *Cell*, **9**, 495.
93 Birnstiel, M., Schaffner, W. and Smith, H.O. (1977), *Nature*, **266**, 603.
94 Schaffner, W., Kunz, G., Daetwyler, H., Telford, J., Smith, H.O. and Birnstiel, M.L. (1978), *Cell*, **14**, 655.
95 Grunstein, M. and Grunstein, J. (1977), *Cold Spring Harbor. Symp. Quant. Biol.*, **42**, 1083.
96 Busslinger, M., Portmann, R. and Birnstiel, M.L. (1979), *Nucleic Acid Res.*, **6**, 2997.
97 Newrock, K.M. and Cohen, L.H. (1978), *Cell*, **14**, 327.
98 Kemp, D.J. (1975), *Nature*, **254**, 573.
99 Lockett, T.J., Kemp, D.J. and Rogers, G.E. (1979), *Biochemistry*, **18**, 5655.
100 McKeown, M., Taylor, W.C., Kindle, K.C., Firtel, R.A., Bender, W. and Davidson, N. (1978), *Cell*, **15**, 789.
101 Vanderkerchove, J. and Weber, K. (1980), *Nature*, **284**, 475.
102 Fyrberg, E.A., Kindle, K.L., Davidson, N. and Sodja, A. (1980), *Cell*, **19**, 365.
103 Tobin, S.L., Zulauf, E., Sanchez, F., Craig, E.A. and McCarthy, B.J. (1980), *Cell*, **19**, 121.
104 Wahli, W., Dawid, I.B., Wyler, T., Jaggi, R.B., Weber, R. and Ruffel, G.U. (1979), *Cell*, **16**, 535.
105 Sim, G.K., Kafatos, F.C., Jones, C.W., Koehler, M., Efstadiadis, A. and Maniatis, T. (1979), *Cell*, **18**, 1303.
106 Jones, C.E. and Kafatos, F.C. (1980), *Nature*, **284**, 635.
107 Mirault, M-E., Goldschmidt-Clermont, M., Atravanis, S., Tsakonas, S. and Schedl, P. (1979), *Proc. Natl. Acad. Sci. USA*, **76**, 5254.
108 Goldschmidt-Clermont, M. (1980), *Nucleic Acids Res.*, **8**, 235.
109 Keller, E.B. and Caluo, J.M. (1979), *Proc. Natl. Acad. Sci. USA*, **76**, 6189.

110 Bird, A.P. (1977), *Cold Spring Harbor Symp. Quant. Biol.*, **42,** 1179.
111 MacGregor, H.C. (1972), *Biol. Rev.*, **47,** 173.
112 Gall, J.G. (1968), *Genetics (Suppl.)*, **61,** 1.
113 Hourcade, D., Dressler, D. and Wolfson, J. (1974), *Cold Spring Harbor. Symp. Quant. Biol.*, **38,** 537.
114 Rochaix, J-D. and Bird, A.P. (1975), *Chromosoma*, **52,** 317.
115 Schimke, R.T., Kaufman, R.J., Alt, F.W. and Kellens, R.F. (1978), *Science*, **202,** 1051.
116 Shields, R. (1978), *Nature*, **273,** 269.
117 Kaufman, R.J., Brown, P.C. and Schimke, R.T. (1979), *Proc. Natl. Acad. Sci. USA*, **76,** 5669.
118 Bostock, C.J. and Clark, E.M. (1980), *Cell*, **19,** 709.
119 Strom, C.M., Moscona, M. and Dorfman, A. (1978), *Proc. Natl. Acad. Sci. USA*, **75,** 4451.
120 Starlinger, P. and Saedler, H. (1976), *Curr. Top. Microbiol. Imm.*, **75,** 111.
121 Bukhari, A.I., Shapiro, J. and Adhya, S. (1977), in *DNA Insertion Elements, Plasmids and Episomes*, Cold Spring Harbor Laboratory, New York.
122 Kleckner, N. (1977), *Cell*, **11,** 11.
123 Schmidt, A., Chernajovsky, Y., Shulman, L., Federman, P., Berissi, H. and Revel, M. (1979), *Proc. Natl. Acad. Sci. USA*, **76,** 4788.
124 Nevers, P. and Saedler, H. (1977), *Nature*, **268,** 109.
125 Cohen, S.N. (1976), *Nature*, **263,** 731.
126 Cohen, S.N. and Shapiro, J.A. (1980), *Sci. Amer.*, **242,** 36.
127 Chou, J., Lemaux, P.G., Casadaban, M.J. and Cohen, S.N. (1979), *Nature*, **282,** 801.
128 Gill, R.E., Heffron, F. and Falkow, S. (1979), *Nature*, **282,** 797.
129 Tu, C-P.D. and Cohen, S.N. (1980), *Cell*, **19,** 151.
130 Rothstein, S.J., Jorgensen, R.A., Postle, K. and Reznikoff, W. (1980), *Cell*, **19,** 795.
131 Ross, D.G., Swan, J. and Kleckner, N. (1979), *Cell*, **16,** 721.
132 Ross, D.G., Swan, J. and Kleckner, N. (1979), *Cell*, **16,** 733.
133 McClintock, B. (1956), *Cold Spring Harbor. Symp. Quant. Biol.*, **21,** 197.
134 Fincham, J.R.S. and Sastry, G.R.K. (1974), *Annu. Rev. Genet.*, **8,** 15.
135 Rasmusan, B., Green, M.M. and Karlson, B.M. (1974), *Mol. Gen. Genet.*, **133,** 237,
136 Finnegan, D.J., Rubin, G.M., Young, M.W. and Hogness, D.S. (1977), *Cold Spring Harbor Symp. Quant. Biol.*, **42,** 1053.
137 Potter, S.S., Borein, W.J., Dunsmuir, P. and Rubin, G.M. (1979), *Cell*, **17,** 415.
138 Stobel, E., Dunsmuir, P. and Rubin, G.M. (1979), *Cell*, **17,** 429.
139 Cameron, J.R., Loh, E.Y. and Davis, R.W. (1979), *Cell*, **16,** 739.
140 Hicks, J. and Strathern, J.N. (1977), in *DNA Insertion Elements, Plasmids and Episomes* (eds. A.I. Bukhari, J.A. Shapiro and S.L. Adhya), Cold Spring Harbor Laboratory, New York.
141 Nasmyth, K.A. and Tatchell, K. (1980), *Cell*, **19,** 753.
142 Williamson, A.R. (1976), *Annu. Rev. Biochem.*, **45,** 467.
143 Hozumi, N. and Tonegawa, S. (1976), *Proc. Natl. Acad. Sci. USA*, **73,** 3628.

144 Brack, C., Hirama, M., Lenhard-Schuller, R. and Tonegawa, S. (1978), *Cell,* **15,** 1.
145 Seidman, J.G., Leder, A., Nau, M., Norman, B. and Leder, P. (1978), *Science,* **202,** 11.
146 Seidman, J.G. and Leder, P. (1978), *Nature,* **276,** 790.
147 Weigert, M., Gatmaitan, L., Loh, E., Schilling, J. and Hood, L. (1978), *Nature,* **276,** 785.
148 Sakano, H., Hupi, K., Heinrich, G. and Tonegawa, S. (1979), *Nature,* **280,** 288.
149 Rabbits, T.H. and Forster, A. (1978), *Cell,* **13,** 319.
150 Matthyssens, G. and Tonegawa, S. (1978), *Nature,* **273,** 763.
151 Lenhard-Schuller, R., Hohn, B., Brack, C., Hirma, M. and Tonegawa, S. (1978), *Proc. Natl. Acad. Sci. USA,* **75,** 4709.
152 Gilmore-Hebert, M., Hercules, K., Komaromy, M. and Wall, R. (1978), *Proc. Natl. Acad. Sci. USA,* **75,** 6044.
153 Gilmore-Hebert, M., and Wall, R. (1979), *J. Mol. Biol.,* **135,** 879.
154 Honjo, T. and Kataoka, T. (1978), *Proc. Natl. Acad. Sci. USA,* **75,** 2140.
155 Rabbitts, T.H., Forster, A., Dunnick, W. and Bentley, D.L. (1980), *Nature,* **283,** 357.
156 Davis, M.M., Calmore, K., Early, P.W., Livant, D.L., Joho, R., Weissmann, I.L. and Hood, L. (1980), *Nature,* **283,** 733.
157 Valbuena, O., Marcu, K.B., Weigert, M. and Perry, R.P. (1978), *Nature,* **276,** 780.
158 Calmane, K., Rogers, J., Early, P., Davis, M., Livant, D., Wall, R. and Hood, L. (1980), *Nature,* **284,** 452.
159 Williams, R.O., Young, R.J. and Majiwa, P.A.O. (1979), *Nature,* **282,** 847.
160 Hoeijmakers, J.H.J., Frasch, A.C.C., Bernards, A., Borst, P. and Cross, G.A.M. (1980), *Nature,* **284,** 78.
161 Cory, S. and Adams, J.M. (1980), *Cell,* **19,** 37.
162 Chikaraishi, D.M., Deeb, S.S. and Sueoka, N. (1978), *Cell,* **13,** 111.
163 Affara, N.A., Jacquet, M., Buckingham, M.E., Robert, B. and Gros, F. (1978), in *Gene Function* Vol. 51, FEBS 12th Meet. Dresden, p. 233.
164 Groudine, M., Holtzer, H., Scherrer, K. and Terwath, A. (1974), *Cell,* **3,** 243.
165 Humphries, S., Windass, J. and Williamson, R. (1976), *Cell,* **7,** 267.
166 Storb, U., Hager, L., Wilson, R. and Putnam, D. (1977), *Biochemistry,* **16,** 5432.
167 Orkin, S.H. and Swetlow, P.S. (1977), *Proc. Natl. Acad. Sci. USA,* **74,** 2475.
168 Swaneck, G.E., Nordstrom, J.L., Kreuzaler, F., Tsai, M.J. and O'Malley, B.W. (1979), *Proc. Natl. Acad. Sci. USA,* **76,** 1049.
169 Tsai, S.Y., Roop, D.R., Tsai, M.J., Stein, J.P., Means, A.R. and O'Malley, B.W. (1978), *Biochemistry,* **17,** 5773.
170 Baker, H.J. and Shapiro, D.J. (1977), *J. Biol. Chem.,* **252,** 8428.
171 Baker, H.J. and Shapiro, D.J. (1978), *J. Biol. Chem.,* **253,** 4521.
172 Guyette, W.A., Matusik, R.J. and Rosen, J.M. (1979), *Cell,* **17,** 1013.
173 Cox, R.F. (1977), *Biochemistry,* **16,** 3433.
174 McKnight, G.S. and Palmiter, R.D. (1979), *J. Biol. Chem.,* **254,** 9050.

175 Minty, A.J., Birnie, G.D. and Paul, J. (1978), *Exp. Cell Res.*, **115,** 1.
176 Terwath, A. and Scherrer, K. (1978), *Proc. Natl. Acad. Sci. USA,* **58,** 2055.
177 Chan, L-N.C. (1976), *Nature,* **261,** 157.
178 Kleene, K.C. and Humphries, T. (1977), *Cell,* **12,** 143.
179 Kleene, K.C. and Humphries, T. (1977), *J. Cell Biol.*, **75,** 340.
180 Ryffel, G.U. and McCarthy, B.J. (1975), *Biochemistry,* **14,** 1379.
181 Young, B.D., Birnie, G.D. and Paul, J. (1976), *Biochemistry,* **15,** 2823.
182 Hastie, N.D. and Bishop, J.O. (1976), *Cell,* **9,** 761.
183 Sures, I., Levy, S. and Kedes, L.H. (1980), *Proc. Natl. Acad. Sci. USA,* **77,** 1265.
184 Busslinger, M., Portmann, R., Irminger, J.C. and Birnstiel, M.L. (1980), *Nucleic Acids Res.*, **8,** 957.
185 Grosschedl, R., Birnstiel, M.L. (1980), *Proc. Natl. Acad. Sci. USA,* **77,** 1432.
186 Simpson, R.B. (1980), *Nucleic Acids Res.*, **8,** 759.
187 Tchurikov, N.A., Zelentsova, E.S. and Georgiev, G.P. (1980), *Nucleic Acids Res.*, **8,** 1243.
188 Coleclough, C., Cooper, D. and Perry, R.P. (1980), *Proc. Natl. Acad. Sci. USA,* **77,** 1422.
189 Sands, M.K. and Roberts, R.B. (1952), *J. Bacteriol.*, **63,** 505.
190 Edlin, G. and Broda, P. (1968), *Bacteriol. Rev.*, **32,** 206.
191 Gallant, J.A. (1979), *Annu. Rev. Genet.*, **13,** 393.
192 Neidhart, F.C. (1966), *Bacteriol. Rev.*, **30,** 701.
193 Stent, G.S. and Brenner, S. (1961), *Proc. Natl. Acad. Sci. USA,* **47,** 2005.
194 Cashel, M. and Gallant, J. (1969), *Nature,* **221,** 838.
195 Cashel, M. and Gallant, J. (1974), *Ribosomes* (eds. M. Nomura, A. Tissières and P. Lengyel), Cold Spring Harbor Monograph Series, p. 733.
196 Pao, C.C., Dennis, P.P. and Gallant, J.A. (1980), *J. Biol. Chem.*, **255,** 1830.
197 Block, R. and Haseltine, W.A. (1974), *Ribosomes* (eds. M. Nomura, A. Tissières and P. Lengyel), Cold Spring Harbor Monograph Series, p. 747.
198 Stark, M.J.R. and Cundliffe, E. (1979), *Eur. J. Biochem.*, **102,** 101.
199 Richter, D. (1980), *Ribosomes: Structure, Function and Genetics* (eds. G. Chambliss, G.R. Craven, J. Davies, K. Davies, L. Kahan and M. Nomura), University Park Press, Baltimore, p. 743.
200 Travers, A. (1974), *Cell,* **3,** 97.
201 Travers, A. (1973), *Nature,* **244,** 15.
202 Leavitt, J.C., Hayashi, R.H. and Nakada, D. (1974), *Arch. Biochem. Biophys.*, **161,** 705.
203 Kung, H-F., Morrissey, J., Revel, M., Spears, C. and Weissbach, H. (1975), *J. Biol. Chem.*, **250,** 8780.
204 Blumenthal, T., Landers, T.A. and Weber, K. (1972), *Proc. Natl. Acad. Sci. USA,* **69,** 1313.
205 Pongs, O. and Ulbrich, N. (1976), *Proc. Natl. Acad. Sci. USA,* **73,** 3064.
206 Silverman, R.H. and Atherly, A.G. (1979), *Microbiol. Rev.*, **43,** 27.
207 Warner, J.R., Tushinski, R.J. and Wejksnora, P.J. (1980), *Ribosomes: Structure Function and Genetics* (eds. G. Chambliss, G.R. Craven, J. Davies, K. Davies, L. Kahan and M. Nomura), Univeristy Park Press, Baltimore, Baltimore, p. 889.

208 Yates, J.L., Arfsten, A.E. and Nomura, M. (1980), *Proc. Natl. Acad. Sci. USA,* **77,** 1837.
209 Lodish, H.F. (1976), *Annu. Rev. Biochem.,* **45,** 39.
210 Chartrenne, H. (1978), *Horm. Cell Reg.,* **2,** 1.
211 Revel, M. and Groner, Y. (1978), *Annu. Rev. Biochem.,* **47,** 1079.
212 Ochoa, S. and De Haro, C. (1979), *Annu. Rev. Biochem.,* **48,** 549.
213 Hindley, J. (1973), *Prog. Biophys. Mol. Biol.,* **26,** 269.
214 Zinder, N.D. (1975), *RNA Phages,* Cold Spring Harbor Monograph Series.
215 Nathans, D., Oeschger, M.P., Polmar, S.K. and Eggen, K. (1969), *J. Mol. Biol.,* **39,** 279.
216 Lodish, H.F. (1968), *Nature,* **220,** 345.
217 Fiers, W., Contreras, R., Duerinck, F., Hageman, G., Iserentant, D., Merregaert, J., Min Jou, W., Molemans, F., Raeymaekers, A., Van den Berghe, A., Volkaert, G. and Ysebaert, M. (1976), *Nature,* **500,** 260.
218 Jeppesen, P.G.N., Steitz, J.A., Gesteland, R.F. and Spahr, P.F. (1970), *Nature,* **226,** 230.
219 Ohtaka, Y. and Spiegelman, S. (1963), *Science,* **142,** 493.
220 Yoshida, M. and Rudland, P.S. (1972), *J. Mol. Biol.,* **68,** 465.
221 Groner, Y., Pollack, Y., Berissi, H. and Revel, M. (1972), *FEBS Lett.,* **21,** 223.
222 Lodish, H.F. (1970), *J. Mol. Biol.,* **50,** 689.
223 Steitz, J.A. (1973), *Proc. Natl. Acad. Sci. USA,* **70,** 2605.
224 Min Jou, W., Haegeman, G., Ysebaert, M. and Fiers, W. (1972), *Nature,* **237,** 82.
225 Robertson, H.D. and Lodish, H.F. (1970), *Proc. Natl. Acad. Sci. USA,* **67,** 710.
226 Sugiyama, T. and Nakada, D. (1968), *J. Mol. Biol.,* **31,** 431.
227 Bernardi, A. and Spahr, P.F. (1972), *Proc. Natl. Acad. Sci. USA,* **69,** 3033.
228 Inouye, H., Pollack, Y. and Petre, J. (1974), *Eur. J. Biochem.,* **45,** 109.
229 Groner, Y., Scheps, R., Kamen, E., Kolakofsky, D. and Revel, M. (1972), *Nature, New Biol.,* **239,** 19.
230 Kolakofsky, D. and Weissmann, C. (1971), *Nature, New Biol.,* **231,** 42.
231 Senear, A.W. and Steitz, J.A. (1976), *J. Biol. Chem.,* **251,** 1902.
232 Goelz, S. and Steitz, J.A. (1977), *J. Biol. Chem.,* **252,** 5177.
233 Weiner, A.M. and Weber, K. (1973), *J. Mol. Biol.,* **80,** 837.
234 Geller, A.I. and Rich, A. (1980), *Nature,* **283,** 41.
235 Rolleston, F.S. (1974), *Sub-Cell Biochem.,* **3,** 91.
236 Palade, G. (1975), *Science,* **189,** 347.
237 Sabatini, S.S. and Kreibich, G. (1976), in *The Enzymes of Biological Membranes,* Vol. 2, p. 531 (ed. A. Martonosi), Plenum, New York.
238 Blobel, G., Walter, P., Chang, C.N., Goldman, B.N., Erickson, A.H. and Lingappa, V.R. (1979), *Symp. Soc. Exp. Biol.,* **33,** 9.
239 Leader, D.P. (1979), *TIBS,* **4,** 205.
240 Davis, B.D. and Tai, P-C. (1980), *Nature,* **283,** 433.
241 Milstein, C., Brownlee, G.G., Harrison, T.M. and Mathews, M.B. (1972), *Nature, New Biol.,* **239,** 117.
242 Blobel, G. and Dobberstein, B. (1975), *J. Cell Biol.,* **67,** 835.

243 Lingappa, V.R., Lingappa, J.R. and Blobel, G. (1979), *Nature,* **281,** 117.
244 Lodish, H.F. and Rothman, J.E. (1979), *Sci. Amer.,* **240** (1), 38.
245 Blobel, G. (1980), *Proc. Natl. Acad. Sci. USA,* **77,** 1496.
246 Pollock, T.J. and Tessmann, I. (1976), *J. Mol. Biol.,* **108,** 651.
247 Lim, L.W. and Kennell, D. (1979), *J. Mol. Biol.,* **135,** 369.
248 Brawerman, G. (1973), *Mol. Biol. Reports,* **1,** 7.
249 Sheiness, D. and Darnell, J.E. (1973), *Nature, New Biol.,* **241,** 265.
250 Brawerman, G. (1976), *Prog. Nucleic Acid Res. Mol. Biol.,* **17,** 117.
251 Perry, R.P. and Kelley, D.E. (1973), *J. Mol. Biol..,* **79,** 681.
252 Marbaix, G., Huez, G., Soreq, H., Gallwitz, D., Weinberg, E., Devos, R., Hubert, E. and Cleuter, Y. (1978), *FEBS Symp.,* **51,** 427.
253 Bard, E., Efron, D., Marcus, A. and Perry, R.P. (1974), *Cell,* **1,** 101.
254 Blobel, G. (1973), *Proc. Natl. Acad. Sci. USA,* **70,** 924.
255 London, I.M., Clemens, M.J., Ranu, R.S., Levin, D.H., Cherbas, L.F. and Ernst, V. (1976), *Fed. Proc. Fed. Am. Soc. Exp. Biol.,* **35,** 2218.
256 Safer, B. and Anderson, W.F. (1978), *CRC Crit. Rev. Biochem.,* **5,** 261.
257 Hunt, T. (1980), *Molecular Aspects of Cellular Regulation,* Vol. 1, *Protein Phosphorylation in Regulation* (ed. P. Cohen), Elsevier/North Holland, Amsterdam, p. 175.
258 Mathews, M.B., Hunt, T. and Brayley, A. (1973), *Nature, New Biol.,* **243,** 230.
259 Legon, S., Jackson, R.J. and Hunt, T. (1973), *Nature, New Biol.,* **241,** 150.
260 Gross, M. and Rabinovitz, M. (1972), *Biochim. Biophys. Acta,* **287,** 340.
261 Farrell, P.J., Balkow, K., Hunt, T. and Jackson, R.J. (1977), *Cell,* **11,** 187.
262 Austin, S.A. and Clemens, M.J. (1980), *FEBS Lett.,* **110,** 1.
263 Amesz, H., Goumans, H., Haubrich-Moree, T., Voorma, H.O. and Benne, R. (1979), *Eur. J. Biochem.,* **98,** 513.
264 Delaunay, J., Ranu, R.S., Levin, D.H., Ernst, V. and London, I.M. (1977), *Proc. Natl. Acad. Sci. USA,* **74,** 2264.
265 Leader, D.P. (1980), *Molecular Aspects of Cellular Regulation,* Vol. 1, *Protein Phosphorylation in Regulation* (ed. P. Cohen), Elsevier/North Holland, Amsterdam, p. 203.
266 Metz, D.H. (1975), *Cell,* **6,** 429.
267 Kimchi, A., Zilberstein, A., Schmidt, A., Shulman, L. and Revel, M. (1979), *J. Biol. Chem.,* **254,** 9846.
268 Kerr, I.M. and Brown, R.E. (1978), *Proc. Natl. Acad. Sci. USA,* **75,** 256.
269 Slattery, E., Ghosh, N., Samanta, H. and Lengyel, P. (1979), *Proc. Natl. Acad. Sci. USA,* **76,** 4778.
270 Lodish, H.F. (1974), *Nature,* **251,** 385.
271 Yau, P.M.P., Godefroy-Colburn, T., Birge, C.H., Ramatshadran, T.V. and Thach, R.E. (1978), *J. Virol.,* **27,** 648.
272 Fernandez-Munoz, R. and Darnell, J.E. (1976), *J. Virol.,* **18,** 719.
273 Trachsel, H., Sonnberg, N., Shatkin, A., Rose, J.K., Lelong, K., Bergmann, J.E., Gordon, J. and Baltimore, D. (1980), *Proc. Natl. Acad. Sci. USA,* **77,** 770.
274 Golini, F., Thach, S.S., Birge, C.H., Safer, B., Merrick, W.C. and Thach, R.E. (1976), *Proc. Natl. Acad. Sci. USA,* **73,** 3040.

275 Carrasco, L. (1977), *FEBS Lett.*, **76**, 11.
276 Gette, W.R. and Heywood, S.M. (1979), *J. Biol. Chem.*, **254**, 9879.
277 Blair, G.E., Dahl, H.H.M., Truelsen, E. and Lelong, J.C. (1977), *Nature*, **265**, 651.
278 Hunt, R.T., Hunter, A.R. and Munro, A.J. (1968), *Nature*, **220**, 481.
279 Lodish, H.F. (1971), *J. Biol. Chem.*, **246**, 7131.
280 Phillips, J.A., Snyder, P.G. and Kazazian, H.H. (1977), *Nature*, **269**, 442.
281 Kabat, D. and Chappel, M.R. (1977), *J. Biol. Chem.*, **252**, 2684.
282 Lockard, R.E. (1977), *Nature*, **275**, 153.
283 Fan, H. and Penman, S. (1970), *J. Mol. Biol.*, **50**, 655.
284 Van Venrooij, W., Henshaw, E.C. and Hirsch, C.A. (1970), *J. Biol. Chem.*, **245**, 5947.
285 Hassell, J.A. and Engelhardt, D.L. (1973), *Biochim. Biophys. Acta*, **324**, 545.
286 Korner, A. (1964), *Biochem. J.*, **92**, 29.
287 Fahmy, L.H. and Leader, D.P. (1980), *Biochim. Biophys. Acta*, **608**, 344.
288 Sherwin, J.R. and Tong, W. (1976), *Biochim. Biophys. Acta*, **425**, 502.
289 Liang, T., Casteneda, E. and Liao, S. (1977), *J. Biochem.*, **252**, 5692.
290 Gross, P.R. (1967), *Curr. Topics Dev. Biol.*, **2**, 1.
291 Huang, F.L. and Warner, A.H. (1974), *Arch. Biochem. Biophys.*, **163**, 716.
292 Lee, S.Y., Krsmanovic, V. and Brawerman, G. (1971), *Biochemistry*, **10**, 895.
293 Rudland, P.S. (1974), *Proc. Natl. Acad. Sci. USA*, **71**, 750.
294 Civelli, O., Vincent, A., Buri, J-F. and Scherrer, K. (1976), *FEBS Lett.*, **72**, 71.
295 Perry, R.P. and Kelley, D.E. (1968), *J. Mol. Biol.*, **35**, 37.
296 Spirin, A.S. (1969), *Eur. J. Biochem.*, **10**, 20.
297 Van Venrooij, W.J., van Eekelen, C.A.G., Jansen, R.T.P. and Princen, J.M.G. (1977), *Nature*, **270**, 181.
298 Bester, A.J., Kennedy, D.S. and Heywood, S.M. (1975), *Proc. Natl. Acad. Sci. USA*, **72**, 1523.
299 Darnborough, C. and Ford, P.J. (1979), *Dev. Biol.*, **71**, 323.
300 Merlie, J.P., Buckingham, M.E. and Whalen, R.E. (1977), *Curr. Topics Dev. Biol.*, **11**, 61.
301 Model, P., Webster, R. and Zinder, N.D. (1979), *Cell*, **18**, 235.
302 Atkins, J.F., Steitz, J.A., Anderson, C.W. and Model, P. (1979), *Cell*, **18**, 247.
303 Beremand, M.N. and Blumenthal, T. (1979), *Cell*, **18**, 257.
304 Narang, F.E., Subrahmanyam, C.S. and Umbarger, H.E. (1980), *Proc. Natl. Acad. Sci. USA*, **77**, 1823.
305 Lawther, R.P. and Hatfield, G.W. (1980), *Proc. Natl. Acad. Sci. USA*, **77**, 1862.
306 Hentschel, C., Irminger, J-C, Bucher, P. and Birnstiel, M.L. (1980), *Nature*, **285**, 147.
307 Spadling, A.C. and Mahowald, A.P. (1980), *Proc. Natl. Acad. Sci. USA*, **77**, 1096.
308 Travers, A.A., Debenham, P.G. and Pongs, O. (1980), *Biochemistry*, **19**, 1651.

309 Spinelli, G., Melli, M., Arnold, E., Casano, C., Gianguzza, F. and Ciaccio, M. (1980), *J. Mol. Biol.,* **139,** 111.
310 Farabaugh, P.J. and Fink, G.R. (1980), *Nature,* **286,** 352.
311 Gafner, J. and Philippsen, P. (1980), *Nature,* **286,** 414.
312 Calos, M.P. and Miller, J.H. (1980), *Cell,* **20,** 579.
313 Potter, S., Truett, M., Phillips, M. and Mahler, A. (1980), *Cell,* **20,** 639.
314 Shimotohno, K., Mitzutani, S. and Temin, H.M. (1980), *Nature,* **285,** 550.
315 Dunnick, W., Rabbits, T.H. and Milstein, C. (1980), *Nature,* **286,** 669.
316 Sakano, H., Maki, R., Kurosawa, Y., Roeder, W. and Tonegawa, S. (1980), *Nature,* **286,** 676.
317 Vanin, E.F., Goldberg, G.I., Tucker, P.W. and Smithies, O. (1980), *Nature,* **286,** 222.
318 Nishioka, Y., Leder, A. and Leder, P. (1980), *Proc. Natl. Acad. Sci. USA,* **77,** 2806.
319 Lacy, E., Hardison, R.C., Quon, D. and Maniatis, T. (1979), *Cell,* **18,** 1273.
320 Fritsch, E.F., Lawn, R.M. and Maniatis, T. (1980), *Cell,* **19,** 959.
321 Lauer, J., Jam Shen, C-K. and Maniatis, T. (1980), *Cell,* **20,** 119.
322 Cordell, B., Bell, G., Tischer, E., DeNoto, F.M., Ullrich, A., Picet, R., Rutter, W.J. and Goodman, H.M. (1979), *Cell,* **18,** 533.
323 Bell, G.I., Picet, R.L., Rutter, W.J., Cordell, B., Tischer, E. and Goodman, H.M. (1980), *Nature,* **284,** 26.
324 Royal, A., Garapin, A., Cami, B., Perrin, F., Mandel, J.L., Le Muer, M., Bregegere, F., Gannon, F., LePennec, J.P., Chambon, P. and Kourilsky, P. (1979), *Nature,* **279,** 125.
325 Heilig, R., Perrin, F., Gannon, F., Mandel, J.L. and Chambon, P. (1980), *Cell,* **20,** 625.
326 Early, P., Rogers, J., Davis, M., Calame, K., Bond, M., Wall, R. and Hood, L. (1980), *Cell,* **20,** 313.
327 Alt, F.W., Bothwell, A.L.M., Knapp, M., Siden, E., Mather, E., Koshland, M. and Baltimore, D. (1980), *Cell,* **20,** 293.
328 Rogers, J., Early, P., Carter, C., Calame, K., Bond, M., Hood, L. and Wall, R. (1980), *Cell,* **20,** 303.
329 Seidman, J.G. and Leder, P. (1980), *Nature,* **286,** 779.
330 Chio, E., Kuehl, M. and Wall, R. (1980), *Nature,* **286,** 776.
331 Lee, A.S., Thomas, T.L., Lev, Z., Britten, R.J. and Davidson, E.H. (1980), *Proc. Natl. Acad. Sci. USA,* **77,** 3259.
332 Ogden, S., Haggerty, D., Stoner, C.M., Kolodrubetz, D. and Schleif, R. (1980), *Proc. Natl. Acad. Sci. USA,* **77,** 3346.
333 Sutcliffe, J.G., Shinnick, T.M., Verma, I.M. and Lerner, R.A. (1980), *Proc. Natl. Acad. Sci. USA,* **77,** 3302.
334 Tamm, I. (1979), *Proc. Natl. Acad. Sci. USA,* **76,** 5750.
335 Laub, O., Jakabovits, E.B. and Aloni, Y. (1980), *Proc. Natl. Acad. Sci. USA,* **77,** 3297.
336 Fraser, N.W., Sehgal, P.B. and Darnell, J.E. (1979), *Proc. Natl. Acad. Sci. USA,* **76,** 2751.
337 Yaoita, Y. and Honjo, T. (1980), *Nature,* **286,** 850.
338 Shoemaker, C., Goff, S., Gilbo, E., Paskind, M., Mitra, S.W. and Baltimore, D. (1980), *Proc. Natl. Acad. Sci. USA,* **77,** 3932.

339 Dhar, R., McClements, W.L., Enquist, L.W. and Vande Woude, G.F. (1980), *Proc. Natl. Acad. Sci. USA,* **77,** 3937.
340 Britten, R.J. and Davidson, E.G. (1969), *Science,* **165,** 349.
341 Davidson, E.H., Hough, B.R., Anderson, C.R. and Britten, R.J. (1973), *J. Mol. Biol.,* **77,** 1.
342 Coggins, L.W., Grindley, G.J., Vass, J.K., Slater, A.A., Montague, P., Stinson, M.A. and Paul, J. (1980), *Nucleic Acids Res.,* **8,** 3319.
343 Davidson, E.G. and Britten, R.J. (1979), *Science,* **204,** 1052.
344 Lee, A.S., Thomas, T.L., Lev, Z., Britten, R.J. and Davidson, E.H. (1980), *Proc. Natl. Acad. Sci. USA,* **77,** 3259.
345 Young, R.A., Hagenbuchle, O. and Schibler, U. (1981), *Cell,* **23,** 451.

Genetic engineering: recombinant DNA technology

14

14.1 GENERAL PRINCIPLES OF GENE CLONING

There are many eukaryotic gene products which would be of considerable use in the fields of medical and agricultural research and technology if they were available in a pure form in sufficient quantities. The most logical way of mass producing these would be to insert the appropriate gene into the bacterial genome so that the high-yielding systems of micro-organisms can be fully exploited. The basic principles of genetic engineering are all based on this biological potential of bacteria. However, the DNA must be introduced into the bacterium in a *vector* which ensures continuance of the ability to express the gene product for a large number of generations. It must also be ensured that the gene is *expressed* in the host bacterium and because of the enormously greater complexity of the animal genome, it is necessary that sophisticated detection techniques are available for analysis of this gene expression. To obtain a fully functional gene product it is necessary to ensure that the intervening sequences in the DNA or transcribed RNA are removed at some stage in the process and also that post-translational modifications such as glycosylation or the removal of peptide sequences can be correctly applied. The potential of this methodology is very considerable and the general principles have been covered in recent reviews [1, 2] and two detailed and wide ranging books [3, 4].

14.2 THE PRODUCTION OF SUITABLE EUKARYOTE DNA SEQUENCES FOR GENE CLONING

14.2.1 The cloning of cDNA

By far the most convenient method of cloning a specific gene is to utilize the ability of reverse transcriptase to copy the mRNA (Chapter 5) and thus form a double-stranded cDNA fragment which may be inserted into the bacterial DNA for cloning [5–7]. However, this technique is dependent on the availability of a purified mRNA (Chapter 5) coding for the protein and therefore restricted to those mRNA molecules which can be isolated in a pure or partially pure form. It is alternatively possible to clone a mixture of cDNA molecules derived from an impure population of mRNA molecules and then select for bacterial populations which produce the required protein. The production of duplex cDNA is shown in Fig. 14.1. The mRNA is copied by reverse transcriptase, usually with

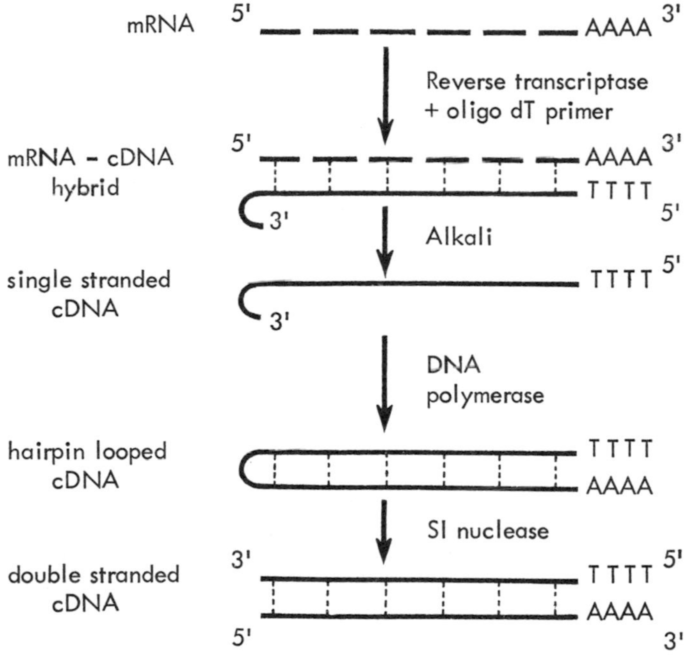

Fig. 14.1 Synthesis of double-stranded cDNA from an mRNA template.

the aid of an oligo (dT) primer and then hydrolysed by alkali (Chapter 5). The single-stranded cDNA formed has the convenient ability to form a transient hairpin which then acts as a primer for DNA polymerase which converts it into the double-stranded form. The advantage of using the cDNA in gene cloning is that tedious and lengthy selection procedures can be minimized. More importantly, it removes the problem of the possible presence of intervening sequences which occur in genomic DNA and which will yield an RNA transcript which cannot be correctly processed by the bacterial host.

14.2.2 The fractionation of genomic DNA

Genomic DNA is generally fractionated by the use of restriction endonucleases (Chapter 6). Clearly, since there may be a restriction endonuclease site in the middle of the required gene, a variety of nucleases can be used for this task so that the likelihood of cutting within the gene is minimized. There are several disadvantages to this method. Firstly, any gene which has intervening sequences will not be correctly expressed in the subsequent cloning procedures. In addition, the screening operation involved is formidable if the cloning procedure is to detect a gene with a small number of copies. If restriction fragments were to average 5000 base pairs in length then of the order of a million different fragments would be expected to be generated from a mammalian genome, only one of which would possess the required gene. This procedure is known as *shotgun cloning*.

DNA may also be fractionated by mechanical shearing [8] but this procedure is less frequently employed.

14.2.3 The chemical synthesis of a gene

Where the gene product is reasonably small, as is the case with some of the genes which code for small polypeptides or tRNA molecules, it is possible to synthesize the gene by chemical techniques which are used to make small oligonucleotides. These are then ligated to form the final gene [9].

14.3 THE SELECTION OF VEHICLES FOR CLONING IN PROKARYOTES

14.3.1 Plasmids

The general biological properties of plasmids have been described in Chapter 4. Small plasmids which carry markers for drug resistance are undoubtedly the most favoured molecules for use as cloning vehicles at this point in time. Ideally, the plasmid must carry a series of target sites for the more commonly used restriction nucleases in the gene which codes for drug resistance. It is then possible to select the host cells into which the plasmids carrying the foreign DNA have been incorporated by virtue of their loss of resistance to the appropriate drug. Thus the best known of the cloning plasmids pBR322 is a small molecule of about four kilobase pairs which carries genes for both ampicillin resistance and tetracycline resistance (Fig. 14.2). The formed gene has the unique Pst I and Pvu I sites in the molecule and the latter unique sites for Hind III,

Bam HI and Sal I. Thus, if the foreign DNA is inserted into the Pst I site, recombinant molecules can be selected by replica plating for their resistance to tetracycline and sensitivity to ampicillin and conversely, if one of the other three sites is used for insertion, the recombinant progeny will be ampicillin resistant and tetracycline sensitive. One of the disadvantages of pBR322 as a cloning vehicle is that its single EcoRI site does not lie in either of the two drug-resistant genes and recombinants created using this widely used restriction nuclease are not readily selected. To circumvent this problem, the plasmid pBR325 has

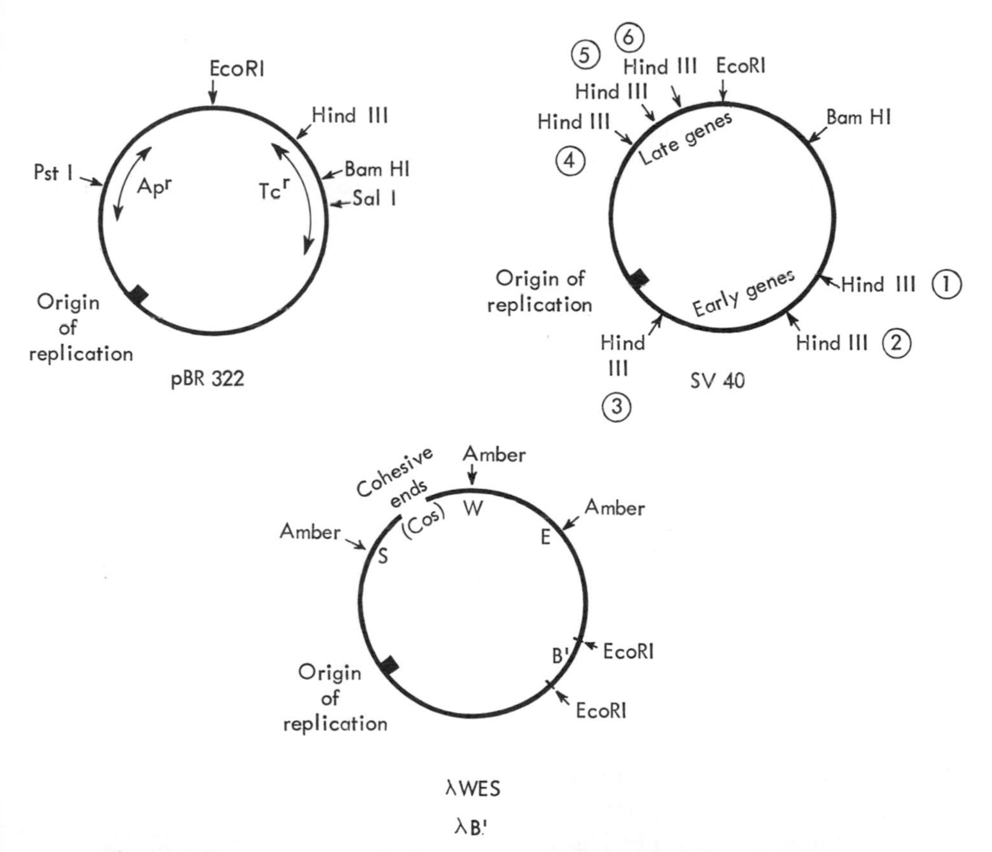

Fig. 14.2 Some common cloning vectors together with their most relevant restriction nuclease sites. In λ WES λ B′, the W, E and S refer to mapped genetic loci whereas the B′ refers to a restriction nuclease fragment which is the product of the reaction of EcoRI with wild-type λ. In SV40, the DNA fragment SVGT5 comprises the larger of the two pieces of DNA obtained by digesting at the Bam HI site and the Hind III 5 site and lacks the distal portion of the late genes which are not necessary for expression of the foreign DNA.

been created. This molecule carries an EcoRI site in a chloramphenicol-resistance gene and recombinants can be selected by their sensitivity to chloramphenicol and resistance to both ampicillin and tetracycline. More recently pBR328, a derivative of pBR325 with unique Pvu II and Bal I sites in the chloramphenicol resistance gene as well as an EcoRI site has been described [11]. Replica plating can be avoided by use of the plasmid pUR2 which has an EcoRI site in the centre of the Z gene of the *lac* operon. Bacteria with inserted sequences are consequently readily identified [52].

Table 14.1 Characteristics of some common plasmids.

	Size (kb)	Marker genes	Single restriction enzyme sites
pMB9	5.2	Tetracycline resistance Immune to Col EI	In T_cR gene Bam HI. Hind III Sal I External EcoRI
pBR322	4.0	Tetracycline resistance Ampicillin resistance	In T_cR gene Bam HI, Hind III, Sal I In Ap^R gene Pst I, Pvu I External EcoRI
pBR325	5.2	Tetracycline resistance Ampicillin resistance Chloramphenicol resistance	In T_cR gene Bam HI, Hind III, Sal I In Ap^R gene Pst I, Pvu I In Cm^R gene EcoRI
BR328	2.9	Tetracycline resistance Ampicillin resistance Chloramphenicol resistance	In T_cR gene Bam HI, Hind III, Sal I In Ap^R gene Pst I, Pvu I In Cm^R gene EcoRI, Pvu II, Bal I
pUR 2	2.7	Ampicillin resistance *Lac* operator and Z gene	In Ap^R gene, Pst I, Pvu I In *lac* genes EcoRI

Another vital feature of plasmids, such as pBR322 and those derived from it, is their ability to replicate under relaxed control, independently of the DNA of the main chromosome, so that many copies may accumulate in each cell. This is a characteristic of the group of plasmids which carry the genes for the colicin antibiotics so that molecules which have been selected to retain this locus are able to replicate extensively in each host cell. The replication relative to the chromosomal DNA is further amplified by the use of chloramphenicol which does not inhibit the synthesis of plasmid DNA so that several thousand copies of each plasmid DNA molecule can be produced in each cell. The production of the pBR322 series of plasmids has recently been reviewed [10].

Plasmid DNA molecules are readily separated from the main bacterial chromosome by buoyant density centrifugation and can be reintroduced into host cells which have first been rendered permeable to DNA by washing in calcium chloride (see p. 62). The properties of the more common plasmids are shown in Table 14.1.

14.3.2 Bacteriophages

The bacteriophage most widely used in genetic engineering is λ bacteriophage. Since the wild-type bacteriophage contains several sites for most commonly employed restriction nucleases in its 47 kilobase pairs, derivative molecules which have only one or two sites have been constructed for this purpose. In addition, amber mutations have been introduced in several sites so that the recombinant bacteriophage is unlikely to grow in any host other than the amber suppressor strain supplied in the laboratory. Thus Charon bacteriophages such as Charon 16A [12] are defective as a result of mutations in the A and B genes and a single EcoRI site in the transduced gene for β-galactosidase. Recombinants can then be selected for their inability to produce colour from a chromogenic substrate of β-galactosidase when grown on a *gal*$^-$ *E. coli* host. In addition, this vector has deletions in the C and E genes (the latter being known as the *nin* deletion) so that the molecule is about 17 per cent shorter than the wild type making the final recombinant molecule able to accommodate up to 10 kilobase pairs of foreign DNA and remain within the size range of the wild-type DNA which is the most suitable for efficient plaque formation. The most commonly used λ bacteriophage for recombinant DNA technology is λgt WESλB′ [13, 14] (Fig. 14.2) which has two EcoRI sites flanking an inverted B gene. It also has the deletions of the C gene and *nin* portion of the E gene and amber mutations in the W, E and S genes. The B gene can be replaced by foreign DNA making this bacteriophage one with replacement rather than insertion vectors. Selection is by virtue of the inability of bacteriophage DNA molecules which lack the foreign DNA to form plaques since the deletion of the B gene together with the C and *nin* deletions yield a molecule too small for viable plaque formation. The foreign DNA molecules with an EcoRI site at either end are then readily inserted into the site previously occupied by the B gene. This vector can accommodate up to 15 kb of DNA. Although the recombinant DNA can be introduced into the bacterial host by itself, a much more effective mechanism is to prepackage the DNA using lysates of mutant bacteriophages which accumulate the appropriate coat proteins in large amounts [15].

While the great majority of bacteriophage recombinant DNA experiments have been performed in λ, the small filamentous bacteriophage M13 has been used more recently [16, 17]. The bacteriophage DNA is single-stranded which means that many of the complications which arise in the sequencing of double-stranded DNA molecules can be circumvented if the DNA to be sequenced is introduced into the double-stranded replicative form of the bacteriophage DNA and then isolated as the single-stranded form from the progeny bacteriophage. Another advantage of M13 is that it does not lyse the host bacterium, but bacteriophage particles are released into the medium in large numbers and are therefore readily isolated.

14.3.3 Cosmids

While it can be seen that λ bacteriophage will accommodate substantial pieces of DNA, there is frequently a requirement for a vector which will incorporate pieces of genomic DNA greater than 15 kilobases. Such vectors are particularly useful in the study of large pieces of eukaryotic chromosomal DNA which encompass a single gene with several large intervening sequences such as the ovalbumin gene, or in the investigation of the regions flanking eukaryote genes. In addition they have a potential role in the construction of simple 'libraries' of the DNA from any particular species of eukaryote. A library is a collection of vectors which contain among them the entire genome of the eukaryote and clearly the larger the pieces which can be incorporated into the vector the smaller the number of components of the library required for screening purposes. Cosmids essentially incorporate some of the features of plasmids and some of λ bacteriophage. They contain the cos sites of λ DNA which encompass the areas immediately around the cohesive termini, known to be essential for correct packaging of the DNA, a drug-resistant gene for selection of recombinant clones and a single restriction endonuclease site in a circular molecule. Since these requirements can be fulfilled by a piece of DNA which is only about 10 per cent of the size of the wild-type DNA molecule, the remaining piece of DNA can be supplied by the foreign material with molecules in the size region of 40 kilobases being readily incorporated. The resultant recombinant molecule is then subjected to the normal λ-packaging system described in 14.3.2 and incorporated into the bacteria, where it replicates as a plasmid, since it lacks the majority of the viral functions [18, 19].

14.3.4 Cloning in eukaryotes

Eukaryotic cells have been investigated as cloning vehicles because of their potential utility in systems which involve the expression of complex eukaryote DNA and may be more readily controlled by eukaryote than by prokaryote cells. Thus one might have anticipated from Chapter 11 that yeast cells would, for example, contain enzymes which could process out the intervening sequences in primary transcripts from genomic DNA with intervening sequences thus permitting the use of genomic DNA rather than cDNA for protein production. However, recent data [51] indicate that although primary transcripts of specific mammalian genomic DNA can be detected within yeast these are not processed to yield mRNAs. Possibly yeast does not have a splicing system capable of coping with mRNA precursors from the genes of higher eukaryotes. Alternatively the transcription products arising from the foreign DNA and the yeast processing enzymes cannot come into contact with one another within the yeast cell for some physiological reason. Also likely is that a complete or near complete transcript may be required for processing (sec also Chapter 11), a particular secondary or tertiary structure being necessary. The specific primary transcripts of mammalian genes so far detected in yeast (e.g. for mouse B-globin [51]) certainly lack both their correct 5′- and 3′-terminal regions, presumably as a result of incorrect initiation and termination by the yeast RNA polymerases. It would, therefore, appear that at least in this respect, yeast has no advantages as a cloning vehicle over bacteria. However, the use of yeast or other eukaryote promoters (and terminators) may well assist the expression of foreign eukaryote DNA [20].

The interest in yeast cells lies partly in their relatively low DNA content and also the presence, in addition to their normal chromosomal DNA complement, of a circular plasmid molecule known as the 2μ plasmid. Eukaryote DNA introduced into this plasmid shows a higher transformation frequency of the cells [21, 22] and allows easy recovery of the cloned plasmid borne in foreign DNA. The plasmid is introduced into cells by incubation with yeast spheroplasts followed by regeneration of the cell walls.

The main vehicle for mammalian cloning experiments is SV40 DNA which has about 5.2 kb of DNA. Functionally this molecule is divided into two separate parts, the 'early' and 'late' regions of DNA (Fig. 14.2). The 'early' genes are required for viral DNA synthesis and the 'late' ones for synthesis of the viral coat proteins. Vectors are therefore selected to retain the early function and those portions of DNA coding for the initiation and processing of the late gene products, the viral coat proteins but to have lost the main body of the late genes which serves a purely structural coding purpose. Thus approximately 20 per cent of the genome can be deleted and can be used to accommodate foreign DNA. The recombinant viral molecule SVGT5 has all these properties and has been shown to act as a viable cloning system for the synthesis of the protein β-globin [24].

The DNA of the hybrid molecule is introduced into the cells by the use of a helper virus (SV40 tsA which has temperature sensitive gene A-protein) which lacks the ability to initiate the synthesis of viral DNA temperatures above 40°C [25]. Transformation can either be with virus or with DNA alone. Plaques are then only formed by cells which have obtained genes from both types of DNA molecule since the vector lacks the viral proteins and the helper virus the ability to perform early functions.

While eukaryote viral systems offer many advantages in the ease with which the DNA can be isolated and the presence of the appropriate initiation and processing functions, it would clearly be better if it were possible to transform cells with such viruses and establish cell lines which could synthesize the required protein in a continuous manner without lysis. This has not yet been achieved.

14.3.5 Genetic transformation of mammalian cells

The ability to introduce foreign DNA or parts or whole eukaryote or viral chromosomes into mammalian cells [53, 54] has played an important role in somatic cell genetics and in the use of mammalian cells as cloning vehicles. Thus transfection (infection of cells with naked DNA) of cells with the SV40 genome or with *in vitro* engineered deletion mutants is playing an important part in determining the region of the genome coding for the cell-transformation capability or required for initiation of transcription of late viral mRNA [55, 56]. As with transformation in prokaryotes it is important to have a marker in order to select for the transformed cells and it has been shown to be possible to cotransform cells in culture with two types of DNA, the first carrying a required gene function and acting as a marker and the second the DNA sequence of interest. The process has most frequently been demonstrated by the use of cells deficient in the thymidine kinase gene. This is only essential when the cells are grown in HAT medium (containing hypoxanthine, aminopterin and thymidine) when the normal pathway of thymidine metabolism is blocked. Cells incubated with DNA which contains a thymidine kinase gene can then be transferred to HAT medium and only those which have acquired the external DNA will grow. It is believed that only a small population of the cells in culture are competent for transformation at any point in time and this technique selects for the appropriate cells. If large amounts of a second type of DNA, containing the genc of interest are also incubated with the cells, then the competent sub-population will take this DNA up at the same time and incorporate it into the genome. Thus it is in theory possible to insert any gene into a population of mammalian cells albeit at a much lower transformation frequency than in bacterial cells [23].

The implication of this general method of eukaryote transformation is clear. It has been shown that processing enzymes are not species specific (see Chapter 11) and thus, it may be anticipated that genomic DNA, which contains inserted sequences, can be inserted into these cells since they will have the potential to process out the intervening sequences from the primary transcription products.

14.3.6 Cloning in plant cells

The prospect of introducing new groups of genes, such as those coding for the enzymes responsible for nitrogen fixation, into plants has aroused great commercial interest. There are two possible vectors at the present time, the plant viruses [26] and tumour-inducing bacteria [27–30]. The former are as yet poor candidates as vectors as they appear never to transform the cells which they infect and the prospect of breeding plants with the new genes is therefore slender as no new DNA is inserted into host chromosomes. The crown-gall bacterium, *Agrobacterium tumefaciens,* which induces tumours in a wide variety of plants by transferring to the cells a large plasmid DNA molecule, offers much greater potential for use as a vector. However, the tumour inducing (Ti) plasmids are difficult to handle because of their comparatively great size (150–200 kb) so that experiments with this molecule are at an early stage of development [30].

14.4 THE INSERTION OF FOREIGN DNA INTO THE CLONING VEHICLE

14.4.1 The use of restriction endonucleases

One of the simplest methods of joining the two different types of DNA molecule is to use a restriction nuclease of the same specificity to produce cohesive ends in both vector and foreign DNA. If the vector is a plasmid this is achieved by using a plasmid which has either a single site for the restriction nuclease, or two sites with an intervening region of DNA which is not required and which can be separated by electrophoresis. When the two types of molecule are mixed and the temperature is reduced sufficiently to favour base pairing of the cohesive ends, a variety of reactions may occur. The plasmids may recircularize or form plasmid dimers, the former reaction being particularly favoured since the ends are physically close to each other in any case. This reaction can be suppressed by the use of alkaline phosphates which removes the 5′-phosphate termini necessary for the joining reaction. The foreign DNA may also reanneal with itself to form linear oligomers. The probability of this can be altered by manipulating the concentration of the DNA. The desired reaction is the annealing of a single piece of the foreign DNA with the DNA of the cloning vehicle. The annealing is then made permanent by use of ligase from either *E. coli* itself [31] or bacteriophage T4-infected *E. coli* [32]. Both require the presence of a 3′-hydroxyl group and a 5′-phosphate group, but the former enzyme uses NAD as cofactor while the latter uses ATP (see Chapter 00). If the phosphate groups have been removed from the vector DNA by alkaline phosphatase treatment the join will still be sealed on one of the two chains of the duplex at each annealing position by virtue of the presence of a 5′-phosphate on the foreign DNA molecule and the final repair can take place once the DNA has been inserted into the host bacterium.

14.4.2 Homopolymer tailing

The homopolymer tailing method is extensively employed in the cloning of cDNA which does not normally have naturally cohesive termini corresponding to any known restriction endonuclease. The enzyme terminal transferase (terminal deoxynucleotidyl transferase from calf thymus [33]) will extend DNA molecules from the 3′-hydroxyl end by the sequential addition of single nucleotide residues derived from a deoxyribonucleotidyl triphosphate substrate (see p. 227). Thus, provided with dATP it will extend, say the foreign DNA molecule with deoxyadenosine residues, and if provided with dTTP it will extend the vector DNA molecule with deoxythymidine residues. When the two populations of molecules are mixed, annealing between the newly acquired A and T residues can take place and the product can be ligated as described in Section 14.4.1. This method has the advantage that self annealing is not a problem (Fig. 14.3) [34, 35]. The single-strand gaps in the new plasmid are sealed by host repair mechanisms.

14.4.3 Blunt-end ligation

DNA ligase from T4-infected *E. coli* is capable of joining two molecules which have not previously self annealed in solution [32]. It has, therefore, the potential of performing the joining reaction between, say a vector molecule and a cDNA molecule which have not previously been treated with the same restriction nuclease, the only requirement being the presence of free double-stranded ends with a 5′-terminal phosphate. However, the efficiency of such a reaction must necessarily be very low

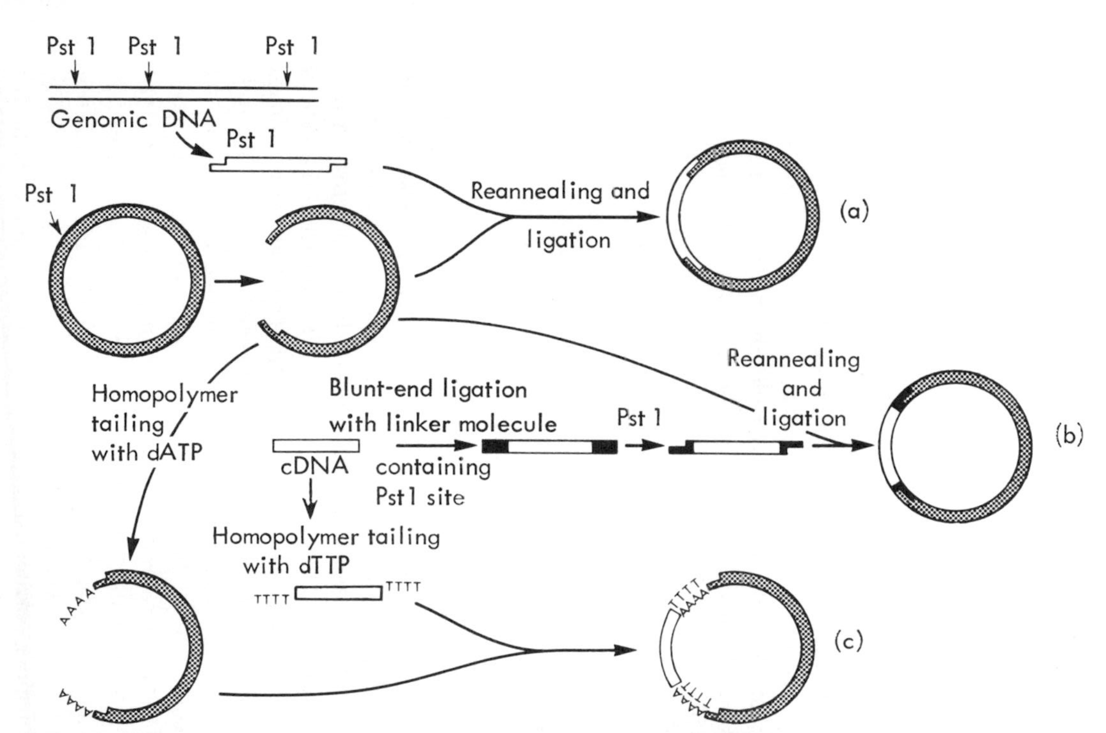

Fig. 14.3 Insertion of foreign DNA into the same site on a plasmid by: (a) direct insertion of genomic DNA; (b) insertion of cDNA using blunt-end ligation and linker molecules; (c) homopolymer tailing. The specificity of Pst I which is:

—CTGCAG—	$\rightarrow$	—CTGCA	G—
—GACGTC—		—G	ACGTC—

gives exposed 3′ ends well suited to homopolymer tailing with terminal transferase.

since it depends on chance, rather than affinity, placing the two ends in the correct juxtaposition. A much commoner use of the blunt end ligating ability of this enzyme is to attach small 'linker' molecules which contain a restriction enzyme cleavage site to either end of the foreign DNA molecule, which is subsequently cleaved with the appropriate restriction endonuclease to produce the cohesive ends required for ordinary ligation [36]. The vector is also treated with the restriction nuclease as described in Section 14.4.1 (Fig. 14.3). Both homopolymer tailing and linker-aided blunt-end ligation have the obvious advantage that the experimenter is not dependent on the fortuitous presence of a suitable restriction endonuclease site which matches one in his vector at either end of the pieces of DNA which he wishes to clone, but rather can select a DNA molecule with the correct properties and insert it into the vector at any one of a number of restriction endonuclease sites by selecting the appropriate linker molecule or by ensuring that the experimental conditions selected for the homopolymer tailing are compatible with type of end produced by the restriction nuclease in the cloning vehicle.

14.5 THE EXPRESSION OF FOREIGN DNA

Although it is possible to insert plasmid DNA into a wide variety of cloning vehicles, the ultimate aim of some experimental procedures in genetic engineering is the actual production of the foreign protein coded for by the DNA. Obvious difficulties occur when eukaryote genomic DNA has been cloned in a plasmid or bacteriophage since the host will lack the processing enzymes required to make the correct mRNA from any high-molecular-weight RNA which may be transcribed from the cloned DNA. However, in addition the correct promoter or terminator region may not be present at the start of the gene and may have to be inserted by further genetic manipulation. Furthermore, the gene may not be in the correct frame for the reading of the genetic code. It may be noted that all the techniques described in Section 4.4. result in two types of inserted DNA molecules, normal and inverted, so that at least 50% of the clones may be expected to have the DNA inserted in the wrong orientation for correct expression.

In order to achieve correct expression it is therefore necessary to have a bacterial promoter at the start of the foreign gene and also preferably to use a technique which produces a series of plasmids differing within a range of a few base pairs in the distance between this promoter and the gene. This should result in the production of some plasmids in the correct reading frame. The primary bacterial promoter used is that of the *lac* operon and the DNA sequence known as Eco R1 UV5 [37] which codes for the regulatory region of the *lac* operon has been widely used in this respect. This can be inserted into either λ [38] or plasmid [39] cloning vehicles. The distance between the promoter and the gene can be adjusted by the pretreatment of either promoter or foreign DNA with homopolymer tailing (which produces cohesive ends of varying numbers of base pairs) or blunt-end ligation and restriction endonuclease cleavage which will add a determined number of base pairs to the ends of the fragment to be incorporated. Alternatively the coding DNA can be inserted in the middle of a known structural gene at the junction between two codons, as in the case of the Pst I site in pBR322 which lies in between two codons of the β-lactamase gene which codes for ampicillin resistance [40], in either case the protein product has attached to it the relevant structural protein under the control of the bacterial promoter in all or in part. Production of the bacterial protein can be minimized by gentle exonuclease treatment of the bacterial DNA prior to annealing with the foreign DNA, so that the appropriate translational signals for the bacterial protein are attenuated without any effect on the translation of the foreign DNA [41].

14.6 THE DETECTION OF VECTORS CARRYING FOREIGN DNA

14.6.1 Selecton by means for a vector marker

Nearly all genetic-engineering experiments incorporate a preliminary selection method involving the loss of a specific gene function such as drug resistance, plaque-forming capacity or synthesis of a specific enzyme such as β-galactosidase. This type of selection method which has already been described in each Section 14.3 results in an enrichment of those clones which have undergone some mutation in the chosen site in the vector. However, further techniques are required to determine the extent to which foreign DNA has been inserted in this site.

14.6.2 Selection for the presence of foreign DNA

In cases in which expression of the foreign DNA is thought unlikely to lead to a viable gene product such as is the case with the cloning of DNA with inserted sequences present, the only method for selection of clones containing the recombinant DNA is by nucleic acid hybridization using a mRNA or a cDNA probe. Such experiments are generally performed by fixation of the DNA from a replica plated colony on a nitrocellulose disc followed by denaturation and hybridization with the probe and autoradiography and comparison with the plate containing the original bacterial clone [42–45] (Fig. 14.4).

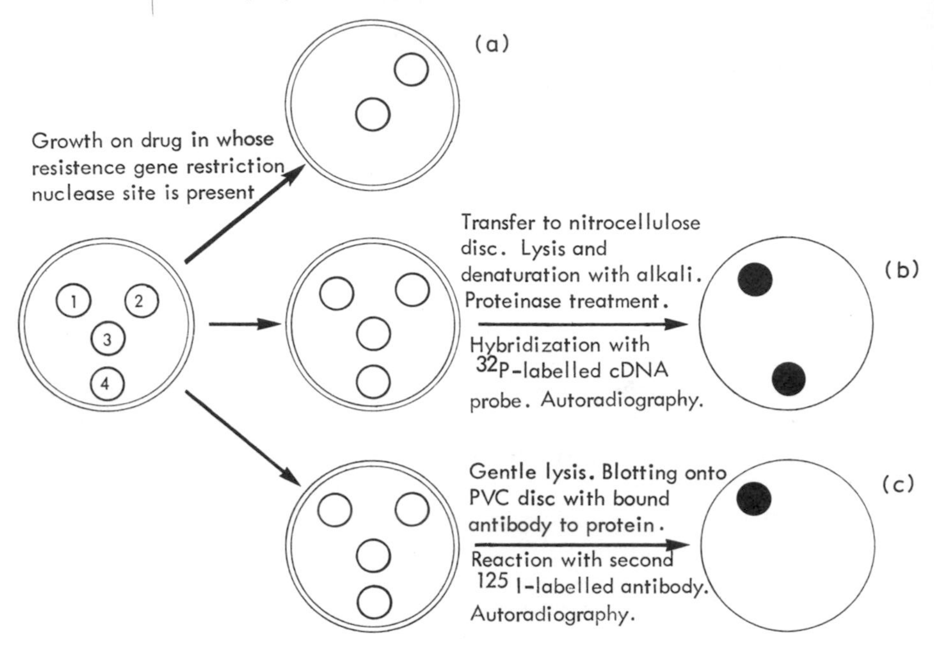

Fig. 14.4 Detection of clones containing required DNA sequences and also making required protein. From (a) it is possible to tell which clones have incorporated DNA into the restriction nuclease cleavage site in the drug resistance gene. From (b) it is possible to tell which clones have incorporated at least part of the DNA complementary to the probe. From (c) it is possible to tell which clones are expressing the incorporated DNA and making the required protein. Only Clone 1 makes the protein in this case though both clones 1 and 4 have incorporated the foreign DNA.

14.6.3 Selection for the presence of foreign protein

In a small number of cases, it is possible to use as host a bacterial auxotroph which requires the presence of the appropriate eukaryote gene function for growth. Viable clones can then be selected and those containing the eukaryote gene separated from spontaneous revertants [46, 47]. However, in the vast number of cases it is necessary to select for the presence of the actual protein coded by the eukaryote gene. This is generally achieved by immunochemical methods which involve a series of 'sandwich' detection methods. The clones are lysed gently so as to release antigen without denaturation and blotted against an antibody to the required protein, which is immobilized on a solid-phase sheet or disc of plastic of polyvinylchloride. The sheet is washed to eliminate non-specific binding and then reacted with more antibody directed against the protein, this time radioactively labelled. After further washing the sheet can be used for autoradiography and the positions of those clones

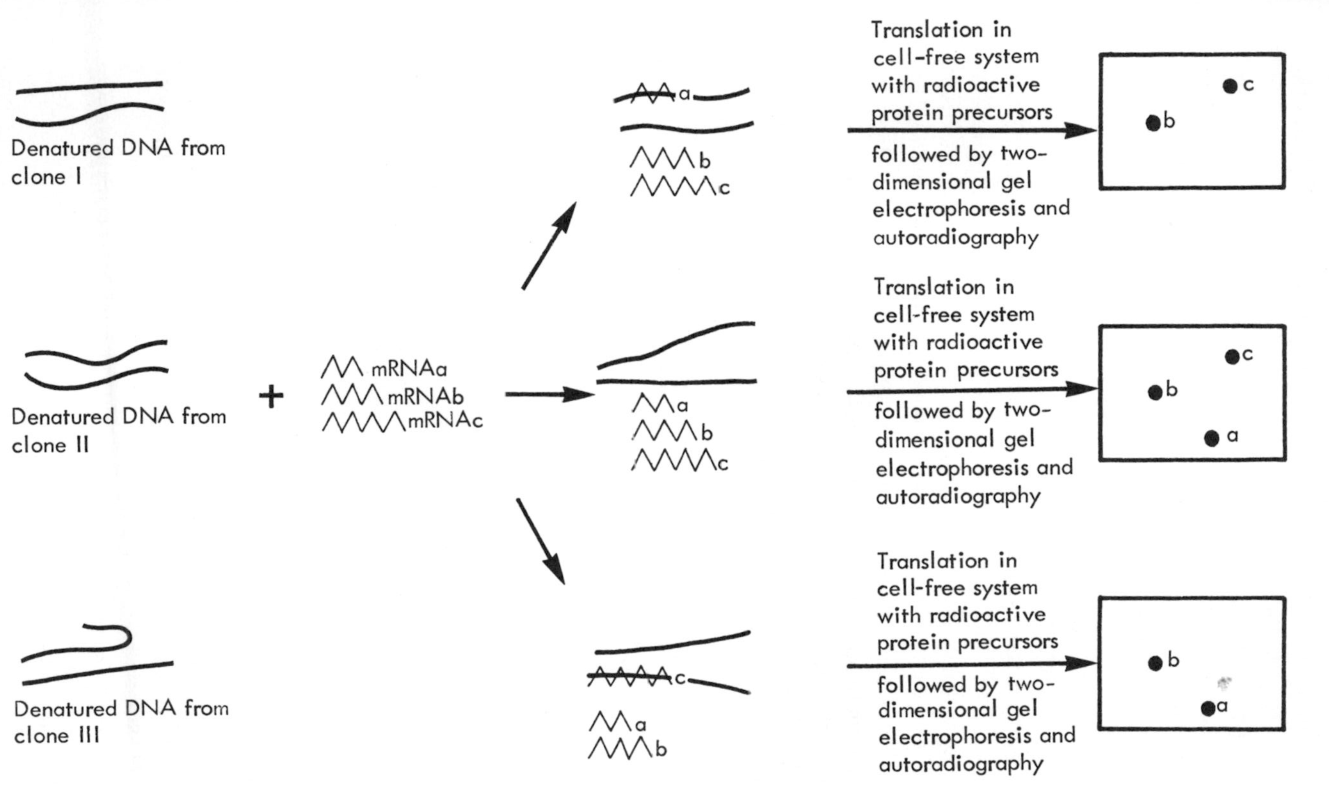

Fig. 14.5 Hybrid Arrested Translation (HART) to detect required protein product from a population of mixed cloned cDNA molecules. Clone I carried the gene coding for protein (a) and therefore it is not synthesized. Clone II carries the gene for none of the three proteins. Clone II carries the gene for protein (c). Immunochemical detection by use of antibodies to the specific protein as shown in Fig. 14.4 can also be employed to improve sensitivity at later stages.

yielding immunochemically active protein compared with the clones on the replica plate [40, 48, 49].

A second method which relies on the detection of the foreign protein essentially depends on the final desired product being absent rather than present in the mixture of polypeptides analysed. In this method, known as *Hybrid Arrested Translation* (Fig. 14.5) [50], the DNA from the plasmid containing the foreign sequence is hybridized to the mixture of mRNA molecules from which the original population of cDNA molecules was produced for cloning. If the cloned DNA contains the desired messenger sequence then it will bind to the appropriate mRNA molecule which will therefore not be available to act as a template in a cell-free translational system. The hybridized nucleic acids are then used to direct protein synthesis and the newly made products are separated by two-dimensional polyacrylamide-gel electrophoresis or identified by immunochemical techniques. All the mRNA molecules which have not hybridized to the cloned DNA direct synthesis of their specific proteins. Any protein missing in one of a series of experiments with several clones will be one which has its sequence encoded in the DNA of that clone. Since genomic eukaryote DNA is unsuitable for cloning in prokaryotes because of its inserted sequences and since in very few cases it is possible to obtain a pure population of mRNA molecules for cDNA synthesis, this method is particularly useful when a population of mRNA molecules requires to be cloned and screened for its gene products.

14.7 GUIDLINES FOR RESEARCH INVOLVING RECOMBINANT DNA MOLECULES

In the majority of countries where experiments involving recombinant DNA are carried out, guidelines for the conduct of such experiments are available in appropriate government publications [57–62].

REFERENCES

1 Gilbert, W. and Villa Komaroff, L. (1980), *Sci. Amer.*, **242,** 74.
2 Boyer, H.W. and Nicosia, S. (eds.) (1978), *Genetic Engineering,* Elsevier, North Holland.
3 Wu, R. (ed.) (1979), *Methods Enzymol.*, **68** (Gen. ed. S.P. Colwick and N.O. Kaplan), Academic Press.
4 Old, R.W. and Primrose, S.B. (1980), *Principle of Gene Manipulation,* Blackwell, Oxford.
5 Verma, I.M. (1977), *Biochim. Biophys. Acta,* **473,** 1–38.
6 Temin, H. and Mizutani, S. (1970), *Nature,* **226,** 1211–1213.

7 Baltimore, D. (1970), *Nature,* **226,** 1209–1211.
8 Wensink, P.C., Finnegan, D.J., Donelson, J.E. and Hogness, D.S. (1974), *Cell,* **3,** 315–325.
9 Khorana, H.G. (1979), *Science,* **203,** 614–625.
10 Bolivar, F. (1979), *Life Sci.,* **25,** 807–818.
11 Soberon, X., Covarrubias, L. and Bolivar, F. (1980), *Gene,* **9,** 287–305.
12 Blattner, F.R., Williams, G.B., Blechl, A.F., Denniston-Thompson, K., Faber, H.E., Furlong, L.A., Grunwald, D.J., Kiefer, D.O., Moore, D.D., Schumm, J.W., Sheldon, E.L. and Smithies, O. (1977), *Science,* **196,** 161–169.
13 Thomas, M., Cameron, J.R. and Davis, R.W. (1974), *Proc. Natl. Acad. Sci. USA,* **71,** 4579–4583.
14 Leder, P., Tiemeier, D. and Enquist, L. (1977), *Science,* **196,** 175–177.
15 Hohn, B. and Murray, K. (1977), *Proc. Natl. Acad. Sci. USA,* **74,** 3259–3263.
16 Meesing, J., Gronenborn, B., Muller–Hill, B., and Hofscheider, P.H. (1977), *Proc. Natl. Acad. Sci. USA,* **74,** 3642–3646.
17 Barnes, W.M. (1979), *Gene,* **5,** 127–139.
18 Collins, J. and Bruning, H.J. (1978), *Gene,* **4,** 85–107.
19 Collins, J. and Hohn, B. (1979), *Proc. Natl. Acad. Sci. USA,* **75,** 4242–4246.
20 Hinnen, A., Hicks, J.B. and Fink, G.R. (1978), *Proc. Natl. Acad. Sci. USA,* **75,** 1929–1933.
21 Struhl, K., Stimchcomb, D.T., Scherer, S. and Davis, R.W. (1979), *Proc. Natl. Acad. Sci. USA,* **76,** 1035–1039.
22 Beggs, J.D. (1978), *Nature,* **275,** 104–109.
23 Wigler, M., Sweet, R., Sim, G.K., Wold, B., Pellicer, A., Lacy, E., Maniatis, T., Silverstein, S. and Axel, R. (1979), *Cell,* **16,** 777–785.
24 Mulligan, R.C., Howard, B.H. and Berg, P. (1979), *Nature,* **277,** 108–114.
25 Yang Chou, J. and Martin, R.G. (1974), *J. Virol.,* **13,** 1101–1109.
26 Shepherd, R.J. (1976), *Adv. Virus Res.,* **20,** 305–339.
27 Zaenen, I., VanLarebeke, N., Teuchy, H., Van Montagu, M. and Schell, J. (1974), *J. Mol. Biol.,* **86,** 109–127.
28 Schell, J. and Van Montague, M. (1977), in *Genetic Engineering for Nitrogen Fixation* (ed. A. Hollaender), Plenum, New York, pp. 159–179.
29 Gordon, M.P., Farrand, S.K., Sciaky, D., Montoya, A.L., Chilton, M.D., Merlo, D.J. and Nester, E.W. (1979), in *Molecular Biology of Plants* (ed. I. Rubenstein), Academic Press, New York.
30 Sciaky, D., Montoya, A.L. and Chilton, M.D. (1978), *Plasmid,* **1,** 238–253.
31 Olivera, B.M., Hall, Z.W. and Lehman, I.R. (1968), *Proc. Natl. Acad. Sci. USA,* **61,** 237–244.
32 Sgaramella, V. (1972), *Proc. Natl. Acad. Sci. USA,* **69,** 3389–3393.
33 Chang, L.M.S. and Bollum, F.J. (1971), *Biochemistry,* **10,** 536–542.
34 Jackson, D.A., Symons, R.H. and Berg, P. (1972), *Proc. Natl. Acad. Sci. USA,* **69,** 2904–2909.
35 Roychoudhury, R., Jay, E. and Wu, R. (1976), *Nucleic Acids Res.,* **3,** 863–877.
36 Scheller, R.H., Dickerson, R.E., Boyer, H.W., Riggs, A.D. and Itakura, K. (1977), *Science,* **196,** 177–180.

37 Backmann, K., Ptashne, M. and Gilbert, W. (1976), *Proc. Natl. Acad. Sci. USA,* **73,** 4174–4178.

38 Charnay, P., Lousie, A., Fritsch, A., Perrin, D. and Tiollais, P. (1979), *Mol. Gen. Genet.,* **170,** 171–178.

39 Mercerau-Puijalon, O., Royal, A., Cami, B., Garapin, A., Krust, A., Gannon, F. and Kourilsky, P. (1978), *Nature,* **275,** 505–510.

40 Villa-Komaroff, L., Efsratiadas, A., Broome, S., Lomedico, P., Tizard, R., Naber, S.P. Chick, W.L. and Gilbert, W. (1978), *Proc. Natl. Acad. Sci. USA,* **75,** 3723–3721.

41 Roberts, T.M., Kacich, R. and Ptashne, M. (1979), *Proc. Natl. Acad. Sci. USA,* **76,** 760–764.

42 Grunstein, M. and Hogness, D.S. (1975), *Proc. Natl. Acad. Sci. USA,* **72,** 3961–3965.

43 Jones, K. and Murray, K. (1975), *J. Mol. Biol.,* **51,** 393–409.

44 Benton, W.D. and David, R.W. (1977), *Science,* **196,** 180–182.

45 Ratzkin, B. and Carbon, J. (1977), *Proc. Natl. Acad. Sci. USA,* **74,** 487–491.

46 Chang, A.C.Y., Nunberg, J.H., Kaufman, R.K., Ehrlich, H.A., Schimke, R.T. and Cohen, S.N. (1978), *Nature,* **275,** 617–624.

47 Ratzkin, B. and Carbon, J. (1977), *Proc. Natl. Acad. Sci. USA,* **74,** 487–491.

48 Skalka, A. and Shapiro, L. (1976), *Gene,* **1,** 65–79.

49 Broome S. and Gilbert, W. (1978), *Proc. Natl. Acad. Sci. USA,* **75,** 2746–2749.

50 Paterson, B.M., Roberts, B.E. and Kuff, E.L. (1977), *Proc. Natl. Acad. Sci. USA,* **74,** 4370–4374.

51 Beggs, J.D., Van den Berg, J., Van Ooyen, A. and Weissmann, C. (1980), *Nature,* **283,** 835.

52 Ruther, V. (1980), *Mol. Gen. Genet.,* **178,** 475–477.

53 Klobutcher, L.A. and Ruddle, F.H. (1979), *Nature,* **280,** 657.

54 Seif, R. and Martin, R.G. (1979), *J. Virol.,* **32,** 979.

55 Subramanian, K.N. (1979), *Proc. Natl. Acad. Sci. USA,* **76,** 2556.

56 Crawford, L.V. (1980), *TIBS,* **5,** 39.

57 *Report of the Working Party on the Practice of Genetic Manipulation,* HMSO Cmnd 6600 (1976).

58 *Second Report of the Genetic Manipulation Advisory Group,* HMSO Cmnd 7785 (1979).

59 *Nature,* **276,** 104–108 (1978).

60 *Genetic Manipulation Advistory Group Notes,* Medical Research Council, London.

60 *Genetic Manipulation Advisory Group Notes,* Medical Research Council, London.

61 *Federal Register,* **45** (2) (1980), 6724–6749, Part VI, Department of Health,

62 *Federal Register,* **43** (247) (1978), 60080–60105, Part VI, Department of Health Education and Welfare, National Institutes of Health, Revised Guidelines Involving Recombination DNA research.

Index